INSECTS *and* POLLUTION

INSECTS *and* POLLUTION

Kari Heliövaara, Ph.D.
Researcher
Finnish Foresr Research Institute
Vantaa, Finland

Rauno Väisänen, Ph.D.
Head
Nature Conservation Research Unit
National Board of Waters and the Environment
Helsinki, Finland

CRC Press
Boca Raton Ann Arbor London Tokyo

Library of Congress Cataloging-in-Publication Data

Heliövaara, Kari.
Insects and pollution / Kari Heliövaara, Rauno Väisänen.
p. cm.
Includes bibliographical references (p.) and indexes.
ISBN 0-8493-6191-5
1. Insects—Effect of pollution on. 2. Insects—Ecology.
3. Pollution—Environmental aspects. I. Väisänen, Rauno. II. Title.
QL496.4.H45 1993
595.7′05222—dc20 92-26120
CIP

Direct all inquiries to CRC Press, Inc., 2000 Corporate Blvd., N. W., Boca Raton, Florida, 33431.

International Standard Book Number 0-8493-6191-5

Library of Congress Card Number 92-26120
Printed in the United States 1 2 3 4 5 6 7 8 9 0
Printed on acid-free paper

AUTHORS

Kari Heliövaara, Ph.D., is a Leading Researcher at the Finnish Forest Research Institute and a Senior Lecturer (applied zoology) at the University of Helsinki. Basically he is a forest entomologist, but his scientific production also deals with ecotoxicological and environmental entomology. Dr. Heliövaara has served as an expert in the scientific and technical cooperation between the USA and Finland, and actively participated in the working group of forest entomology in the Nordic countries. He is married, has two sons, and lives in Helsinki.

Rauno Väisänen, Ph.D., is the Head of the Nature Conservation Research Unit in the Board of Waters and the Environment, and a Senior Lecturer (zoology) at the University of Helsinki. He started his scientific career in systematics, but has later been more interested in conservation ecology and biogeography. He has served in several governmental committees and working groups on nature conservation and environmental issues. Dr. Väisänen is the vice-chairman of the Finnish Lepidopterological Society and the chairman of the Lepidoptera Group of the WWF Finland. He is also a member of the IUCN Species Survival Commission. He is married and lives in Helsinki.

PREFACE

In this book we have attempted to provide an overview of both the direct and indirect effects of pollution on insects and to discuss the ecological and economic consequences of these changes at the individual, population, and ecosystem levels. The book is primarily intended for advanced undergraduates, graduate students, and researchers in entomology and environmental sciences, as well as for civil servants concerned with insects or ecosystems. This book can also be used to gain a better understanding of human impact on the global ecosystem.

We have reviewed the published literature dealing with insects and pollution. The first chapters review the studies on pollutant-induced changes in insects classified according to their trophic position, taxonomy, and developmental stage. These changes are considered in different spatial and temporal scales, climatic and vegetational zones, as well as in different habitats. The effects of different pollutants are described in both terrestrial and aquatic ecosystems, in two comprehensive chapters. Subsequent chapters consider the effects of pollutants on insect physiology, ecology, and evolution, updating and synthesizing the available material. The last part of the book discusses the ecological and economic consequences of pollution-induced changes in insects.

We decided to act as authors of the whole volume instead of editing texts from several authors. This challenging alternative speeds up the procedure and makes a more coherent approach possible. It also decreases overlapping information in different chapters. However, due to the multiapproach structure of the book, some overlap is unavoidable. We preferred this structure that probably makes the text easier to use for applied readers and hope that general readers do not find it too sectioned.

The literature used in compiling the text was partly obtained from abstract services. We have carried out an exhaustive survey of the topic by collecting information from *Water Resource Abstracts, Biosis Previews, NTIS, Pollution Abstracts, Zoological Record, Life Science Collection, Pascal,* and *Agricola.* However, since the information is widely scattered among a wide range of sources in different countries, we cannot be sure that all the essential information has passed through our hands. The text of the book is based on the data and descriptions derived from the original papers. Although this field of science is rather young, we have laid particular emphasis on the most recent literature. We attempt to present well-documented facts on which we have based our interpretations. The documentary text is followed by short conclusions at the end of each chapter. Our own bias is on research; consequently, in several chapters in this book we emphasize the need for further entomological and ecotoxicological research.

The compilation and writing of this book was made possible by the research performed by a large number of scientists. We are indebted to all those authors who are cited in this review.

We are particularly indebted to Mr. Reijo Lindström, who collected most of the literature from libraries and did much to promote the whole procedure in many ways, and to Mr. Björn-Olof Olsson for careful assistance in the text processing. Ms. Arja Hyvärinen, Ms. Liisa Ikävalko-Ahvonen, Mr. Vilke Pursiheimo, and Ms. Marita Rosengren helped in the literature survey. We thank Ms. Sari Elomaa, Mr. Risto Hiukka, Ms. Auli Immonen, and Ms. Outi Nummi for technical assistance. The English language of the first version was checked by Mr. John Derome. We acknowledge support from the Finnish Forest Research Institute, Vantaa, Finland, which also provided the working facilities, and from the National Board of Waters and the Environment, Helsinki, Finland.

We are grateful to several colleagues for valuable discussions or for commenting on the manuscript: W. Al-Houty (Safat, Kuwait), R. O. Butowsky (Moscow, Russia), I. M. Campbell (West Hill, Ontario), W. H. Clements (Fort Collins, CO), V. Clulow (Sudbury, Ontario), J. S. Coleman (Syracuse, NY), R. L. Crunkilton (Columbia, MO), C. Dahl (Uppsala, Sweden), A. Davis (Canberra, Australia), L. M. Dosdall (Vegreville, Alberta), H. Ellenberg (Hamburg, Germany), M. Eyre (Newcastle-upon-Tyne, U.K.), S. A. Frenzel (Logan, UT), E. Führer (Vienna, Austria), R. Haack (East Lansing, MI), S. Hågvar (Ås, Norway), F. P. Hain (Raleigh, NC), A. S. Heagle (Raleigh, NC), H. Hokkanen (Helsinki), J. K. Holopainen (Kuopio, Finland), S. P. Hopkin (Reading, U.K.), A. Jansson (Helsinki), P. Kosalwat (Gainesville, FL), Z. Kovats (Windsor, Ontario), D. P. Kreutzweiser (Sault Ste. Marie, Ontario), S. Larsson (Uppsala, Sweden), S. Laurema (Helsinki), M. Lodenius (Helsinki), W. J. Mattson (East Lansing, MI), C. P. McCahon (Research Triangle Park, NC), K. Mikkola (Helsinki), S. Neuvonen (Turku, Finland), M. A. Novak (Albany, NY), S. J. Ormerod (Cardiff, U.K.), Z. Przybylski (Rzeszów, Poland), S. Rundgren (Lund, Sweden), J. Sarvala (Turku, Finland), D. W. Schloesser (Ann Arbor, MI), G. H. Schmidt (Hannover, Germany), J. M. Scriber (East Lansing, MI), N. M. van Straalen (Amsterdam, The Netherlands), I. Teräs (Helsinki), J. Trumble (Riverside, CA), S. Turunen (Helsinki), A.-L. Varis (Helsinki), S. Warrington (Hatfield, U.K.), G. R. Williams (West Hill, Ontario), J. Witter (Ann Arbor, MI), and L. W. Wood (Albany, NY).

This book includes material reproduced from original sources in a wide range of journals and other publications. We are deeply grateful to the authors, learned societies, and publishers, who granted permission for reproduction.

Finally, we wish to thank our families for support, patience, and judiciousness.

TABLE OF CONTENTS

Chapter 1

INTRODUCTION

I. BACKGROUND AND OBJECTIVES

Man-induced pollution is not a new phenomenon, but has gradually grown from a local nuisance into a global menace.[753] It has caused changes in insects as well as in ecosystems and human economy through insects. Up until the 19th century, pollution was mainly a local phenomenon without any marked impact at the ecosystem level. However, sewage spoiled small lakes and rivers, affecting their fauna. In medieval cities, urban waste combined with poor hygiene created favorable conditions for rats and human fleas (*Pulex irritans*). The fleas spread the bubonic plague caused by the bacterium *Yersinia pestis*. Epidemics of the black death swept over Europe and the Middle East in the 14th century. As the vector of the disease, fleas had a decisive effect on the human population and political situations. Bubonic plague is still widely distributed, the greatest number of cases occurring each year in Vietnam. It also occurs in South America, Africa, Asia, and even in North America.[1575]

The industrialization of Europe in the 1800s created entirely new pollution problems. The burning of coal for energy caused air pollution in the form of sulfur dioxide and soot. Populations of the peppered moth (*Biston betularius*) started to darken in England in the mid-1800s.[829] Changes in insect population densities near polluting industries have been known for at least 150 years. Counts of *Epinotia tedella* larvae on Norway spruce around a German iron foundry between 1832 and 1833 showed that this tortricid moth was seven times more abundant in the fume zone than in nonpolluted adjacent areas.[322]

As a whole the study of insects in relation to pollution is a relatively new field of science that has advanced rapidly during the past few years. Nevertheless, the extensive information is very diffuse and scattered in the literature; relatively little attention has been paid to insects in environmental studies, and few studies in ecological entomology have investigated the influence of environmental pollution. Due to the essential role of insects in both natural and managed ecosystems, the changes caused by pollution are not restricted to the insects themselves, but also include detrimental effects on biodiversity, ecosystem dynamics, and human welfare.

The main goals of this book are

1. To describe the direct and indirect effects of pollution on insects
2. To describe the secondary effects on ecosystems of pollution-induced changes in insects

Consequently, this book strives to give a general overview of the diverse impacts of pollution on insects representing different taxa, developmental stages, and trophic levels, living in a range of terrestrial and aquatic habitats in different climatic and vegetational regions. The effects of man-induced pollution on insects have both significant ecological and economic consequences. We have tried to explain the role of insects in ecosystems where pollution has affected both the role of insects and the whole structure and functioning of ecosystems.

II. CONCEPTS AND DEFINITIONS

A. INSECTS

The Insecta is conventionally regarded as a class of Arthropoda. The Insecta, with its approximately 0.8 million described (and many more undescribed) species, is not only the biggest class in the animal kingdom, but its species richness exceeds that of all the other animal groups combined. Insects are predominantly terrestrial, but there are many freshwater and relatively few marine species. Insects are dominant components of many food webs. In recent years insects have often been used as indicators of ecological conditions and environmental quality. Insect fossils are available from the Lower Devonian period onwards.[672]

The Arthropoda also includes the myriapods (centipedes, millipedes, etc.), crustaceans (crabs, shrimps, woodlice, etc.), and arachnids (spiders, mites, scorpions, etc.). The Insecta is traditionally considered as one of the five classes in the suphylum Uniramia, the others being Chilopoda, Diplopoda, Pauropoda, and Symphyla, which are collectively known as Myriapoda.[1177] The Insecta and the Myriapoda apparently form a monophyletic group (including species descended from a single-stem species of which they are the sole living descendants). This group is also called the Tracheata or Atelocera.[672] These animals have mandibles, one pair of antennae, and other appendages that are primitively unbranched. Insects have a tracheate respiratory system and a body divided into a head with two antennae, a thorax with six legs and up to two pairs of wings, and an abdomen. However, there is considerable variation in morphology.

The success of insects can obviously be partly attributed to the evolution of flight, which has improved dispersal, escape from predators, and access to food and optimal environmental conditions. The evolution of a wide range of feeding habits that facilitate utilization of nearly all natural organic substances has surely contributed to the success of insects.[577] The major radiation in insect herbivores has been in the taxa associated with angiosperms. These plants and insects have been coevolving since the Mesozoic period, and their interactions have resulted in the great proliferation in species in both groups.[418] The generally high level of organization of insect sensory and neuromotor systems, their short generation time, and small size have apparently also considerably contributed to such success.[994]

In the present book we treat all traditional groups of this class as insects. Such orders as Protura, Diplura, Thysanura, and Collembola have been included, although all scientists do not regard them as insects because these groups may have evolved independently.[979] Spiders, crustaceans, mites, etc. are definitely not insects, but there is much pollution-related information on these arthropod groups, which can probably apply to insects as well. However, these groups have been included only in cases where corresponding knowledge on insects is lacking or scarce.

B. POLLUTION

It is problematic to define pollution. There are many ways of classifying pollutants, and the appropriate choice depends on the purpose.[699] A pollutant may be defined as a substance that occurs in the environment, at least in part, as a result of man's activities and which has a detrimental effect on living organisms.[1075] Freedman[502] wrote that pollution occurs when a chemical is present in the environment at a concentration that is sufficient to have a physical effect on organisms and thereby cause an ecological change. He considered that environmental contamination by a potentially toxic agent does not necessarily connotate pollution, since it is possible to measure very low concentrations of toxic agents by means of modern analytical chemistry. The ecological effects of pollutants depend, of course, on exposure and dose.[772,1075,1149]

Pollution can also be defined as the introduction of man-made substances and energy (e.g., heat, radiation, and noise) into the environment that are likely to harm man, living organisms, or the ecosystem, or damage structures and amenities. This definition covers almost all aspects, ranging from contamination of rivers by sewage, pesticide residues in living organisms, noise near factories and airports, industrial atmospheric pollution causing damage to buildings, and radiation dangers.[1214] Pollution has also been defined as the process of overloading the Earth's ecosystems with damaging materials or waste energy.[753] Pollution causes perturbations in ecological systems. A perturbation is a deviation or a displacement of a system's structure or function from its normal operating range.[1149] A perturbation may have a subsidy effect (such as increased biomass production) or a stress effect.[722] In this book we understand pollution in a very broad sense and omit only a few marginal phenomena.

Natural emissions of sulfur and ash from volcanoes or forest fires can pollute the atmosphere,[502] and some fish contain mercury levels toxic to man that have resulted from natural accumulation in the oceans or artificial lakes. It could thus be argued that natural pollution also occurs,[1214] but we have decided to exclude it and concentrate on anthropogenic pollution.

We do not discuss the contamination of foodstuff, such as grain, caused by insects. So-called biological pollution is also beyond the scope of this book, i.e., the direct or indirect adverse effects of exotic species on indigeneous ones. The invasion of ecosystems by exotic or alien biological organisms is a growing problem resembling, to some degree at least, chemical pollution as regards to its detrimental effects on the original biota. Such biological pollution is well documented,[38,59,432,1069,1623] but it receives only passing mention here. Alien insects and other animals have even resulted in the reduction of native insect populations, especially on islands. For example, Zimmerman [1673] noted that the introduction of predatory fish species into Hawaii was a factor probably contributing to the extinction of the dragonfly *Megalagrion pacificum* on Oahu. Introduced ant species have had adverse effects on the insect faunas of Hawaii, Bermuda, and the Galapagos Islands.[332,949,1674] Rats may have been the primary cause of the extinction of the stick insect *Dryococelus australis* on Lord Howe Island.[1124] The impact of exotic species, such as goats, rabbits, or pigs, is often indirect through the destruction of vegetation and the habitats of insects.

Some pollutants at ambient levels do not have any apparent direct effects on insects and other living organisms, but they do alter the physical and chemical environment and thus affect the ability of species to survive. Carbon dioxide, a trace gas in the atmosphere, with a natural concentration of about 0.03%, has probably the most far-reaching effects. Carbon dioxide emitted through the combustion of fossil fuels is increasing CO_2 levels in the atmosphere by about 0.2% each year. According to some predictions, this will elevate global temperatures and subsequently affect other aspects of the climate, too.[186,410,1192] The result would be radical changes in the distribution of species throughout the world.

C. BIOLOGICAL AND ENVIRONMENTAL CONCEPTS

Prior to approaching the subject matter, it is necessary to discuss some basic concepts. Although the concept of an individual does not cause confusion in insects, it is not always so clear with respect to their host plants. Modular organisms make the definition of an individual more complex. For example, according to Harper[620] a tree is a population of suborganismic units (modules), each of which has a characteristic time of birth and death and a discrete lifespan of its own. Watson[1570] elaborated this definition and stated that a tree is a population of modules with associated stems and roots.

A group of individuals in a given area can be called a population and may be defined in several ways. A more precise definition was provided by Mayr,[1007] who defined the local population or deme as the group of potentially interbreeding individuals at a given locality. All the members of a local population share a single gene pool. Thus, a population can be regarded as a group of individuals of the same species who live together at the same time and place. A metapopulation, a kind of "population of populations," is a system of local populations that interact via individuals moving between populations.[611,613]

According to the multidimensional species concept, species are groups of interbreeding natural populations that are reproductively isolated from other such groups.[1007] Species are essentially distinguished, not by any difference in form or structure, but by not interbreeding. Species change with time; they are not constant in their properties, unlike the chemist's elements or the physicist's abstraction from natural phenomena.[1075]

The populations of different species that exist in the same area form a community. Whittaker[1601] defined a natural community as an assemblage of plants, animals, bacteria, and fungi, that live in an environment and interact with one another, forming a distinctive living system with its own composition, structure, environmental relations, development, and function. In practice only a limited range of species is usually studied. For this reason the use of the term community has sometimes been considered misleading or confusing. The assemblages of plant and animal populations that make up ecological communities may have well-defined spatial boundaries that separate them from other communities. These boundaries can usually be recognized by rather abrupt changes in the dominant species in the community or in the physiographic structure of the landscape.[126]

The term environment denotes the whole surroundings of an individual organism: both the abiotic components, such as the air, soil, and water; and conspecifics, as well as other animals and plants. The abiotic environment constitutes the habitat for individual, population, and community,[1602] although the term biotope is sometimes used for the environment in which the community exists.[1510] A biological community with its abiotic environment constitutes an ecosystem. Biodiversity is the total variety of genetic strains, species, and ecosystems.[753]

The term ecotoxicology was introduced by Truhaut in 1969[1499] as an extension of toxicology, the science of the effects of poisons on individuals. Toxicological studies are concerned with single organisms, while ecotoxicology is concerned with the effect on populations and ecosystems.[1075,1361] One branch of ecotoxicology attempts to establish risks in various constituents of ecosystems. Responses of individual organisms to pollutants are to be extrapolated to higher organization levels.[800]

A compartment is usually defined as an ecosystem unit that contains a quantity of material or energy.[502] In recent years data from laboratory experiments have often been analyzed by compartmental models. In this context a compartment is defined, more strictly, as a mass of pollutant that has uniform kinetics of transformation or transport and whose kinetics are different from those of all the other compartments. A whole animal is envisaged as consisting entirely of compartments, the peripheral compartments all being linked to the central compartment, but not with each other.[1075]

Moriarty[1075] defined bioconcentration as the increase of pollutant concentration from water when passing directly into species. Bioaccumulation has a similar meaning, but it indicates the combined intake from food as well as from water, whereas biomagnification indicates the increase in concentration of a pollutant in animal tissue in the successive members of a food chain. In this book the units expressing concentrations of pollutants are used in standardized SI-units, as far as possible. However, this is not always possible, since the original conditions during studies were not given.

A microcosm is a manmade, miniaturized ecosystem designed for experimental studies on the effects of perturbations at the community level, whereas manipulated enclosure experiments are often called mesocosms. The relevance of microcosms to populations is uncertain,[943] and it has been argued[746] that microcosms are more suited to studies on the "metabolism" of a community, i.e., when investigating the rates of transfer of energy and nutrients.[879] The same question of relevance to field situations applies to all other measurements made on a microcosm, but it is sometimes suggested that tests of "community metabolism" will indicate the impact of a pollutant on "ecosystem health."[546] It is meaningful to consider the health of an individual organism, but the definition of a healthy ecosystem is less obvious.[1075]

The systematic use of biological states and effects to gain environmental information is called biological monitoring.[1259] Bioindication is a biotic process that responds significantly to changes in an affecting-state variable. A bioindicator provides insight on the condition of a system, because its behavior is an integrated response to its total biotic and abiotic environment.[1259]

A bioassay is the quantitative estimation of the intensity or concentration of a biologically active environmental factor, measured via a specific biological response under standardized conditions.[502] Critical load is a quantitative estimate of the exposure to one or more pollutants, below which significant harmful effects do not occur to specified sensitive elements of the environment, according to present knowledge.[1127] Bull[201] has recently reviewed the definitions of critical loads and critical levels. At its simplest, the critical value (load or level) is exceeded when the concentration or load causes harmful effects on a receptor, but this definition is often difficult to apply. "Harmful" effects may be difficult to define, and the effects may be dependent upon the developmental stage of a species, or they may result from mixtures of pollutants. In addition, critical loads for specific parts of an environment may differ from each other. Occasionally other stresses[1179] may increase the sensitivity of species to pollutants. At present there is far too little information available about ecosystems for this concept to be applied safely.

Chapter 2

STUDIES ON INSECTS IN RELATION TO POLLUTION

I. INTRODUCTION

The number of insect species is enormous, although most species still remain to be described[1004] (Table 1). The proportion of undescribed species is especially large in Coleoptera, Diptera, and Hymenoptera, but even in Lepidoptera at least 40% of the species have been estimated to be undescribed.[674] Insects have been used in attempts to detect, measure, and interpret environmental disturbance resulting from pollution by fertilizers, pesticides, nutrient enrichment, sewage and chemical waste disposal, radioactivity, and acid deposition.[921] Insects are particularly useful because they are ubiquitous and diverse and possess many readily measurable biological characteristics. Several species have been used as test animals in studies concerning the effects of chemicals on living organisms.

Reviews dealing with insects and pollution have been rather rare, most of them being either very general or limited to some narrow aspects of the field. Alstad et al.[27] reviewed the effects of air pollutants on insect populations. Since this paper a tremendous amount of information has been published on this topic. The effects of metals in terrestrial invertebrates, especially soil fauna, have been relatively well covered recently.[109,706,802,1508] Various effects of pesticides on nontarget invertebrates have also been reviewed.[207,327,433,559,1474] The mechanisms and interactions between air pollution and terrestrial insects, forest pests in particular, have also been considered in several papers.[77,517,589,660,661,914,1029,1284,1599] However, scattered records of the pollution-induced effects on insects are common byproducts of environmental studies. Since this field of science is young and developing rapidly, the number of new papers and undergoing projects seems to increase almost exponentially.

The effects of pollution on insects in aquatic ecosystems differ considerably from those recorded in terrestrial ones. In contrast to terrestrial ecosystems, pollutants transported by water are often directly toxic and may abruptly change the composition of invertebrate communities or even cause the deaths of entire fauna. A large database now exists on the toxicity of pollutants to aquatic species, both singly and in combinations.[853,1203,1359] Recently, Muirhead-Thomson[1091] extensively reviewed the impact of pesticides on stream fauna, covering, among others, aquatic insect groups of running waters. Respectively, the review by Clements[274] covered the community responses of stream organisms to metals. The toxicity of water pollutants to aquatic animals, with special emphasis on methods for measuring lethal toxicity, has been reviewed by

TABLE 1
Estimated Number of Described Species in Different Insect Orders

Order	Common name	Number of species
Thysanura	Silverfish and bristletails	580
Diplura	Diplurans	660
Protura	Proturans	325
Collembola	Springtails	6,000
Ephemeroptera	Mayflies	2,000
Odonata	Dragonflies and damselflies	4,950
Plecoptera	Stoneflies	1,550
Grylloblattodea	Rock crawlers	20
Orthoptera	Grasshoppers, crickets, etc.	12,500
Phasmida	Stick insects or walking sticks	2,000
Dermaptera	Earwigs	2,000
Embioptera	Webspinners	150
Dictyoptera	Cockroaches and mantises	5,500
Isoptera	Termites	1,900
Zoraptera	Zorapterans	22
Psocoptera	Booklice and barklice	1,100
Mallophaga	Biting or chewing lice	2,675
Anoplura	Sucking lice	250
Heteroptera	True bugs	23,000
Homoptera	Leafhoppers, aphids, etc.	32,000
Thysanoptera	Thrips	4,000
Neuroptera	Alder flies, snake flies, lacewings, etc.	4,600
Mecoptera	Scorpion flies, etc.	350
Lepidoptera	Butterflies and moths	146,300
Trichoptera	Caddisflies	4,500
Diptera	Two-winged flies	85,000
Siphonaptera	Fleas	1,370
Hymenoptera	Bees, wasps, ants, etc.	103,000
Coleoptera	Beetles, incl. stylopids	290,000

Abel.[1] Water acidification has been reviewed by Sutcliffe and Hildrew,[1452] Baker et al.,[74] and Kauppi et al.[819] Crunkilton and Duchrow[333] reviewed the effects of oil on aquatic invertebrates.

Numerical changes in the insect species observed in the field are difficult to interpret in relation to causal factors, since insect populations fluctuate widely and basic data on the extent of natural changes are usually lacking. Insect populations may fluctuate due to life-history phenomena, intraspecific and interspecific interactions, and unpredictable abiotic stress factors. Field evaluations of agricultural pests are further hampered by annual alterations in agricultural practices. For some forest pests there are, however, long-term records available that make such interpretations easier.[1284]

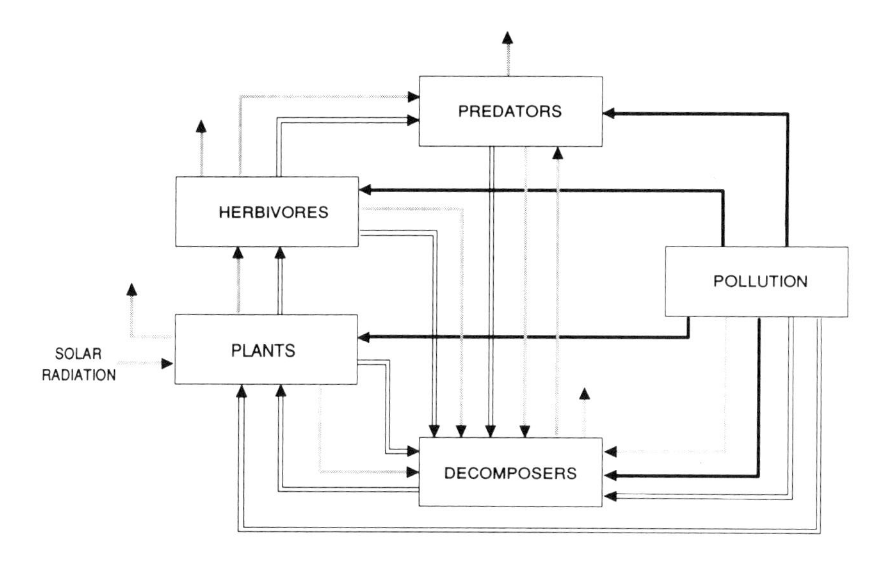

FIGURE 1. A model of the effects of pollutants on an ecosystem. Boxes represent organisms of a particular trophic level; black arrows represent the toxic and directly harmful effects of pollutants. Gray arrows indicate the one-way flow of energy, and white arrows indicate the cycling of nutrients. All energy received daily from the sun is finally converted to heat and lost from the ecosystem. Nutrients can be recycled indefinitely.

II. TROPHIC GROUPS

A. HERBIVORES

More than 90% of the Earth's biomass is plant biomass, and the amount of phytomass is directly related to vegetational zones.[1547] Insects are the major herbivores in many natural ecosystems. Consumption of annual plant production by herbivorous insects is usually less than 10%, but in Australia the estimates are higher, and in some habitats 80% of the plant material may be ingested by insects. However, much of the food eaten by insect herbivores is rapidly returned to the decomposers through feces.[1483]

Vascular plants acquire carbon via their leaves, and water and nutrients via their roots; these resources are translocated throughout the plant (Figure 1). Air pollutants like sulfur dioxide, nitrogen dioxide, and ozone can directly damage plant tissues and disrupt normal patterns of resource acquisition and allocation.[1497] These disruptions can, in turn, potentially influence the plant's ability to defend itself against pests and pathogens.[914] The response of plants to such atmospheric pollutants is extremely variable and is dramatically controlled by genetic factors, plant age, and health, as well as environmental conditions. The dose required to produce acute injury varies widely with the type of pollutant and vegetation. In addition, field symptoms of air pollution injury are not highly specific, and the causalities are often difficult to interpret.[1398] General quantitative tendencies for the suppression of root growth, increased leafiness, and reduced flowering and fruiting are obvious among different plant species and growth forms. A similar qualitative trend in decreased nitrogen concentration exists in leaves, but in general, the carbohydrate, lipid, and protein concentrations of tissues in stressed plants do not follow any predictable pattern. The dose dependence and response time varies with species and exposure.[914]

Stress-induced changes in plant quality may or may not improve insect performance.[791,905] Terrestrial insects that feed on the living tissues of higher plants are virtually confined to the orders Coleoptera, Lepidoptera, Heteroptera, Homoptera, Diptera, Orthoptera, Hymenoptera, Thysanoptera, Phasmida, and Collembola.[1446] Insect responses to stressed plants are affected by both the mode of resource exploitation and the insect life-history parameters.[791] The aspects important in the way herbivores exploit plants include feeding guild, plant tissue specialization or specialization to particular developmental stage(s) of plants (e.g., unexpanded, expanding, fully expanded, or senescent leaves), and behavioral and physiological specialization. Important life-history parameters include inherent growth and developmental rates, number of developmental stages, intrinsic reproductive output and mode of reproduction, voltinism and diapause requirements, host alteration, and insect aggregation and dispersal.[791]

Larsson[905] reviewed the possibility of differences in stress responses between herbivorous forest insects belonging to different feeding guilds, including gall-formers, chewers, miners, suckers, and cambium feeders. Presumably, if individual insects respond to host-plant stress by alterating their performance (e.g., increase in survival, fecundity, or growth rate), there may be substantial potential for population change.[27] However, processes counteracting such effects (e.g., density dependent predation, dispersal, intraspecific competition) may, of course, exist at the population level.[905] Insects belonging to different guilds are likely to respond to stress-induced changes in their food resource in different ways. For example, when the effects of a large, coal-fired, electric power station on insect fauna were studied in the Krasnoyarsk region, Siberia,[1660] the reported changes included an increase in the damage frequency to birch (*Betula*) and aspen (*Populus*) leaves by sucking insects (aphids, cicadas, bugs) and a decrease in that by gnawing insects (larvae of moths, sawflies, adults and larvae of leaf beetles, and weevils) in proportion to the increase in pollution level.

Experimental studies on the response of gall-forming insects to stressed trees are scarce. Gall-forming sawflies performed less well on stressed willow cuttings.[264,1230] Several gall formers have been reported to prefer large-sized plant tissues as oviposition sites,[9,514,1236,1598] possibly because establishment success and/or larval survival is better in such tissues.[1236] Since trees respond to environmental stress by reducing cell growth, resulting in smaller buds, shoots, and leaves, it is likely that stress to the host plants has a negative effect on gall-forming insects.[905]

Most stress experiments have been carried out with chewing insects, as shown, for example, by the reviews of Whittaker and Warrington[1599] and McNeill and Whittaker.[1029] In general, the performance on stressed trees has been the same as that on control trees.[905] In several cases insects performed less well on stressed trees, while in some studies insects preferred stressed tissues.[771,909] One explanation for the lack of positive response in chewing insects may be the fact that, in general, they do not separate chemical fractions in their food as efficiently as other insects.[905] If secondary compound concentrations increase in stressed trees[539,998] and if chewers are negatively influenced by such a change,[907] this might counteract the possible positive impact of increased concentrations of soluble nitrogen.[905]

Bultman and Faeth[203] demonstrated that experimentally induced drought stress in oaks had no effect on the survival of leafminers. The long-term field study of Münster-Swendsen[1101] showed that lower-than-normal precipitation in early summer resulted in an increase in the larval weight of the spruce needle miner *Epinotia tedella* (Tortricidae) feeding in late summer. The delayed response of this moth suggested that it was mediated by changes in host-plant quality. Larsson[905] proposed a mechanism by which miners could respond positively to plant stress. Some mining insects may avoid defensive substances that are compartmentalized in their host plants. Several species feeding on conifer needles never come into contact with the terpenoid substances in the resin ducts, since they feed only on the mesophyll tissue.[119] Consequently, miners may be able to take advantage of the increased nitrogen levels in stressed trees, while being less adversely affected than chewers by a possible increase in the concentrations of defensive compounds.[905]

Experiments dealing with sucking insects on woody plants have been carried out with aphids, and the stress effects have often been evaluated by measuring aphid densities rather than the responses of individual insects. Although these studies only took intraspecific population effects into account, it seems clear that aphids generally respond to host stress more positively than do other folivorous insects.[905] Their food, e.g., phloem sap, is low in nitrogen,[382] and noxious compounds are possibly less abundant in vascular tissue than in leaf tissue.[1262] Stress-induced, increased soluble-nitrogen levels may therefore affect the growth and reproduction of aphids more than leaf-eating insects. Due to the fast reproductive response of many aphid species, even comparatively short periods of stress may cause substantial population changes.[905,1599]

Bark beetles are the most widely studied organisms among cambium feeders. In a number of experiments where tree stress level was manipulated or the natural variation in the level of stress was known, the colonization success of naturally occurring bark beetles was found to increase when the trees were subjected to stress.[469,908,1099,1253,1552] These studies support the idea that tree stress is important for bark beetles attacking living trees. Bark beetles depend upon a resource that is normally resistant, i.e., vigorous trees cannot be colonized by most bark beetle species at endemic population levels.[1252]

Environmental stress renders trees less resistant inasmuch as the flow of oleoresin is reduced.[469] The induced response to bark beetle attack may be less effective under stress conditions.[262] Larsson[905] concluded that bark beetles are sensitive to stress-induced changes in food quality, but via a different mechanism than other insects. Some changes in symbiotic fungi associated with scolytids have been recorded in relation to the pollution level.[663]

Much of the complexity of terrestrial food webs is due to the large number of insects associated with angiosperm plants. One of the insect groups closely related to plants is pollinators. This group includes species ranging from primitive beetles and flies to butterflies and moths and to bees, which are the most important pollinators.[453,836,1239] The wide-scale application of insecticides has resulted in a decrease in pollinators[356,834,1211] or pollinating activity[57,780,1643] in many areas.

There are several other groups of insect herbivores, but little is known about their relationship with pollution. They include agricultural pests that have been controlled with chemicals. The seeds of many plants are susceptible to heavy destruction by insects. There are numerous insects that feed on fruit and cause a decrease in their cosmetic value, at the minimum. Root-feeding insects include a few serious pests of crop plants.

B. PREDATORS

Predators are animals that prey upon other animals. During their life span predators kill and consume many animal-food items. Parasitoids differ from other predators in that their offspring usually consume the prey after the death of the parental generation. A parasitoid requires and eats only one animal in its life span, but may be responsible for killing many.[1231] A typical species of insect herbivore is attacked by five to ten species of parasitoids.[643,1004] Several predators are beneficial to man because they are natural enemies of insect pests. Well-known examples include several Carabidae and Coccinellidae (Coleoptera), Chrysopidae (Neuroptera), and Syrphidae (Diptera). Among parasitoids several Diptera (Tachinidae) and Hymenoptera (Ichneumonidae, Braconidae, Aphelinidae, Aphidiidae, Encyrtidae, etc.) are important beneficial species.[628]

The susceptibility of insects to insecticides generally increases in the order: herbivorous pest — predator — parasitoid.[433,1098] Parasitoids are often also considered to be sensitive to pollutants.[27] However, parasitoids may also be more tolerant to contaminants than their hosts, as has been demonstrated, for example, in the hymenopterous parasitoids of leaf-mining agromyzids (Diptera).[990] Of course, predators can be affected indirectly via depletion of the prey populations. Long-term studies on the effects of pesticides on beneficial predatory insects have demonstrated marked effects on some predators, but hardly any in others.[207] These differences may be associated with exposure and susceptibility to pesticides, as well as with the dispersal ability and prey availability.

The different parasitoid species of *Rhyacionia buoliana* (Lepidoptera, Tortricidae) did not react in the same way to increased pollution levels in Scots pine stands of a French forest exposed to sulfur dioxide and fluorine.[1529] Generally, ectoparasites were susceptible to atmospheric pollution. In contrast, the distribution of larval endoparasites between the areas supporting different levels of population differed greatly among the species. As pointed out by Heagle,[646] pollution may cause population decreases of obligatory parasites and thus benefit facultative parasites.

C. DECOMPOSERS

The habit of feeding on dead and decaying organic matter is known in most insect orders. The rich decomposer fauna utilizing dead wood, litter, carrion, dung, etc. in terrestrial and freshwater systems primarily consists of insects.[1460] Several insect species feeding on detritus actually eat saprophytic microorganisms (e.g., fungi) in the decaying organic matter.[1483] Detritivorous insects usually assimilate less than 10% of the ingested material, most being returned to the soil or water substrate.[338] The intestines of termites contain symbiotic bacteria or protozoa that break down the cellulose of woody litter, and termites are therefore more efficient at assimilating food.[1645] The energy channeled through decomposer-based food webs is often greater than that channeled through autotroph-herbivore food webs.[1483] In some ecosystems fire can often replace decomposers and bring about very rapid mineralization of the accumulated litter.[1547]

Research into the effects of toxic compounds on soil-inhabiting organisms is still very limited compared to allied research on aquatic organisms.[1508] In forests the decomposition of organic matter in soil litter provides the nutrients needed to keep a forest productive. The soil litter effectively binds metals (e.g., Pb and Zn), and high concentrations may accumulate. Sites heavily contaminated with metals have smaller populations of decomposer organisms such as fungi, earthworms, and arthropods.[109,129]

Microarthropods (Acari and Collembola) usually account for about 20% of the total soil animal biomass in temperate coniferous forests.[110,1193] Only 10% or less of the carbon dioxide produced during decomposition has been attributed to soil animals,[1190,1193] but microarthropods may still have a crucial impact on litter decomposition. They may increase the loss of mass by over 70% more than microbes.[1534] However, the effects reported depend on the year, soil type, and microhabitat. Seastedt,[1353] summarizing 15 standardized studies, concluded that the average mass lost due to microarthropods is about 23%.

The effects of soil animals on element fluxes have been investigated in several ecosystem studies. Acceleration of decomposition rates in the presence of various invertebrates has been reported in experiments in different field conditions as well as in laboratory microcosms.[31,292,1525] Ammonium ions are often found to be mobilized from the litter and mor/humus layers as a result of animal activity, but the extent to which the different soil invertebrates contribute to mineralization varies. Lumbricid earthworms are often the most important group quantitatively. The impact of the fauna on mineralization depends on population density and soil temperature, and a correlation between NH_4^+-N mobilization and faunal biomass has been demonstrated.[31] If this mobilization is expressed per gram of animal, springtails (Collembola) seem to have a relatively larger effect on nitrogen mineralization than do other soil invertebrates.[31]

The loss of litter, resulting from respiration, fragmentation, decomposition, and leaching of soluble compounds, is usually greater in the presence of microarthropods than mineralization due to grazing, which may stimulate the activity of microorganisms.[1353] However, this tentative conclusion may have to be reconsidered once the effects of microarthropods on different elements and litter of different degree of decomposition are better documented.[110]

Metal pollution around smelters has been shown to adversely affect abundance and species numbers of the soil fauna[106,107,109,112,113] and soil microorganisms.[804,1132,1508] The subsequent reduction in the biological activity of the soil might have implications for decomposition processes,[110] as evidenced by the accumulation of litter,[503,1307,1444] reduced[744] or enhanced[65] loss of litter, decreased enzyme activities,[1505] and changes in mineralization rates.[757,1506] Strojan[1444] observed that the numbers of microarthropods and the loss of litter decreased with increased metal deposition.

Bengtsson et al.[110] designed a laboratory microcosm experiment to assess the relative significance of enchytraeids and microarthropods on the loss in mass and on mineralization, under two metal deposition levels in coniferous forest soil. Soil columns were prepared with litter, mor, and mineral soil, and the indigenous microorganisms from two field sites approximately 300 and 8000 m from a brass mill. The presence of animals increased carbon (C) and nitrogen (N) mineralization and enhanced the leaching of dissolved organic C and nutrients by 20 to 30%. All of the reduced rate of mass loss due to metal pollution and 20 to 35% of the reduction in the amounts of organic C and inorganic nutrients in the leachate were explained by reduced animal activity. The calculated decomposition rate constants predicted a delay in decomposition in the metal-polluted soil by a factor of 10. An 18% difference in mass loss due to a doubling of the background concentration of metals was detected by the microcosm method.

A decrease in the number and activity of soil invertebrates has several ecological consequences. The amount of food available for birds and other wildlife feeding on invertebrates is reduced. The decomposition rate may be reduced, since decomposers break up litter and bring it into close contact with microorganisms, as well as digest some of it directly. A reduced decomposition rate may disrupt nutrient cycling and result in an increase through accumulation in the amount of soil litter. Litter decomposers tend to contain higher metals concentrations than other groups of organisms in a contaminated ecosystem and may be an important source of contaminants in food chains.[129]

III. INSECT TAXA

A. COLLEMBOLA

Springtails (Collembola) are small, wingless hexapoda that often have an abdominal furcula, a jumping apparatus. Most species live in soil, leaf litter, and refuse with high humidity. They are also not uncommon in vegetation, and a few species are aquatic. Springtails occur commonly in marine vegetation along the shores of oceans and seas. Several springtail species abound in and on snow in the temperate regions, while certain species occur even in the harsh Arctic and Antarctic conditions. Springtails are among the most numerous of forest soil invertebrates, especially in mor-type profiles. Through their feeding activity, springtails and other soil animals contribute to the decomposition of litter and soil formation.[1460] Due to their high abundance and productivity, these insects are regarded as important links in biomass transfer in the forest soil.[443,1193,1432]

As a group, Collembola have not shown any pronounced density variation along the pollution gradients investigated by a number of authors.[108,109,1104,1444,1618] An extensive study was carried out on the effects of metals on Collembola in the Gusum area of Sweden.[108] Species numbers, density, and diversity showed a bell-shaped distribution with a peak at intermediate metal concentrations. The vertical distribution of springtails was dependent on soil metal concentrations. At the most polluted (Cu, Zn) site, the species assemblage was predominated by *Folsomia fimetarioides*, and the species diversity was the lowest.[1496]

Many springtail species are fungivorous and tend to discriminate between fungal species.[114,116] As fungal biomass and species composition are affected by metal contamination,[1132,1308] fungus-feeding Collembola in metal-polluted soils will presumably be influenced by changes in their food source. Species that graze selectively on metal-sensitive fungal species would be at a disadvantage compared to those preferring metal-tolerant fungi.[1496]

Whether springtails transfer significant amounts of pollutants along the food chain depends on the pertinent assimilation and excretion mechanisms of these substances.[1436] For a forest-floor population of *Orchesella cincta*, van Straalen et al.[1438] calculated that the food-chain transfer of lead was very small compared to the flux of lead through consumption and defecation. Given the relatively high resistance of Collembola to lead,[799,801] the effects springtails have on the cycling of metals may be more important than the detrimental effects of intoxication.[1436]

Collembola, e.g., *Folsomia candida*, *Cyphoderus* sp., and *Xenylla* sp., have been used as test organisms when assessing the toxicity of contaminants to soil fauna.[95,806,1409,1433,1482] The effects of metals on growth, survival, and the number of eggs have been studied especially in *Onychiurus armatus*.[108,109,112,115,116] Van Straalen and de Goede[1435] demonstrated that there was a great difference between the effects on individual Collembola and the effects at the population level deduced from the individual effects. Woltering[1639] considered that the mechanisms of population regulation, which may compensate for chemical perturbations, are the keys to population level performance. This was supported by field observations: populations of *Orchesella cincta* were found in habitats where the cadmium level in their food was far above the no-effect level for female growth.[1435]

In nature Collembola suffer from high predation,[443] which mainly affects the juvenile stages.[1431] The impact of cadmium toxicity on growth became overt in relatively few adults only, and fertility was slightly affected. This did not significantly decrease the productivity of the population as a whole. The high natural mortality from sources other than soil pollution acted as a compensating mechanism. Kooijman[863] stated that in predation-controlled populations, moderately toxic effects on survival often do not manifest themselves in the species assemblage, unlike toxic effects on reproduction. Consequently, the presence of springtails in polluted soils does not mean that toxic effects are absent.[1435]

B. EPHEMEROPTERA

Mayflies (Ephemeroptera) spend almost all of their life cycle in the water where they are important food for many fish.[181] They are basically herbivores, feeding on algae and plant debris. The nymphs breathe by means of tracheal gills. Both the adults and nymphs are characterized by two or three long tail-like abdominal filaments. The short-living adults have one or two pairs of wings. Unlike other insects, mayflies also moult after attaining the winged state. The premature winged stage is called subimago.

Mayflies are known to be sensitive to acidification, exhibiting either reduced density, diversity, or richness at sites with a low pH.[23,101,471,511,711,718,774,955,1196,1378,1451,1495] *Baetis rhodani* and *Leptophlebia marginata* are among the species most investigated.[538,678,1015,1154,1164,1249] Mayflies often have a higher drift rate in response to lowered pH than do other benthic invertebrates.[711] Baetidae are also sensitive to pesticide applications.[559] Distribution of *Hexagenia* nymphs has been studied in relation to oil pollution.[689]

Mayfly larvae have been regarded as useful indicator organisms of the quality of monitoring aquatic environments.[47,508,901,1475] Reduced mayfly numbers have been commonly observed in streams below urban and industrial areas. For example, no mayflies were found in a polluted 20-km section of the St. Marys River below Sault Ste. Marie, Ontario, Canada, while mean densities up to 457 per square meter (mainly pollution-sensitive *Hexagenia* spp.) were found in sections with better water quality (Figure 2).[1337] The larva of *Baetis thermicus* is one of the aquatic organisms most tolerant to metal pollution.[635,1448,1456] *Hexagenia limbata* with burrowing benthic nymphs has been used in bioassays to determine the toxicity of sediments.[543]

C. ODONATA

The nymphs of dragonflies (Odonata) are aquatic, while the winged adults are usually found in a range of terrestrial habitats in addition to the vicinity of waters. The greatest part of the life cycle with a series of gradually growing nymphal stages is spent in an aquatic environment. Many species live in still water, but there are running-water and bog species as well. Both the nymphs and adults are carnivores. Almost all species are diurnal as adults.

Water pollution, including eutrophication caused by the discharge of nutri-

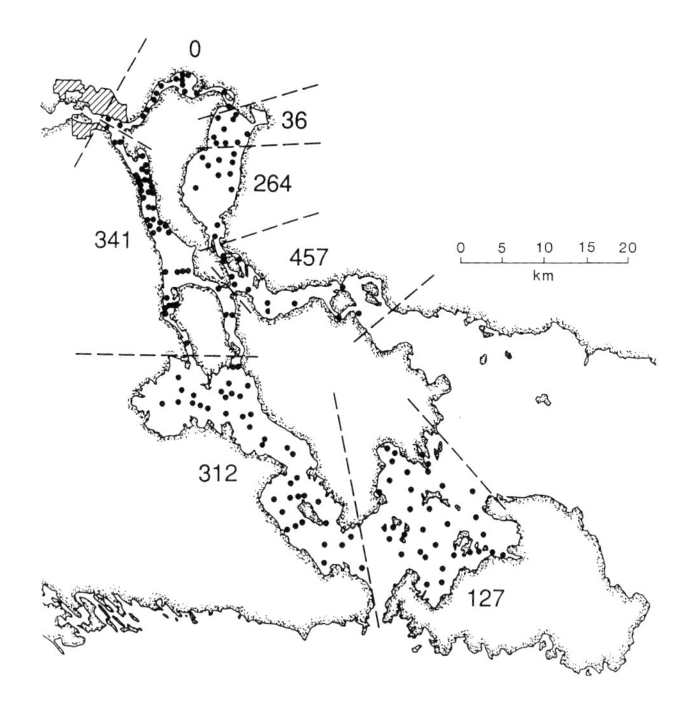

FIGURE 2. Mean density (number per square meter) of all mayflies (Ephemeroptera) at sampling stations (black dots) in seven geographical sections of the St. Marys River, Ontario, Canada, downstream from Sault Ste. Marie. Section with no mayflies was heavily polluted by industrial wastes. (From Schloesser, D. W., *J. Great Lakes Res.*, 14, 227, 1988. With permission.)

ents from agricultural areas, sewage effluent, and other sources, is considered a significant threat to several dragonfly species in central Europe,[861,1271,1584] Britain,[1362] and Australia.[685,1613] Drought-reduced water clarity, eutrophication, and algal blooms probably decrease the feeding success of preying nymphs. The application of deltamethrin in the control of tsetse flies caused a catastrophic drift of Gomphidae in Zimbabwe.[559] Pyrethroid insecticide applications markedly affected Anisoptera and Zygoptera in Nigeria.[1390]

D. PLECOPTERA

Stoneflies (Plecoptera) have aquatic nymphs, and most species live in running waters with stony and gravelly bottoms. Both nymphs and winged adults are important items of the diet of many fish. Many species in Europe and North America have a northern distribution and live on mountain streams. Both adults and nymphs have a pair of tail-like abdominal cerci. The nymphs have accessory gills, but obtain much of their oxygen directly through the body surface. The nymphs feed mainly on algae and mosses, but there are also carnivorous species.

Stoneflies are known to be intolerant of water pollution and oxygen deficiency.[901] For this reason drastic losses of species have been reported from many districts.[774] However, in general, stoneflies seem to be more tolerant to acidification than mayflies.[774,1038,1249]

E. ORTHOPTERA

Grasshoppers, including locusts and a number of cricket species (Orthoptera or Saltatoria) with their well-known stridulating voices, are predominantly tropical, and many species live in arid open areas such as meadows and steppes. They are primarily herbivores, but may also eat animal material. Grasshoppers (Acrididae) alone account for 20 to 30% of the biomass of arthropods in some ecosystems in Europe. They are important prey for numerous birds and other vertebrates, and thus they often play an important role in the food chains. Grasshoppers can also be destructive, causing huge annual crop losses. The likes of infamous locust plagues described in the Bible still continue to cause widespread famine in many developing countries.[1514]

In highly stressed agricultural areas in Europe, the percentage and absolute numbers of grasshoppers have fallen drastically during the last few decades. Investigations using biotests under simulated conditions have shown that acridids cannot tolerate insecticides or herbicides and are very sensitive to heavy metals and fertilizers in the soil (Figure 3).[1338,1339,1341]

In Romania the repeated chemical campaigns to exterminate mosquitoes along the coast of the Black Sea are reported to have resulted in the disappearance of several Mediterranean species of Orthoptera, such as *Decticus albifrons*, *Platycleis escalerai*, and *Saga campbelli*.[847]

F. HETEROPTERA

True bugs (Heteroptera) are hemimetabolous insects with piercing mouth parts adapted for sucking the juices of plants or animals. Bugs have diverse structures and habits. Most species are terrestrial, but the order also includes aquatic and semiaquatic forms such as pond skaters (Gerridae). Terrestrial species are usually herbivorous, but there are also predators and blood-sucking ectoparasites of vertebrates.

Exposure to high levels of nitrogen dioxide and sulfur dioxide, singly [463] or in mixtures,[462] can directly enhance the growth and reproduction of seed-eating plant bugs. In general, little is known about the effects of pollution on Heteroptera.[103,135,652,766-768,942,1313] Bendell[103] studied the effects of acidification on Gerridae in Canada. He suggested that reduced food resources and toxic effects on the eggs may have affected some species. Jansson[766] observed, using both sampling and underwater recording equipment, that the distributions of *Micronecta griseola*, *M. minutissima*, and *M. poweri* (Corixidae) in Lake Päijänne, Finland, showed a strong correlation with the results of studies on the eutrophication and pollution of the lake. *M. griseola* was found only in waters that were clearly eutrophicated, *M. minutissima* in waters with at least signs of eutrophication, and *M. poweri* in waters ranging from somewhat eutrophicated

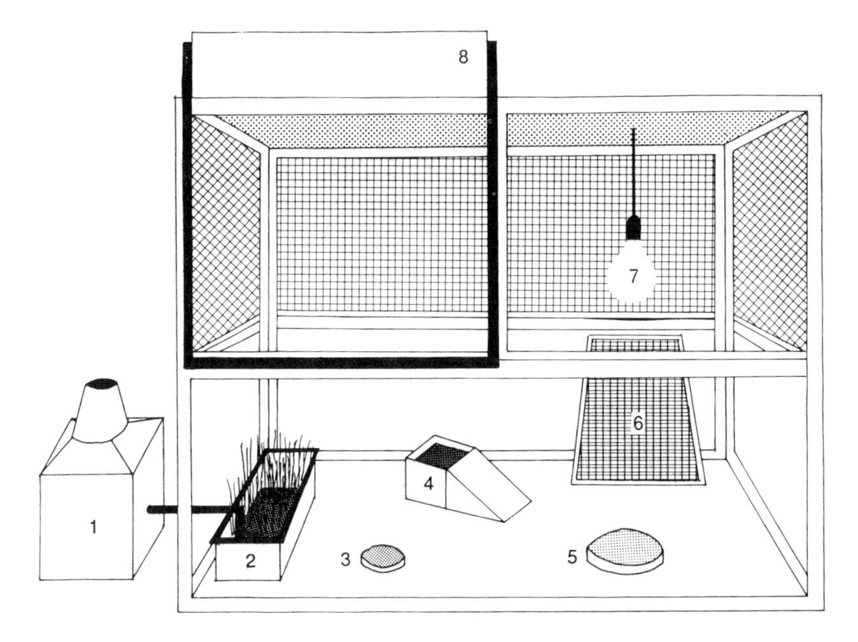

FIGURE 3. Diagram of a terrarium used for growing grasshopper (Orthoptera, Acrididae) larvae and allowing egg laying by adults and for testing chemicals. (1) Water flask with a communicating pipe, (2) hydroculture system, (3) feeding plate, (4) pot for egg deposition, (5) Petri dish with wet sand, (6) nylon gauze for molting, (7) radiator for lighting and heating, and (8) glass door. (From Schmidt, G. H., *Agric. Ecosyst. Environ.*, 16, 175, 1986. With permission.)

to oligotrophic. Jansson[767] also observed that there were only small differences in the average catches in waters ranging from oligotrophic to eutrophic, although the species composition changed completely. He was even able to present a classification of water quality based on corixids. The results from a 10-year study on Micronectinae[768] gave results that were very similar to those from limnological investigations on water quality in another Finnish lake.

Lis[942] observed a decrease in the numbers of bug individuals and species, along with an increase in pollution levels, around a zinc smelter in Poland. Heliövaara and Väisänen[652] studied in detail the distribution and abundance of the pine bark bug (*Aradus cinnamomeus*, Aradidae) in the surroundings of a factory complex emitting, for example, sulfur dioxide and metals (Figure 4). The results showed that the bug was almost completely absent in the immediate vicinity of the complex, reached the outbreak level in the zone only about 1 to 2 km from the pollutant sources, and decreased gradually with decreasing pollution levels.

Biesiadka and Szczepaniak[135] studied the occurrence of water bugs in the strongly polluted and eutrophicated Lake Suskie in northern Poland. They concluded that a distinct impoverishment of species richness and the large numbers of four bug species were due to pollution. The occurrence of only 14 species indicated a notable reduction in the bug fauna of the lake. The absence

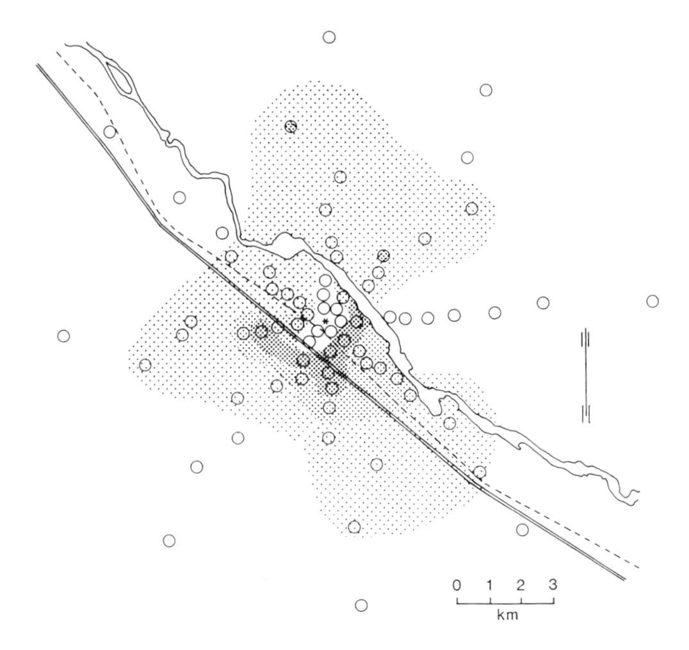

FIGURE 4. Zones of the pine bark bug (*Aradus cinnamomeus*, Heteroptera, Aradidae) densities in samples taken around an industrial complex (asterisk) at Harjavalta, Finland, emitting, for instance, sulfur compounds and metals. Density categories from white to dark 0–0.50, 0.51–1.75, 1.76–3.50, and 3.51–12.2 bugs (per 0.01 m^2 bark surface), respectively. (From Heliövaara, K. and Väisänen, R., *J. Appl. Entomol.*, 101, 469, 1986. With permission.)

of many common species of the genera *Micronecta*, *Plea*, *Ranatra*, *Mesovelia*, and *Microvelia* was striking. The Corixidae accounted for 98.6% of the individuals collected.

Studies on the side effects of large-scale eradication programs against tsetse flies in Africa showed that Corixidae, Nepidae, Notonectidae and Naurocoridae (e.g., *Laccocoris limigenus*), and Belostomatidae (*Sphaerodema nepoides*) were sensitive to pyrethroid applications.[559,1390]

G. HOMOPTERA

The order Homoptera includes various hemimetabolous insects with sucking mouth parts, such as cicadas, spittle bugs, leaf hoppers, scale insects, aphids, and psyllids. Only aphids have been extensively studied in relation to pollution, though there are also some records of adelgids and coccids on trees.[1284] Often, air pollution seems to increase the number of aphids.[175,387,1029,1030,1599] Aphids are also the most serious pests of crops among the Homoptera. They damage plants directly by removing material from the phloem vessels and act as vectors in the transmission of viral diseases.

Population changes of aphids have been attempted to relate to pollution gradients in several studies. Wentzel[1586] reported an unusually high abundance of the spruce gall aphid *Sacchiphantes abietis* in a pine forest surrounding a

FIGURE 5. Chamber experiment with *Phyllaphis fagi* (Homoptera, Drepanosiphidae) in an ozone-polluted area, showing an aphid cage fixed on a beech leaf.[177] (Photo by W. Flückiger.)

brickworks. The trees that were obviously damaged by hydrogen fluoride (HF) emmissions had far more aphid galls than did the healthy looking trees away from the pollutant source. Flückiger et al.[486] observed high population levels of *Aphis pomi* adjacent to a motorway. The abundance of the aphid decreased with increasing distance from the road. Wiackowski[1603] noted very large populations of aphids in the vicinity of a factory emitting sulfur dioxide. Flückiger et al.[486] found increased numbers of aphids near a motorway. Only few papers have reported less numerous aphid populations in polluted areas.[122,524,1530]

Several authors have studied experimentally the effects of pollutants on aphids (Figure 5). For example, Dohmen et al.[390] studied the effects of air pollution on aphids in a fumigation experiment. They found increased aphid (*Aphis fabae*) growth rate on plants (*Vicia faba*) that had been fumigated with sulfur dioxide (SO_2), nitrogen dioxide (NO_2), or ambient London air. Bolsinger and Flückiger[154] compared the population growth of aphids (*Aphis fabae*) on a woody plant (*Viburnum opulus*) in ambient air near a motorway and that in charcoal-filtered air at the same location. Their results also indicated an increased aphid growth rate in the polluted air. Dohmen[387] studied the effect of ambient Munich (Germany) air on aphid growth rate. Rose bushes were fumigated with either ambient urban air or charcoal-filtered air. The plants were then infected with aphid *Macrosiphon rosae* nymphs. The mean relative growth rate of aphids feeding on roses in ambient air was about 20% higher than those feeding in the filtered atmosphere. Further experiments using agricultural aphid pests have shown that the mean relative growth rate of aphids increased on plants previously fumigated with either SO_2 or NO_2, compared with those maintained under identical conditions in charcoal-filtered, clean air.[715] The results also suggest that NO rather than NO_2 may be the most important factor in mediating the interaction.[716]

Aphids can detect a wide range of substances in plant sap[225,1345] and on leaf surfaces[805] and show increased restlessness when exposed to toxic chemicals taken up from the soil and translocated by their hosts.[152]

H. LEPIDOPTERA

Lepidoptera form a large insect order comprising butterflies and moths. They are economically important because the larvae of a number of species cause serious damage to crops and other plants. On the other hand, silkworms, which are larvae of *Bombyx mori* and other saturniid moths, have been in domestic use for thousands of years. They were originally utilized in ancient China, and the industry is said to have been founded in 2640 BC.[1584] The aesthetic value of butterflies is not insignificant, and the loss of butterflies, resulting in a colorless summer, would have the same symbolic value as the absence of birds in Rachel Carson's book, *Silent Spring*.[235] The decline and disappearance of butterflies and moths has often been attributed to pollution.

Despite what is said in the above, butterflies and moths have received relatively little attention in ecotoxicological research. However, there is one aspect of pollution that has been a subject of several intensive studies: industrial melanism. The peppered moth *Biston betularius* (Geometridae) is the classic example of how an organism can respond to industrial pollution[829,934] (Figure 6). The first black specimen (f. *carbonaria*) was caught in 1848 in Manchester, and although it has only one generation a year, by 1885 about 98% of the population in the original area were melanics.[268] Since then the black form has spread rapidly over large areas. By the early 1900s it occurred throughout the northwestern parts of continental Europe, and nowadays it even occurs in Scandinavia.[398,1048,1051]

The North American equivalent of *carbonaria* is the form *swettaria* of the subspecies *Biston betularius cognatarius*. It was first recorded in Philadelphia in 1906. Subsequent sightings were made in New Jersey in 1920 and Chicago in 1935, but not in New York City until 1948. However, its spread was so rapid that by 1961 more than 90% of the population in parts of Michigan belonged to the *swettaria* form.[125,1172,1173]

The effects of air pollutants on Lepidoptera are poorly understood. The population density of the pine bud moth *Rhyacionia buoliana* (Tortricidae) increased in the young Scots pine (*Pinus sylvestris*) stands of the Roumare Forest, France, exposed to heavy atmospheric pollution by sulfur dioxide and fluorine.[1528,1529] The periodically recurring oak damage caused by *Tortrix viridana* (Tortricidae) was highest in the vicinity of an ironwork near Krakow, Poland.[1144] Acidic precipitation and other forms of air pollution have destroyed woodlands to such an extent in certain areas of Czechoslovakia and Germany

FIGURE 6. Color forms of the peppered moth *Biston betularius* (Lepidoptera, Geometridae): the light morph f. *typica* (upper row), the dark morph f. *insularia* (middlle row), and the black morph f. *carbonaria* (lower row); males on the left, females on the right. (Photo by K. Mikkola.)

that the survival of butterflies is unlikely, owing to the adverse changes in the habitats and host plants.[888] According to Weidlich,[1577] several species are threatened in heavily polluted areas in eastern Germany, due to water pollution (*Archanara dissoluta*, *Chilodes maritima*), the application of agricultural chemicals (e.g., *Chazara briseis*, *Spialia sertorius*, *Phyllodesma tremulifolia*, *Gastropacha quercifolia*, *Pachythelia villosella*, *Cucullia asteris*, *C. verbasci*, *Oria musculosa*, *Hoplodrina superstes*, *Luperina nickerlii*, *Theria rupicapraria*, *Lycia pomonaria*), and industrial air pollution (*Cucullia verbasci*, *Hoplodrina superstes*, *Lycia pomonaria*). The effects of electric lighting on nocturnal moths were recently reviewed by Frank.[493]

The apollo butterfly *Parnassius apollo* (Papilionidae) has declined over parts of northern and central Europe lying to the north or east of the main industrial areas in Europe, where the prevailing winds are from the southwest.[299] The decline of *P. apollo* in Norway may be attributed to acidic precipitation. Its decline in Sweden and Finland since the 1950s may also be associated with pollution (other causes include climatic and land-use changes). Stable populations seem to be able to survive in localities with calcareous bedrock. This raises the hypothesis that acidic precipitation and subsequent metal pollution via the food plant were responsible for the decline in other localities with a lower pH buffering capacity. Nuorteva[1142] reported that the levels of iron and aluminium in the host plant in the area where *P. apollo* had disappeared were clearly higher than in the areas where it still exists.

However, Bengtsson et al.[111] could not find any correlation between bedrock characteristics and the metal content (Al, Mg, Mn, Fe, and Cd) of its host plant, *Sedum telephium*. Feeding experiments with young larvae of *P. apollo* did not result in significant biomagnification of the analyzed metals from the plants to the animals. However, feeding on plants from different localities resulted in significant differences in larval growth and mortality, but overall, the relationship with air pollution is still obscure. Metal concentrations of moths have been investigated also in several other papers.[665,811,1144,1204]

The toxicity of insecticides has been investigated in several butterfly species, e.g., *Pieris brassicae* (Pieridae).[360,361,1379] The application of insecticides may have caused the decline of several species. In Romania, for instance, the great peacock moth *Saturnia pyri* (Saturniidae) has reportedly declined due to the intensive chemical treatment of fruit trees, and the death's head hawk-moth *Acherontia atropos* (Sphingidae) has almost disappeared due to the chemical control campaigns against the Colorado beetle *Leptinotarsa decemlineata* (Chrysomelidae) on the potato.[862] Furthermore, the aerial applications of insecticides to control *Lymantria dispar* (Lymantriidae), *Tortrix viridana* (Tortricidae), and occasionally *Thaumetopoea processionea* (Notodontidae) have lead to a drastic decline in the oak-feeding *Catocala* species, *C. conversa*, *C. diversa*, and *C. dilecta* (Noctuidae). Other species, like *Catocala sponsa*, *C. promissa*, *Catephia alchymista* (Noctuidae), *Hoplitis milhauseri* (Notodontidae), *Minucia lunaris* (Noctuidae), *Eriogaster rimicola* (Lasiocampidae), and *Phalera bucephaloides* (Notodontidae), have also been reported to become more rare. In Australia, mosquito control has been considered to be a threat to *Hypochrysops epicurus* (Lycaenidae).[685] Nevertheless, there is still little evidence to show that insecticides would have had more than an occasional effect on butterlies in Europe at least.[1243,1479] The spruce budworms (*Choristoneura* spp.) (Tortricidae) are serious defoliating pests of conifers in North America. They have been controlled using the aerial spraying of DDT and fenithrothion and other organophosphorous insecticides. Application of these insecticides has had considerable effects on nontarget fauna in both terrestrial and aquatic ecosystems.[1091]

I. TRICHOPTERA

Caddisflies (Trichoptera) are the only order of holometabolous insects with primarily aquatic young stages. The short-lived adults are mostly nocturnal and resemble dull-colored moths. The larvae usually feed on algae and organic debris, but a few species are carnivorous, and a few others herbivorous. Larvae of several species bear a caddis case to protect their soft abdomen, while some others spin silken nets among the aquatic vegetation. Caddisflies seem to be more abundant in temperate waters than in tropical waters. They are one of the key orders in the trophic structure of streams, displaying a great ecological diversity, with species from a wide range of habitats, and representatives from all major feeding strategies (grazers, shredders, collectors, and predators). The larvae form an important component of the diet of adult and particularly nestling riverine birds and many fish species, while the adults are favored by insectivorous bats.[1014]

The Trichoptera are relatively tolerant to pollution,[281,1485,1554,1616] although there are also sensitive species.[205] Larvae of Hydropsychidae frequent erosional habitats and filter suspended particles from the water, using silken nets. Genera like *Macronema* primarily tend to consume fine organic particles, while others like *Cheumatopsyche* derive most of their energy from larger particles and small invertebrates. Because several organic contaminants can be adsorbed on small particles, differences in contaminant concentrations among taxa may provide a measure of the relative distribution of contaminants among the various food categories.[263]

Malicky[967-969] and Chantaramongkol[243] have recommended light-trapping of Trichoptera as a method for assessing water quality in large rivers. Phenology and the effect of meteorological conditions on the catching success of light trapping were discussed by Waringer.[1553] The most pollutant-resistant caddisfly species in central Europe (e.g., in the Danube) was apparently *Hydropsyche contubernalis*. In southern Europe it may be replaced by *H. modesta*, which was the only species in heavily polluted sections of the River Tevere below Rome.[243]

J. DIPTERA

The two-winged flies (Diptera) are characterized by only one pair of membraneous wings and hind wings that are reduced to club-shaped halteres. The mouth parts of Diptera are suctorial and frequently adapted for piercing. The flies are a large insect order. They are well represented everywhere, including Arctic regions. The order includes a great variety of herbivores, predators, and decomposers. Diptera play an important role in both terrestrial and, as larvae, in aquatic ecosystems. Some species act as pollinators, e.g., *Forcipomyia* spp. (Ceratopogonidae) in cocoa (*Theobroma cacao*), while other species are agricultural or veterinary pests. Many blood-sucking species carry veterinary and especially human diseases. Some large groups of flies, such as Tachinidae, are mainly parasitoids of other insects. Diptera have also played a considerable scientific role. The common housefly (*Musca domestica*) is one of the most thoroughly studied insects. The drosophilids, for example, the banana fly (*Drosophila melanogaster*), have been used as important tools in genetics and evolutionary biology.

An increase in the numbers of Diptera near an industrial plant emitting fluorine compounds has been reported, but the increase may have been connected with the effects of other anthropogenic factors such as the presence of waste dumps and rotting organic matter.[818] Chloropidae and other Diptera have also been studied near a phosphate fertilizer factory in Germany.[69,71]

Benthic invertebrates, such as chironomid larvae, are important components of many freshwater systems. In some areas larval populations of up to 100,000 per square meter have been reported. Species of the family Chironomidae, especially the tribe Chironomini, are well known for their resistance to various environmental perturbations, e.g., the presence of organic pollution, low dissolved oxygen concentrations, and high or low pH values.[23,63,64,619,638,640,810,1083,1251,1328,1377] Most chironomid larvae live in the bottom substrate and are often abundant, especially in eutrophic lakes or ponds. They are important food items for fish in lakes and rivers and in summer months can represent up to 80% of the food of certain fish in some areas.[197] Therefore, anything affecting these aquatic insects can alter the stability of aquatic communities.[866,867] Many chironomid species inhabit aquatic environments polluted with metals,[631,1585,1627,1662] and several investigators have studied their metal tolerance.[275,276,876,1116,1449,1605,1614,1615,1659] *Chironomus riparius* was suggested as a suitable test species to assess both acute and chronic toxicity to pollutants,[1009,1013,1181,1182] and *C. tentans* was demonstrated to be useful in sediment toxicity bioassays.[543] Clean waters are dominated by larvae of the subfamily Orthocladinae, and polluted waters by Tanypodinae. Tanypodinae dominance may be indicative of metal pollution in a nonacid environment.[1604]

The predatory *Chaoborus* larvae (Chaoboridae) have recently been investigated with respect to water acidification[442,1518] and pesticide contamination.[603,605,606] Dixidae were considered suitable bioindicators of disturbances in lotic ecosystems.[1475]

Mosquitoes (Culicidae) and other blood-sucking flies may carry dangerous diseases and have been major targets of chemical control campaigns in the tropics. Malaria continues to be one of the most serious public health problems in many regions of the world. According to recent estimates, about 100 million cases occur every year. The mosquito vector has become more resistant to insecticides.[1514] Mosquito larvae can be used in toxicity tests. For example, the toxic effects of pollutants alter the phototactic behavior of *Aedes aegypti*, a species that has been used for research in laboratories all over the world.[391,1373]

Black flies (Simuliidae) are blood-sucking pests of humans and domestic animals in temperate and Arctic areas, which discourage tourism and outdoor recreation and interfere with lumbering, mining, and building activities during spring and summer.[1039] In the tropics their main importance is in their role as vectors of human diseases, such as river blindness or onchocerciasis caused by a parasitic filarial worm in Central and South America, Africa, and the Arabian peninsula.[1514] Pesticides have been widely used to control black flies living in streams as larvae. They include the application of methoxychlor in New York and the use of DDT and methoxychlor in Canadian rivers, as well as the application of the organophosphorous insecticide Temephos (Abate) in the Volta River Basin, West Africa, and Guatemala to control onchocerciasis. These campaigns have caused some marked effects on nontarget insects.[1091,1355]

Tsetse flies (*Glossina* spp.) inhabit over 10 million km^2 in tropical Africa between the Sahara and Kalahari deserts. They spread nagana in livestock and sleeping sickness in man, the causal agent being a parasitic protozoan *Trypanosoma* spp.[24] Attempts to eradicate the tsetse fly and trypanosomiasis have resulted in the slaughter of 1.3 million game animals, extensive bush clearance, and, since the 1950s, the application of insecticides (DDT, lindane, dieldrin, endosulfan, and deltamethrin). The side effects of the tsetse fly control campaigns have been considerable.[995,1091] Many of the pioneer investigations on the effects of the aerial control of tsetse flies were conducted in association with campaigns against the East African game tsetse, *Glossina morsitans*, *G. pallidipes*, and *G. swynnertoni*, which largely live in semiarid savannah woodlands and thickets. The extensive control measures were later extended to western Africa, where the riverine species *G. palpalis* and *G. tachinoides* are associated with dense riverine gallery forest and extensive river systems.[1091] However, the residual application of highly persistent organochlorine insecticides by helicopter has recently been replaced largely by sequential ultralow-volume spraying with very low concentrations of endosulfan or pyrethroids[559] and by control with sterile males. Grant[559] reported on the methods and materials used for monitoring the impact of tsetse control on nontarget insects.

K. HYMENOPTERA

The members of the order Hymenoptera are of special importance for human welfare. Honeybees (*Apis mellifera*) have a long history as domestic insects, dating back to ancient Egypt, producing honey and beeswax and being effective commercial pollinators. Many other species of Hymenoptera, such as the alfa-alfa leaf-cutting bee *Megachile rotundata*, the alkali bee *Nomia melanderi*, and bumblebees, pollinate crops and natural plants and, hence, play a key role in both agricultural and natural ecosystems. Most species of the order are small inconspicuous parasitoids that destroy vast numbers of agricultural and forest pests. Some parasitoids (e.g., *Encarsia formosa*) have been used in biological control. Social ants have interested people at least from the time of Aesop. They are, indeed, important predators of pest insects. Some species of wasps sting and may even cause human deaths. Herbivorous species include a few serious pests such as a number of pine sawflies (Diprionidae) and gall wasps (Cynipidae).

Honeybees may traverse a radius of several miles from their hives and contact innumerable surfaces while collecting nectar, pollen, propolis, and water. In the process, the worker bees may become contaminated with surface constituents that are indicative of the type of environmental pollution in their particular foraging area.[1081] Due to food change other individuals in the hive become susceptible. In fact, honeybees have been studied in respect to radioactive,[544,1080] PCB,[1081] insecticide,[578,1527] fluoride,[376,1001,1003,1005] and toxic-element contamination.[184,1545] Honey has also been analyzed as a possible indicator of metal pollution.[1491] Insecticides used in the vicinity of beehives have also been found in bees[1082] and honey.[781] Bees may even collect insecticide capsules as pollen (e.g., parathion product Penncap-M). The effects of insecticides on bees have been widely investigated.[57,81,359,1428] Insecticide applications have also reduced the numbers of wild bees and bumblebees.[834,1210,1211,1643]

Pesticides usually have a greater adverse effect on parasitoids than on herbivores. Elzen[433] made a useful review on the sublethal effects of pesticides on parasitoid species. The effects of metal contamination on parasitoids has also been studied (e.g., *Pimpla turionellae* on *Galleria mellonella*;[1168,1169] *Glyptapenteles liparidis* on *Lymantria dispar*[1128]). The parasitoid species have also been considered sensitive to dust.[27] On the other hand, Wentzel and Ohnesorge[1587] observed a massive increase in the population density of the herbivorous sawfly *Pristiphora abietina* around a charcoal-producing factory, far away from its normal feeding areas.

Czechowski[343] reported that ants that are associated with aphids occurred in the largest numbers in areas with a high traffic density. In Poland, studies along an industrial pollution (mainly SO_2) gradient showed that both the size of ant colonies (*Myrmica* sp. and *Lasius niger*) and the biomass of the ant workers were reduced.[1191] Reduced number and size of ant hills (*Formica rufa*) have also been recorded in heavily polluted areas.[818] In Czechoslovakia, populations of *Formica* spp. have been observed to decrease in forests affected by atmospheric pollution. This has been attributed to the accumulation of metals derived from their food.[1421] Ants are suggested as useful aids in monitoring the effects of metal pollution on forest soil biota.[107] Ants are also susceptible to pesticides.[1188]

L. COLEOPTERA

Coleoptera is not only the largest of all insect orders, but also has the greatest number of species of any group of living organisms. The exceptional success of beetles may be largely due to their tough elytra and hard cuticle. Not surprisingly, beetles can be found in all available habitats, and they utilize a wide variety of food sources. Although most beetles can fly, they are primarily insects of the ground and vegetation. They include various predators (e.g., ladybirds, many ground beetles) and decomposers (e.g., carrion beetles, dung beetles, and various wood decomposers), as well as a huge number of herbivores, some of which are serious pests (e.g., several members of the families Chrysomelidae, Scolytidae, Curculionidae, Cerambycidae, and Buprestidae).

Considering their species richness, abundance, and great significance for ecosystems, beetles have received relatively little attention in ecotoxicology. In 1929 Evenden[446] described a heavy attack of bark beetles on trees damaged by sulfur dioxide in the vicinity of smelters. In the 1970s over 1000 trees were examined for ozone damage and infestation and mortality from bark beetles in the San Bernardino Mountains of California. Stark et al.[1420] observed that the living crown ratio of conifers decreased, and the occurrence of bark beetle infestation increased, as the degree of oxidant damage caused by the Los Angeles urban complex increased. More detailed studies showed that the reduction in oleoresin exudation pressure, quantity, rate of flow, and increased propensity of oleoresin to crystallize, and the reduction in phloem and sapwood moisture content, all correlated with photochemical atmospheric pollution in ponderosa pine (*Pinus ponderosa*) and consequently increased the damage caused by the western pine beetle (*Dendroctonus brevicomis*) and mountain pine beetle (*D. ponderosae*).[285,286] The effects of metal pollution on bark beetles have recently received some attention.[663,1531,1532] Among agricultural pests, the effects of host-plant exposure to air pollutants on the Mexican bean beetle (*Epilachna varivestis*) have been studied intensively.[246,256,726-728]

Dunger et al.[407]reported certain changes in the numbers of different carabid species that were dependent on systematic group and habitat preferences in the valley of the River Neisse affected by industrial pollution. Kleinert[850] found that the carabid species richness was lower in an area affected by magnesite dust emissions. At study sites smaller, diurnal and winged species predominated. Jepson[777] provided a theoretical discussion on the dynamics of pesticide side effects on nontarget invertebrates, especially Carabidae, within arable cropping systems. The long-term effects of pesticides on carabid beetles in cereal crop fields have been monitored at Boxworth Experimental Husbandry Farm in Huntingdonshire, U.K.[207] Tachyporinae and aphid-specific predators had low susceptibility to long-term effects, and some carabids, such as *Bembidion obtusum* and *Notiophilus biguttatus*, had high susceptibility. The long-term indirect effects were considerable in several carabid species, such as *Agonum dorsale*, *Bembidion lampros*, *Trechus quadristriatus*, *Harpalus rufipes*, and *Pterostichus melanarius*, although they were not very susceptible to direct effects. The effects of some herbicides on carabids have also been studied.[195,258] Some carabids and staphylinids have been used as test organisms in toxicity tests,[95,628] and elmids as bioindicators in aquatic ecosystems.[1476]

Among the Coleoptera the melanism of the two-spot ladybird *Adalia bipunctata* has received much attention in relation to air pollution, although other factors such as climate also affect the melanism.[117,323-325,849,1054,1086,1358]

M. OTHER ORDERS

There is little information available about the other insect orders in respect to pollution. Soil-dwelling orders such as Protura and Diplura are exposed to similar pollutant stresses as Collembola. Diplurans are small, wingless, soil- or detritus-living insects with two abdominal cerci. High levels of cadmium have been found in some species. This may be associated with the fact that removal of the intestinal surface in these animals is only partial and takes over four to five molting episodes to be completed.[1437]

Termites (Isoptera) are basically tropical, soft-bodied, winged or nonwinged insects. They live in colonies with different castes and may build impressive nests. Most species feed on wood and may be serious pests, while some species feed on grass and fungi. The wood-eating termites have mutualistic bacteria or protozoans in their digestive system that break down the cellulose. Much attention has been paid to the chemical control of termites. Termites are supposed to be of profound importance in the environmental cycling of metals in the tropics. In Zimbabwe, worker termites feeding on plant material and soil contained 5000 μg/g nickel and 1500 μg/g cromium. Other castes fed with saliva had much lower concentrations, since most of the metals are removed from the saliva.[1610]

The bird lice or biting lice (Mallophaga) are small, wingless parasites primarily associated with birds, while sucking lice (Anoplura) are associated with mammals. The detrimental effects of agricultural and industrial chemicals on their host vertebrates are obviously reflected in these poorly known insects as well, but there are little, if any, data on such secondary effects.

IV. DEVELOPMENTAL STAGES

A. EGG

The biology of insect eggs has been reviewed, for example, by Sander[1319] and Hinton.[690] Insect eggs are normally surrounded by a vitelline membrane and chorion and an internal layer of wax. The chorion is a complex structure consisting of layers of proteins and lipoproteins.[245] These layers might chemically react with copper and prevent it from entering the interior of the egg. Egg number is correlated with body size in Diptera, but not in Hymenoptera or Coleoptera, whereas egg size and ovary volume are correlated with body size in all three orders.[124]

Pollutants may directly affect the eggs of terrestrial insects. In experiments the eggs of the European pine sawfly (*Neodiprion sertifer*, Hymenoptera, Diprionidae) were subjected to simulated acid precipitation. Egg viability analyses showed that the proportion of hatched first instar larvae increased with increasing acidity. Although the precise mechanism remained unclear, it was evident that direct acid-induced changes took place in the eggs themselves.[667,668] Pollutants may also affect insect populations by disturbing the synchrony between hatching and phenology of the host plant. This was suggested as the explanation of the reduced numbers of *Aphis frangulae* (Homoptera, Aphididae) on *Frangula alnus*.[524]

Studies on chronic toxicity of copper to the aquatic midge *Chironomus decorus* (Diptera, Chironomidae) showed that larval stage was the most sensitive, while eggs were least sensitive. At concentrations that caused sublethal effects on larvae, the development and hatchability of eggs were not affected.[867]

B. LARVA

The larval development of insects has been reviewed, for instance, by Chen,[248] Jungreis,[807] and Reynolds.[1275] In Apterygota (Thysanura, Diplura, Protura, and Collembola), metamorphosis is slight or absent. The young insects are smaller and lack reproductive organs. In Exopterygota or hemimetabolous insects, larvae (nymphs) are structurally similar to adults, but the wings and reproductive system are undeveloped. Their cuticular exoskeleton must be shed, and a larger one constructed, the biomass increasing two- to fourfold. Each molt (ecdysis) terminates an instar, and there are commonly from five to seven instars, the final one being of longer duration and terminated by metamorphosis to the adult.[553] In Endopterygota or holometabolous insects, the larvae differ greatly from their adults as regards outer structure, musculature, nervous system, tracheation, and many internal organs. The larval stage is adapted to invade diverse habitats and utilize a range of food resources.[553]

Most studies on the effects of pollutants on insects primarily deal with larvae, especially in aquatic species. Differences have been observed between honeybee adults and larvae in their susceptibility to similar levels of the insecticides carbofuran or dimethoate. The results of Davis et al.[359] suggested that long-term exposure of honeybee larvae to insecticide-contaminated diets at concentrations not immediately lethal to worker adults may cause significant latent damage to the colonies. They studied carbofuran and dimethoate at concentrations sublethal to adult worker honeybees (<1.25 μg/g food) on larval growth and pupation success in the laboratory. Honeybee larvae initially exposed to either insecticide gained weight at a slower rate than did control larvae, and died relatively early. Insecticide concentrations of 1.25 μg/g rarely disrupted the growth of larvae first exposed at an age of 72 to 96 h. Concentrations sublethal to adults, on occasions, reduced the mature weight of larvae first exposed at an early age (44 h). There were reduced numbers of potentially viable pupae when these larvae were exposed to carbofuran at 1.25 μg/g or dimethoate even at 0.3 μg/g.[359]

In chironomids the life stage most sensitive to pollutants is the larva. The larval stage most sensitive to the effects of nickel was the first instar larva, followed by the second instar larva in the case of *Chironomus riparius*.[1226] In the case of *C. tentans*, the most sensitive life stage to copper was the second instar larva.[1116] Giesy and Hoke[541] recommended that a 10-day assay starting with the second instar larvae of *C. tentans* should be used in a screening evaluation of sediment toxicity.

Biochemical and structural changes in the integument may explain the differences in the behavior of copper and cadmium between developmental stages in the grasshopper *Chorthippus brunneus*. In the second and third instar the integument is a flexible, waxy membrane, whereas in the fourth instar and adult animals the integument is more protective and shows increased sclerotization.[734]

C. PUPA

Holometabolous insects or Endopterygota have a pupal stage. The pupa is a quiescent, nonfeeding stage often adapted to survival in unfavorable environmental conditions. Histolysis of larval tissues predominates during the early period, whereas histogenesis of adult structures occurs in the later period. The duration of the pupal stages often ranges from 4 to 14 days, but may be greatly prolonged by lower temperatures or by a diapause stage.[553]

Little attention has been paid to pupae with respect to pollution. The pupal stage is generally well protected against the effects of pollutants. High concentrations of metals and other pollutants have been reported in pupae,[656,658] and pupal or cocoon size has been measured in relation to pollution.[657,666] Pollution may also cause prolongation of pupal development.[662]

D. IMAGO

The adult stage of an insect is specialized to reproduction and often also to dispersal. Thus, it may be heavily affected by pollutants. For example, industrial melanism has been reported only in adult insects.

Adult size is usually considered to be directly associated with fecundity. Since pollutants affect insect size,[657,666] they also indirectly affect the number of offspring produced. Females of the sawfly *Neodiprion sertifer* reared on metal-polluted needles oviposited an average of ten eggs less than females reared on unpolluted needles (69 vs. 79 eggs). However, the effect of reduced female size was masked by the higher egg viability in the former group. In both groups, larger individuals produced relatively fewer viable eggs.[667]

While a large proportion of studies on terrestrial insects deal with adults, only few aquatic studies consider adult insects. For example, studies have seldom investigated the use of adult aquatic insects as pollution bioindicators.[263,273,869,999]

V. CONCLUSIONS

Among the four or five trophic levels commonly present in many terrestrial and freshwater food chains, insects occupy the critical middle links.[1483] Most investigations dealing with the effects of pollution on terrestrial insects have concentrated on herbivores, especially aphids and a few defoliators. However, the results cannot be generalized to other trophic levels or even to other feeding guilds among herbivorous insects, since there are fundamental differences in exposure routes and biological uptake. It is necessary to obtain more basic information on other trophic levels as well because the ecosystem structure and functions are often the final objects of environmental studies. For example, far too little attention has been paid to parasitoids and other predators. Although insects play a crucial role in decomposition processes and the mineralization of organic matter, only a small fraction of the literature is concerned with soil insects, and most of these papers are concentrated on metal contamination. Research on insect-fungus and insect-plant-fungus relations in polluted environments has hardly begun.

Although species should clearly be the basic taxon in ecotoxicology, some generalizations seem to be possible within higher taxa with relatively uniform ecology and physiology. Several insect orders are poorly known, and some of them have not been investigated in relation to pollution at all. On the other hand, even closely related insect species at the same trophic level may differ considerably from each other in physiology and behavior, so that the lumping of species into genus groups or even higher taxonomic groups should be avoided.

Immature stages have received too little attention, except for larvae/nymphs of some aquatic species. Nevertheless, altered mortality or developmental failures in eggs and young larval instars induced by pollution may be of great significance to population dynamics of insect species. Developmental changes are often linked with changes in timing, e.g., growth rates, hatching dates, or duration of immature stages, which may be overlooked in laboratory experiments, but may be of utmost importance in the field.

Chapter 3

EFFECTS OF POLLUTION IN DIFFERENT ENVIRONMENTS

I. SPATIAL AND TEMPORAL SCALES

To understand the effects of pollution on insects and ecosystems as a whole, it is necessary to view the patterns and processes on the appropriate spatial and temporal scales. Investigations of biological systems should be carried out at various hierarchical levels in order to ensure that lower-level driving forces and higher-level regulating processes are included.[90]

Recently, Eberhardt and Thomas[412] described and classified methods for designing environmental studies. The main problem identified by them was the inadequate sample sizes used in field experimentation. They defined eight categories of techniques for field studies, in terms of the nature of control exerted by the observer, by the presence or absence of a perturbation, and by the domain of study (Figure 1):

1. Experiments with replication were preferred when feasible. Confirmation of the fact that two experimental outcomes are indeed different depends on randomization and replication to provide a measure of variability in units treated alike.
2. In experiments without replication, cost or circumstances prohibit replication, whereas in intervention analysis, the intervention is not subject to control by the investigator. Intervention analysis measures the effect of known perturbation in a time series.
3. Sampling for modeling is an efficient form of experimentation for parameter estimation in specified nonlinear models.
4. Observational studies contrast selected groups from a population.
5. Analytical sampling provides comparisons over the entire population. Instead of experimentation in intact, functioning ecosystems, parts of systems or model ecosystems such as microcosms, are normally used, but any real confirmation will depend on combining the experimental data with observations of real-world systems. Major emphasis should be put on the observational process and the design of the sampling effort.
6. Descriptive sampling is an efficient estimation of means and totals.
7. Sampling for pattern provides a description of spatial pattern over a selected region. It may produce, for instance, a contour map.

The ability to detect patterns is a function of both the extent and grain. Extent is the overall area encompassed by a study, and grain is the size of the individual units of observation, e.g., quadrats. Finer-scale studies may reveal

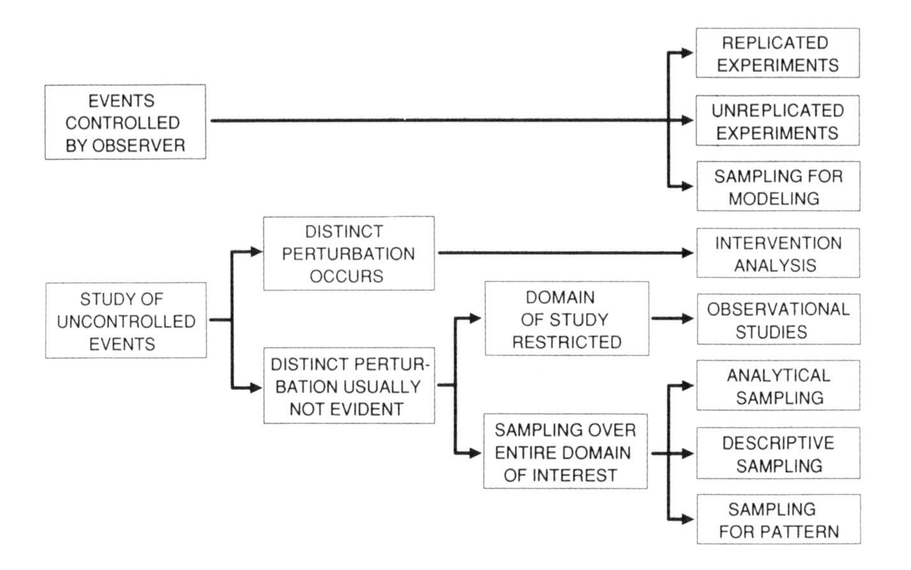

FIGURE 1. A classification of environmental field studies. (From Eberhardt, L. L. and Thomas, J. M., *Ecol. Monogr.*, 61, 53, 1991. With permission.)

in greater detail the biological mechanisms underlying patterns, but generalizations are more likely to emerge at broader scales.[1607]

For example, acidic precipitation affects soil invertebrates in forests at different scales. There are regional and local differences between forests, but stemflow-induced soil acidification creates a more complex mosaic. Stemwater running down the trunks of beech (*Fagus sylvatica*) has an acidifying effect on the soil near the base of the trunk.[458,1635] The deposition of acidifying substances may be two to four times higher close to the stem compared to the stand in general. The increase in soil acidity was observed to amount to 50% after 15 years, and only small further recovery occurred over the next 10 years.[459] Wolters[1640] reported that soil fauna is able to respond to such changes (Figure 2).

The effect of environmental pollution on insect communities is difficult to evaluate because of the diversity of insect species and the natural fluctuations in their populations. Levin[928] and Moriarty[1075] suggested that there was a need for toxicological studies on the functioning of ecosystems, rather than merely investigations on the effects of chemicals on individuals or on populations of particular species.

Ecological succession is the process and pattern of change in the biotic community over time. Change can occur at any site, since organisms arrive at different times, and once established, they change the living conditions for themselves and other organisms, benefiting some and forcing others into local decline.[1233] As the vegetation changes, e.g., when a plant responds to pollution, the animals respond to this change. Animals modify the assemblages of plant

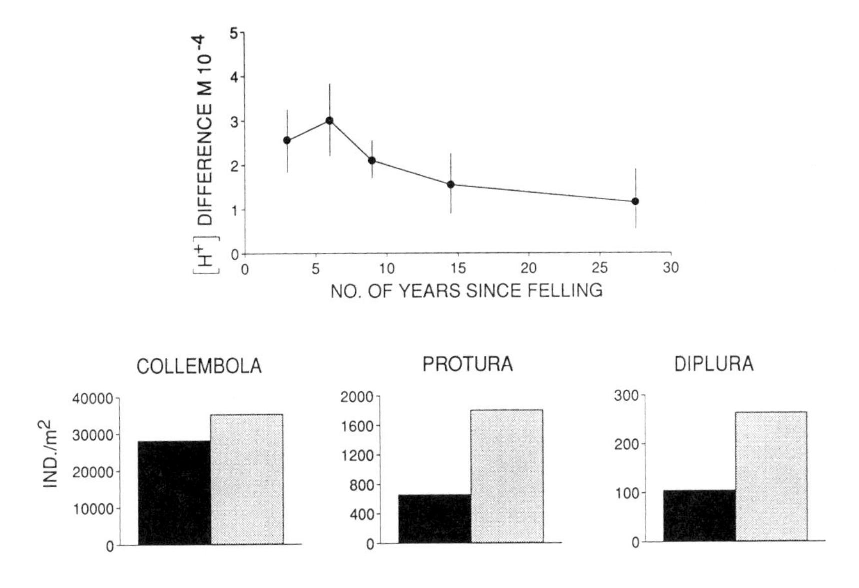

FIGURE 2. Effects of stemflow-induced soil acidification on soil fauna. Upper part shows the difference in H^+ concentration between proximal and distal soil samples in the most evident stemflow transect around each beech stump, related to the number of years since felling in a beech (*Fagus sylvatica*) forest in Sweden. Mean ± 95% confidence interval for each class is based on 15–18 stumps (five in the year class 25–30). (Redrawn from Falkengren-Grerup, U. and Björk, L., *Environ. Pollut.*, 74, 31, 1991. With permission.) Columns of the lower part shows the densities of Collembola, Protura, and Diplura close to the trunks of beech (shaded; pH $H_2O < 4$) and far from the stems (black; pH H_2O about 5.5).[1640]

species by affecting location, reproduction, and survival, the mechanisms whereby mature communities persist, and the role of other major ecosystem processes.[1233]

The baseline data required for making postpollutant evaluations on ecosystem structure and function are largely lacking. This is especially true of insects. The available data mainly concern individual species, not the broad interactions involved in ecosystem analysis.[1497] Bordeau and Treshow[160] drew the following tentative generalizations about ecosystem responses to air pollution, which are probably applicable not only to plants, but also to insects. Acute exposure sufficient to kill dominant species, such as has occurred around point sources, tends to shift succession back to an earlier stage. This is generally accompanied by a decrease in species richness. Chronic exposure, as exemplified by regional pollution at locations more remote from point sources, brings about gradual modifications of the ecosystem structure to adapt to the new conditions. Species are replaced by more tolerant ones that are better able to compete in the new environment. Steeply graded distributions of species' relative abundances are commonly seen in polluted environments and in early successional communities, with a handful of dominant species accounting for most of the individuals present.[1004]

II. CLIMATIC AND VEGETATIONAL ZONES

A. ARCTIC

Arctic vegetation is widely distributed along the northern parts of Eurasia and North America, as well as on several islands, such as Iceland and Greenland. The Arctic zone is underlain by permafrost. The Arctic areas are treeless, and large areas are glaciated. Compared to the other vegetational zones, the ecosystems are relatively simple. They are characterized by short, intense growing seasons and low decomposition rates.[1120] Although the Arctic areas are poor in insect species, insects may play an important role both in terrestrial and aquatic ecosystems. The proportion of Diptera, including many aquatic species of Chironomidae, Culicidae, Simuliidae, is high in the Arctic insect fauna.[347] Generalist Diptera are dominant pollinators in the Arctic, where few bee species persist.[348] Insects have evolved the ability to perceive environmental cues that signal the approach of regular, geographically widespread seasonal changes. They can respond to these cues by undergoing specific physiological, morphological, and behavioral modifications (diapause syndrome) that prepare them for the approaching adverse conditions.

Although the Arctic areas are sparsely populated and there is little industry, several widely distributed pollutants have reached these ecosystems, and some volatile compounds, such as HCB and toxaphene, are often detected at concentrations similar to those in source regions.[773] However, metals originating from anthropogenic emission sources are found in lower concentrations in the Arctic regions of Greenland, Iceland, and Svalbard than on the Scandinavian mainland.[1309] The Arctic areas are apparently sensitive to environmental contamination.[773,1120] Increasing mining and oil industry are apparent threats to the biota. The effects of oil production on insects were studied at Prudhoe Bay in Arctic Alaska, which is the center for oil and gas exploration as well as a focal point for crude oil production. One source of contaminants was the drilling fluids seeping from reserve pits.[1590] Unlike other invertebrate groups, dipterans, particularly chironomids, increased in abundance in ponds that received direct discharges of reserve-pit fluids. A shift in invertebrate community composition appeared to occur in ponds receiving reserve-pit discharges. Such shifts towards increasing numbers of dipterans and diminished numbers of other macroinvertebrate taxa and lowered species diversity have also been documented in temperate streams affected by oily effluents.[1648]

Arctic ecosystems may be the ecosystems most sensitive to climate change. Climate models show that a greater increase in mean annual temperature will occur in these regions compared to lower latitudes.[609] Arctic vegetation would undergo widespread changes. Owing to the presence of permafrost, the active layer of the Arctic soil is shallow, and the top 10 cm of the soil contains most of the root and rhizome systems that in these ecosystems represent the major component of the living biomass.[1460] Up to 90% of the carbon dioxide (CO_2) evolving from the soil is derived from root and rhizome respiration.[136] Tundra soils also contain large quantities of organic matter that, being primarily located in the permafrost, is normally unavailable to decomposers. The following scenario emerges from the available data: as atmospheric CO_2 concentration and temperature rise, the growing season lengthens, decomposition of organic matter increases sharply, nutrient availability increases, net CO_2 uptake increases, and transpiration increases as a result of the higher temperature and increased conductance.[89] After a while, however, the water table recedes, photosynthesis and net ecosystem productivity decrease, and the system becomes a CO_2 source, and a positive feedback loop would be established.

The Antarctic continent is almost totally covered by ice. Terrestrial life is virtually confined to the exposed 1% of the land area. The insect fauna includes chironomid midges, springtails, feather lice on birds, and sucking lice on seals, as well as bird fleas. The terrestrial ecosystems in the Antarctic have been considered ideal test beds for investigating environmental impacts resulting from human activities.[910]

B. BOREAL, NEMORAL, AND MEDITERRANEAN

Most forms of pollution, including the overuse of nitrogenous fertilizers, pesticides, gaseous industrial emission, or oil spillage, occur in the developed and more industrialized countries in the temperate region of the world. Pollution is by no means a new phenomenon in these areas. Two hundred years ago the River Thames in London was an open cesspool. During the 1800s there was considerable industrial pollution of the atmosphere and rivers in much of the developed world. Although the emphasis in this book is largely on temperate species, it is partly an artifact reflecting the lack of data from the tropics and elsewhere.

The boreal zone supports coniferous forests with some deciduous tree species and an understory of a few common vascular plants, mosses, and lichens. Terrestrial ecosystems have two contrasting spatial elements: the enormous size and relative homogeneity of the predominantly forested areas, and the heterogeneity on the scale of local habitats.[349] Local differences are due to both edaphic differences and the successional stages of forests and peatlands. Aquatic habitats are also diverse and support large populations of Chironomidae, Culicidae, Simuliidae, and other aquatic insects. Important insect habitats in the boreal zone include the canopies of trees, understory plants, successional habitats in disturbed areas, surface litter, dead wood, the soil, and fresh waters. The boreal soil fauna differs from that of more southern areas by the absence or reduction of several groups of insects and other invertebrates. Oribatid mites and fly larvae are, hence, especially important for the fertility of northern forest soils.[349] Some herbivorous insects are also characteristically boreal, including diprionid sawflies and scolytid bark beetles. The population dynamics and especially outbreaks of forest insects have received much attention, to some extent in respect to air pollution.[77,261,517,589,660,661,914] Pesticides have been used for the control of some forest pests, such as the spruce budworms in North America.[1322]

In seasonal environments insects are often unable to survive the winter unless a specific developmental stage has been reached. Dormancy enables insect survival during adverse seasons and times the appearance of active stages to coincide with suitable conditions. In many boreal species a small portion of the population remains in hormonally controlled diapause for more than one year.

The Nemoral zone is less homogeneous and can be divided into smaller sectors, e.g., on the basis of continentality or aridity. It is usually characterized by diverse deciduous forests. Food chains are relatively complex, and plants are susceptible to a variety of insect pests. The litter on the ground is broken down by microorganisms, fungi, and bacteria. Insects and other arthropods, but mainly earthworms, feed on the litter and render it more accessible to microorganisms.

Most entomological studies on the effects of pollution have been conducted primarily in densely populated and industrialized nemoral areas, especially in North America or Britain and central Europe. However, serious pollution problems also exist in Japan as well as in eastern Europe and Russia. For example, the highest concentrations and deposition levels of sulfur compounds in Europe occur in southwestern Poland and neighboring regions in eastern Germany and Czechoslovakia. About 50% of the forest cover was affected in the former territory of East Germany in 1987. Soils contaminated by pollutants include 25% of the agricultural land. The middle or lower courses of virtually all large rivers in Bulgaria are heavily polluted by heavy metals, nitrate, oil derivatives, or detergents.[750]

Mediterranean areas (including the Mediterranean, Californian, Chilean, Capensic, and Australian floristic biomes) have also suffered from these pollutants and are dealt with here, although they could be classified as the subtropical winter rain region.[1547] Most of the studies in such areas come from California, where much attention has been paid to ozone and acidic precipitation. Being important agricultural areas, the use of pesticides is also widely distributed.

It is generally concluded by a variety of computer projections that climate change due to global warming will be relatively greater at higher latitudes.[609] Thus, boreal and Arctic species may ultimately be in greater jeopardy from climate change, although tropical systems may be more diverse and are currently seriously threatened by habitat destruction. An attempt to map climate-induced changes in world biotic communities projected that high-latitude communities would be particularly stressed;[435] the boreal forest area, for example, was projected to decrease by 37% in response to global warming of 3°C. In response to a doubling of CO_2 concentrations, Murphy and Weiss[1102] suggested a probable decline in the boreal habitat, caused by vegetation moving upslope in the Great Basin mountain ranges. It was estimated that this reduction in habitat would result in the loss of approximately 44% of mammalian species and 23% of butterfly species, on average, in the different mountain ranges.[1192]

C. SUBTROPICAL AND TROPICAL

The principal part of the global forest biomass, about 50%, is found in tropical forests, which are characterized by a large number of tree species and huge diversity of insects. The balance between nutrient immobilization and mineralization is often delicate in the tropics. In relatively nutrient-poor systems, such as the Amazonian rain forest, the highly efficient recycling of nutrients from decomposing leaf litter back into the vegetation is considered critical for the maintenance of primary productivity. Mycorrhizae are also important in nutrient cycling. In many areas tropical soils are considered sensitive to acidification.[1019,1294] More and more attention has been paid to environmental pollution in subtropical and tropical areas.[479,916,1019,1043,1293,1294]

Biodiversity is not evenly distributed, and there really are conservation hot spots, which are consequently also risk areas of special concern in respect to potential pollution damage. Although these areas vary from one group of organism to another, it is clear that some tropical areas account for a very large proportion of the total biodiversity. This is illustrated by the following example on butterflies.

Collins and Morris[299] analyzed the distribution of endemic species of swallowtail butterflies (Papilionidae) throughout the world, adopting political boundaries in their analysis (Figure 3). The results showed that the five countries (Indonesia, Philippines, China, Brazil, Madagascar) with the highest swallowtail species endemism included 54% of the world total, and in the further five countries (India, Mexico, Taiwan, Malaysia, Papua New Guinea) the total increased to 68%. There were seven other countries (U.S., Cuba, Ecuador, Colombia, Australia, Andaman and Nicobar, Jamaica) with at least three endemic species. Similar analyses of critical faunas in smaller areas could probably be used in environment impact assessment on a smaller scale as well. However, when dealing with small areas with little endemism, the use of rarity[452] and species richness as complementary or supplementary criteria may be convenient.

Pest problems caused by herbivorous insects are most serious in the tropical areas. The indirect effects of pollutants on pests are poorly known, but the extensive use of agricultural chemicals has caused serious damage in nontarget insects and, via insects, also at higher trophic levels. Pesticides have been used, for example, in Africa on a wide scale since the 1950s when ground-applied DDT dust and aerially applied BHC (lindane) smoke eradicated the tsetse fly *Glossina pallidipes* from South Africa in 1954. In the more recent tsetse fly control campaigns, dieldrin, endosulfan, and synthetic pyrethroid insecticides have been used. Indiscriminate aerial spraying of high doses has caused serious damage to a wide variety of other organisms, including many pollinating insect species.[995]

So far other forms of environmental pollution in the tropics are mostly localized, but this does not mean it is absent. For instance, some 1100 kg of arsenic oxide were discharged daily in smoke from gold mining camps in Ghana. In Zambia about 1250 kg of SO_2 were emitted daily in some areas of

FIGURE 3. An analysis of critical faunas of swallowtail butterflies (Lepidoptera, Papilionidae) in the world. The map shows the smallest combination of countries (or comparable areas) in which all the species are represented at least once. Black portion of the pies indicate the proportion of endemic species of the total number of swallowtail species in each of these countries. The size of the pies indicates the total number of species in each country as follows: <12, 12–24, 25–49, 50–100, >100 species.[299]

the copper belt.[1214] In developing countries pollution results not only from industrial effluents and the use of pesticides, but also from forest and savanna fires.[334,1652]

III. HABITATS

A. FORESTS

Forest with closed canopies currently cover about 29 million km^2.[753] They are major stores of carbon, and they moderate local climates. Together with wooded grasslands and shrublands, forests cover 53 million km^2 or some 40% of the Earth's land surface.[753] About 82% of the plant biomass on land is wood in forests.[1547] Temperate forests cover 1.8 billion ha, and the temperate forest ecosystems produce 200 to 400 tons per hectare per year of biomass. With the primary productivity of 5 to 20 tons per hectare per year, temperate forest ecosystems rank third behind only tidal zones and rain forests.[1398] The variety of forest forms include evergreen and deciduous, coniferous and broadleaved, wet and dry, closed canopy forest, and open woodland. The major forests of the world have been classified into tropical hardwood forests, temperate hardwood forests, and coniferous forests, but there is much diversity within each of these broad groups.[871] Temperate forest ecosystems are located in the zone of maximum air pollution because of their extensive distribution throughout the zone of primary urbanization and industrialization in the world.[1398]

Trees regulate most biotic interactions in forests by steering ecosystem processes such as succession. Factors capable of causing injury, disease, and mortality in forest systems are called stresses.[1398] Mattson[996] reviewed the role of insects in forest ecosystems. Trees subjected to stress undergo physiological changes and are frequently attacked by insect herbivores. The stresses may be natural, e.g., microbial infection, climatic extremes (e.g., drought), nutrient deficiencies, and age; or anthropogenic, such as soil erosion, poor harvesting practices, and perhaps air pollution. Stress decreases tree vigor and increases the susceptibility of trees to some insect pests as well as parasitic fungi, bacteria, nematodes, and viruses. Several factors often operate simultaneously (Figure 4).[261,589] Insects and other arthropods play important roles in the structure and functioning of forest ecosystems. The release of nutrients from organic compounds is accomplished by a large number of soil microbes and animals. The detrimental effect of high population densities of certain insect species can cause widespread forest destruction. Well-known examples include the Douglas fir tussock moth (*Hemerocampa pseudotsugata*), the gypsy moth (*Lymantria dispar*), the eastern and western spruce budworms (*Choristoneura* spp.), and southern and western bark beetles (*Dendroctonus* spp.) in North America, and spruce bark beetle (*Ips typographus*) and pine sawflies (*Neodiprion sertifer*, *Diprion pini*) in Europe.

Forest insect epidemics are often initiated in mature, slow-growing forests; during epidemics it is usually the least vigorous trees of all age classes that are attacked.[589] There are several hypotheses explaining insect outbreaks in forest

FIGURE 4. Damaged Norway spruce (*Picea abies*) forest on the mountains (Sudetes) in Poland. The planted trees, originating from lower altitudes and stressed by cold winters and pollutants from brown coal burning, were first attacked by *Zeiraphera diniana* (Lepidoptera, Tortricidae), after which they were injured by snow and storm. Increased amount of dead wood provided breeding material, which caused an outbreak of *Ips typographus* (Coleoptera, Scolytidae). (Photo by M. Nuorteva.)

ecosystems.[126,712,1028] Outbreaks are highly individual phenomena, and only detailed investigations of the host plant, the insect herbivore, its natural enemies, and environmental conditions permit predictions of future outbreaks of insect pests.

Fungi play an important role in forest ecosystems. Many of the decomposition processes involved in the mineralization and recirculation of nutrients are carried out by fungi. Tyler[1507] showed that the macrofungal flora in Sweden was characterized by great differences in the chemical properties of the soil (base saturation and organic matter content of the humus horizon). Sporophore production of most mycorrhiza taxa was distinctly highest in the most acidic soils. Thus, soil acidification is likely to affect both the species composition and sporophore production of fungi and, consequently, insect species associated with them (e.g., Diptera, Mycetophilidae).

Forest trees form ectomycorrhiza with fungi. This symbiosis is regarded as a prerequisite for a sufficient supply of minerals or even their survival in poor soils. Gehring and Whithan[532] demonstrated that the herbivore-mycorrhizae-host plant interaction differs between resistant and susceptible trees. In their study on the pinyon pine (*Pinus edulis*), herbivores reduced the mycorrhizal levels of the trees by an average of 33%. Thus, the effects of pollutants on any of these interactive components may be reflected in the other components as well. Nitrogen deposition, especially, can be detrimental to mycorrhizas.[484]

The biggest threat to forest death is often regional pollution that can change forests across national boundaries.[1395-1398,1480] Tree or forest decline is a widespread phenomenon in both Europe and North America,[138,529,589,1238,1381,1480,1497] but forests in China, Korea, and Japan[916] have also suffered from air pollution. Over 100 years ago forests began to die in the vicinity of industrial plants, which were then appearing in Europe. Pollution control was an unknown concept. Tree decline starts with chlorosis of the older leaves, often leading to premature defoliation. The younger needles of conifers turn yellow and are shed. The shoot tips finally die back to produce a thinning canopy. The symptoms include loss of foliar biomass, loss of feeder-root biomass, and decreased annual increments in height and diameter. Tree vigor is weakened, and the growth rate is reduced. Death of the whole tree may take several years.[496,589,1497] The effects of forest decline may be unexpected, as described by Treshow and Anderson.[1497] In Brazil the lush tropical forest of the Serra do Mar above Cubatao was completely destroyed by fluoride pollution from an industrial source in the Moji River Valley. This, combined with torrential rainfall, resulted in hundreds of mudslides over an extensive area.

There is ample circumstantial evidence to show that air pollution can be the determinative factor causing forest decline.[1346,1480] There are at least six hypotheses[589,1348,1511] that invoke air pollution as the cause of the tree decline syndrome:

1. The acidification-aluminium toxicity hypothesis states that the acidification of forest soils, resulting from air pollution, leads to toxic concentrations of soluble aluminium and necrosis of the fine roots.[37]
2. The ozone hypothesis postulates that excessive concentrations of ozone lead to foliar loss and other growth-reducing effects. It is based on field observations and controlled-exposure studies of foliar symptoms and high ozone concentrations.
3. The magnesium deficiency hypothesis holds that leaching of magnesium from foliage is accelerated by ozone or frost damage to the cuticles and cell membranes, resulting in the yellowing of conifer needles at high elevations.
4. The excess nitrogen hypothesis states that the nitrogen supply to the trees is in excess as a result of increased atmospheric nitrogen deposition. Nitrogen may increase growth and the demand for other essential nutrients, leading to subsequent deficiencies in these other elements; inhibit or cause necrosis of mycorrhizas; increase susceptibility to frost; increase susceptibility to root diseases; alter the root/shoot ratios; or change patterns of nitrification, denitrification, and nitrogen fixation.
5. The hypothesis concerning the transport of growth-modifying organic substances holds that synthetic organic compounds contribute to the symptoms of forest decline by changing the balance of growth regulators.

6. The general stress hypothesis states that air pollution leads to a decrease in photosynthesis and a diversion of photosynthate from mobile carbohydrates to less mobile secondary metabolites. This, in turn, leads to a reduced energy status and an increased susceptibility of the trees to other stress factors.[589]

In addition to air pollution, other factors may contribute to, or be the primary cause of, tree decline. These include wind in association with wood-decaying fungi, pests and pathogens, climatic and other natural physical stress, and forest successional processes. All the forest-decline hypotheses assume that the decline is caused by stress. The metabolic response of trees to stress is the breakdown and mobilization of nitrogen compounds and the translocation of nitrogen from the stressed tissues. White[1596] suggested that stressed plants are a more suitable food source for invertebrate herbivores because stress causes an increase in the tissue content of soluble nitrogenous compounds and, hence, an increase in the nutrient value of the tissue to herbivores. White[1596] cited a number of publications that demonstrated that exposure to airborne pollutants increases the content of free amino acids in damaged tissues and predisposes plants to attack by insects. Mobilization of nitrogen and the associated increase in insect herbivores only occurs at intermediate levels of exposure.

There are numerous studies on the effects of air pollution on insects,[27,77,79,122,165,175-177,247,516,517,644,646,651-668,1029-1031,1079,1219-1221,1284,1366-1371,1385,1472,1530,1531,1586,1587,1599,1631] but causal relationships have usually remained uncertain, since the evidence is mostly correlative. The majority of these investigations were conducted in forest ecosystems subjected to particulate and/or gaseous pollutants. Several investigations on the impact of acidic precipitation *per se* on plant-insect interactions and soil fauna have been conducted in forest ecosystems.[320,583,586,587] By affecting tree health, air pollutants are likely to affect the interaction between trees and insect pests. They do so not only by exerting an effect on the tree, but also by affecting insect growth rate, dispersal, fecundity, pheromone detection, host finding, and mortality.[27,589,657,662,666-668] Previously innocuous species may occasionally become pests.[589,1370,1371]

For example, Sierpinski[1369] reported several cases of insect attacks in forests weakened by chronic industrial air pollution in Poland. He suggested that some pest species could be used as bioindicators of air pollution: namely *Zeiraphera griseana* (Tortricidae), *Cephalcia falleni* (Pamphiliidae), and *Pristiphora abietina* (Tenthredinidae) on spruce; *Dreyfusia nordmannianae* (Adelgidae) on fir; *Taeniothrips laricivorus* (Thysanoptera) on larch; and *Exoteleia dodecella* (Gelechiidae) on pine. Air pollution may increase pests both directly and indirectly as a consequence of increased illumination within the tree stand. Sierpinski[1369] anticipated an insect succession: attack by insects preferring dense stands thins the forest and results in more intensive penetration of industrial gases. When more trees are weakened, they are then attacked by species preferring more open stands.

According to Anisimova,[44] some bark beetles, especially *Tomicus piniperda* and *T. minor*, increased in an area near an aluminium plant. In the heavily polluted zone, the crowns of trees taller than 10 m were totally damaged. Damage to young stands by curculionids increased as well. Mozolevskaya and Pechenzhskaya[1084] studied forests affected by gaseous wastes from metallurgical plants. Bark beetles (*Tomicus piniperda*, *Ips typographus*) played the major role in the desiccation of tree stands. Anisimova and Sokov[45] noted that the increased annual litter fall from trees in polluted forest areas created long-term foci with conditions favorable for xylophagous insects. Plantations gradually being killed by industrial pollution become a source of insect invasion that is dangerous to adjacent, viable tree stands. The numbers of curculionids and cerambycids (*Monochamus* spp.) in the heavily polluted zone were considerable. Bogdanova[150] observed increased numbers of *Tomicus piniperda*, *T. minor* (Scolytidae), *Phaenops cyanea* (Buprestidae), and *Monochamus galloprovincialis* (Cerambycidae) in pine trunks in the vicinity of an industrial plant near Novosibirsk, Siberia.

These patterns may arise either from a decreased efficiency of the natural enemies controlling pest populations[175,664,1529,1530] or from decreased resistance of the host plants.[176,390,727,771,1596] Furthermore, as Alstad et al.[27] correctly emphasized, it is difficult to make correlations with a parameter that already has a large variance independent of the association of interest; indeed, forest insects often exhibit large fluctuations in population size irrespective of pollution.

Prior to heavy air pollution, forest ecosystems in national parks and the ecosystems of deciduous and coniferous mountain forests in eastern Europe were rather resistant to outbreaks of pests.[230] In the 1980s, however, outbreaks of phytophagous insects, such as *Cephalcia falleni*, the larch bud moth *Zeiraphera diniana*, and fir budworms, occurred in central European mountain forest stands, including the ecosystems of national parks in Poland. This is considered to be associated with a weakening of the tree stands due to air pollution.[230] The degree of air pollution in all of the park was moderate or strong.[1632] Zabecki[1670] investigated the insects attacking firs in the Ojców National Park, Poland. As the dying-off process progressed, firs were attacked by insects zonally, starting from the tops of trees most seriously damaged by industrial air pollution. Major roles were played by *Pissodes piceae* (Curculionidae), *Pityokteines spinidens*, *P. curvidens*, *Pityophthorus pityographus*, *Pityokteines vorontzovi* (Scolytidae), and the family Siricidae. Some xylophagous insects, such as *Sirex juvencus* (Siricidae) and *Serropalpus barbatus* (Serropalpidae), attacked fir trunks in early stages of dying off, before the trees were attacked by cambium feeders.

Although air pollution *per se* will not create conditions conducive for bark beetle outbreaks, it may disorganize forest management, creating conditions for permanent bark beetle problems, as shown by the experiences in Poland and Czechoslovakia.[261] For example, *Pityogenes chalcographus* (Scolytidae) may have become more abundant in areas of pollution damage. This may be associated with air pollution, but inadequate forest hygiene and control measures and insufficient thinning of dense younger stands are likely to have also contributed.[261]

B. AGRICULTURAL AREAS

A number of elements at elevated concentrations are commonly regarded as contaminants of agricultural soil. Anthropogenic sources of contamination include industrial activities and top dressing with wastes such as sewage sludge or liquid pig manure.[1514] The employment of chemical fertilizers in agriculture dates back to the early 1800s when nitrates were imported from Chile and sulfate ammonium was produced as a byproduct of coal gas manufacture. The use of nitrates, phosphates, and potash has been extensive during the 20th century. The application of DDT and related pesticides was introduced on a worldwide basis only after World War II. The use of fertilizers, insecticides, herbicides, and fungicides on agricultural land has caused serious pollution problems in waters and other ecosystems. Agriculture, as well as forestry, has also caused considerable water pollution by suspended sediments.[555] However, this topic is more related to soil erosion and beyond the scope of this book.

There are fewer studies on the effects of air pollution on phytophagous insects in nonforest habitats, including agricultural systems. Detailed studies conducted by Hughes et al.[725-728] and Warrington[1557,1558] described cause-and-effect relationships between pollutant-induced changes in crop plants and the response of economically important insect species. The pollutant used in these studies was sulfur dioxide, a precursor of sulfuric acid formation in precipitation. Little is known about the interactive effects of acid precipitation on economically important plants and insects in agricultural ecosystems.[1427]

Intensive farming is designed to concentrate a large number of animals in a limited area, and this may have adverse consequences on the environment essentially because of the concentration and accumulation of animal excreta. In addition to the specific problems associated with fecal contamination, there are other concerns derived from the use of chemical compounds in feed, with the consequent environmental risks. These compounds are essentially antibiotics, sulfonamides, furans, quinoxalin derivatives, imidazoles, and copper. The use of waste containing such drugs as fertilizer may have an adverse effect on ground-nitrifying bacteria and cause disruption of the purification processes, or it may have toxic effects on aquatic and terrestrial ecosystems.[960] Macrì and Sbardella[959] studied the acute toxicity of nitrofurazone and furaltadone on *Musca domestica*. Furanic drugs, particularly furazolidone, are widely used in medicated feed. Macrì et al.[960] showed that this compound was toxic to *Culex pipiens molestus* (Culicidae) larvae.

The application of sludge to land is increasingly viewed as a practical and economical means of recycling this waste material. Sludge from industrialized cities may be contaminated with a wide spectrum of toxic chemicals, including polychlorinated biphenyls (PCBs), pesticides, and heavy metals.[337] Consequently, sludge application to land may have detrimental effects on the biota.

The trace element of most concern is cadmium, since it is toxic, readily absorbed by plants, and accumulates in the tissues of foraging animals. The effects of sewage sludge on invertebrates are relatively poorly known. Decreases in the populations of dragonflies have been reported in aquatic ecosystems.[1569] Soil invertebrates may be exposed directly to sludge particles or gases from freshly applied sludge. Metals, such as cadmium, were shown to be concentrated by earthworms inhabiting soils amended with municipal sludge.[729]

Huhta et al.[730] studied the animal succession in test plots made of different mixtures of digested, activated, or limed sewage sludge and crushed bark. Flying insects (Coleoptera, Diptera), and the phoretic nematodes and mites transported by them, colonized fresh mixtures in a few days and reproduced rapidly in the beginning. Springtails (Collembola) increased within some weeks and remained important throughout the succession. The community at the early stages of succession was typical of dung. Even in the oldest test material, the community differed considerably from that in the adjacent arable soil.

Culliney and Pimentel[336] demonstrated marked reductions in fecundity and survival of green peach aphids (*Myzus persicae*) on collard plants (*Brassica oleracea*) growing in soil treated with chemically contaminated sludge, as compared to aphids on plants growing either in soil treated with uncontaminated sludge or soil conventionally fertilized. Reduced plant growth and increased aphid restlessness were observed in the contaminated-sludge treatment. Culliney and Pimentel[337] also observed that green peach aphids preferred plants grown in sewage sludges to unfertilized plants. The severe reduction in aphid growth observed on some of the treated plants was possibly associated with the presence of phytotoxic contaminants.

Hughes et al.[729] grew cabbage and collards in soil amended by municipal sludge. Cadmium accumulated at significantly higher concentrations in the leaves of both cabbage or collards grown in the sludge-soil mixture. Cabbage loopers (*Trichoplusia ni*, Noctuidae) feeding on the sludge-grown plants accumulated cadmium from the contaminated foliage. Hughes et al.[729] concluded that the concentrations of metals present in the plant tissue did not inhibit their feeding habits or cause toxicity. Concentrations of the di- and polyamines putrescine, spermidine, and spermine were higher in the leaves of sludge-grown plants and cabbage loopers that had fed upon these plants. Increases in the concentration of the diamine putrescine and glutathione in plants have been reported as possible indicators of various forms of plant stress. The cabbage loopers had a larger final size, but developed more slowly on the sludge-grown collards than on the control plants. Differences in plant nutrition and leaf temperature may have caused the changes observed in insects on these plants.[729]

Agricultural production can be significantly affected by insect pests or by crop diseases carried by vector insects. Consequently, insecticides have been widely used in agricultural areas, but their effects have not been restricted to target and nontarget organisms in the fields.[559,1091] Long-term studies in agricultural areas have indicated declines in the population densities of, for instance, Cryptophagidae, Lathridiidae, Staphylinidae, Lonchopteridae,

Lepidoptera, Parasitica, and Symphyta, as well as spiders and several Carabidae.[5,207,778] Fungicide applications seem to have been important in directly depleting the fungal food resources for mycetophagous species.[779] The long-term effects are associated with the spatial (proportion of fields sprayed) and temporal (frequency of application) scale of pesticide applications.[777,779]

C. GRASSLANDS AND OTHER OPEN AREAS

Treshow and Anderson[1497] reviewed the effects of air pollution on plants in deserts, heaths, and grasslands. For example, annual and herbaceous perennial plants, grasses, cacti, and some shrubs are almost completely absent near copper smelters in the Sonoran desert.[1642] Ecotoxicological studies on copper and cadmium in a contaminated grassland ecosystem in England[733-737] showed that the flora was of low diversity and was dominated by metal-tolerant populations of *Agrostis stolonifera* and *Festuca rubra*. Invertebrates showed significant elevation of total body metal concentrations, with marked seasonal variation.

In southwestern Poland the plant species composition of grasslands subjected to heavy pollution was highly impoverished: only three species accounting for 75 to 97% of the plant biomass. The soil water was very acidic, and the biomass of the soil fauna decreased. Above-ground insects were small, assimilated less efficiently, and thus needed to consume more plant matter.[751]

Great changes have taken place in plant community structure on the heathlands of the Netherlands in recent years, due to elevated nutrient levels.[650] The grasses *Molinia coerulea* and *Deschampsia flexuosa* expanded at the expense of *Calluna vulgaris* and other heathland species. Many grasslands are given applications of artificial fertilizers, e.g., nitrogen compounds. While increasing meadow productivity, these fertilizers alter the species composition, tending to increase the proportion of grasses while decreasing legumes and forbs.[751] This may have serious direct and indirect effects on insect fauna (see Chapter 4, Section III.C). The effects of pasture improvement procedures (including applications of nitrogen fertilization) and subsequent pesticide (chlorpyrifos) use on the nontarget ground beetle and spider fauna of seminatural upland grasslands were studied in Northumberland, U.K.[1310] Species composition of improved sites were poor, and the frequency of use of the pesticide appeared to be an important factor influencing the species composition.

Deserts are arid regions where the potential evaporation is very much higher than annual precipitation. These regions cover 22% of the Earth's land surface, but contain only 0.8% of the total terrestrial plant biomass.[1547] Their insect fauna are scarce and apparently have been only slightly affected by pollution, primarily from oil production and mining industries. After the Gulf War, the desert ecosystem in Kuwait suffered from heavy pollution due to burning of 950 oil wells which produced soot and several oil lakes. Various desert insects including dragonflies and tenebrionid beetles were killed when visiting oil lakes resembling water lakes. However, sand termites and ants seemed to survive in their below-ground chambers near the soot and oil spills.[11a]

D. URBAN AREAS AND ROADS

Urbanization encourages human overcrowding and slum conditions. In such conditions some harmful insects can increase in numbers due to the excess of organic waste. Some insects, like cockroaches, may mechanically transport pathogens. Fleas transported by rats, mice, and other mammals are potential vectors of typhus and plague. Even the common house fly can be a vector of amoebic dysentery, typhoid, and cholera.[216,1575] In most slum areas there is inadequate sanitation, an accumulation of refuse and excreta, and numerous open cesspits, all of which lead to a proliferation of filth flies (*Fannia*, *Musca*, *Calliphora*, *Lucilia* spp.). As reviewed by Nelson,[1119] inadequate housing, the necessity of storaging water, an inadequate and polluted water supply, and the lack of sewage and waste disposal create substrates for vectors such as *Aedes aegypti*, *Culex pipiens*, *Anopheles* spp. (Culicidae), *Phlebotomus papataci* (Psychodidae), *Triatoma barberi*, and synanthropic flies. The mosquito *Culex quiquefasciatus* breeds in polluted waters and is a vector of the filarial worms that cause the disease bancroftian filariasis. The incidence of this disease is increasing throughout much of the tropics as a result of increased urbanization. In much of Africa this species has become the most common man-biting mosquito in the towns.[1214]

Urban ecosystems are not random samples of species, but systems molded by general ecological and evolutionary rules.[495] The number of insect species is usually reduced under the influence of urban pressure.[1139] For instance, park areas in Poland are characterized by a relatively low but constant density of elaterid beetles (30 to 70 ind. per m^2) in comparison to rural meadows (more than 100 ind. per m^2).[1140] Ant species with wide habitat niches dominate in urban areas. Species that succeed in colonizing are relatively free from predators and other natural enemies in urban areas and can thus exhibit very individual-rich populations.[1524]

Lichens are intolerant of atmospheric pollution.[470] Almost all species disappear from heavily polluted areas, sulfur dioxide being probably the main pollutant responsible for their decline. Although several insect species are associated with lichens, there seems to be little information on the indirect effects on insect fauna caused by the disappearance of lichens. André[40] alleged that changes in corticolous microarthropod communities will parallel that of epiphytic lichens during succession. Gilbert[545] showed that the herbivore fauna living on tree trunks, especially psocids, were heavily affected by pollutants, whereas the predators were less affected. The decline of *Alcis jubatus* (Geometridae) has been attributed to the loss of epiphytic lichens on trees in northern Europe.[660] In Sweden *Cleora arenaria* (Geometridae) feeding on epiphytic lichens of trees has been reported to have declined.[417] In addition, industrial melanism is, of course, usually associated with the sensitivity of tree trunk lichens and algae to sulfur dioxide and other pollutants.[829]

The verges of roads and motorways provide a complex habitat for a large number of insects. The impact of roads on terrestrial [154,155,175,176,218-220,485,486,1104,1219-1221,1241,1415,1416] and even aquatic invertebrates[1392] has received some attention. Insects close to roads may also be influenced by changed microclimate.[175,485] Przybylski[1241] surveyed winter wheat, meadows, and orchards along a gradient from a motorway. The number of aphids and Heteroptera was highest near the motorway, but the the total number of arthropods decreased with increasing proximity. Especially the numbers of spiders, ladybirds, lepidopteran larvae, clickbeetles (Elateridae), Hymenoptera, and weevils were highest at longer distances from the road. Port and Thompson[1221] showed that aphids and the larvae of *Euproctis similis* and *Phalera bucephala* (Lepidoptera) abounded near a motorway. The increased abundance of the aphid *Aphis pomi*[486] and *Aphis fabae*[154,155] near a motorway has also been demonstrated. An experiment with different deicing salt (NaCl) concentrations showed no effect on any life-history parameter of the aphid *Rhopalosiphum padi*. However, salt may affect the absorption of nitrogen oxides (NO_x) by soils, its subsequent oxidation, and its availability to plants.[1415]

E. WATERS AND WETLANDS

Aquatic environments of insects include lakes, ponds, running waters, springs, and marine ecosystems. In relation to pollution, it is the whole drainage basin, not just the body of water, that should be considered in environmental studies.[1148] Recently, water pollution levels have decreased in some North American and European waters. However, this is not the global trend. In eastern Europe and in Russia, as well as in several developing countries, the situation is unsatisfactory. The distribution of industry and human population as well as hydrological conditions affect the consequences. Since Chapter 5 deals with water pollution, only some notes are given here.

Plant nutrient levels in waters are often elevated by the urbanization process and discharge of nutrients from agricultural areas (as well as by discharges from agricultural areas). Phosphorus and nitrogen losses from urbanized watersheds may be two to ten times greater than those from forested watersheds. These nutrients may stimulate algal growth in urban streams, leading to changes in aquatic food webs. Levels of metals, toxic organic compounds, and road salt, all of which may be damaging to aquatic life, are greatly enhanced in streams draining urbanized areas. Urbanization of a watershed may significantly alter stream water quality even in the absence of direct industrial or municipal discharges. Suspended sediment loads from streams draining urban areas are often much greater than those from nearby forested watersheds. Suspended sediments from construction activities can adversely affect aquatic life through habitat elimination under heavy loading or by interference with feeding under lighter stress. Other factors associated with urbanization may also alter the hydrology of watersheds. The increase in the impervious surface area and drainage networks has resulted in a substantial increase in the proportion of rainfall that is rapidly discharged from the watershed as direct runoff and streamflow. The volume of flood flow increases, and there is a decrease in low flow volume during nonstorm periods. The decrease in low flow discharges decreases the available stream habitat, increases the likelihood that the stream may dry up, may increase diurnal temperature fluctuations, and increases the concentration of pollutants due to a lack of dilution. The increase in flood flow causes enhanced erosion and scouring.[796]

Most freshwaters in densely populated and industrialized areas like central Europe are more or less polluted, but even such large water bodies as Lake Baikal in Siberia are threatened if pollution continues at present rates.[752] However, the effects on insect fauna of pollutants are often pronounced in small lakes and ponds. For example, the North American prairie ponds may be particularly vulnerable due to the extensive agricultural use of insecticides that may contaminate nontarget ecosystems near cropland and have a serious impact on the natural fauna.[566,1077,1360,1572,1573]

Rapid changes in stream discharge, and the resultant pH fluctuation, may make stream biota more susceptible to acidification than the biota of lakes.[594] The pollution of rivers is affected by both the amount of pollutants and the water volume. In Africa the River Nile carries relatively little water, but runs through densely populated areas, while the River Congo, which is the world's second largest in terms of discharge into the sea, runs mainly through sparsely populated areas.[1214] Large rivers in industrialized countries are generally strongly influenced by pollution, regulation, or both. The Rhine is no exception to this. Its catchment area comprises densely populated parts of Switzerland, France, Germany, and the Netherlands. Waste-water treatment plants have improved oxygen concentrations in river water considerably in recent years, but the pollution of the Rhine by toxic substances is still a serious problem. The Rhine transports large amounts of heavy metals as well as organic pollutants. Only a part of the total organic load can be characterized by chemical analysis. Biological indicators are needed to demonstrate the combined effects of the toxic sustances in Rhine water.[1516]

Wallace and Hynes[1542] suggested that river macroinvertebrates respond to toxic pollution in three phases: an initial catastrophic phase during the initial passage of the chemical, when substantial mortality and downstream drift of susceptible taxa occur; a reversion phase, when pesticide concentration and mortality rates decline, but chronic effects occur; and a recovery phase, when successful recolonization occurs. The severity, extent, and duration of toxic effects will determine the impact on the macroinvertebrate community, while recovery depends on the availability of recolonization sources. Moreover, although many taxa return relatively quickly, recovery in severely affected populations may take several years.[182,601,739,1091,1195,1263]

Wetland ecosystems, including floodplains, freshwater marshes, mangrove swamps, peatlands, and estuaries, play a central role in the water cycle. For example, they absorb nutrients and retain sediment.[753] Coastal wetlands may become susceptible to inundation due to sea-level upheaval caused by potential global warming. The fauna of the small bogs of central Europe are threatened by vegetational changes caused by acidification or eutrophication. Extensive areas of the boreal zone are covered by communities on peaty soils. However, there is little information about the effects of pollutants on insects in such habitats.

F. MOUNTAINS

Mountainous regions cover over 20% of the Earth's surface, but only 10% of the world's people live in them.[753] The mean annual temperature decreases with altitude, and mountainous altitudinal belts are about 100 times narrower than the vegetational zones on plains.[1547] Mountains are split up into very small climatic units (e.g., sunny and shady slopes, snowless and snow-covered sites). Temperature inversions and cold air pools play important roles and can cause a reversal of the order of altitudinal belts[1547] and influence the effects of pollution on plants and insects. Combined with metropolitan areas (e.g., Mexico City), orographic characteristics may cause serious pollution problems.

Climatological differences may result in changes in the vigor of host plants or in the population dynamics of insects in mountainous regions. For example, in the southern Appalachians, latitudinal variation in Fraser fir (*Abies fraseri*) mortality was attributed to atmospheric deposition, and tree sensitivity was suggested to have increased by the balsam woolly adelgid (*Adelges piceae*, Homoptera).[590]

IV. CONCLUSIONS

Pollution is not evenly distributed in the world. However, the studies on the effects of pollution on insects or on ecosystems mediated by insects are still more concentrated to some limited areas of temperate Europe and North America. More research should be directed to the effects of pollutants in tropical and subtropical ecosystems with high species richness that is often restricted to the remaining isolated patches of natural vegetation, e.g., nature reserves. Furthermore, the pollution control is often less effective in developing countries, and due to the considerable pest problems, the use of potentially dangerous pesticides is common. On the other hand, even the remote Arctic areas are not safe from long-distance pollution. These simple ecosystems may be susceptible to relatively small pollutant-induced perturbations, e.g., in decomposition processes. The Arctic areas are also suitable for field experiments with controls utilizing the low local-pollution levels.

The majority of the information on pollutant-induced changes in insects originates from waters, cultivated forests or agricultural areas, or from the laboratory. For example, little is known about the changes in coastal marshlands or mangroves, steppes, or deserts. The forest ecological data is rarely based on studies in untouched forests, so the effects of pollution on the wood-decomposing system or on the conservation value of the forest (e.g., compared to the effects of silvicultural practices) are poorly understood. The results from boreal or nemoral forests are not readily applicable to tropical forests.

Forest decline in Europe is apparently associated with a loss in tree vigor due to exposure to air pollutants. Symptoms of decline vary by species and region, and if air pollutants are responsible, it is possible that a number of interacting agents are involved. Mechanisms of decline are unknown, although several hypotheses have been developed. There is some evidence that air pollution increases the susceptibility of trees to insect pests. Air pollutants may alter plant-herbivore interactions so that a formerly innocuous insect may become a pest.

In agricultural habitats there is an increasing need to predict the ecological consequences of deliberate releases of chemicals (insecticides, herbicides, fungicides, fertilizers, etc.). The principal entomological question is the acceptable reduction of insect populations and the quality of their repeatedly treated habitat.

Scaling is a fundamental factor in both planning studies and interpreting the results. The traditional scales used in environmental studies dealing with vertebrate animals or vascular plants are usually inappropriate for insects. Furthermore, even among insects different scales are needed for different species, e.g., springtails and butterflies. The distribution of pollutants may be mosaic-like and involve mosaics in different scales (e.g., regional level, local level affected by vegetation type and topography, insect habitat level affected by microclimate and single trees). The most important measure is apparently the dispersal or recolonization rate of the insect species in relation to the spatiotemporal distribution of pollutants.

Chapter 4

POLLUTION IN TERRESTRIAL ECOSYSTEMS

I. INTRODUCTION

The effects of pollutants in terrestrial and aquatic ecosystems are dealt with separately, since there are profound differences between terrestrial and aquatic food chains. At most trophic levels in terrestrial systems, food is the principal source of persistent insecticides and many other pollutants, whereas aquatic insects may also acquire residues directly from the abiotic environment.[1075]

Terrestrial ecosystems are contaminated by a wide range of pollutants originating from different sources and having very different distribution. Insects are affected directly or indirectly by gaseous pollutants such as sulfur dioxide, oxides of nitrogen, ozone, carbon dioxide, and fluoride. Since acidic precipitation is associated with the emissions of reducing molecules, it is dealt with after sulfur and nitrogen compounds. The effects of climate change on insects are discussed after carbon dioxide, a major greenhouse gas. Unlike most gaseous pollutants (not to speak about acidification or climate change), metals, agricultural chemicals, industrial chemicals, and fuels usually affect insects locally around pollutant sources. Terrestrial insects are also influenced by radiation, heat, and light from anthropogenic sources.

The most important gaseous pollutants are sulfur dioxide (SO_2), hydrogen sulfide (H_2S), oxides of nitrogen (NO_x), ammonia (NH_3), carbon monoxide (CO), carbon dioxide (CO_2), methane (CH_4), ozone (O_3), and peroxyacetyl nitrate (PAN). The last two compounds are formed in the atmosphere by photochemical reactions.[502] Hydrocarbon and elemental mercury vapors, as well as small particulate material that also contains toxic elements, are atmospheric pollutants, also. Jones[794] stated that, effectively, all soils in industrialized countries have become contaminated with selected trace substances (notably Pb, Cd, PAHs, PCBs, and polychlorinated dibenzo-p-dioxins and -furans) at levels above their historical background, as a result of aerial inputs.

Several classes of sampling described in the previous chapter (Section I) offer important gains of efficiency in environmental field studies.[412] Figure 1 suggests one approach to distinguishing the objectives of four classes of sampling in terrestrial environment (e.g., soil contamination). Although dealing with more general aspects of pollution, it emphasizes the need to set a single set of objectives also for an entomological study when planning field investigations. Descriptive sampling is an "inventory" method, sampling for pattern can produce a contour map, analytical sampling provides a basis for comparisons, and sampling for modeling focuses on parameter estimation for a specific model.

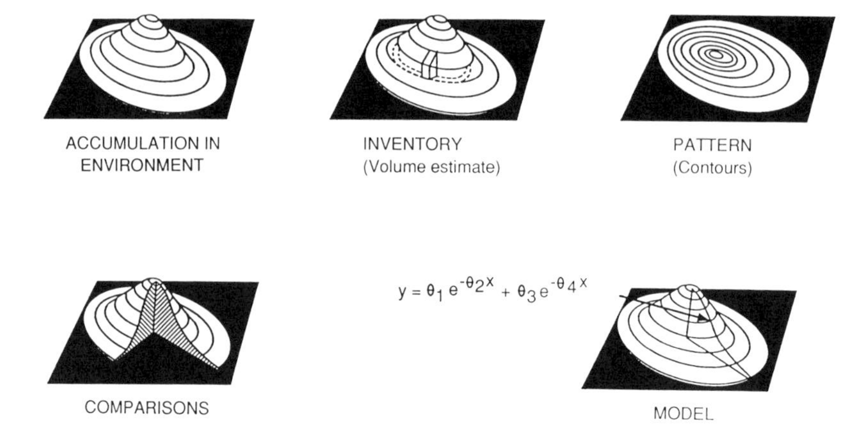

FIGURE 1. The four categories of sampling. The diagrams at the right are hypothetical concentrations in soil around a point source: here suggested as a smokestack. In practice it is likely that the elliptical concentration pattern might be more strongly skewed in the direction of the prevailing winds. (From Eberhardt, L. L. and Thomas, J. M., *Ecol. Monogr.*, 61, 53, 1991. With permission.)

Suspended particulate matter is one of the most relevant atmospheric pollutants. Approximately half of the anthropogenic emissions are derived from fossil fuel combustion from both industrial and agricultural sources.[1514] Particulate emissions may be almost totally controlled in some countries, but older plants and facilities in developing countries often lack such technology.[1497] Small particulate material or dust has a desiccating effect that may be harmful or even lethal to insects. Highly absorptive dusts that absorb the wax-lipid secretion may cause rapid desiccation and death in even large insects.[411] The small parasitic wasps are said to be especially sensitive to dust.[27] Outbreaks of the black pineleaf scale (*Nuculaspis californica*) on ponderosa pine (*Pinus ponderosa*) have been attributed to the sensitivity of its eulophid parasitoid to dust.[413] In contrast, for example, Kulfan[890] found only small differences in the abundance of herbivorous Lepidoptera between sites of high, low, and very low dust deposition around a cement factory in the Little Carpathians in Slovakia. Only the goat willow (*Salix caprea*) was considerably less infested at the most polluted site.

Correlative evidence suggests that air pollution influences the distribution and abundance of terrestrial insect species, although there is still little reliable information about causality.[27,77,261,516,517,589,660,661,1079,1284,1370,1371,1395,1396,1599] There are several mechanisms whereby air pollutants can affect the population dynamics of insects. Führer[517] classified three types of insect outbreak patterns along an emission gradient. The first type, which is common in heavily polluted areas, involves species that have either a high tolerance to the pollutant or a life history that avoids direct exposure to the contaminant. Outbreaks of these species may be due to a reduction in plant resistance or in competitors or predators that are sensitive to the pollutants. The second type occurs in areas of moderate-to-low pollution: species that either have an exposed way of life habit or are highly sensitive to the pollutant, but have a protected way of life. Their increase may result from the elimination or reduction of competitors or predators. The third type comprises species that commonly attack stressed plants, but are scarce in areas exposed to air pollutants. These species may be highly sensitive to air pollutants and have an exposed life history. Thus, they are unable to utilize the stressed host species within a heavily or moderately polluted area.

Invertebrate numbers usually decrease in the immediate vicinity of industrial plants. For instance, the average density of soil arthropods was 35,000 and 161,000 individuals per square meter at distances of 1 and 40 km, respectively, from a metallurgical plant.[1444] Some resistant herbivorous insects may increase in numbers. Reported cases include *Exotelia dodecella* (Lepidoptera, Gelechiidae), *Rhyacionia buoliana*, *Retinia resinella* (Tortricidae),[247,818,1369] and several aphids.[1029,1599] Lesniak[927] suggested that pollution causes changes in the vegetation, which in turn creates favorable conditions for an increase in phytophagous species. Simultaneously, the number of other insects (including saprophagous species and predators) decreases. Near industrial plants, bryophytes and dwarf shrubs are often very scarce, which may affect insects associated with them.

II. SULFUR COMPOUNDS

A. BACKGROUND

The most important sulfur-containing pollutants are sulfur dioxide (SO_2) and hydrogen sulfide (H_2S). Sulfur dioxide is produced naturally by volcanic emissions and forest fires. However, most of the SO_2 in the atmosphere comes from anthropogenic sources; 146–187 $\times$ 10^6 MT are emitted yearly.[502] In contrast, only a small fraction of the H_2S is emitted from anthropogenic sources. One half of the total anthropogenic emissions of SO_2 is due to the burning of fossil fuels.[1068] Total annual emissions have increased from 5 million tons in 1860 to more than 180 million tons in the mid-1980s.[502]

B. CHANGES IN PLANTS

Sulfur dioxide is phytotoxic, but hydrogen sulfide usually occurs at low concentrations that do not affect plants. Access of SO_2 is mainly via stomata. Short-term exposure may cause tissue damage such as necrosis, while more chronic injury can be seen as premature abscission of foliage or chlorosis in leaves. Long-term expose to low-level SO_2 leads to loss of yield, even though clear symptoms in the foliage would be missing. The loss of yield may also be due to general soil acidification caused by sulfuric acid and other phytotoxic substances. Air pollutants such as SO_2 have been considered the most likely factors involved in forest decline.[529,589,1238,1346]

The visible symptoms in plants caused by SO_2 were recently reviewed by Treshow and Anderson.[1497] Air pollution also causes biochemical changes in plants. Plants commonly respond to SO_2 pollution by an increase in the total soluble nitrogen and free amino acid content in the tissues.[176,315,390,550,966,1246,1599] Decreased nitrogen concentrations, after exposure to SO_2, were reported in the leaves of soybean (*Glycine max*) and pea (*Pisum sativum*)[1326] and in American elm (*Ulmus americana*) seedlings.[302] Increased free amino acid concentrations have been reported in SO_2-fumigated bean (*Phaseolus vulgaris*) leaves[550] and SO_2-fumigated jack pine (*Pinus banksiana*) needles.[966] After an 8-week fumigation at SO_2 concentrations of 55 nl/l, pea leaves showed a 7.3% increase in soluble-protein content, but at higher SO_2 concentrations of 185 and 370 nl/l, soluble-protein levels decreased by 7.0% and 16.5%, respectively.[1326] Significant increases in amino acids and amides were found in the shoots of pea plants continuously fumigated with 300 nl/l SO_2 for 18 d.[1237] These studies showed that the biochemical response of peas to SO_2 may be complex, causing some nitrogen compounds to increase while others decrease. SO_2 can increase the amino acid content of fumigated plants, even at low concentrations, as was shown in experiments with the Mexican bean beetle (*Epilachna varivestis*) on Pinto beans.[727] In Scots pine (*Pinus sylvestris*) shoots, total free amino acid concentration and the concentrations of ornithine, lysine, histidine, and arginine were significantly negatively correlated with the distance from a pulp mill, while in Norway spruce (*Picea abies*) only glycine, ornithine, lysine, and histidine had significant negative correlation.[808]

Reducing-sugar concentrations were increased by moderate SO_2 fumigation in jack pine[966] and red kidney bean,[870] but at higher SO_2 doses, reducing sugars decreased in the kidney bean. Starch levels in the leaves of red kidney bean followed the same trend.[870] The changes in leaf carbohydrate concentrations may alter plant susceptibility to high and low sugar diseases.[1520] Levels of glucose and fructose were significantly increased in Scots pine and Norway spruce needles near an industrial plant emitting mainly SO_2, while there were no changes in sucrose and starch concentrations.[808]

Environmental factors modify the response of a plant to SO_2 in several ways. Conditions that allow greater stomatal opening, such as high temperatures or high light intensity, can cumulatively increase plant susceptibility to SO_2 by increasing pollutant uptake.[871] After SO_2 is absorbed, environmental conditions may alter tissue sensitivity to the pollutant or affect the rate of detoxification of the pollutant directly. There may also be indirect effects on sensitivity as a result, for example, of the effects on respiration rate and sugar content.[569] Conditions leading to rapid growth may allow for faster recovery once the plant is removed from the polluted area.[1131] Winner and Mooney[1629] described SO_2 resistance as the result of interplay between the ecological, morphological, and physiological characteristics of a plant.

The sensitivity of plants to SO_2 exposure varies greatly. In part, the variation appears to be related to shade tolerance: slow-growing, shade-tolerant trees of late forest succession stages are more tolerant to SO_2 than are pioneer species.[1513] The expression of pollution injury and inhibition of growth under various postfumigation regimes after fumigation with SO_2 also varies among plant species.[1129] Postfumigation temperature has little effect on the amount of injury to leaf tissue. The effects of postfumigation temperature regime on the response of the growth rate to SO_2 are small, but there is positive correlation between the effects of temperature on relative growth rates and the effects of temperature on growth inhibition caused by SO_2. Temperature can affect the mechanisms of either SO_2 avoidance or tolerance, or both, in different plant species. Thus, generalizations about the influence of exposure temperature on the resistance of plants to SO_2 may be inappropriate.[1129,1130]

C. NEGATIVE EFFECTS ON INSECTS

Several studies have been performed on the toxicity of gaseous SO_2 to insects. Sulfur dioxide can be harmful to a number of insects, in high concentrations and even sometimes at ambient levels.[1284] The emission of sulfur contaminants reduced the efficiency of pollinating bees and, consequently, fruit production in orchards near an industrial plant in Poland. Bees clearly avoided flowering trees near the factory.[1240] Flight activity and brood-rearing activity were reduced in bees (*Apis mellifera*) exposed to sulfur dioxide. The total colony weight gain of honeybees was reduced by 14-week SO_2 fumigations, which may have been as low as 150 nl/l within the body of the hive.[686] Sulfur dioxide-induced changes in plants resulted in decreased feeding rate, inactivity, nonuniform growth, delayed cocooning, and cuticular softening in silkworm larvae.[891]

The response of different arthropods to SO_2 exposure seems to be highly variable. Ginevan and Lane[547] reported significantly reduced feeding activity and survival and increased development times of *Drosophila melanogaster* when exposed to 400 nl/l SO_2. Petters and Mettus,[1197] on the other hand, did not observe any effect on reproduction in a parasitic wasp (*Bracon hebetor*) exposed to SO_2. In the milkweed bug (*Oncopeltus fasciatus*), growth and reproduction were stimulated by SO_2 exposure.[462,463] The density of ground beetle populations correlated negatively with the deposition of sulfuric acid from a kraft paper mill in Canada.[504,506] A decrease in the number of soil invertebrates caused by the emission of SO_2, NO_2, and other harmful substances from a rubber shoe factory was reported by Arakelian et al.[50] A significant reduction in the number of grasshoppers was observed on mixed-grass prairie plots during a 3-year fumigation period with SO_2 in an open fumigation system. Monthly mean values for SO_2 on the 0.52-ha plots were <10, 28, 52, and 87 ppb, and the respective densities of grasshoppers 2.9, 2.4, 1.9, and 1.8 individuals per hectare.[1026]

D. EFFECTS ON THE BEAN BEETLE

The Mexican bean beetle (*Epilachna varivestis*) is an agricultural pest of the soybean (*Glycine max*). Increased susceptibility of greenhouse-grown soybeans to the Mexican bean beetle was demonstrated after plant exposure to SO_2 (Table 1).[728] The beetles clearly fared better on foliage that had been exposed to SO_2 than on untreated leaves. The larvae developed faster and grew larger,

TABLE 1
Effects of Sulfur Dioxide on Insects in Experimental Studies

Species	Host plant	Dose	Response	Refs.
Orthoptera	Grasses	10, 28, 52, 87 nl/l per month	62% decrease in density	1026
Homoptera				
Drepanosiphum platanoides	*Acer pseudoplatanus*	3, 40 nl/l, 5 days	No change[a]	1029
Phyllaphis fagi	*Fagus sylvaticus*	40, 100 nl/l, 5 days	22.2% increased growth rate[b]	1029
Rhopalosiphum padi	*Triticum aestivum* *Hordeum vulgare*	100 ppb, 7 days	Increased MRGR[c]	1030
R. padi	*Triticum aestivum*	100 nl/l, 3–4 days[d]	35% increase in MRGR	715
	Hordeum vulgare		40% increase in MRGR	715
Aphis fabae	*Vicia faba*	0. 80, 150, 250 ppb, 7 days	Dome-shaped response in MRGR, 9% difference	1030
A. fabae	Artificial diet	150 ppb, 4 days	No change	1030
A. fabae	*Vicia faba*	100 ppb, 7 days	Increased MRGR[c]	1030
A. fabae	Artificial diet	143 μg/m^3, 3 days	No change	390
A. fabae	*Vicia faba*	0.09, 0.12, 0.15, 0.23 ppm, 7 days	Dome-shaped response, 10.8% increase in MRGR at 0.12 ppm[e]	389
A. fabae	*Vicia faba*	100 nl/l, 3–4 days[d]	9% increase in MRGR	715
Brevicoryne brassicae	*Brassica oleracea*	100 nl/l, 7 days	Increased MRGR[c]	1028
B. brassicae	*Brassica oleracea*	100 nl/l, 3–4 days[d]	10% increased MRGR	715
Elatobium abietinum	*Picea sitchensis*	100 ppb, 0.5, 3, 16, 48 h	Dome-shaped response in MRGR, 19.4% difference	1030
E. abietinum	*Picea sitchensis*	25 nl/l, 2 months	Threefold (well-watered trees) or twofold increase (drought-stressed) in population density	1559
Myzus persicae	*Brassica oleracea*	100 ppb, 7 days	Increased MRGR[c]	1030
M. persicae	*Brassica oleracea*	100 nl/l, 3–4 days[d]	7% increase in MRGR	715

Table 1 (continued)

Species	Host plant	Dose	Response	Refs.
Acyrthosiphon pisum	*Pisum sativum*	0–300 nl/l, 4 days	11% increase in MRGR at 90–100 nl/l, decreased MRGR at 220 nl/l	1557
A. pisum	*Pisum sativum*	100 ppb, 7 days	Increased MRGR	1030
A. pisum	*Vicia faba*	100 ppb, 7 days	Decreased MRGR	1030
A. pisum	*Vicia faba*	100 nl/l, 3–4 days[d]	12% decrease in MRGR	715
	Pisum sativum	100 nl/l, 3–4 days	11% increase in MRGR	
Metopolophium dirhodum	*Triticum aestivum* *Hordeum vulgare*	100 ppb, 7 days	Increased MRGR[c]	1030
M. dirhodum	*Hordeum vulgare*	9, 21, 32, 43 ppb, summer	Twofold increase in MRGR	1030
M. dirhodum	*Triticum aestivum*	100 nl/l, 3–4 days[d]	15% increase in MRGR	715
	Hordeum vulgare	100 nl/l, 3–4 days[d]	20% increase in MRGR	715
Macrosiphum albifrons	*Lupinus* sp.	100 ppb, 7 days	Increased MRGR[c]	1030
M. albifrons	*Lupinus* sp.	100 nl/l, 3–4 days[d]	35% increase in MRGR	715
Sitobion avenae	*Triticum aestivum*	100 ppb, 7 days	19–21% increase in MRGR	1030
S. avenae	*Hordeum vulgare*	9, 21, 32, 43 ppb, summer	Fourfold increase in growth rate	1030
S. avenae	*Triticum aestivum* *Hordeum vulgare*	21, 46, 57 nl/l, 9 months	Threefold increase in population density	28
S. avenae	Artificial diet	100 ppb, 4 days	No change	1030
S. avenae	*Triticum aestivum*	100 nl/l, 3–4 days[d]	28% increase in MRGR	715
	Hordeum vulgare	100 nl/l, 3–4 days[d]	32% increase in MRGR	715
Cinara pilicornis	*Picea abies*	Mean 75–430 μg/m^3, 2 months	Threefold increase in population density	700
Lepidoptera				
Bombyx mori	*Morus* sp.		Decreased feeding rate, inactivity nonuniform growth, delayed cocooning, cuticular softening	891

Table 1 (continued)
Effects of Sulfur Dioxide on Insects in Experimental Studies

Species	Host plant	Dose	Response	Refs.
Diptera				
Drosophila melanogaster		400 nl/l	Reduced feeding activity	547
Hymenoptera				
Apis mellifera		150 nl/l in the hive, 14 weeks	Reduced honey gain	686
Coleoptera				
Epilachna varivestis	*Glycine max*	524 μg/m^3, 7 days	Increased growth and fecundity	726, 728
E. varivestis	Pinto bean	390 μg/m^3, 7 days	Preference for treated plants, no effects in fecundity	727

[a] A slight increase in May (13.2%), decrease in September (–9.2%).
[b] Highest increase in June (26.3%), lowest in September (2.8%).
[c] 5–30% increase in MRGR, highest in cereal aphids.
[d] Plants previously exposed to SO_2 for 7 h.
[e] Less-pronounced effect with increasing plant age.

whereas female fecundity was increased in terms of the number of females laying eggs, the number of eggs laid per female, the percent fertility of the eggs, and possibly the age at which egg laying began. All of these effects would tend to create higher population levels of the beetle.

In studies on the longevity and fecundity of adult Mexican bean beetle,[848,945] a higher proportion of the females fed on reproductive soybeans oviposited, and these ovipositing females laid more eggs per female than females fed on vegetative soybeans. Longevity was not affected by the phenological stage of the plant, and in choice experiments the beetles preferred leaves from the reproductive plants over those from the vegetative plants. The females reacted to the SO_2-fumigated plants in a similar way as to mature plants. These similarities suggest that the action of SO_2 may, in part, be to induce physiological changes similar to those occurring at maturation. However, the precise mechanisms by which SO_2 alters host-plant suitability are not known.

Hughes et al.[725] reported that developmental time from egg to adult among Mexican bean beetles was consistently shortened when insects were reared on detached leaves compared with those reared on whole plants. However, the response of the Mexican bean beetle to SO_2 fumigation was the same for detached soybean leaves and whole plants.[726] Chiment et al.[256] found a direct relationship between the increase in Mexican bean beetle pupal weight and the increase in foliar glutathione concentrations with increasing amounts of SO_2 and concluded that glutathione was being utilized by the insect as a food source.

Plant exposure to SO_2 may exacerbate the pest status of Mexican bean beetles by affecting the period of time during which the beetles begin to reproduce on soybeans in the field.[728] The beetle does not live on young soybeans, but thrives on the older plants later in the season. The vegetative soybean plants are obviously not suitable hosts for the beetle, whereas plants in reproductive growth stages are. If exposure to SO_2 makes the young plants suitable for colonization, damage may possibly be more severe in a more susceptible stage of the plant.

Hughes et al.[728] used a computer simulation model to estimate the impact of SO_2-induced changes on Mexican bean beetle populations. When the predictions were compared, using the egg-laying rate and mortality values for females that were fed on either control leaves or fumigated leaves, the model showed substantially higher insect populations as a result of SO_2-induced effects after only one generation. Forty-one days after the simulation began, the treated population had 3.5 times more adults and 5.5 times more larvae per hactare than the control population; 10 days later this difference had increased to 5 times more adults and 17 times more larvae per hactare. Although the density-dependent effects of predation would tend to reduce these differences, the predictions suggested that even much smaller changes in the egg-laying rate and egg mortality could have a significant impact on field populations.

E. EFFECTS ON APHIDS

Populations of some aphids may be enhanced by gaseous atmospheric pollutants (Table 1).[389,1029,1030,1284,1599] Numerous field observations have been made about increased aphid densities on plants in areas where air pollution is high, SO_2 usually being the major air pollutant.[154,164,175,1241,1530,1586] In a number of studies pollutants were filtered from ambient air[177] or controlled levels of gases were added to ambient air.[387,390,1557] The influence of the treatments on the growth rates of the aphids was then followed in either enclosed or open-top chambers. McNeill et al.[1031] observed increased numbers of *Sitobion avenae* and *Metopolophium dirhodum* on wheat and barley that had been fumigated with low concentrations of SO_2 in an open field experiment.

Increased development of aphid populations in woody plants stressed by air pollutants has been reported in several studies.[156,177,387,1121] On Scots pine (*Pinus sylvestris*) exposed to sulfur and fluoride pollution, Villemant[1530] found increases in aphid populations of *Cinara pini*, *C. pinea*, and *Protolachnus agilis*, but the opposite trend was observed for *Schizolachnus tomentosus* and *Pineus pini*. In contrast, Wiackowski[1603] found *S. tomentosus* to be common around factories emitting sulfur and dust. Near factories emitting sulfur and heavy metals, *Cinara* spp. were more numerous in heavily and moderately polluted pine stands than in lightly polluted ones.[655] In an experiment carried out near a pulp mill emitting mainly sulfur dioxide, numbers of *Schizolachnus pineti* on Scots pine did not increase compared to controls, while the reproduction of *Cinara pilicornis* on Norway spruce (*Picea abies*) was significantly faster at 0.2 and 0.5 km from the pollution source than at more distant sites.[701] On Norway spruce, a gall aphid, *Sacchiphantes abietis*,[1371,1473] and the spruce shoot aphid, *Cinara pilicornis*, are common in polluted areas.

A 2-year study on the occurrence of aphids (*Sitobion avenae*) on spring wheat in an area polluted by sulfur compounds suggested that aphid damage to spring wheat was rather light.[1242] In both years the number of aphids on spring wheat ears was higher in the polluted zone than in the control zone, both in the full earing stage and the grain milk-ripe stage of the crop. The number of aphids in wheat plantations located near the sulfur-processing factory was greater than in control stands. Regardless of the intensity of aphid attack in a given year, the number of aphids on wheat was considerably greater near the factory than in the control stands.

Air pollution can modify the relationship between aphids and their host plants in favor of the aphid. The response of the grain aphid *Sitobion avenae* populations to elevated SO_2 levels, and the mechanism behind such a response, were investigated in a controlled field study using an open-air fumigation system.[28] The performance of the grain aphid was monitored after exposing cereal crops to elevated SO_2 (mean levels 0.02 to 0.06 μl/l) over three seasons. Winter wheat (*Triticum aestivum*) was grown during the first season, whereas winter barley *(Hordeum vulgare)* was cultivated in the following two seasons. The results suggested that the aphid population levels were enhanced by exposure to SO_2. However, natural-enemy populations were not affected significantly by these SO_2 levels, nor did they increase with enhanced prey levels. Analysis of the plant material suggested that changes in the food quality of the plants with respect to the aphids was a factor possibly causing enhanced aphid performance.

The effects of a range of SO_2 concentrations on aphid-host plant relationships are not well known. The mean relative growth rate (MRGR; see Chapter 6, Section II.B) of aphids (*Macrosiphon rosae*) feeding on plants in ambient city air was significantly increased, compared with the MRGR of aphids in a filtered atmosphere.[387,390] McNeill et al.[1030] showed that when potted seedlings of Sitka spruce (*Picea sitchensis*) had been prefumigated with 100 nl/l SO_2, the MRGR of the aphids (*Elatobium abietinum*) was increased, with a peak in the response after 3 h of fumigation. Warrington[1557] investigated the dose response of the pea aphid (*Acyrthosiphon pisum*) to SO_2 in controlled conditions. He found a linear increase in the MRGR of aphids feeding on SO_2-fumigated plants relative to control aphids feeding on plants supplied with charcoal-filtered air. MRGR increased with increasing SO_2 concentration up to a peak of 11% above the controls, which occurred at SO_2 levels between 90 and 110 nl/l. A factor in the decline of the pea aphid at higher SO_2 concentrations may be partly attributable to the toxic effect of the gas on the insect.[1557] McNeill and Southwood[1028] observed reduced growth of the aphids (*Acyrthosiphon pisum*) on *Vicia faba* exposed to either 100 ppb SO_2 and 100 ppb NO_2 or ambient air, compared to the performance on clean air controls. Further investigations are needed to separate possible toxic effects of SO_2 on the growth of the pea aphid from changes in the pea plant to which the aphid responds through an altered growth rate. It would be important to identify the critical nutrients that vary with SO_2 dose. The aphid MRGR has a considerable potential in identifying changes in host-plant response to pollution.[1557]

Chamber studies on aphids (*Acyrthosiphum pisum*) have covered a wider spectrum of SO_2 levels and have shown that the relationship between relative growth rate and SO_2 level tails off above concentrations of 0.1 μl/l, and levels above 0.2 μl/l actually cause a decrease in growth rate.[1557] These adverse affects, however, were observed with relatively high SO_2 concentrations. The sort of levels commonly found in the field are more consistent with those that enhance aphid growth rates.

F. RESPONSE MECHANISMS

The amount of sulfur ingested by insects fed on plants depends on the food type and its sulfur content, as well as on the amount of gaseous sulfur compounds in the atmosphere, that can penetrate the body through the cuticle and respiratory system.[1282] Some phytophagous insects are highly tolerant,[878] but an excess of sulfur compounds usually affects the activity of a number of enzyme activities, both NAD and FAD dependent.[575]

Larval mortality in the early instars of the gypsy moth (*Lymantria dispar*) increased, their development period was prolonged, and their mass decreased already at an atmospheric SO_2 concentration of 0.2 mg/m^3.[1357] Migula[1044] showed that rearing the satin moth (*Leucoma salicis*) in increased concentrations of SO_2 led to an increased activity of several cytoplasmic and mitochondrial enzymes.

Some insect herbivores (e.g., the Mexican bean beetle, *Epilachna varivestis*) prefer SO_2-fumigated leaves.[727,728] However, larval growth was found to be dose dependent and reached a maximum at intermediate concentrations. Reduced growth at higher concentrations might be attributed to the reduced foliar uptake of pollutants as a result of reduced stomatal conductance, increased internal leaf resistance to pollutant uptake, reduced food quality of the leaf, or a toxic effect of the pollutant on the insect.[725]

The mechanism involved in changes in the success of the insect is believed to be mediated via the host plant.[390,914] Subdamaging levels of SO_2 can cause changes in plants that affect insects.[727,728] In the cases studied so far, the changes favored the success of the insect. If the reactions of the Mexican bean beetle to SO_2-affected plants are the same in the field as in the laboratory, damage caused by this insect and subsequent crop losses could be greater in polluted areas. If such a relationship can be extended to other pests and crops, then the total impact of the pollutant-plant-insect interaction on crop losses or cost of crop protection might be quite significant. Consideration of this interaction is important in integrated pest management.

Field observations suggest that aphid populations are enhanced by elevated SO_2 levels. The cause of such an enhancement could be due to a decreased performance of the natural enemies or increased performance of the aphids, or perhaps a combination of the two. The available evidence favors the former explanation.[28] The natural enemies of the cereal aphids (*Sitobion avenae*) did not respond to enhanced aphid populations to any significant extent.[28] The parasitoids exhibited a slight negative density dependence, suggesting that their numbers remained fairly constant in the face of increasing aphid numbers. There was no evidence that SO_2 had a detrimental effect on the two specialist predators, while two of the generalists displayed a slight negative relationship with SO_2 level, suggesting that they were adversely affected directly (although it is not known whether this was due to mortality or habitat choice). There is no evidence, however, that these slight adverse effects on one subset of natural enemies were responsible for the rise in aphid numbers.

Dohmen et al.[390] found that 3-day fumigation with SO_2 at 400 $\mu g/m^3$ (148 nl/l) had no significant effect on the growth of *Aphis fabae* on artificial diets and concluded that the enhanced growth of the aphid on *Vicia faba* in response to SO_2 fumigation was mediated entirely via the host plant. Warrington[1557] suggested that some physiological changes had taken place in the pea plant, that the aphids (*Acyrthosiphon pisum*) were responding to these changes, and that these changes may be directly proportional to the dose of SO_2. Uptake of SO_2 by the plant is obviously proportional to dose.

Analysis of plant material showed an increase in soluble nitrogen with SO_2 exposure, which is consistent with the idea that nitrogen is the limiting factor for aphid populations.[438] Increased soluble nitrogen levels in a plant usually increase growth and reproduction of herbivorous insects, e.g., aphids.[997] A positive correlation between aphid fecundity and the foliar concentration of soluble nitrogen of their host plant has been demonstrated.[439,622] Pea varieties susceptible to pea aphid (*Acyrthosiphon pisum*) attack contain higher concentrations of free and total amino acids than do resistant varieties.[61] The reported increase in aphid populations and growth rates in response to air pollution are most probably the result of changes in the nitrogen metabolism of the host plant.[1557] Subnecrotic levels of SO_2 enhanced populations of the grain aphid (*Sitobion avenae*) in the field.[28] Their material suggested that the mechanism for such an increase was through an increase in the availability of plant soluble nitrogen for the aphid. The resulting enhanced aphid populations may be an important factor through which SO_2 could affect crop productivity.

An increase in needle free amino acid concentrations of Norway spruce (*Picea abies*) also with HF-induced injury has been reported.[758] The concentration of free amino acids has also increased in plants subjected to other air pollutants[155,156,177,966] or other environmental stresses.[1596] Translocation of free amino acids as a result of the degradation of leaf protein, as well as changes in amino acid metabolism,[914] increases the levels of available nitrogen in the phloem. This has been proposed as a mechanism explaining the increase in sucking insects on stressed plants.[288]

Secondary plant compounds may act as feeding deterrents for needle-eating herbivores on conifers.[240,1279] However, the responses of phenolic, monoterpene, and resin acid concentrations to air pollution are still obscure.[808] Terpenes[1299] and phenolic compounds[400] are feeding deterrents for aphids in grasses. Changes in the terpene production of conifers have been observed in environments polluted by SO_2 and F.[922] The increased release of volatile monoterpenes might be due to the degradation of lipids in needles damaged by air pollution. The composition of lipids in conifer needles is strongly influenced by fluoride compounds,[1677] but their effect on plant secondary compounds is not known.[700]

G. PEST DAMAGE ON PLANTS

The changes caused by air pollutants, e.g., SO_2, may create conditions favorable for the development of some harmful insect species, especially aphids, feeding on plant fluids. Similar phenomena have been observed both in agricultural[1241,1242] and forest areas.[259] The effect of air pollution on plant-insect relationships has been favorable on insect herbivores on many occasions.[390,652,700,725,1241,1415,1502,1557]

Already Evenden[446] reported a heavy attack of bark beetles on trees damaged by SO_2 in the vicinity of ore smelters. An increase in bark beetle populations is a characteristic of conifer stands damaged by SO_2,[164,447,663,1332] although there are cases where exposure to pollution has not increased herbivores.[1367] Sulfur dioxide-induced changes in spruces improved the success of spruce sawflies feeding on needles.[1586,1587] Increased densities of the pine shoot moth (*Rhyacionia buoliana*) and a tineid moth (*Exoteleia dodecella*) have been attributed to similar changes in trees.[1472]

Insect damage may increase the susceptibility of plants to atmospheric pollutants.[1560] In the experiments by Berge,[122] needle injury on conifers exposed to high levels of SO_2 occurred rapidly after the onset of an insect infestation, while branches treated with insecticides remained free of injury. Gehring and Whithan[532] showed that herbivores may reduce mycorrhizal levels of trees. The interaction between plants, mycorrhizae, and herbivores are still poorly investigated, especially in relation to air pollution.

The aphid *Elatobium abietinum* is perhaps the most important pest of Sitka spruce (*Picea sitchensis*). The response of spruce to *E. abietinum* includes a decrease in leader extension and root dry mass. Short-term exposure to SO_2 increased the MRGR of *E. abietinum.*[1030] Over a 2-month period the aphid population increased by a factor of 250 on the watered, SO_2-fumigated plants compared with 80 times on watered unfumigated plants.[1559] The densities were comparable with those recorded in the field in moderate-to-severe infestations.[366] The population increase was maintained throughout the 2-month period and was not alleviated by increased intraspecific competition. Damage to Sitka spruce was broadly proportional to the number of aphids present; the enhanced number of aphids on the SO_2-fumigated trees resulted in a significantly greater reduction in tree growth than for trees with aphids in ambient air. Drought alone had a relatively small effect on the aphids; the slight increase in mean numbers was not significant.

The effect of low concentrations (25 nl/l) of SO_2 alone on the Sitka spruces was not significant, but there was a significant interaction between SO_2 and the aphids. Aphids increased more in numbers and caused greater reduction in leader length and root dry mass than on the trees growing in ambient air.[1559] A similar observation has been reported for pea plants.[1561]

Drought and SO_2, separately, caused reductions in certain plant growth parameters, but no direct interactions between drought and SO_2 were found, except for a further decrease in root dry mass (24%) in Sitka spruces.[1559] Drought did, however, make the spruces less attractive to the aphids than the well-watered trees exposed to SO_2, although they were still more acceptable than trees not exposed to pollution, whether watered or not. Consequently, the effect of low concentrations of SO_2 on the aphids was more important than the effects of drought, although the latter was severe and reduced tree growth significantly.

Sitka spruce trees simultaneously exposed to green spruce aphids and SO_2 pollution are likely to suffer greater damage than would be caused by either factor separately.[1559] Drought alone had little effect on aphid numbers, and the combination of drought and aphid attack was not additive. This was probably because the effects of drought alone and aphids alone were so severe that there was insufficient plasticity in the trees for further reductions in growth to occur.

This was also the case when aphids, SO_2, and drought were combined; the only significant interaction was in root growth.

H. COMBINED EFFECTS OF STRESS FACTORS

The response of four aphid species to short-term fumigation with SO_2 and NO_2 showed that both of these pollutants enhanced the performance of aphids on previously fumigated plants.[715] Dohmen[389] demonstrated experimentally that fumigation of *Vicia faba* with 0.15 ppm SO_2 (400 $\mu g/m^3$) or 0.2 ppm NO_2 (400 $\mu g/m^3$) during a 7-day period caused changes in plant metabolism, which resulted in higher growth rates of the aphid *Aphis fabae* feeding on these plants. Fumigation of *V. faba* with 0.085 ppm O_3 during 2 or 3 days caused decreased aphid growth on the fumigated plants. The result was reversed by higher ozone concentrations or through the presence of NO_x during ozone fumigation. Ambient air comprising a mixture of pollutant gases had an even stronger enhancing effect on aphid performance than did single gases. Thus, the growth of *A. fabae* on field bean plants was significantly higher in ambient summertime London air than in charcoal-filtered air. Similarly, the growth of *Macrosiphon rosae* on rose bushes was improved in ambient summertime Munich air; the increase in growth rate averaged about 20%.[389]

Holopainen et al.[700] monitored the development of spruce shoot aphid (*Cinara pilicornis*) populations in natural and artificial infestations on Norway spruce (*Picea abies*) seedlings exposed to air pollutants applied both singly and in mixtures as gaseous SO_2, NaF (30 mg/l F), and $Ca(NO_3)_2$ or $(NH_4)_2SO_4$ in aqueous solutions (200 mg/l N) in an experimental field. All the pollutants and their combinations significantly increased the numbers of aphids per seedling. Four apterous females were transferred to spruce seedlings growing in containers on the same plots. After 4 to 5 weeks, aphid numbers were significantly higher in the fluoride treatment as well as in the combined fluoride, nitrogen, and SO_2 treatment. The pollution treatments had no significant effect on shoot growth. There were no significant differences in the concentrations of free amino acids in shoot stems between the control and fluoride treatment. However, the relatively low concentration of arginine in the fluoride treatment at the end of the growing season might indicate disturbances in the nitrogen metabolism of the spruce seedlings.

The results of Holopainen et al.[700] showed that all the pollutants or their combinations had a stimulating effect on aphid numbers. One explanation could be the additional water given to the seedlings as precipitation, which would increase seedling vigor and suitability for aphids during very dry periods in June. In the combination treatments, SO_2 and the spraying of aqueous pollutants seemed to have a synergistic effect on aphid success. Neuvonen and Lindgren[1121] also observed that dry periods and pollution stress have interactive effects on aphid (*Euceraphis betulae*) reproduction. According to their results simulated acid rain increased aphid reproduction on birch when the weather was drier than normal, but no effect was observed when precipitation was above normal.

III. NITROGEN COMPOUNDS

A. BACKGROUND

The air in industrial areas usually contains a mixture of active polluting agents. It is therefore not easy to specify the contribution of each component in damaging trees. In northern regions especially, both the physiological condition of the trees and the weather conditions exhibit several special features that apparently have some effect in a polluting situation. Gaseous nitrogen pollutants include ammonia (NH_3), nitrous oxide (N_2O), nitric oxide (NO), and nitrogen dioxide (NO_2). In the atmosphere, NO is oxidized to NO_2 mainly by ozone. The combustion of fossil fuels is the largest anthropogenic source of nitrogen oxides (NO_x).

Nitrogen is also a major component of fertilizers. These compounds are often plant nutrients that in moderate quantities increase plant growth. In excess, however, they may cause growth disturbances. Ammonia emissions originate from cattle, pig, and poultry breeding, fur farming, artificial fertilizers, and industry. Nitrogen compounds can also indirectly cause soil acidification.

An increased nitrogen input due to atmospheric deposition or fertilization can lead to eutrophication of terrestrial ecosystems.[429] Applications of up to 50 kg nitrogen per hectare per year usually result in only slight changes in the vegetation, but above this level the numbers of dicotyledonous plant species, as well as insects, decline. Applications of about 200 kg/ha of nitrogen in Eastern Länder, Germany, reduced the insect fauna in the soil by about one third and the number of insects in the vegetation by three quarters.[751]

Eutrophication may have a major impact on insect fauna even though it causes relatively small changes in the vegetation. Perhaps one of the best examples of the effects of a small change in vegetation on insects is the disappearance of the large blue butterfly *Maculinea arion* (Lycaenidae) from Britain in the 1970s.[1479] The butterfly has an obligate parasitic relationship with the ant *Myrmica sabuleti*. Sufficiently large populations of *M. sabuleti*, needed to support a colony of *M. arion*, only occurred on warm sites with a very short turf. In slightly taller turf, *M. sabuleti* was replaced by another ant species, *M. scabrinodis*, which is an unsuitable host. Although this process, which led to the disappearance of the butterfly, was attributable to a cessation of grazing, similar effects could also be caused by eutrophication.

Van Wingerden et al.[1626] studied the influence of temperature on the duration of egg development in west European grasshoppers (Acrididae). The duration of embryonic postdiapause development was strongly temperature dependent in all the species studied. They postulated that the increase of vegetation height in grasslands subjected to eutrophication will lead to lower maximum temperatures near the soil surface. Consequently, egg hatching will be delayed, and thermophilous species disappear.

B. CHANGES IN PLANTS

Nitrogen pollution may contribute to forest decline.[589,1480] As a result of the effect of ammonium sulfate aerosols, gaseous ammonia, and nitrogen oxides, pine stands in the neighborhood of a nitrogen plant in Pulawy, Poland began to die.[1368] Large-scale damage has been noted in the pines (*Pinus sylvestris*) and birches (*Betula pubescens*) surrounding a plant producing nitrogen fertilizers in Oulu (65°N), Finland.[642] The symptoms on Scots pine (and also of the birch) indicated that the damage had been initiated in winter conditions. Although the visible damage did not appear until the spring, laboratory experiments showed that it had already occurred in the winter. Large amounts of fine fertilizer dust discharged from the fertilizer plant during winter accumulated in the needles of the test trees placed near the pollutant source. Dust deposition occurred under certain weather conditions during the winter, the dust often being wet and therefore readily adhering to the branches of the trees.

Deposition of fertilizer dust in the vicinity of the nitrogen fertilizer plant clearly increased the growth rate of the pines and the size of their needles. The needles became more susceptible to desiccation during the winter. The fine dust blocked the stomatal slits and depressions, thus disturbing gas exchange in the spring and even during the summer unless rainfall washed the dust off the needles.[642] Exposure to NO_2 either increased leaf nitrogen in tomato plants[1498] or had no effect.[1471] Ammonia and NO_2 emissions may increase foliar concentrations of nitrogen in plants, but near pollutant sources, SO_2 emissions may contribute to foliar nitrogen concentrations.[1025]

C. EFFECTS ON INSECTS

Exposure to high levels of NO_2 and SO_2, singly[463] or in combinations,[462] can directly enhance the growth and reproduction of seed-eating plant bugs. In the vicinity of a nitrogen plant in Pulawy, Poland, insects attacked about 35% of the dead and dying trees damaged by nitrogen compounds (ammonium sulfate, gaseous ammonia, and nitrogen oxides).[1368] The greatest increase was observed in the activity of the beetles *Pissodes piniphilus* (Curculionidae), *Monochamus galloprovincialis* (Cerambycidae), *Acanthocinus aedilis* (Cerambycidae), and *Tomicus piniperda* (Scolytidae). An increase in the bark beetles *Ips sexdentatus*, *Orthotomicus laricis*, and *Ips acuminatus* was also reported. Air pollution along a motorway had a major impact on nearby plants and was assumed to change the plant-insect balance to the detriment or favor of phytophagous insects.[1221,1242] Aphid outbreaks have been observed on motorway verges in other studies.[175,486]

A fumigation experiment conducted next to a busy motorway, revealed a significant increase in the population development of aphids (*Aphis fabae*) on two main hosts (*Viburnum opulus* and *Phaseolus vulgaris*) in ambient air compared to filtered air[155] (Figure 2; Table 2). The prevailing air pollutants, mainly NO (404 $\mu g/m^3$, daily mean) and NO_2 (210 $\mu g/m^3$), seemed to have a strong impact on the nitrogen metabolism of the fumigated plants. This was reflected by the increased organic nitrogen concentrations in the foliage and

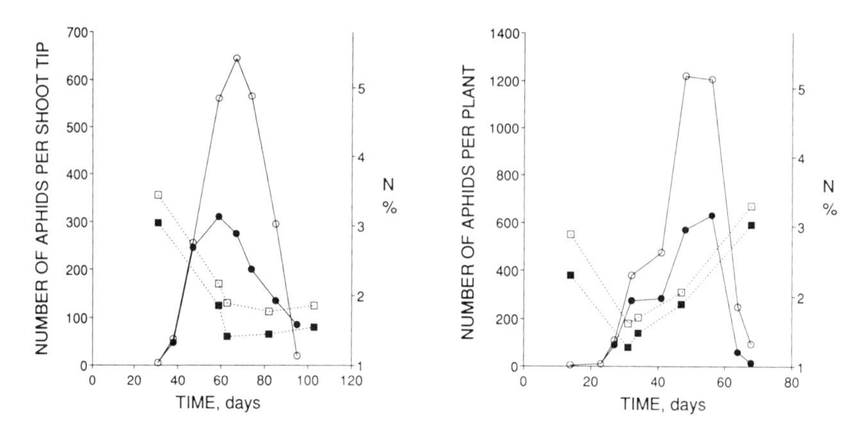

FIGURE 2. Population development of *Aphis fabae* (Homoptera, Aphididae) (solid lines) and foliar nitrogen content (percent of dry matter) of host plants (dotted lines) on a motorway. Treatment of *Viburnum opulus* (left) and *Phaseolus vulgaris* (right) in filtered air (filled symbols) or ambient air (empty symbols) of continuous fumigation. Horizontal axis shows the number of fumigation days. (From Bolsinger, M. and Flückiger, W., *Entomol. Exp. Appl.*, 45, 237, 1987. With permission.)

total amino acids in the phloem sap. It is thus believed that the plant-insect interaction becomes unbalanced when plants are submitted to ambient air on motorways, owing to changes in the nutritional quality for aphids, indirectly induced by the enhanced nitrogen uptake from ambient air by the host plant. Recent experiments suggest that the effect on aphid performance was more strongly linked to NO, rather than NO_2, concentrations.[716]

Air pollutants along a motorway significantly enhanced aphid development at the roadside, and the main promoting factor was thought to be NO_x.[155] In the experiment mentioned above, both plant species, during certain periods, had higher organic nitrogen levels in their leaves and increased amounts of free amino acids in the assimilate transportation system when submitted to ambient air. The lowered stomatal diffusive resistances might have eased the uptake of pollutants from the motorway air. The fumigation of plants with NO_2 was followed by intensified reduction of dissolved NO_2 to ammonia and subsequent conversion to amino acids, resulting in higher protein concentrations. NO_x is assimilated into organic compounds when plants suffer from a shortage of inorganic nitrogen in the soil.[1418] This was verified by the organic nitrogen concentration in *Phaseolus* seedlings grown on a N-deficient solution. Despite the fact that all the plants were supplied with sufficient plant nutrients, the plants grown in ambient air obviously metabolized more free amino acids. The beneficial effect of additional incorporated NO_2 on plant nutrition was reviewed by Wellburn.[1581]

The increased organic nitrogen concentrations in the foliage and phloem of host plants growing in the ambient air of motorways represents a considerable improvement in nutritional quality for herbivorous insects.[1596] Large populations of defoliating insects have been observed on roadside shrubs.[1220] These

TABLE 2
Effects of Nitrogen Dioxide on Insects in Experimental Studies

Species	Host plant	Dose	Response	Refs.
Homoptera				
Euceraphis punctipennis	*Betula pubescens*	9, 57, 105, 157 nl/l, 5 days	Dome-shaped response, 30.2% increase in MRGR at 105 nl/l	1029
Rhopalosiphum padi	*Triticum aestivum*	100 nl/l, 3–4 days[b]	38% increase in MRGR	715
	Hordeum vulgare	100 nl/l, 3–4 days[b]	30% increase in MRGR	715
R. padi	*Triticum aestivum*	NO: 53 nl/l, NO_2: 29 nl/l, 42 days[c]	24% decrease in MRGR	716
R. padi	*Triticum aestivum*	NO: 68 nl/l, NO_2: 32 nl/l, 84 days[c]	25 or 30% decrease in MRGR	716
Aphis fabae	Artificial diet	200 nl/l, 4 days	No change	1030
A. fabae	*Vicia faba*	100 ppb, 7 days	Increased MRGR	1030
A. fabae	*Vicia faba*	200–220 ppm, 7 days	7.7% increase in MRGR	389
A. fabae	*Viburnum opulus* *Phaseolus vulgaris*	NO 404 μg/m^3, NO_2 210 μg/m^3[(a)], 60 days	Twofold increase in population density	154,155
A. fabae	*Vicia faba*	100 nl/l, 3–4 days[b]	9% increase in MRGR	715
A. fabae	*Vicia faba*	NO: 87 nl/l, NO_2: 41 nl/l, 42 days[c]	17.6% decrease in MRGR	716
A. pomi	*Crataegus* sp.	500–1200 μg m/l^3, 4 weeks[a]	4.4-fold increase in population density	176
Elatobium abietinum	*Picea sitchensis*	100 nl/l, 3 days	18–58% increase in MRGR	1029
Acyrthosiphon pisum	*Vicia faba*	100 ppb, 7 days	Increased MRGR	1030
A. pisum	*Vicia faba*	100 nl/l, 3–4 days[b]	10% decrease in MRGR	715
A. pisum	*Vicia faba*	NO: 87 nl/l, NO_2: 41 nl/l, 42 days[c]	15% increase in MRGR	716
Metopolophium dirhodum	*Triticum aestivum* *Hordeum vulgare*	100 ppb, 7 days	Increased MRGR	1030
M. dirhodum	*Triticum aestivum*	100 nl/l, 3–4 days[b]	17% increase in MRGR	715
	Hordeum vulgare	100 nl/l, 3–4 days[b]	14% increase in MRGR	715
M. dirhodum	*Triticum aestivum*	NO: 45 nl/l, NO_2: 38 nl/l, 42 days[c]	29% decrease in MRGR	716
	Hordeum vulgare	NO: 42 nl/l, NO_2: 34 nl/l, 42 days[c]	29% decrease in MRGR	716
Macrosiphon rosae	*Rosa* sp.	4–15 to 280 μg/m^3[(c)]	20% increase in MRGR	387

Table 2 (continued)

Species	Host plant	Dose	Response	Refs.
Macrosiphum albifrons	*Lupinus* sp.	100 nl/l, 3–4 days[b]	75% increase in MRGR	715
Sitobion avenae	*Triticum aestivum*	100 ppb, 7 days	Increased MRGR	1030
	Hordeum vulgare			
S. avenae	*Triticum aestivum*	100 nl/l, 3–4 days	25% increase in MRGR	715
	Hordeum vulgare	100 nl/l, 3–4 days	36% increase in MRGR	715
Cinara pilicornis	*Picea abies*	$Ca(NO_3)_2$, $(NH_4)_2SO_4$, 200 mg/l N	Twofold increase in population density	700

[a] Ambient and filtered air of a heavily frequented motorway.
[b] Plants previously exposed to NO_2 for 7 h.
[c] Urban air of Munich or London; SO_2 and O_3 also present.

outbreaks were related to higher nitrogen concentrations in the infested plants. Previous experiments[1221] had shown that predators play a minor role in determining aphid population density, but the authors supposed that a combination of a harsh microclimate and air pollutants and especially NO_x should be considered as a factor promoting development. An increased MRGR was reported in *Macrosiphon rosae* nymphs when fumigated with ambient urban air (Munich, Germany) in exposure chambers.[387] A study performed on the central reservation of a motorway revealed a strongly increased population of *Aphis pomi* on *Crataegus* sp.[176] Experiments showed the same effect with different plant and aphid species.[154] These investigations indicated that polluted air strongly promoted insect infestation on host plants.

There is much evidence to show that the air pollutants along motorways can indirectly increase the potential of herbivorous insects due to strongly increased organic nitrogen concentrations in the foliage and phloem, which is an important aspect in insect nutrition.[155] Nitrogen pollution should therefore be regarded as a further, and increasingly eminent, factor that can affect plant-insect relationships.

IV. ACIDIC PRECIPITATION

A. BACKGROUND

Atmospheric acidification results from the emission of several reducing molecules during the combustion of fossil fuels and related photochemical processes, which reinforce each other (synergism) in a complex manner. The anthropogenic release of sulfur and nitrogen oxides has increased the acidity of precipitation 10 to 100 times compared to preindustrial levels.[1398] Deposition of acidic material can be "dry" (particulate matter) or "wet" (molecules washed out through precipitation), and the pattern of deposition varies according to the seasonal direction of the prevailing winds and the source of pollution. The effects of atmospheric acidification vary from local episodes of high toxicity to widespread, low-level, long-term insidious poisoning.[554] In more northern latitudes a fairly high proportion of the acid molecules are deposited and accumulated in snowfall. When the snow melts in spring, there can be a sudden and very threatening pulse of acidic water entering streams and lakes.[554,1333] The acidification of soils is of considerable regional significance. Jones[794] concluded in his review that atmospheric inputs have enhanced the rate of soil acidification over large areas of Scandinavia and elsewhere. In general, forest soils are considered to be more vulnerable to the effect of acidic precipitation than are agricultural soils.[1398]

Acidic precipitation has caused widespread environmental effects in Europe and North America. Acid aerosols in the form of mists and fogs form at high elevations, condensing on the vegetation and causing "leaf burn" and bud disorders; conifers are especially vulnerable in this respect. Over the last century hydrogen ion concentrations in the soil have been growing; once the latent buffering potential of a soil has been exceeded, an acidic layer is formed

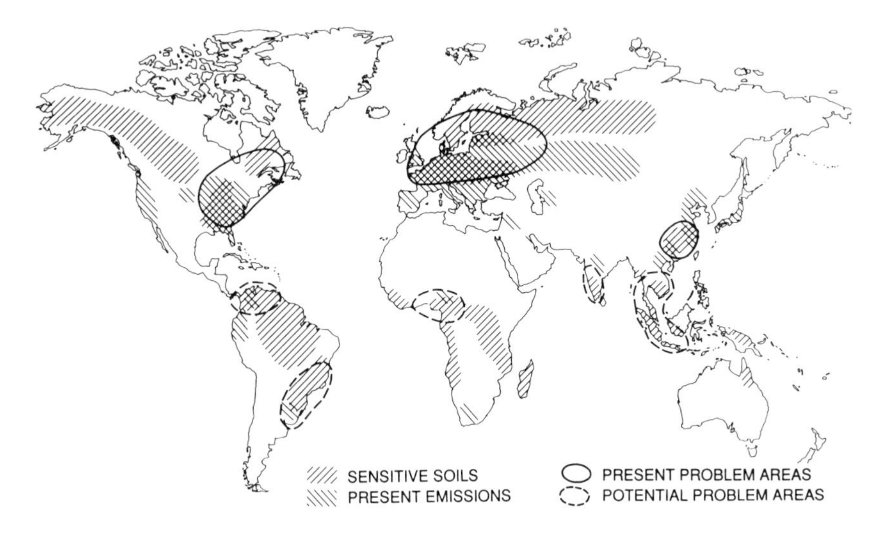

FIGURE 3. Schematic map showing regions that have acidification problems, and regions where, based on soil sensitivity, expected future emissions, and population density, acidification might become severe in the future. (From Rodhe, H. and Herrera, R., *Acidification in Tropical Countries*, John Wiley & Sons, Chichester, U.K., 1988. With permission.)

that damages root systems (principally through the leaching of essential nutrients, including potassium, magnesium, and calcium, and the release of toxic metal ions such as aluminium, manganese, and cadmium), leading to the death of trees (especially conifers, which have shallow rooting systems) and the inhibition of natural regeneration by acid-sensitive plants.[554]

Acidic fogs may have a beneficial as well as a detrimental effect on plants. Wet deposition of pollutants on foliar surfaces can either act as fertilizers or gradually cause cellular collapse and the formation of necrotic lesions.[1363] Deleterious effects, such as lesion development, weathering of cuticular wax, foliar leaching, and modification of physiological responses, have been reported. These effects have been reproduced in short-term exposure to acidic fogs of pH 2.3 or lower, under laboratory conditions.[557] Because ambient fogs in the Los Angeles basin have an acidity ranging from pH 1.7 to 3.0, significant foliar necrosis leading to yield losses in crop plants is not unusual.[696]

Until relatively recently, recognition of the acidification problems has been confined to North America and Europe. Nevertheless, several other regions are likely to experience acidification problems in the future. These potential areas, identified on the basis of soil sensitivity and predicted pollutant emission rates, are illustrated in Figure 3.[1293,1294,1514]

B. DIRECT EFFECTS ON INSECTS

The direct effects of acidic precipitation on terrestrial invertebrates have received little attention. Migula[1044] reported that the change in environmental pH resulting from acidic rainfall mixed with dust and gaseous pollutants caused inhibition of enzymatic activity in the satin moth (*Leucoma salicis*).

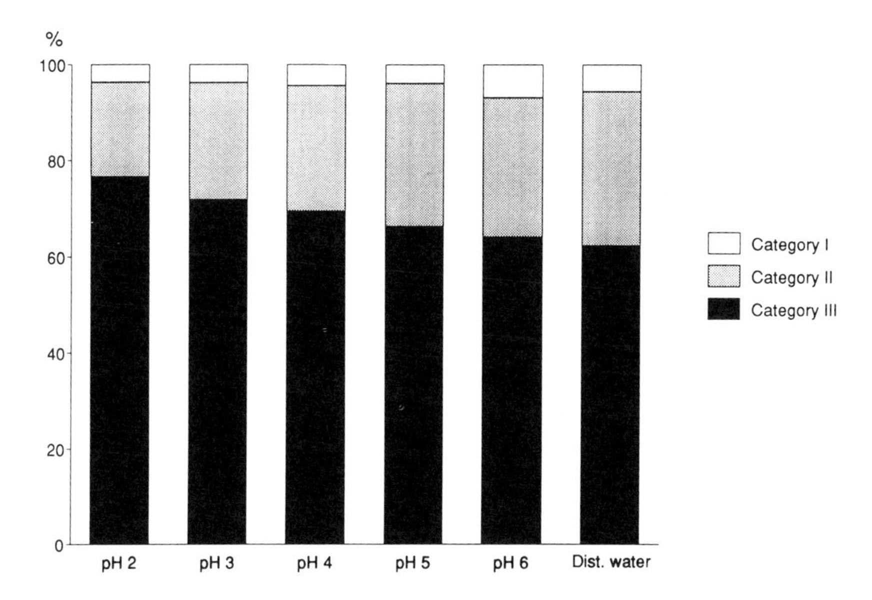

FIGURE 4. Increasing egg survival of *Neodiprion sertifer* (Hymenoptera, Diprionidae) with increasing acidity of the spraying treatment. The three categories used in the experiment were as follows: (1) Developmental failure of the embryo not due to the withering of the pine needle. The contents of the egg have dried out during embryonal development, or the development of the embryo has stopped or been prolonged. In some cases the fully developed larva has not hatched successfully from the egg. (2) Developmetal failure of the egg due to withering of the needle. The egg itself appears to be normal, but the embryo has died in an early stage, owing to physical disturbances caused by the withering of the needle. (3) Egg development has been successfully completed, and a normal larva has hatched from the egg pocket. (From Heliövaara, K., Väisänen, R., and Varama, M., *Entomol. Exp. Appl.*, 62, 55, 1992. With permission.)

Heliövaara et al.[668] studied the effects of simulated acid precipitation on the egg viability of the European pine sawfly (*Neodiprion sertifer*). Overwintering egg clusters on Scots pine (*Pinus sylvestris*) needles were sprayed with a mixture of sulfuric and nitric acid (1:1, pH 2, 3, 4, 5, 6, distilled water, natural precipitation) for a month in April–May. The proportion of hatched first instar larvae increased with increasing acidity (Figure 4). Although the better survival of sawfly eggs on pine needles in more acidic conditions may be partly due to changes in needle physiology, the results on eggs reared without needles on Petri dishes were similar, indicating direct acid-induced changes in the eggs themselves.

Gunnarsson and Johnsson[574] studied the effects of acidic precipitation on the growth rate of a spruce (*Picea abies*) living sheetweb spider (*Pityohyphantes phrygianus*) by spraying water of different acidity on spiders in experimental vials. No increasing difference in growth between pH 4.0 and the control was found during the 2.5-month experiment. Gunnarsson and Johnsson[574] concluded that the indirect effects of air pollution on spider communities[270,573] are probably more important than are toxic effects.

C. EFFECTS ON SOIL FAUNA

Several experiments have demonstrated that the artificial acidification of soils results in decreased decomposition rates. This effect has been documented by directly measuring the decrease in organic material over time and by measuring the evolution of CO_2.[580]

Hågvar et al.[588] investigated the impact of artificial acidification of the forest floor on insects and spiders, as reflected by catches in pit-fall traps and hatching traps. The abundance of two *Carabus* species (Coleoptera, Carabidae) and one lycosid spider (*Trochosa terricola*) was significantly reduced on the most acidified plots, while one theridiid spider (*Euryopis flavomaculata*) and one staphylinid beetle (*Astilbus canaliculatus*) were caught in higher numbers on those plots. Hågvar et al.[588] suggested that changes in population densities might be related to the reduction in moss cover on the most acidified plots.

Norwegian long-term studies on the effect of artificial acidification and liming in coniferous forest have revealed characteristic effects on the soil fauna.[2,3] Studies were performed for Protozoa, Rotifera, Nematoda,[1419] Enchytraeidae,[2] and microarthropods (Collembola and Acari).[579,585,586] Both acidification and liming reduced the abundance of Protozoa, Rotifera, Nematoda, and Enchytraeidae. Acidification did not affect the total number of Acari or Collembola, but liming decreased both of these groups. Liming decreased the total number of Acari and Collembola. Esher[444] found that the number of Oribatei, Prostigmata, and Astigmata increased with decreasing pH after artificial acid precipitation in pine forests. Recent laboratory studies in microcosms indicate that acidification may influence soil microarthropods via competition mechanisms.[581]

Wolters[1640] found a marked increase in the abundance of Collembola, Protura, and Diplura in the soil (mull) around beech trees in a forest on limestone. The soil close to the trunks of the beech trees was strongly acidified (pH $H_2O < 4$) by the high input of protons via stemflow. Differences in soil acidity on the different sides of the trees were correlated with the distribution of Apterygota. Collembola and Protura reached their highest densities in the most acidic soil, while Diplura only occurred in the soil with the higher pH value.

D. EFFECTS ON PLANT-INSECT INTERACTIONS

The effects of acidic precipitation *per se* on plant-insect interactions are poorly known (Table 3).[320,583,585-587,1399] Surprisingly little is known about the interactive effects of precipitation, acidity, and economically important insects even in agricultural ecosystems.

The larvae of the black cutworm (*Agrotis ipsilon*, Noctuidae) damage corn (maize, *Zea mays*) by cutting young plants at the soil level, which can result in significant economic losses. Stinner et al.[1427] investigated the effects of simulated acidic precipitation on interactions between this pest and corn. They observed an increase in approximate digestibility and a decrease in the efficiency of digested food conversion in the high-acidity treatments of a Petri dish experiment with black cutworms. This result suggested that simulated acidic precipitation had a stressing effect on food utilization by black cutworm larvae. However, no indication of stress effects were discerned in pot experiments with more closely approximated field conditions. In the pot experiments with sixth instar larvae on two-leaf corn, high acidity appeared to act as a subsidy for the larvae, which grew larger and developed faster in the pH 2.8 and 4.2 treatments than in the pH 5.6 treatments.

The results of the above-mentioned Petri dish and pot experiments suggested that simulated acidic precipitation can affect the quality of corn plants to the extent that some significant effects on insect herbivores are observable.[1427] However, no response was found in the nitrogen concentrations of corn leaves in the pot experiments. Other components of food quality, such as feeding deterrent compounds, may have therefore had an effect on the sixth instar black cutworms in the two-leaf corn experiments. Cracker and Bernstein[319] observed that corn leaves have a high buffering capacity against acidic precipitation. The mechanism of this buffering capacity is not known, but their results suggest that the internal disruption and release of cell contents may be involved in neutralizing the hydrogen ion component of acidic precipitation. Reese and Field[1267] suggested that the structurally complex tissue of intact corn plants may be an important factor in defense against attack by black cutworms. If the buffering mechanism against acidic precipitation in corn leaves involves internal disruption of cells, then acidity may have caused the breakdown of the structurally complex tissue and thus improved the food quality for the sixth instar larvae.[1427]

Trumble and Hare[1500] studied the acidic-fog-induced changes in *Phaseolus lunatus* as the host plant of *Trichoplusia ni* (Lepidoptera, Noctuidae). Plants exposed to fogs of greater acidity had significantly higher total nitrogen concentrations compared to the controls. Considerable evidence is available suggesting that the growth of plants exposed to acidic fogs may be enhanced by the increased availability of nitrogen and sulfur.[466,1363] However, no improvement in *T. ni* larvae weight was found, although nitrogen is often a limiting factor for insects.[693,997,1106,1108,1464] Results for other lepidopteran herbivores feeding on deciduous trees suggested, however, that nitrogen ceased to be limiting at higher levels of availability.[1477] Alternative explanations were proposed for the lack of increased weight gain of the larvae on plants with higher levels of total nitrogen, soluble protein, or specific free amino acids.[1500] The nitrate form of nitrogen may be unsuitable for *T. ni*. Furthermore, test plants from the pH 2.0 and 2.5 fogs had more necrotic areas, which reasonably could be expected to have interfered with palatability or phagostimulation. Larvae on these plants would spend more time in transit between acceptable or palatable regions, and less time feeding, than larvae on foliage with minimal damage. The changes in nitrogen content and quality were at least partially offset by increases in other compounds produced in response to damage that deterred feeding and/or assimilation by the larvae. Furthermore, since *T. ni* is a generalist feeder, the variation in nitrogen form and concentration between stressed and unstressed plants within *P. lunatus* may be small relative to the variation among species within the broad host range of *T. ni*.

TABLE 3
Effects of Acidity on Terrestrial Insects in Experimental Studies

Species	Host plant/habitat	Dose	Response	Refs.
Collembola	Soil	H_2SO_4, pH 2–6, 25 or 50 mm, monthly during May–September	Increased abundance[a]	580,582
Homoptera				
Euceraphis betulae	*Betula pendula*	H_2SO_4, pH 3.5, 1–2 times weekly during 3 successive summers	40–100% increase in offspring number[b]	1121
Phyllaphis fagi	*Fagus sylvatica*	$H_2SO_4/HNO_3/HCl$, pH 2.6 or 3.6, 2 months, 2 times per week[c,d]	Decreased population density	177
Aphis fabae	*Phaseolus vulgaris*	$H_2SO_4/HNO_3/HCl$, pH 3.6, 2 weeks, 1-2 times per day[c,d]	Decreased population density	177
Eulachnus agilis	*Pinus sylvestris*	H_2SO_4/HNO_3, pH 2.99 5 weeks, 0.5 l, 5 days per week	20% increase in MRGR	840
Schizolachnus pineti	*Pinus sylvestris*	H_2SO_4/HNO_3, pH 2.99 5 weeks, 0.5 l, 5 days per week 2 months, pH 5.7	12% increase in MRGR	840
Cinara pini	*Pinus sylvestris*	H_2SO_4/HNO_3, pH 2.99 5 weeks, 0.5 l, 5 days per week	Twofold increase in MRGR	840
Lepidoptera				
Agrotis ipsilon	*Zea mays*	H_2SO_4/HNO_3, pH 2.8, 4.2, 5.6, 2 times weekly	Larger larvae, faster development at pH 2.8 and 4.2	1427
Trichoplusia ni	*Phaseolus lunatus*	H_2SO_4/HNO_3, pH 2, 2.5, 3, 6.3–6.5, 2 h	Increased larval weight at pH 3, no change in feeding preference or egg survival	1500
Spodoptera exigua	*Apium graveolens*	H_2SO_4/HNO_3, pH 2, 2 h	Larval survival reduced 50–8 %[e]	1503

Table 3 (continued)

Species	Host plant/habitat	Dose	Response	Refs.
Hymenoptera				
Neodiprion sertifer	*Pinus sylvestris*	H_2SO_4, pH 2.5, 6, 7 years, 50 mm, 5 times per year[f]	Lower adult weight at low and high pH	66
N. sertifer	*Pinus sylvestris*	H_2SO_4/HNO_3, pH 3, 4, 60 mm per month, 2 successive summers	Twofold increase in larval survival at pH 3[g]	1122
N. sertifer	*Pinus sylvestris*	H_2SO_4/HNO_3, pH 3, 4, 5 mm, 3 times per week, 3 successive summers	No effect in growth rate or cocoon weight	1123
N. sertifer	*Pinus sylvestris*	H_2SO_4/HNO_3, pH 2, 3, 4, 5, 6, 1 month, 18 times	14% increase in egg survival	668
Coleoptera				
Carabus spp.	Soil	H_2SO_4, pH	Reduced population density	583, 588
Astilbus canaliculatus	Soil	2–6	Increased population density	
Xanthogaleruca luteola	*Ulmus* sp.	H_2SO_4/HNO_3, pH 4, 9 weeks[h]	Decreased number of eggs laid	599

[a] Decreased number at pH 2; between-species differences.
[b] Response obtained in four cases; no response in other four cases.
[c] Ambient air dominated by O_3, and filtered air.
[d] 0.1 mg/l H_2O_2 was added to the pH 3.6, and 1 mg/l to the pH 2.6.
[e] Mortality caused by psoralens (linear furanocumarin) following acidic fog episodes.
[f] High pH = pH 6 + added $CaCO_3$.
[g] Reduced susceptibility to viral disease.
[h] SO_2 (0.2 ppm), O_3 (0.1 ppm), or SO_2 + O_3 (0.2 + 0.1 ppm) also present.

The most important economic question is whether current levels of ambient acidity can affect the amount of damage caused by insect herbivores on crops. Herbivory data from the pot experiments indicated that acid precipitation, especially precipitation with ambient levels of acidity, had no significant effect on the total number of corn plants cut by black cutworms. Stinner et al.[1427] concluded that acidic precipitation may have greater effects on herbivorous insects that inhabit the foliage of agricultural ecosystems than on soil-inhabiting insects, such as the black cutworm.

The great water dock *Rumex hydrolapathum* (Polygonaceae) is the host plant of the large copper butterfly *Lycaena dispar batava* (Lycaenidae) in the Netherlands. Bink[137] studied the influence of acid stress (pH 3.5 to 7.5) in the host plant of the butterfly. The pupal weight of the butterfly was highest on plants growing at the pH 5.5 to 6.5. The host plant had the highest nitrogen concentration in the same pH range.

It has been proposed that acidic deposition superimposes on natural stress factors and that a significant increase in forest damage following warm and dry years characterizes the forest decline in central Europe.[1511] Atmospheric deposition may influence trees by impairing photosynthetic capacity and by reducing carbohydrate production and nutrient retention.[1023] Repeated exposure to acidic precipitation may lead to acidification of the leaf surfaces or alter the leaf surfaces by eroding the cuticle.[697,1422] Most studies on the effects of acidic precipitation per se on plant-insect interactions have been conducted in forest ecosystems.[320,582,583,585,586,588]

The pine bud moth (*Exoteleia dodecella*) caused severe damage to Scots pine forests in southernmost Norway. The outbreak occurred in the same areas of Norway that receive the highest levels of acidic air pollutants from the industrialized parts of western and central Europe. As the attack cannot be explained by local pollution, it is possible that acidic air pollutants of distant origin weakened the trees, creating favorable conditions for the insect.[587] In Poland the epidemics of the pine bud moth are most intense and persistent in areas that are uninterruptedly influenced by gaseous emissions high in sulfur compounds. Consequently, Hågvar et al.[587] suggested that long-range acidic air pollutants might have weakened the health of the trees, creating conditions favorable for the insect.

Hall et al.[599] treated cuttings from two clonally propagated elm (*Ulmus*) hybrids with ozone, SO_2, or simulated acid rain in order to investigate the effects of these pollutants on the susceptibility of the elms for elm leaf beetle, *Xanthogaleruca luteola* (Chrysomelidae). Plants were subjected 7 h per day, 5 days per week for 9 weeks in open-top chambers to 0.1 ppm ozone, 0.2 ppm SO_2, 0.1 ppm ozone plus 0.2 ppm SO_2, or charcoal-filtered air. An acid rain treatment (pH 4.0) of about 1.3 mm was performed weekly in rain simulation chambers. Elm leaf beetles were fed on foliage harvested from the trees subjected to the treatment combinations. The fumigation treatments had little direct effect on preoviposition period and fecundity. However, the beetles fed

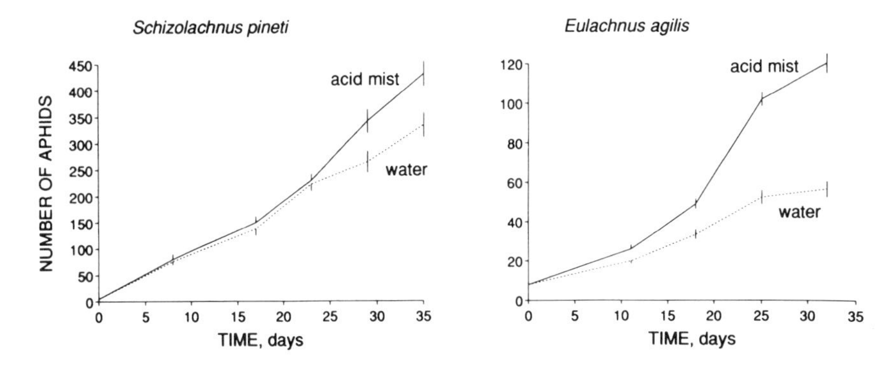

FIGURE 5. Population growth of *Schizolachnus pineti* and *Eulachnus agilis* (Homoptera, Lachnidae) on trees sprayed with either deionized water or acid solution. Vertical bars indicate standard errors. (From Kidd, N. A. C., *J. Appl. Entomol.*, 110, 524, 1990. With permission.)

on foliage subjected to the acid rain laid significantly fewer eggs than the beetles fed on untreated foliage.

The results of Kidd and Thomas[841] demonstrated that acid mist can promote the individual and population growth rates of the aphid *Schizolachnus pineti* in the laboratory. Kidd[840] studied the responses of pine-feeding aphids (*S. pineti*, *Eulacnus agilis*, *Cinara pini*) to acid-mist treatment of their host tree (*Pinus sylvestris*). The performance of all three aphid species was significantly enhanced in the presence of simulated acid mist (Figure 5). Nymphal growth rates generally increased by between 10 and 20% on the acid-treated trees, although an increase of 100% was recorded for *C. pini*. Improvements in population growth rates (34 to 89%) under acid treatment, however, tended to be much greater than for nymphal growth rates. The damage exhibited by the trees in the experiments appeared to be enhanced by the acid treatment. Chlorotic mottling was also significantly enhanced by *E. agilis*. This species is known to cause defoliation of *P. sylvestris*, the first symptom prior to needle shedding being chlorotic mottling.

Neuvonen and Lindgren[1121] studied the effect of simulated acid rain (dilute sulfuric acid) on the reproduction and survival of the phloem-feeding aphid *Euceraphis betulae* on silver birch (*Betula pendula*) in Turku, southern Finland. The results were ambiguous, but showed that artificial wet deposition could decrease the resistance of trees to aphids. In four of the bioassays, the aphids produced 40% to over 100% more progeny on birches watered with dilute sulfuric acid (pH 3.5) than did the control trees. In four other cases the aphid performance did not differ between the treatments. An aphid reproduction index pooled over the whole study was significantly higher on the acid-treated birches than on the control birches. The reproduction of aphids on acid-treated birches was enhanced when precipitation was below the long-term average, suggesting an interaction between the stress caused by acid treatment and dry periods.

Austarå and Midtgaard[66] investigated the effects of acidification and liming on the individual weights of the European pine sawfly (*Neodiprion sertifer*) in Norway. They reported that *N. sertifer* larvae reared on pines growing on soil treated earlier with sulfuric acid or limestone had significantly lower dry weights than those reared on trees growing in soil of intermediate values. Neuvonen et al.[1123] studied the effect of artificial acid rain (H_2SO_4 and HNO_3) on the quality of Scots pine needles as food for *Neodiprion sertifer* larvae in northernmost Finland. They found no consistent effects of simulated acid rain on the performance of large *N. sertifer* larvae, but it should be noted that their bioassays did not consider survival and/or growth of young larvae, which may be the developmental stage most susceptible to changes in foliage quality.[907,1596] Consequently, these experiments did not totally exclude the possibility that acidic irrigation may affect foliage quality in a way that is important for the population dynamics of *N. sertifer*.

Some plants, such as the celery (*Apium graveolens*), contain the phototoxic linear furanocumarins psoralen, bergapten, and xanthotoxin.[1501] These compounds generally exhibit increased toxicity in the presence of ultraviolet (UV) radiation. Increased production of these compounds in *A. graveolens* can be induced by a variety of environmental stresses. The high-nitrogen acidic fogs of the type and duration occurring in the Los Angeles basin can generate exceptionally high levels of linear furanocumarins in celery.[375] Trumble et al.[1503] demonstrated that *Scodoptera exigua* (Lepidoptera, Noctuidae) larvae feeding on celery with high linear fumarocumarin concentrations following acidic-fog episodes will suffer substantial additional mortality, with the resulting reduction in survival likely to produce economic consequences.

In addition to affecting plant-herbivore relationships, acidic precipitation may affect insects indirectly through their diseases or parasites. Fischer and Führer[472] have demonstrated that acidic precipitation can also affect insect populations indirectly by reducing their parasites (see Chapter 6, Section III.C). Neuvonen et al.[1122] reported that simulated acid rain reduced the susceptibility of the European pine sawfly (*Neodiprion sertifer*) to its nuclear polyhedrosis virus. Trumble et al.[1503] showed that acute incidence of acidic fogs as low as pH 2.0 will not reduce the efficacy of *Bacillus thuringiensis* against *Scodoptera exigua* if the bacteria are not physically washed from the leaves of celery. Furthermore, antagonistic interactions with linear furanocumarins (see above) that would reduce the effectiveness of *B. thuringiensis* in the field were not observed.

V. OZONE

A. BACKGROUND

Ozone (O_3) concentrations in the troposphere are increasing and becoming more pervasive. Ozone is the most important phytotoxic air pollutant in the U.S. and is becoming a major concern on a worldwide scale.[887] Comparison of global distribution of potential photochemical smog regions and forest

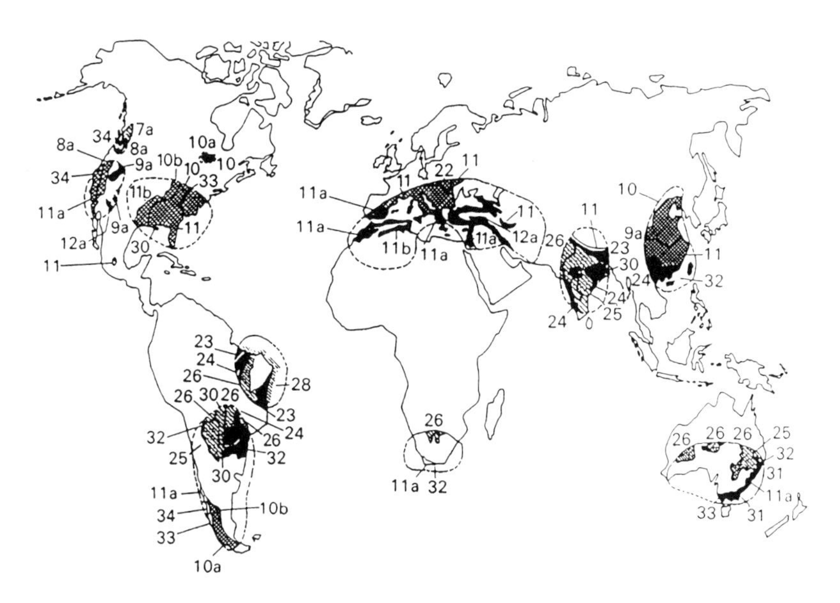

FIGURE 6. Global distribution of potential photochemical smog regions and forest ecosystem distribution patterns. Cool mixed forest and woodlands: (7a) mountain snow forest, (8a) cordilleran, (9a) montane, (10a) beech. Cold deciduous forest: (10) central broadleaf, (10b) grassland border. Southern mixed forest: (11) deciduous and evergreen forest, (11a) hard-leaf and scrub savanna, (11b) grass woodland. Dry woodland: (12a) alpine evergreen. Montane complex: (22) alpine meadow-forest-scrub. Tropical ecosystems: (23) rain-leaf evergreen forest, (24) drought deciduous mixed forest, (25) dry woodland savanna complex, (26) scrub woodland savanna complex, (28) mangrove swamp. Special limited-area ecosystems: (30) alluvial forest-herbage complex, (31) temperate broadleaved evergreen, (32) subtropical seasonal rain forest, (33) temperate or cool mixed rain forest, (34) giant evergreen forest. (From Kickert, R. N. and Krupa, S. V., *Environ. Pollut.*, 68, 29, 1990. With permission.)

distribution patterns have revealed important forest/woodland geographic regions where little or no O_3 data exist even though the potential threat to the forests in those regions appears to be large within 35 years from now (Figure 6).[837] Krupa and Manning[887] reviewed the formation and effects of ozone on vegetation. Ozone is present both in the troposphere and the stratosphere. Tropospheric ozone is mainly produced via photochemical reactions involving precursors generated by natural processes and, to a much larger extent, by human activity. There is evidence for a tendency towards increasing tropospheric ozone concentrations. However, tropospheric ozone is known to account for only 10% of the vertical ozone column above the Earth's surface, while the stratosphere accounts for the remaining 90%. Losses of stratospheric ozone apparently occur due to the updraft of ozone-destroying pollutants generated by both natural processes and by human activity. These losses can result in substantial changes in the amount of UV radiation reaching the Earth's surface.

The occurrence of regional-scale photochemical smog in areas downwind of urban centers in many parts of the world is of great concern. Ozone is formed in the troposphere when nitrogen dioxide (NO_2) is converted into nitric oxide (NO) by the action of sunlight. The freed oxygen atom reacts with oxygen molecules (O_2) to form ozone (O_3). Normally, the reaction is reversible, and a state of equilibrium exists between ozone, nitrogen dioxide, and nitric oxide. When automobile exhaust products are present, they react with nitric oxide and stop the back reaction, so that ozone accumulates. Ozone and other photochemical oxidants can be transported hundreds of kilometers from areas where their precursors originate, so that ozone is not really a local problem.[887,1497] Widespread ozone pollution in the Southern Hemisphere, mainly originating from biomass burning in Africa, was recently deduced from satellite data.[479]

As ozone, peroxyacetyl nitrate (PAN) is formed in the atmosphere by photochemical smog reactions. It is the most abundant compound of peroxyacyl nitrates. These major pollutants typically affect the lower leaf surface of exposed plants. PAN can interfere with photosynthesis and other basic metabolic reactions of plants.[570,1497] Relatively high concentrations of PAN have been reported from California, but its effects on the growth and reproduction of plants is still poorly known, not to speak of its impact on insects.

B. CHANGES IN PLANTS

Tropospheric ozone is highly phytotoxic. Ozone enters plants mainly through open stomata during the day. Ozone symptoms in plants have been reported from North America and Europe. Exposure to ozone can result in both acute injury and chronic injury, or it may affect growth and yield, with or without visual symptoms.[570,887,1273,1497,1548] Common symptoms of acute foliar ozone injury include flecking and stippling and those of acute injury pigmentation, chlorosis, and premature senescence. Warm temperatures, sunlight, high relative humidity, good nutrition, adequate soil moisture, and other factors are necessary for ozone injury to occur.[887] Chronic effects are of great concern in terms of both crops and forests.[887] In the U.S. ozone causes more plant damage than any other air pollutant.[887,1502]

A number of experimental techniques are available for evaluating the chronic effects of ozone on plants. Of these field evaluation techniques, open-top chambers are the most frequently used for evaluating the chronic effects of ozone on crops. Acute effects can be evaluated with less complexity through the use of biological indicator plants. The numerical modeling of such effects is less complicated than establishing numerical cause-and-effect relationships for chronic effects.[887] There are considerable differences among plant species in their sensitivity to ozone.[1497] The presence of other kinds and forms of pollutants in the ambient atmosphere, and the incidence of pathogens and pests, confound the acute or chronic responses of plants to ozone. The resulting complex interactions and joint effects on plants are poorly understood.[887]

The effects of ozone on carbon balance and growth of individual plants can be quantified on the basis of concentration, external dose, or uptake.[1272] Several mathematical models have been published representing the effects of ozone on crops and native vegetation.[837] According to Reich,[1272] agricultural

FIGURE 7. Chambers used to study the effects of ozone on plants, insects (e.g., aphids), and plant-insect interactions in California.

crops are the most sensitive to ozone, with hardwoods intermediate, and conifers least sensitive. However, there are considerable differences among different tree species in their pollutant sensitivity, and different models may give controversial results.[837] Changes in water exchange processes throughout the hydrological cycle can be used as early warning indicators of forest responses to ozone.[837]

Ozone interferes in several ways with the basic metabolic processes of plants.[157,1497] The physiological changes reportedly decrease photosynthesis in several species.[1166,1273] Long-term exposure of plants to low concentrations of ozone affects photosynthesis and results in reduced translocation of photosynthates from shoots to roots.[305] This may adversely affect mycorrhizae,[820] enhance root senescence,[976] and subsequently affect yields due to reduced uptake of water and minerals. Upon entering leaves through the stomata, ozone reacts with plant membranes and cellular structures such as chloroplasts, resulting in structural damage and, in some cases, the onset of premature senescence, which may lead to the release of soluble nitrogenous compounds from plant structural forms on foliar tissues.

C. EFFECTS ON INSECTS

Attempts have recently been made to experimentally determine the insect response to ozone and to study the mechanisms causing any changes (Figure 7; Table 4). While it is now generally accepted that SO_2 tends to increase growth rates and promote a population increase in many herbivorous insects

TABLE 4
Effects of Ozone on Terrestrial Insects in Experimental Studies

Species	Host plant	Dose	Response	Refs.
Homoptera				
Phyllaphis fagi	*Fagus sylvatica*	0.04 ppm (78 μg/m^3) (max 240 μg/m^3), 1 month[a]	Ca. 57% population increase	177
Eucallipterus tiliae	*Tilia x europaea*	5.72 nl/l, 5 days	No change	1029
Chaitophorus populicola	*Populus deltoides*	397 μg/m^3 (0.20 ppm), 5 h	No change in performance	295
Rhopalosiphum padi	*Hordeum vulgare*	70, 120, 170 nl/l		
		24–27. June	4.5, 26.2, and 24.4% change in MRGR	
		22–25. July	6.4, 13.1, and 14.7% change in MRGR	1558
		4–7. August	–3.1, 6.2, and 14.2% change in MRGR	
Aphis fabae	*Phaseolus vulgaris*	0.04 ppm (78 μg/m^3) (max 240 μg/m^3) 1 month[a]	Ca. 57% population decrease	177
A. fabae	*Vicia faba*	0.085 ppm, 2 or 3 days	3.4–13.1% decrease in MRGR[b]	389
A. rumicis	*Rumex obtusifolius*	34–206 nl/l, 4 or 8 days	–19.9–+13.6% change in MRGR, not dose-related[c]	1600
Elatobium abietinum	*Picea sitchensis*	100 nl/l, 4–96 h, continuous	No response	1029
E. abietinum	*Picea sitchensis*	100 nl/l, 24–96 h, episodic	No response	1029
Acyrthosiphon pisum	*Pisum sativum*	21–206 nl/l, 4 or 8 days	–6.5–+24.0% change in MRGR, not dose-related	1600
Schizolachnus pineti	*Pinus sylvestris*	100 nl/l, 4–96 h, continuous	Decreased performance at 48 nl/l	1029
S. pineti	*Pinus sylvestris*	100 nl/l, 24–96 h, episodic	No response	1029
Cinara pilicornis	*Picea sitchensis*	100 nl/l, 4–96 h, continuous	No response in old shoots	1029
C. pilicornis	*Picea sitchensis*	100 nl/l, 24–96 h,	No response[d]	1029

C. pini	*Pinus sylvestris*	100 nl/l, 4–96 h, continuous	No response	1029
C. pini	*Pinus sylvestris*	100 nl/l, 24–96 h, episodic	No response	1029
Lepidoptera				
Keiferia lycopersicella	*Lycopersicon esculentum*	280 ppb, 3 h, 2 or 4 times	100% increased larval survival, 10.6% faster development, no change in longevity or fecundity	1502
Danaus plexippus	*Asclepias curassavica*	150–183 nl/l, 16–17 days, 7 h/day	O_3-treated leaves preferred by 3rd instar larvae, no preference with 4th instar larvae, relative growth and consumption rates of 5th instar larvae greater on fumigated plants	157a
D. plexippus	*Asclepias syriaca*	165–178 nl/l, 14–19 days, 7h/day	No preference with 3rd instar larvae, 4th instar larvae preferred control leaves, relative growth and consumption rates of 5th instar larvae greater on fumigated plants	157a
Lymantria dispar	*Quercus alba*	90, 150 nl/l	Preference for O_3 treated foliage, decreased palatability in intermediate levels	771
Coleoptera				
Rhynchaeus fagi	*Fagus sylvatica*	Monthly mean 14–101 $\mu g/m^3$, 2 yrs, maximal hourly mean 92–227 $\mu g/m^3$[a]	Preference for O_3-treated leaves	687
Epilachna varivestis	*Glycine max*	2.5, 5.0, 7.8, and 11.4 pphm, 37 days, 16 treatments	Preference for O_3-treated leaves	440

Table 4 (continued)

Species	Host plant	Dose	Response	Refs.
E. varivestis	*Glycine max*	30, 60 nl/l, 3 weeks, 7 h/day	No feeding preference, increased defoliation, faster development	246
Gastrophysa viridula	*Rumex obtusifolius*	70 nl/l, 24 days, 7 h/day	Larger egg batches, higher survival, and productivity, lower food consumption	1600
Plagiodera versicolora	*Populus deltoides*	393 $\mu g/m^3$ (0.20 ppm), 5 h	Unexposed plants preferred for oviposition, reduced fecundity	294

[a] Urban air of the city of Basel, Switzerland.
[b] 15.8% increase in MRGR when 0.03–0.04 ppm NO_2 was present.
[c] Greater MRGR on new leaves.
[d] 63% increase in MRGR at 24 h on new shoots. Decreased MRGR at high temperatures (23°C).

and that this effect is mediated via the food plant, the evidence for ozone is more questionable.[246,294,440,771,1502] The consequences of ozone fumigation on associated insect herbivores cannot yet be predicted.[1600] Exactly the same treatment can produce opposite responses in similar insects feeding on different food plants, or in different types of insects feeding on the same food plant. This simply reflects the conflicting evidence about how plants respond to ozone fumigation.

Ozone air pollution may alter the interactions of insects and pathogens with their host plants. This has been shown by studies on changes in insect feeding and oviposition behavior due to ozone stress of the host plants; by studies on changes in insect growth, development, survivorship, and fecundity; by studies on the performance of plant pathogens on ozone-stressed plants; and by studies on community level responses of pests to ozone fumigation of plants.[294] However, not all insects are influenced by host-plant exposure to ozone,[294,428] while others are.[790,1502] Thus, it is not possible to predict the outcome of pollution-induced changes in the resistance of plants to insects and pathogens, and subsequent changes in plant productivity and reproduction.

Ozone-weakened pines in California, Virginia, and Mexico become more susceptible to invasion by bark beetles.[368,1057,1382] The bark beetle infestation of ozone-damaged ponderosa pine (*Pinus ponderosa*) was studied in the San Bernardino Mountains of southern California. Both the western pine beetle (*Dendroctonus brevicomis*) and the mountain pine beetle (*Dendroctonus ponderosae*) frequently attack trees with advanced symptoms of pollution injury. Pollution-damaged trees not yet infested by bark beetles had reduced oleoresin exudation pressure, yield, and rate of flow, and increased rate of resin crystallization. All these characteristics were associated with increased susceptibility to bark beetles. However, the infested trees did not permit good brood production. Thus, the pollution injury may result in a more susceptible, but less suitable, host. Moreover, in spite of the heavy infestation of individual pollution-damaged trees, there was no noticeable increase in the number of infested trees.[285,286,1058,1420]

Prolonged exposure of house fly populations to high ozone concentrations caused insignificant variations in growth and survival, but apparently stimulated oviposition.[94] When fumigated for a prolonged time with high concentrations of ozone, flies (*Drosophila*, *Musca*, *Stomoxys*) showed a 15% inhibition of egg hatching, but no differences in larval or pupal stages, whereas adults showed stimulated oviposition, an increase in the number of eggs laid, and an increase in the adult population compared to the controls.[929] Such effects were not found in fire ants or cockroaches.[930]

Most studies have concentrated on chewing insects and have generally shown a preference for, and increased growth rate on, fumigated foliage compared with control foliage when fumigation was ceased before the insects were released onto the foliage. Endress and Post[440] reported that Mexican bean beetle adults (*Epilachna varivestis*) preferred to feed on excised soybean leaf discs (*Glycine max*) in the laboratory when the leaves had been prefumigated with air enriched by concentrations of ozone between 25 nl/l and 114 nl/l. The effect was detectable up to 23 days after fumigation had ceased, and the response was dosage dependent. Similarly, Jeffords and Endress[771] demonstrated that gypsy moth larvae (*Lymantria dispar*) favored ozonated foliage excised from white oak (*Quercus alba*). Moderate prefumigation by ozone (90 nl/l) made the leaves less palatable, and high concentrations (150 nl/l) made them more palatable, to moth larvae.

Chappelka et al.[246] studied the effects of ozone on adult feeding preferences of the Mexican bean beetle and on its larval growth and development in soybean. The favored hosts of the beetle are *Phaseolus* spp., but the species is also an important pest of soybeans. Soybeans were fumigated with ozone levels of between 0 and 40 nl/l in unfiltered open-top chambers and then transferred to charcoal-filtered chambers into which the beetles were released. Defoliation of the plants was positively related to the degree of prefumigation, and the development time of the beetles decreased, and larval weight increased, when the beetles were fed on O_3-fumigated foliage. Hughes et al.[726,728] reported similar findings with SO_2-treated soybeans.

Trumble et al.[1502] evaluated the influence of ozone on interactions between the tomato pinworm (*Keiferia lycopersicella*), a specialist herbivore of economic significance in the U.S., and its host plant. Survival of pinworm larvae developing on ozonated foliage was significantly greater than larvae fed control plants. Survival was more than doubled for tomato pinworm fed plants exposed to two fumigations. A significant decrease in survival was observed for tomato pinworms developing on foliage exposed to four fumigations, as opposed to two fumigations. Whether such an effect was due to contact toxicity, irritation, or some other cause warrants additional investigation. The adverse effects of ozone apparently decreased upon entering the leaf-mining stage, as the counts of mines on the foliage equaled the numbers of larvae exiting from leaves. No differences were detected in either percent pupation or successful adult emergence from pupal cases. Although tomato pinworm larvae feeding on ozonated plants survived better, no corresponding improvements in other fitness parameters were observed; significant differences were not detected for male-to-female ratios, female longevity, or fecundity. Developmental rates for larvae feeding on ozonated tomato plants were faster by more than 32 h than larvae feeding on control plants. Mean times for emergence from the plants were 75.9 h for larvae fed control foliage, 55.5 h for tomato pinworm fed on plants exposed to two ozone fumigations, and 49.4 h for larvae fed tomatoes fumigated four times. These values represent increases in overall larval development rates of 1.3, 7.1, and 10.6%, respectively, as compared to expected larval developmental time predicted at 26°C. The increase in developmental rate may have occurred in response to ozone-induced stomatal closure in tomato foliage (which would elevate temperature through a decrease in evaporative cooling).

Experiments on a chewing insect, *Gastrophysa viridula* (Coleoptera, Chrysomelidae), feeding on *Rumex obtusifolius* showed that gravid females feeding on ozone-fumigated plants laid more eggs over a shorter period of time than did beetles feeding on control plants.[1600] Hatching was not affected, but survival through the three larval stages to pupation was significantly better on fumigated plants, and total population production was 50% greater on fumigated, compared with the control, plants. The leaf area consumed by the developing larvae for every milligram of beetle produced was, however, over twice as great on the control as on the fumigated plants. It was considered likely that fumigated leaves were relatively more nutritious than control leaves, or less well defended.

Jones and Coleman[790] examined the feeding and oviposition preference of the imported willow leaf beetle, *Plagiodera versicolora* (Coleoptera, Chrysomelidae), for two clones of the eastern cottonwood (*Populus deltoides*) that had been exposed to a single acute dose of ozone or charcoal-filtered air. This ozone dose had no significant effect on the growth of the plant, nor did it produce any visible injury on the bioassayed leaf material.[297] The same dose of ozone had effects on cottonwood resistance to some members of the insect and pathogen community, but not others.[294,295,297] Such a dose significantly changed the feeding and oviposition preference of the willow leaf beetle (Figure 8). Both *P. versicolora* larvae and adults preferred to feed on, and adults consumed more of, the ozone-stressed foliage than the controls, when presented with a choice. In contrast, females preferred to oviposit on the unstressed foliage. The changes in preference occurred consistently and to about the same degree in two cottonwood clones, in three different years, and in experiments using leaf discs, attached leaves, and whole plants.[790]

Air pollution stress on plants does not necessarily lead to better insect performance. Although foliage consumption by *P. versicolora* was greater on cottonwood foliage exposed to air pollution, insect development and survivorship did not change, and beetle fecundity decreased. The ultimate result of ozone fumigation was that the reproductive fitness of *P. versicolora* decreased on ozone-exposed cottonwoods, contrary to predictions made by Hain,[589] Führer,[517] and Dohmen.[387] The feeding preference of the willow leaf beetle for ozone-fumigated cottonwood plants is therefore not an indicator of the quality of foliage for beetle performance or an indicator of population fitness. It has been asserted that the increased feeding preference of some insects for ozone-exposed plants[440,771] may be an indicator of increases in their performance and population sizes in ozone-polluted areas,[589] but this assertion is not necessarily correct.[294]

Sap-feeding insects, and particularly aphids, have often been used in atmospheric pollution studies because they show a rapid and easily measurable response to physiological changes in the food plant. In the case of ozone, however, there have been few studies, and these have produced contradictory results. Braun and Flückiger[177] reported that population growth of *Aphis fabae* on *Phaseolus vulgaris* was inhibited by ozone (40 nl/l), but that of *Phyllaphis fabae* on *Fagus* was stimulated. Acute ozone exposure (200 nl/l) produced no effect on survivorship, reproduction, or development of the aphid *Chaitophorus*

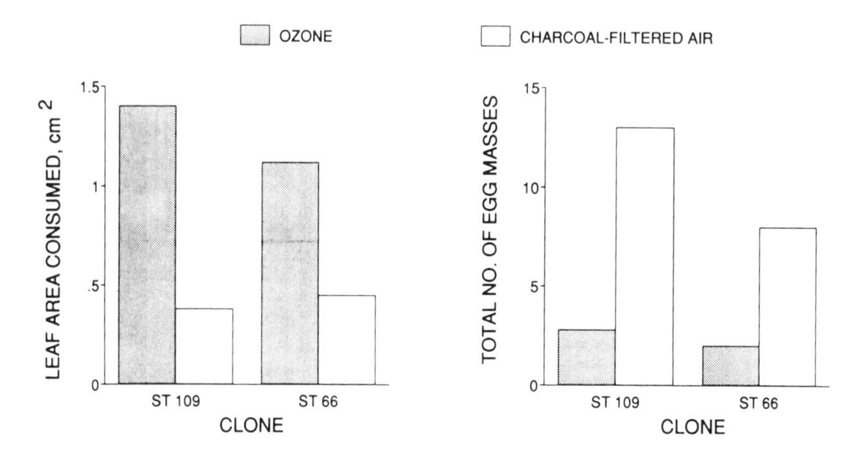

FIGURE 8. Mean leaf area consumed by adult *Plagiodera versicolora* (Coleoptera, Chrysomelidae) in choice tests with two cottonwood (*Populus deltoides*) clones, shown as consumption after 7 days (left), and oviposition by females on these clones, given as total number of egg masses (right), in dual-choice leaf disc assays. Cottonwood clones (ST 109, ST 66) were treated with ozone (shaded) or charcoal-filtered air (white). (From Jones, C. G. and Coleman, J. S., *Oecologia,* 76, 51, 1988. With permission.)

populicola feeding on eastern cottonwood trees.[295] Elden et al.[428] found that the population growth of the pea aphid on alfa-alfa did not change as a result of preinfestation exposure of the plants to ozone.

Dohmen[389] found that fumigation of *Vicia faba* plants with 85 nl/l ozone caused decreased growth of the black bean aphid (*Aphis fabae*), while higher concentrations stimulated aphid growth. This was explained by accelerated senescence of the host plant at these concentrations. Warrington[1558] studied the effects of ozone on the growth rate of cereal aphids (*Rhopalosiphum padi*) on spring barley grown in greenhouses supplied with charcoal-filtered air. Early instar larvae of the cereal aphid were caged on leaves for 3 days in June, July, and August. Ozone did not affect aphid survival, but the highest concentrations of ozone (25 and 36 nl/l) appreciably increased the mean relative growth rate of the aphids, the effect being larger early in the growing season.[1558] Whittaker et al.[1600] studied the effect of ozone fumigation on two aphid species, *Acyrthosiphon pisum* on *Pisum sativum* and *Aphis rumicis* on *Rumex obtusifolius.* They did not find any clear evidence of a general effect on the MRGR of aphids exposed to concentrations of ozone between 21 and 206 nl/l.

Furthermore, Dohmen[389] found that the presence of NO_2 (formed as a result of NO reacting with O_3) in combination with ozone during fumigation could enhance aphid growth, even though ozone alone caused the opposite effect. The differential concentrations of NO and NO_2 within the cabinets might account for the variability in aphid MRGR and complicate the effect of ozone fumigation in some of the pea aphid experiments. When concentrations of NO_x contaminants were low, the growth rate of *A. rumicis* on ozone-fumigated plants was not only significantly higher on new growth compared with old growth, but actually increased on new growth and decreased on old growth, relative to the controls.[1600] This effect may occur because of an increased turnover rate of fumigated leaves, i.e., the duration of individual leaves on the plant is reduced. However, it will almost certainly mean that aphids feeding on actively growing plant shoots will benefit from an ozone episode.

If increased insect survival and developmental rate on ozonated plants proves applicable to other agricultural systems, programs dependent on developmental models driven by temperature will underestimate both the stage of development and population size. Trumble et al.[1502] suggested that the exposure of host plants to ozone is an important developmental factor for phytophagous insects. Increased developmental rates may affect the success of biological control agents. For example, even though an increased density of the tomato pinworm may allow increases in the most common parasitoid species, the time period for most successful parasitization would be shortened to less than the reported 48-h period.

D. RESPONSE MECHANISMS

Gaseous air pollutants induce biochemical changes in plants. Increased concentrations of amino acids and other nitrogen compounds have been reported following exposure to ozone.[1187,1490] Nitrogen is often a limiting nutrient in the growth of herbivorous insects, and increased availability of nitrogen frequently improves insect growth and reproduction.[1028,1229] Ozone can affect the nutritional status of host plants for insects or change their morphology, making them more or less palatable or suitable as habitats and sites of reproduction.

Hughes and Laurence[724] stated that host vulnerability to insect selection can be altered by changes in the physical characteristics of a plant chlorosis, stippling, etc. or by alterations in plant metabolism, which may influence insect discovery. Although there was a significant, positive correlation between visible ozone injury and Mexican bean beetle defoliation percent, the data of Chappelka et al.[246] indicated that alterations in host preference occurred primarily due to nonvisible changes with increasing ozone concentration. The mechanism(s) responsible for the increased preference of the Mexican bean beetle to ozonated foliage is unknown. Host selection by the beetle is, however, influenced by increases in leaf sugar concentrations and host volatile compounds.[62] Ozone has been reported as causing increases in the level of soluble sugars in the foliage[83] and the release of host volatile compounds.[1292]

Increased foliage consumption by the imported willow leaf beetle (*Plagiodera versicolora*) on ozone-exposed plants might be a compensatory mechanism used by the beetle to cope with the reduced foliage quality of fumigated cottonwoods.[294] Compensatory consumption occurred in a number of insects reared on host tissue of lower nutritive quality.[564,997,1376] For example, soybean loopers (*Pseudoplusia includens*, Noctuidae) increased their foliage consumption in order to obtain adequate nitrogen, on plants exposed to carbon dioxide, when the carbon/nitrogen ratio of these plants increased.[938] The quality of nitrogen compounds in tomato and cottonwood leaves changed, respectively, as a result of ozone fumigation.[1502]

The data of Coleman and Jones[295] did not support the hypothesis of White,[1596] Dohmen,[387] and Führer[517] that aphid performance will directly increase on plants exposed to the environmental stress of air pollution. The ozone dose used by Coleman and Jones[295] was obviously an adequate environmental stress, since the performance of *Plagiodera versicolora* changed on fumigated cottonwoods of the same two clones.[297] It is probable that different ozone doses are required to affect the response of different insect species. Coleman[293] hypothesized that aphids would prefer leaves at the developmental stage of the sink-to-source transition, because they are anatomically, physiologically, and biochemically the most suitable resource for insects and pathogens. Hartnett and Bazzaz[627] suggested that aphid leaf preference was a function of the assimilative value of a leaf to the plant.

The imported willow leaf beetle (*Plagiodera versicolora*) and the cottonwood leaf rust (*Melampsora medusae*) decreased on cottonwoods exposed to ozone, but the performance of the cottonwood aphid (*Chaitophorus populicola*) and the leaf spot fungus (*Marssonina brunnea*) was apparently unchanged.[297] Ozone fumigation of cottonwoods has the potential to change the dynamics of the pest species of the plant. *Chaitophorus populicola*, *Marssonina brunnea*, *Melampsora medusae*, and *Plagiodera versicolora* all feed on cottonwood in the same geographic area, all feed on similar leaves preferentially, and all feed and reproduce throughout the growing season of the eastern cottonwood.[295] Thus, there is a potential for resource competition to occur between these four pests.[295] *P. versicolora* larvae use methylcyclopentanoid monoterpene "defenses" to repel cooccurring herbivores.[1261] If *C. populicola* populations are held at low levels because of competition with *M. medusae* or *P. versicolora* for cottonwood leaf resources, acute ozone exposure of cottonwoods could result in the increased availability of leaf tissue for *C. populicola* and subsequently increase the pest potential of this aphid. This is because ozone exposure of cottonwoods decreases the reproductive performance of the other two pest species that use the same leaf resource.

Coleman and Jones[294] hypothesized that in the field, acute exposure of cottonwoods to ozone can reduce the total egg production of the imported willow leaf beetle (*Plagiodera versicolora*) through a short-term reduction in oviposition rates, and this could affect overall population dynamics. The decrease in the fecundity of the imported willow leaf beetles on ozone-fumigated cottonwoods was quite striking.[294] Raupp and Denno[1260] suggested that *P. versicolora* adults withhold egg mass rather than oviposit on unsuitable willow foliage. Jones and Coleman[790] showed that control foliage was preferred by *P. versicolora* for oviposition over ozone-treated foliage. Female beetles on ozone-treated plants made their first oviposition at a much later date than those on control plants. Females probably withheld their eggs. Inhibition of oviposition apparently ceased within 10 days after insects were placed on the plants. Despite the short duration of differences in beetle oviposition rates between treatments, the total number of eggs oviposited after 2 weeks was still significantly greater on control plants, being equivalent to about one whole generation of insects.

Changes in the nutritional quality of the host can alter the growth and development of an insect feeding on it.[724] Ozone fumigation can cause increases in the levels of free amino acids, soluble proteins, soluble sugars, phenolic compounds, and other metabolites[83,1488,1502] that can be utilized by insects as food sources.[724] The implications of increased concentrations of readily available forms of nitrogen for insect herbivore development are considerable.[997,1596] Nitrogen availability affects such basic life processes as growth rate, survival, and reproductive capacity. White[1596] speculated that improved host quality resulting from an increase in free amino acid concentration following an episode of plant stress would greatly enhance survival of newly enclosed larvae. White[1596] proposed that the breakdown of insoluble proteins to smaller, more soluble compounds was a general response to plant stress and adaptive to the extent that the increase in free amino acid concentration stimulated seed production during stressful periods. While this may be true of some stresses (e.g., drought), Trumble et al.[1502] suggested that the increases in soluble nitrogen concentrations following ozone exposure are more likely to be simply the direct consequence of the chemical reactivity of ozone with plant proteins and amino acids.[321,1085]

Since the biological effects of feeding on ozonated foliage were more evident for tomato pinworm (*Keiferia lycopersicella*) larvae than adults, Trumble et al.[1502] suggested that either the nutritional suitability of the foliage was improved for the young larvae, as suggested for stressed plants by White,[1596] or the plant's defensive system was being adversely affected. Analysis of total nitrogen suggested that ozone-damaged foliage should be less nutritious than untreated leaves by 1 week following exposure to ozone fumigations. However, total nitrogen analysis includes nitrogen from structural proteins and nonprotein amino acids and the nitrogen incorporated in defensive compounds.[615] Significant enhancement of soluble protein concentrations for up to 2 days postfumigation provided an increase in the amount of readily assimilable nitrogen available for the larvae, which is a plausible explanation for the observed improvement in larval survival.[1502] Similar increases in soluble protein concentrations have been documented for beans exposed to ozone,[96] but reports in the literature have not been consistent.[1487] The concentrations of 10 of the 20 amino acids determined from ozonated tomate foliage differed significantly from those in the control leaves.[1502] The reductions in methionine and tyrosine immediately following fumigation were anticipated from earlier reports,[242,1085] and significant reductions in glutamine and proline were reported by Trumble et al.[1502] Of the other amino acids that reportedly react with ozone, cysteine and cystine were only nominally present in tomato foliage, and histidine was not significantly affected. In contrast to previous studies,[321,1037] Trumble et al.[1502] did not detect significant increases in the total pool of free amino acids. However, a trend toward higher concentrations of total free amino acids in ozone-treated plants was evident. The concentrations of readily assimilable nitrogenous compounds proved to be considerably better indicators of host suitability than were the results of total nitrogen analysis.

The results of Endress and Post[440] indicated that the metabolic changes in the host plant induced by ozone can be stable for long periods of time.[246] Mexican bean beetle adults (*Epilachna varivestis*) still preferred ozonated soybean foliage over controls 23 days after the last exposure. Comparisons between amino acid concentrations in tomato immediately following fumigation and those after 7 days postfumigation may explain the variable reports of amino acid fluctuations in the literature.[1502] While the proportions of some amino acids remained approximately the same on both sampling dates, the relative amounts of others (serine, glutamine, histidine, and valine) increased considerably. The tomato plant may be stimulated to compensate for ozone-induced reductions in certain amino acids, such as glutamine, or the leaves are entering a premature senescence causing nitrogen compounds to be converted into soluble forms available for reallocation. The general increase in free amino acid concentration in both ozonated and control plants possibly reflected a general mobilization of energy reserves from the leaves to the flowers or immature fruit.[1502] Regardless of the cause of differences in amino acid concentration between ozonated and control plants, the time elapsed between ozone fumigation and leaf sampling can affect the quantities of amino acids, soluble protein, and total nitrogen measured.

The implications of ozone-induced changes in amino acid concentrations are potentially significant for insect populations. Of the amino acids considered necessary for insect growth,[344] methionine and valine concentrations differed significantly in ozonated foliage. Since growth of the tomato pinworm was not inhibited by the transient reduction in methionine concentration, and valine concentrations increased, the changes in these amino acids were not limiting. Relatively small changes in free amino acid concentrations can have significant negative effects on insect development. Potentially critical changes in free amino acid concentrations occurred in the key supplementary amino acids, all of which increased significantly in ozonated tomato foliage. Alanine, glycine, aspartic acid, and serine have been reported as important growth factors for the silkworm (*Bombyx mori*), while proline was documented as semiessential for the silkworm[748] and critical to the development of many dipterans.[513] Since there were major changes in concentrations of these amino acids, increased availability of supplementary amino acids probably accounted for at least some of the improved survival rate demonstrated for tomato pinworms fed on ozonated foliage. Trumble et al.[1502] suggested that exposure to ozone affected nitrogen, soluble protein, and total free amino acid concentrations independently.

The amino acid composition of the phloem plays an important role in aphid growth. The amino acid pattern in a phloem exudate of *Phaseolus vulgaris* and *Viburnum opulus* was changed qualitatively and quantitatively by motorway air pollutants, and this pattern promoted aphid growth and development when simulated in artificial diets.[156] In *P. vulgaris*, however, the amino acid composition in the phloem exudate was not changed significantly by the rural fumigation treatments. The reduced aphid growth rate on *Vicia faba* after ozone fumigation was also mediated via the plant.[389]

Of the secondary plant compounds (in commercial tomatoes) that have proven antibiotic properties against lepidopterans, only chlorogenic acid has an exposed double bond susceptible to rapid oxidation by ozone.[1502] Since chlorogenic acid production can be an induced response occurring after stress caused by insects,[430] oxidation should prevent chlorogenic acid levels from increasing as long as contact with ozone is frequent. Trumble et al.[1502] found that the observed increase in survival and developmental rates of tomato pinworm larvae feeding on ozonated tomato foliage was due to a complex of factors, the most important of which were nutritional. Analyses of ozonated foliage at 0, 2, and 7 days following fumigation showed a transient, but significant, increase (18 to 24%) in soluble protein concentration. Significant changes were observed in at least ten specific amino acids, some of which are critical for either insect development or the production of plant defensive chemicals. Chlorogenic acid production in tomatoes may be induced by exposure to ozone, but the plant response occurred too late to affect the enhanced growth and survival of the tomato pinworm.

E. INCREASED PEST DAMAGE TO PLANTS

Plant health is the product of interaction between the plant and the physical and chemical climatology and pathogens and pests.[887] In addition to causing visible symptoms on leaves and reducing the growth and yield of plants, ozone may have many other subtle and interactive secondary effects. If these are not considered when assessing the effects of ozone on plant yields, then misleading or incomplete results will be obtained.[887] Since air-pollution episodes typically occur over broad areas of crop production, the cumulative effects on herbivores at the population level may be quite significant. Thus, potential changes in plant physiology due to stress or direct injury resulting from air pollutants may have more serious consequences than surveys of direct economic losses have suggested.[1502] For example, as a result of increased Mexican bean beetle (*Epilachna varivestis*) preference and larval growth and development, ozone may have a serious economic impact on soybean, in addition to its direct phytotoxic effects.[246]

It is reasonable to ask what impact such ozone-induced changes in behavior could have on the distribution of the insect and the damage it produces. The imported willow leaf beetle (*Plagiodera versicolora*) is a major pest of cottonwoods (*Populus* spp.) and willows (*Salix* spp.). Its feeding and oviposition can occur together on the same cottonwood plant and on the same leaves, and on the same or similar leaves on other hosts.[790] Multiple-choice whole-plant experiments suggested that *P. versicolora* would actively feed on cottonwoods that are susceptible to ozone stress, following an acute ozone episode, but female adults might well then disperse to other plants for oviposition.[790] Those other plants could either be clones that are not susceptible to ozone stress, or plants of the same genotype that exhibit less of a response to ozone stress due to local conditions. The light and moisture conditions affect the dose of ozone taken up by plants. Considerable variation has been observed in the amount of visible injury to older leaves of cottonwood of the same clone within the same fumigation chamber.[790]

Three possible consequences of this change in insect distribution were proposed by Jones and Coleman.[790] First, trees showing the greatest susceptibility to ozone stress could experience the greatest amount of beetle (*Plagiodera versicolora*) damage immediately following the ozone exposure. However, the subsequent generation(s) of beetles that would have normally been oviposited on these plants would be oviposited elsewhere, and the ozone-susceptible plants should subsequently be exposed to less beetle damage. Second, those trees that are growing near susceptible plants, but are not themselves susceptible to ozone stress, might gain initial benefit from a reduction in beetle consumption as the adults moved to stressed plants. However, they would then receive a subsequently greater oviposition load following the acute ozone episode as beetles moved away from stressed plants to oviposit on them instead. This greater egg load might subsequently lead to greater damage to ozone-resistant plants, an unexpected consequence of being resistant to stress.[790] Third, it is also possible that there might be alterations in the structure of the insect and pathogen species assemblage on cottonwood, and the damage to the plant that these other organisms cause, following an acute ozone episode.[790] The cottonwood aphid (*Chaitophorus populicola*) and the leaf spot fungus (*Marssonina brunnea*) are unaffected by the same ozone dose that affects the beetle,[294] while cottonwood leaf rust (*Melampsora medusae*) reproduction is reduced at these doses.[297] These organisms might experience changes in leaf resource availability on stressed plants and adjacent unstressed plants because of changes in beetle consumption and distribution. Such changes could occur irrespective of whether or not these organisms are directly affected by ozone stress to their host plant.[790]

The consequences of ozone stress to a plant on the behavior of a cottonwood insect are not simple or immediately obvious. It is clear that attempting to deduce the consequences of changes in insect behavior to both the insect and the plant cannot be accomplished from studies that address only the feeding preference of insects on stressed vs. unstressed plants.[790] It has been assumed that if a given insect or pathogen species does not exhibit significant changes in performance on plants exposed to air pollution, there is no reason to expect air-pollution-induced changes in the population dynamics of that organism.[27,428] This view does not account for the fact that most plants are susceptible to a large number of insects and pathogens and that these organisms may interact directly with each other or through the host plant. Different pests will respond differently on cottonwoods exposed to the same acute ozone dose.[297] Ozone exposure of eastern cottonwood apparently changes the resource quality of cottonwood leaves in a different way for different pests, and this will directly or indirectly result in changes in the structure of the cottonwood pest species assemblage.[295]

The long-term effects of ozone-plant-insect and/or pathogen interactions in relation to crop yields are poorly known.[887] Ozone can directly or indirectly affect the course of development of plant diseases. Ozone generally increases the incidence of diseases caused by nonobligate fungi, particularly necrotrophic fungi. Manning et al.[977] demonstrated that infection of potato leaves was increased when they were injured by ozone. Fehrman et al.[461] found that preinoculative exposures to ozone generally increased the incidence of cereal leaf pathogens. Ozone generally decreases the incidence of diseases caused by obligate fungi, such as rusts and powdery mildews.[647] Dohmen[388] showed that preinoculative exposure of young wheat plants to ozone reduced the effectiveness of the inoculum of brown rust. Ozone-weakened pines in California and Virginia may also become more susceptible to root rots caused by *Heterobasidion annosum* and *Verticicladiella procera*.[1057,1382]

VI. FLUORINE COMPOUNDS

A. BACKGROUND

Fluoride is widespread in natural ecosystems, but in excess it may be harmful for plants and animals. Polluted air provide the main source of elevated fluoride levels in plants. This pollution may originate from heating of clays, rocks, coal, or ores containing fluoride. Fluoride emissions have been largely controlled in the U.S. and Europe, but the costly technology is not applied universally.[1497]

For example, rural industries have developed rapidly in China in recent years and have resulted in air pollution that has serious impacts on agriculture, horticulture, and forestry. Among the types of pollution that have contributed to the damage of many plant species is airborne fluorine (F) emitted from brick kilns, phosphate fertilizer plants, coal, and other industries. In the silkworm (*Bombyx mori*) industry, damage caused by the ingestion of F-polluted mulberry leaves has been widespread in Jiangsu and Zhejiang provinces.[1549]

Toxicological effects of prolonged fluoride intake are well known in humans, cattle, and sheep.[42] Information about wild animals is gradually accruing, but the relationships between environmental fluoride and fluoride toxicosis are still poorly known. Andrews et al.[42] found evidence of dental fluorosis, loss of enamel color, and the banding of incisor teeth in the field vole (*Microtus agrestis*), but not in the common shrew (*Sorex araneus*).

B. CHANGES IN PLANTS

Weinstein[1578] and Treshow and Anderson[1497] reviewed the plant symptoms and responses caused by fluoride. In addition to leaf necrosis and chlorosis, fluoride causes metabolic changes in plants. Changes in keto acids, organic acids, amino acids and amides, free sugars, peroxidase, DNA and RNA, phosphorus, and starch and nonstarch polysaccharides have been reported in plants exposed to fluorides. Fluorides are known to have an effect as enzyme inhibitors, and this may be associated with the observed changes in plants.

An increase in free amino acid concentrations in the needles of Norway spruce (*Picea abies*) with HF-induced damage has been reported.[758] No increase in free amino acids in the spruce stems of the current year's shoots was observed in fluoride treatments after two growing seasons.[700] However, it was not known whether the levels of free amino acids were increased in the fluoride-treated seedlings during active growth in June. The relatively low concentration of arginine in the F treatments at the end of the growing season might indicate disturbances in the nitrogen metabolism of spruce seedlings. Arginine concentrations are normally lower during active growth than during dormancy.[408] According to Wellings and Dixon,[1582] arginine is more abundant in sycamore leaves previously infested with sycamore aphids than in uninfested leaves. Lower concentrations of this essential amino acid were found in spruce foliage infested by aphids than in uninfested foliage.[475] *Cinara pilicornis* did not reproduce in laboratory on spruce cuttings that had high concentrations (about 750 μmol per 10 μg fresh weight) of arginine. This amino acid may be an indicator of the nutritive value of the host plant for conifer aphids.[700]

C. EFFECTS ON INSECTS

Field studies concerning the effects of fluoride on insects were reviewed by Alstad et al.[27] Populations of silkworms,[891] honeybees,[1001,1003,1087] some bark beetles (*Pityokteines* sp.),[1200] and the European pine shoot moth (*Rhyacionia buoliana*)[974] were reported to be lower than normal in areas polluted by fluoride. Populations of some herbivores, such as the pine bud moth (*Exoteleia dodecella*), increased in areas polluted by fluoride in southern Norway.[587] Among invertebrates collected near an aluminium smelter plant, scavengers tended to have the highest concentrations of fluoride, followed by predators, omnivores, and herbivores.[213]

Fluorine and its compounds may enter the insect body not only with food, but also in the course of respiration. Feeding experiments with fluoride resulted in both negative and positive changes in *Tribolium* beetles, depending on the concentration,[782] and in changes in mortality and fecundity of *Drosophila* flies, with some variation among fly strains.[536,537] Fluorine enters the body of worker bees exclusively through respiration and may accumulate to reach the critical concentration of 110 mg/kg (in the control, approximately 10 mg/kg), resulting in the death of the bees.[183]

There are conflicting opinions about whether fluoride is accumulated in food chains.[27] Mexican bean beetles (*Epilachna varivestis*) reared on HF-fumigated bean foliage showed delayed development, reduced growth, and lower fecundity compared to the controls. However, less than 0.1% of the fluoride ingested by beetles on fumigated foliage was retained, the rest being excreted with the feces.[1578] Hymenoptera, as a group, tend to accumulate relatively high concentrations of fluoride compared to other kinds of insects.[376] However, the main route of exposure (respiration, food grooming) of fluoride to Hymenoptera has not been identified,[27] and the trap-nesting wasps may have a much lower exposure to flouride than species such as honeybees, which forage for pollen. Recently flouride measured in sawfly (*Diprion pini*) larvae was considered to be due to surface contamination and to the flouride contained within the gut contents, but no flouride was detected in the pupae.[355a] High fluoride concentrations have been found in the larvae and adults of some bark beetles.[233]

The incidence of damage caused by bud pests in young pine stands rose markedly as fluoride pollution increased.[247] The numbers of insect groups, such as the aphids, increased considerably.[247,1368,1371] Thalenhorst[1473] found a positive correlation between the incidence of Norway spruce (*Picea abies*) infested with a gall aphid (*Sacchiphantes abietis*) and the concentration of fluoride-containing air pollutants. He also reported a histologically visible defensive reaction against insertions by the aphid. Charles and Villemant[247] studied the numbers of aphids in relation to the degree of fluoride pollution. Aphid numbers changed from 3218 individuals per 1 m^2 of leaf surface in the light pollution zone and 5679 in the medium zone to 7312 individuals in the heavy pollution zone.

On the other hand, Pfeffer[1200] found decreased populations of bark beetles (*Pityokteines* spp., *Cryphalus* spp.) and weevils (*Pissodes piceae*) in eastern European fir stands damaged by fluoride, while populations of the aphids (*Dreyfusia piceae*, *D. nusslini*) increased. There was no relationship between the foliar injury on ponderosa pine caused by fluorides and the population density of the diaspidid scale insect (*Nuculaspis californica*).[300,414] *Cryptoccus* scale insects associated with the fungus *Nectria* on beeches showed density variation suggestive of a complex interaction with fluoride and other air pollutants in France.[370]

Beyer et al.[132] collected trap-nesting wasps at eight sites at distances of 1.2 to 33.0 km from an aluminum smelter producing high emissions of fluoride on Kentucky Lake in Tennessee. The wasps belonged to several families, but had similar nesting habits. All seek out holes in wood where they construct cells partitioned with soil or debris. Cells are provisioned with prey, an egg laid, and the cells and hole sealed. The plant emitted more than 3000 tons of fluoride per year when operating at peak capacity. Although the emissions were very high compared to other industrial plants in the U.S., they seemed to have no discernible effect on the populations of trap-nesting wasps. The degree of fluoride pollution was unrelated to the relative densities of the wasps and to the number of their cells provisioned with prey.

Laboratory experiments carried out by Katayev et al.[818] have shown that hydrogen fluoride can increase larval mortality and the duration of the developmental period and decrease larval and pupal size and female fecundity in the moths *Lymantria dispar* and *Orgyia antiqua* (Lymantriidae). In another experiment,[1061] needles of Norway spruce (*Picea abies*) that had been subjected to different levels of pollution from an aluminium factory were used as food for the nun moth (*Lymantria monacha*) larvae reared both in the laboratory and in

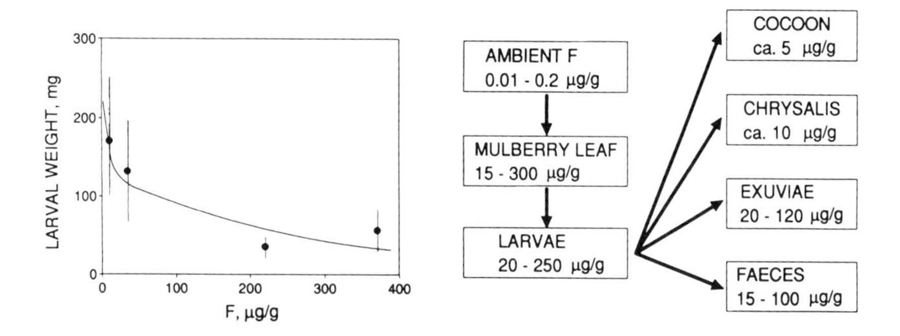

FIGURE 9. Dry weight of sixth instar larvae of *Lymantria monacha* (Lepidoptera, Lymantriidae) in relation to fluoride concentration in their food (left), and the range of fluoride concentrations in air, mulberry leaves, and different parts of the silkworm (*Bombyx mori*, Lepidoptera, Saturniidae) (right). (From von Mitterböck, F. and Führer, E., *J. Appl. Entomol.*, 105, 19, 1988; and from Wang, J. and Bian, Y., *Environ. Pollut.*, 52, 11, 1988. With permission.)

field conditions. Elevated concentrations of fluoride (up to 365 ppm F) resulted in reduced larval weight, increased mortality (by up to 75%), and delayed development (Figure 9). Reduced larval growth rate resulted in a decrease (by up to 55%) in pupal weight. The maximum factor of fluoride accumulation from the food was 7.3, which was found in larvae that died during the fifth instar. Fluoride concentrations in the pupae indicated that accumulation did not exceed a factor of 3.8. A level of 1400 to 1500 ppm F content of larvae is considered to be lethal. Lower concentrations have been found in surviving pupae.

Several papers have been published on the effects of fluoride on the feeding of silkworms (*Bombyx mori*), rates of growth, and mortality.[133,134,520,521,741,891-893,1064,1244,1549-1551] The toxicity of F-containing mulberry leaves to silkworm larvae depended less on the total quantity of fluoride ingested than on the kind or form of the fluoride compounds provided. Fluorides depressed the feeding rates of the larvae and caused softening of the cuticle, reductions in growth rate, and increased mortality.

Wang and Bian[1549] examined the absorption and accumulation of fluoride from ambient air by mulberry leaves, and its transfer and effects on silkworm development (Figure 9). Fluoride was absorbed from the atmosphere by the mulberry leaves and transferred to the silkworm, soil, water, and back to the atmosphere. A close correlation was found between the atmospheric fluoride concentration at different distances from a brick kiln and the amount of fluoride accumulated in mulberry leaves. Old leaves accumulated more fluoride than young leaves. When the concentration of fluoride in the air exceeded 1.5 mg/ dm^2, the fluoride concentration of the mulberry leaves was more than 30 ppm. The threshold concentration of accumulated fluoride that was toxic to silkworm larvae was above 30 ppm, and the lethal dose between 120 and 200 ppm. A fluoride concentration of up to 30 ppm had no apparent effect on silkworm mortality, but above 30 ppm and below 50 ppm, the mortality rate was more than 30%.

Fluoride-polluted mulberry leaves inhibited the growth and development of silkworms. A fluoride concentration of 20 to 30 ppm reduced the wet mass of silkworm larvae by about 15% after 6 days. Leaves containing 50 to 60 ppm fluoride reduced growth by more than 60% after 6 days, demonstrating a nonlinear relationship between fluoride accumulation and larval growth.[1549] Leaves containing more than 80 ppm fluoride severely inhibited silkworm cocoon production. Cocoon weight appeared to be more sensitive to fluoride than was the wet mass of the larvae. Leaves containing 15 to 20 ppm produced cocoons that were 25% lighter than the controls. There was a close correlation between silkworm cocoon development and the fluoride concentration of the leaves or the fluoride concentration of mulberry leaves at different distances from a pollution source.[1549]

Earlier observations[1473,1530] suggested that fluoride pollution especially promotes the growth of aphids on conifers. According to Thalenhorst,[1473] Norway spruce (*Picea abies*) has a histologically visible defensive reaction against gall aphids, and fluoride exposure reduces this reaction, thus improving aphid survival. Fluoride, alone or in combination treatments, caused brown and chlorotic needle tips, indicating a strong physiological response to fluoride,[700] while SO_2 and nitrogen alone caused only a few visible symptoms.[1657]

Detailed studies providing data on the distribution of fluoride in contaminated seminatural ecosystems are few. This is especially the case with the contamination derived from the soil, with little or no input of atmospheric fluoride. The hazards that fluoride poses to grazing cattle are well documented, but the pathways of fluoride into wild animals are still unclear.[365] Extremely high soil fluoride levels are present in association with fluorspar waste, largely present as CaF_2. After revegetation these very high levels of soil fluoride are reflected as elevated concentrations in the vegetation, invertebrates, and small mammals.[42]

Katayev et al.[818] investigated the effects on invertebrates of fluorine pollution from an aluminium plant in the former Soviet Union. Fluorine ions form toxic compounds that act both as contact and sometimes also as systemic poisons in the soil and litter. The numbers and biomass of beetles, arachnids, and myriapods declined markedly near the industrial plant. The response of Heteroptera and Homoptera differed from other insect groups. The Heteroptera were primarily represented by the families Miridae and Plataspidae. Among the Homoptera, species of Aphidoidea and Fulgoroidea dominated, and the families Aphrophoridae and Cicadellidae were abundant. These groups reached their maximum density and biomass in the immediate vicinity of the aluminium plant. The increase in the numbers of sucking insects in this zone was significant in all cases.[818] According to Katayev et al.,[818] the numbers of the ant *Formica rufa* decreased markedly near the aluminium plant. The density and the average volume of the ant hills diminished as the distance from the plant became shorter. In addition, the average mass of the worker ants decreased.

Andrews et al.[42] studied the distribution of fluoride in the abiotic and biotic components of the grassland ecosystem established on a revegetated fluorspar tailings dam. High total soil fluoride concentrations (10 mg/g) in the metalliferous fluorspar tailings was reflected by elevated concentrations in the standing live vegetation (0.3 to 1 mg/g), plant roots (c. 6 mg/g), plant litter (c. 4 mg/g), total body concentrations of invertebrates (0.4 to 4 mg/g), and the common shrew *Sorex araneus* (0.14 to 0.25 mg/g). Levels in the vegetation on the tailings dam were very high and were higher than has normally been found in atmospherically polluted sites, although the latter can be spatially and temporally very variable.

Fluoride concentrations in all invertebrate groups trapped at the fluorspar tailings dam were markedly elevated.[42] Detritivores had the highest fluoride concentrations, reflecting the very high concentrations in litter and soil. The concentrations at the tailings dam increased from May to October, and at the control site they decreased. In both instances, the invertebrate fluoride levels were not generally related to vegetation levels. When the data for the invertebrate groups were combined from each sampling period, distinct differences between taxa were evident. The data on the soil invertebrates were generally consistent with the view that fluoride concentrations are related to feeding strategy, i.e., the concentration in carnivores is higher than that in herbivores, but lower than that in detritivores.[213,376,828] The high value for *Lumbricus* earthworms at the tailings dam was almost certainly due to the gut contents. Ground beetles (Carabidae) and spiders (Araneae) are mainly carnivorous and/or scavengers, both feeding on invertebrate prey, yet their fluoride concentrations were different, with the Araneae having considerably higher values at both sites. This may be because of unexplained differences in their respective prey fluoride levels or their longevity, or differences in surface contamination, or differences in their fluoride metabolism.

VII. CARBON OXIDES

A. BACKGROUND

Bazzaz[89] recently reviewed the response of natural ecosystems to the rising atmospheric carbon dioxide (CO_2) levels. Evidence from many sources shows that the concentration of atmospheric CO_2 is steadily rising,[153] and this rise is strongly correlated with the increase in the global consumption and burning of fossil fuels.[1304] Controversy continues as to whether the biosphere is presently a source or a sink for carbon.[692,714] However, most scientists seem to agree that rising CO_2 levels will have substantial effects on the biosphere. Because CO_2 is a greenhouse gas, an increase in CO_2 levels in the atmosphere may influence the Earth's energy budget. General circulation models predict increased global warming and substantial shifts in precipitation patterns.[1343] However, some scientists[817] have questioned the predictions of these models. Regardless of changes in temperature and other climate variables, increasing CO_2 levels can influence ecosystems through direct effects on plant growth and development.[89]

Recent reviews have been made of the response of crops[339,842,1536,1637] and managed forests and trees.[410,873] Bazzaz[89] concentrated on the response of natural vegetation to elevated CO_2 (and some of the predicted climate change) and addressed the CO_2 response of individual plants at the physiological level and the consequences of such a response at the population, community, and ecosystem levels. Most of the first findings on the physiological and allocational response to CO_2 were obtained with agricultural crops. Much of the initial work on plants from natural ecosystems has tested the variation among species in these responses.

B. CHANGES IN PLANTS AND VEGETATION

Plant biologists have long been aware of some of the effects of high CO_2 levels on plants, and greenhouse growers have used CO_2 amendment to increase plant yield. Many plant and ecosystem attributes will be directly or indirectly influenced by elevated CO_2.[1441]

Information about the response of grasslands to elevated CO_2 is limited.[89] Smith et al.[1394] compared the response of four grass species from the Great Basin. High CO_2 resulted in increased growth, especially basal-stem production, in the C_3, but not in the C_4, species. Work with blue grama (*Bouteloua gracilis*), an important native perennial in the same region, showed that biomass and leaf area were greatly enhanced at elevated CO_2 levels, which is unusual for a C_4 plant.[1283] Carbon dioxide concentration differentially influenced the growth of six plant species from the short annual grasslands found on serpentine soil in California when the plants were grown individually.[1620] In this low-stature community with a very short growing season and nutrient limitation, competitive networks and adaptation can develop and dampen the CO_2 effects.[89]

In temperate forests only a few studies have examined the response of tree species in a community context, and fewer still in competitive situations.[1619] Seedlings of the dominant species of a floodplain forest community and of an upland deciduous forest community were grown in competition under ambient and elevated CO_2 concentrations. Photosynthetic capacity (rate of photosynthesis at saturating CO_2 and light) tended to decline as the CO_2 concentration increased. Stomatal conductance also declined with an increase in CO_2. Nitrogen and phosphorus concentrations generally decreased as CO_2 increased. The overall growth of both communities was not enhanced by CO_2, but the relative contribution of individual species to the total community biomass changed in a complex way and was also influenced by light/CO_2 interactions.

Reekie and Bazzaz[1266] studied competition and the patterns of resource use among seedlings of tropical tree species under ambient and elevated CO_2, using five relatively fast growing, early successional species from the rain forest of Mexico. Elevated CO_2 only slightly affected photosynthesis and overall growth of the individually grown plants, but greatly affected mean canopy height. Carbon dioxide had marked effects on species composition, with some species decreasing, and others increasing, in importance. There were also some differences among species in allocation to roots and in the timing of such allocation. The results suggested that competition for light was the major factor influencing community composition and that CO_2 influenced competitive outcome largely through its effects on canopy architecture.

C. EFFECTS ON INSECTS

It has been hypothesized that high CO_2 levels and the resulting high availability of photosynthates will enhance root growth and root exudation in the soil.[1442] These will, in turn, influence plant nutrition by enlarging the soil volume utilized by the roots and by increasing mycorrhizal colonization, nodulation, and nitrogen-fixing capacities.[898,952] Several authors have predicted that the rate of litter decomposition may be lower in high CO_2 environments.[1442,1619] These predictions are based on the finding in the majority of studies that the C/N ratio of tissues grown under elevated CO_2 levels increases, and on experimental evidence that tissue with high lignin and low nitrogen concentrations decay slowly.[1035]

A CO_2-enriched atmosphere may affect the growth or physiological state of plants in a way that affects species interactions and community composition.[939] Elevated CO_2 concentrations within the range predicted by global change scenarios are unlikely to influence herbivores directly. However, the tissue quality of plants grown under high CO_2 environments may be altered, thereby indirectly affecting insect performance.[89,1442] Other important nutritional factors, such as foliar carbon-based allelochemical and fiber concentrations, do not seem to be affected by elevated CO_2 conditions,[455-457,787,937] and foliar water content does not change in any consistent way in higher CO_2 atmospheres.[456,939] Too few systems have been examined to make any general statements about these patterns.

Plants grown in enriched CO_2 environments generally have lower foliar nitrogen concentrations, a limiting nutrient for insect herbivores,[997] than those grown in ambient conditions.[456,938] In turn, both the behavior and performance of herbivorous insects may be modified in response to the reduction of foliar nitrogen concentrations. To compensate for the lower nitrogen concentrations, insect herbivores feeding on high-CO_2-grown foliage increase their consumption rate by 20 to 80% compared to the larvae feeding on low-CO_2-grown tissue.[455,787,937,938] Presumably to compensate for these lower foliar nitrogen concentrations, both early and late instar lepidopteran larvae consume additional elevated-CO_2-grown foliage.[455,938]

Despite the increased consumption, insect herbivore performance on high-CO_2-grown plants is often poorer than on low-CO_2-grown plants. Lepidopteran larvae reared on enriched CO_2-grown foliage may experience a reduction in fitness-related parameters such as increased mortality, reduced pupal mass, or longer development times.[10,455-457] Slower growth might reduce insect

TABLE 5
Effects of Carbon Dioxide on Terrestrial Insects in Experimental Studies

Species	Host plant	Dose	Response	Refs.
Homoptera				
Bemisia tabaci	*Gossypium hirsutum*	350, 355, 511, 662 μl/l	No response	217
Lepidoptera				
Junonia coenia	*Plantago lanceolata*	350, 700 μl/l	Longer larval period	457
J. coenia	Artificial diet	350, 700 μl/l	No response	457
Pseudoplusia includens	*Glycine max*	350, 500, 650 μl/l	Increased feeding decreased growth rate	938,939

herbivore fitness in the wild, due to an increased exposure to predators and parasitoids[1234] and a decrease in the likelihood of their completing development in seasonal environments.[250]

The experiments conducted thus far only permit a limited interpretation of how future, enriched CO_2 atmospheres may affect natural plant-insect herbivore systems,[457] primarily because insects have rarely been permitted to develop directly on plants growing in different CO_2 environments. In most of the studies the insects were raised in separate growth chambers, often in Petri dishes, under ambient CO_2 conditions (Table 5). Moreover, wild strains of both plants and insects were rarely investigated.[457]

Fajer et al.[457] attempted to determine how an enriched CO_2 atmosphere could affect interactions among plants and herbivorous insects in the wild. The offspring of buckeye butterflies, *Junonia coenia* (Nymphalidae), from crosses between wild- and laboratory-raised butterflies, were reared directly on a major host plant, *Plantago lanceolata* (Plantaginaceae), grown from seeds collected at the same site as the wild-collected butterflies. Both plants and insects were reared in either ambient (350 μl/l) or enriched (700 μl/l) CO_2 atmospheres. Previous work had shown that *P. lanceolata* grown under enriched CO_2 atmospheres have significantly larger root and shoot masses than those grown under an ambient CO_2 atmosphere. Plant nutritional qualities were analyzed, and the duration of the entire larval stage and pupal mass was measured. In addition, *J. coenia* larvae from a laboratory stock were fed on artificial diets and raised in either ambient or enriched CO_2 conditions in order to compare the direct effects of CO_2 concentrations on insect herbivore performance.

Carbon dioxide treatment affected buckeye larval growth (Figure 10), presumably because plants reared in high CO_2 environments had lower nitrogen concentrations, whereas other nutritional factors known to influence the performance of insect herbivores were not significantly affected by CO_2 treatment.[457] Differences in development rate probably did not result from the direct effects of high-CO_2 conditions on buckeye physiology: when reared on artificial diets, early instar larvae from different CO_2 treatments attained similar masses.

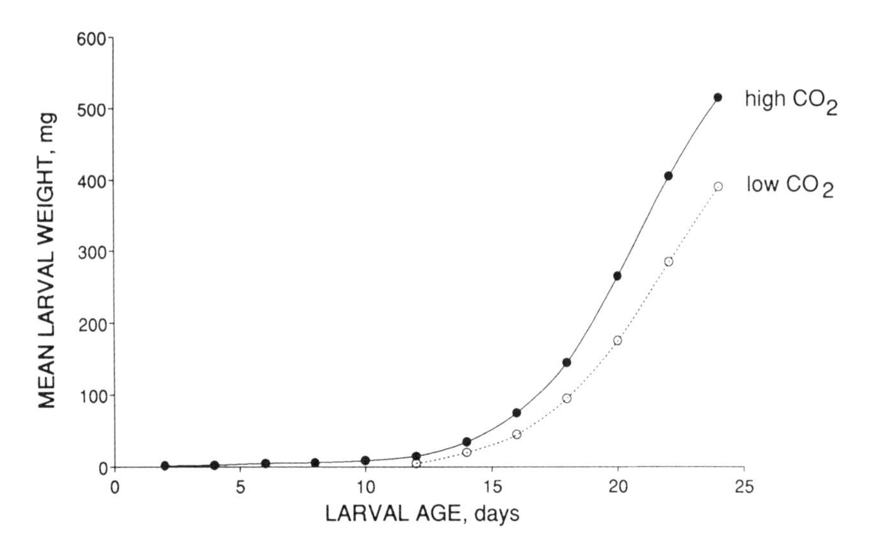

FIGURE 10. Growth of larvae of the buckeye butterfly *Junonia coenia* (Lepidoptera, Nymphalidae) on *Plantago lanceolata* grown at ambient and elevated CO_2 concentration. (From Bazzaz, F. A., *Ann. Rev. Ecol. Syst.*, 21, 167, 1990. With permission.)

Although elevated-CO_2 environments can affect insect herbivore performance by lowering the nutritional quality of the host plants, direct effects on insect development and fecundity may not result from exposure to anticipated future concentrations of atmospheric CO_2 (e.g., 700 µl/l). Instead, severe physiological effects have only been observed for insects reared in extremely high CO_2 concentrations (i.e., 50,000 µl/l, or 5% atmospheric concentration by volume[1125]), which will not occur globally.

In approximate agreement with previous results, the performance of buckeye larvae was reduced on high-CO_2-grown plants.[457] Larvae required longer times for pupation when reared on high-CO_2-grown *P. lanceolata*, which in the field could lead to fitness reductions resulting from increased exposure to predators and parasitoids[1234] and/or from a diminished likelihood of completing development on the same host in climatically limited environments.[1580] However, if higher temperatures during the insect growing season accompany rising atmospheric CO_2 concentrations,[1343] buckeye butterflies may not take longer to reach pupation in a high-CO_2 world. The weights of pupae of neither sex of the buckeye butterfly were affected by CO_2 treatment.[457] Thus, since female egg production and male spermatophore size are often correlated with pupal weight,[151,253] high-CO_2 conditions would probably not affect individual reproductive fitness.

VIII. CLIMATIC CHANGE

A. BACKGROUND

Climate is expected to change on a global scale due to increasing concentrations of atmospheric greenhouse gases (carbon dioxide,

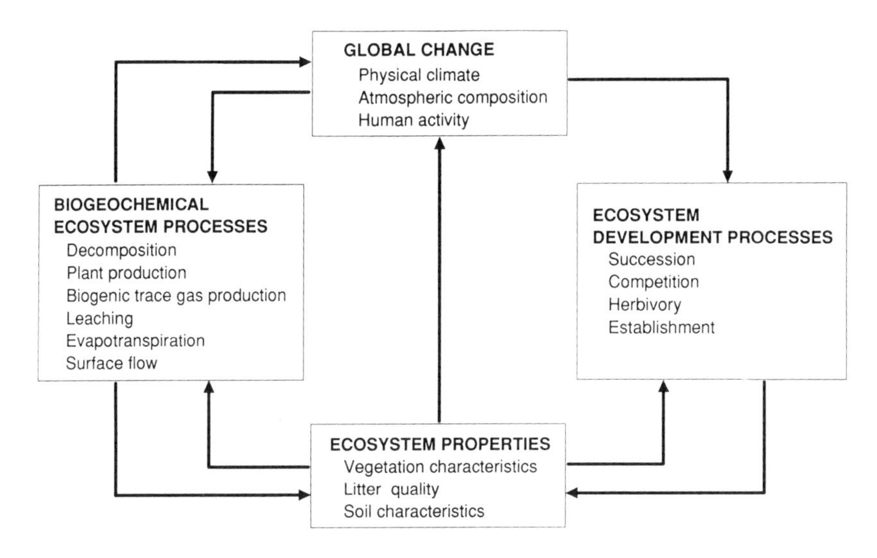

FIGURE 11. Global change-terrestrial ecosystem interactions. Effects of global change on terrestrial ecosystems will directly feed back to the global environment through the biogeochemical processes. Indirect feedback is observed through changes manifested in ecosystem properties brought about by changes in the ecosystem development processes. (From Ojima, D. S. et al., *Ecol. Appl.*, 1, 316, 1991. With permission.)

chlorofluorocarbons, nitrous oxide, methane, and ozone).[186] The response of terrestrial ecosystems to the climate weather system is dependent on the spatial scale of the interactions between these systems and the temporal scale that links the various components (Figure 11).[1151] The last 100 years have already seen an increase in the global mean temperature of between 0.3 and 0.7°C. Although a direct relationship between changes in the concentrations of greenhouse gases and temperature is complicated by numerous factors, this rise in global temperature is consistent with the observed increasing concentrations of carbon dioxide and the other greenhouse gases.[89,186,410]

Rising concentrations of CO_2 and other anthropogenic polyatomic gases may raise global average temperatures substantially.[608] Even if the production of all greenhouse gases stopped today, it would be likely that the concentrations in the air are now high enough to result in an ecologically significant warming after a brief lag period. Because there are uncertainties both in the methodology of the scenarios and in our understanding of the nature of climatic change, scenarios for climate cannot be taken as predictions of future changes. They can, however, be used for looking at the impacts of different magnitudes and distributions of climatic change on environment and society.[759]

It is expected that within the next 40 years greenhouse trace gases in the atmosphere will reach a concentration equivalent to double the preindustrial concentration of carbon dioxide. Various predictive models suggest that an elevated concentration of greenhouse gases may result in an increase of the Earth's temperature of between 1.5 and 4.5°C or more. Warming of less than 1°C would have substantial ecological effects.[1192]

The magnitude and order of variation of the climatic change, as well as the changes in the individual seasons, will vary with latitude. The largest increases in temperature are expected to occur in regions of high latitude.[186,608] Because of the time lag caused by the thermal inertia of the oceans, some of this warming will be delayed by 30 to 40 years beyond the time that a doubling equivalent of carbon dioxide is reached, but substantial warming could occur soon.[1285]

Global warming will be associated with regional and local changes in average temperature, in the distribution of hot and cold periods, and in a number of other chemical and physical variables, including precipitation, evaporation rates, sea level, and soil and water chemistry.[1192] For example, the summer and winter scenarios project warmer temperatures throughout Europe as a result of a doubling of carbon dioxide levels, with the largest increases occurring in the winter period.[186] The increase in winter temperature could be as much as 6°C in parts of northern Europe. Summer temperatures will increase by less than 2°C for most of Europe, while an increase of more than 4°C is projected during the winter period over Scandinavia and in the north of the Russia Briffa et al.[180] used tree-ring data to reconstruct the mean summer (April–August) temperature for each year from AD 500 to the present in Fennoscandia. Summer temperatures have fluctuated markedly on annual, decadal, and century time scales. Comparison of the expected trends in summer temperature with those that have occurred over the past 1481 years suggested that summer warming induced by greenhouse gases may not be detectable in this region until after the year 2030. The summer and winter scenarios project increased precipitation over the northern half of Europe and less precipitation over much of southern Europe.[186]

B. CHANGES IN VEGETATION

Past natural climate changes have resulted in large-scale geographical shifts in the ranges of species, alterations in the species composition of communities, and species extinction. If the predicted greenhouse effect does occur, natural ecosystems will respond in similar ways as in the past, but the effects will be more severe because of the fast rate of the projected change. Warming by 3°C would present natural systems with a warmer world than has been experienced in the past 100,000 years; 4°C would make the earth its warmest since the Eocene period, 40 million years ago. Such a warming would not only be large compared to recent natural fluctuations, but it would be perhaps 15 to 40 times faster than previous natural changes. Such a rate of change may exceed the ability of many species to adapt.[1192]

The effects of climate change would be pronounced in temperate forests and Arctic areas, where temperature increases are projected to be relatively large.[89,1192] Range retractions of trees will be proximally caused by temperature and precipitation changes, an increase in fires, changes in the ranges and severity of pests and pathogens, changes in competitive interactions, and additional effects of nonclimatic stresses such as acidic precipitation and low-level ozone. Serious changes may occur in tropical forests as well, particularly due to changes in rainfall. Ocean systems may show a similar shift in species ranges and community composition if a warming of ocean water or an alteration in the patterns of water circulation occur. For example, recent El Niño events demonstrated the vulnerability of primary productivity and species abundances to changes in ocean currents and local temperatures.[404,549]

Peters[1192] concluded that global warming during the next 50 to 100 years may lead to the disruption of natural communities and the extinction of populations and species. Climate change and habitat destruction together would threaten many more species than either factor alone. Peters[1192] also stressed that extreme events, like droughts, floods, blizzards, and hot or cold spells, may have more effect on species distributions than average climate per se.[858]

C. EFFECTS ON INSECTS

When temperature and rainfall patterns change, species' ranges will also change. Tropical rain forest species of *Drosophila* appear vulnerable to a temperate increase of as small as 2°C, at which point there would be the extinction of certain species.[1179] A key question is whether the dispersal capabilities of most species prepare them to cope with the coming rapid warming.[1192] The density of natural populations of certain species of Lepidoptera have been observed to increase during extended periods of warm, dry weather.[809,1133,1595] Even small temperature changes of less than 1°C within this century have been reported to result in substantial range changes. The white admiral butterfly (*Ladoga camilla*) and the comma butterfly (*Polygonia c-album*) greatly expanded their ranges in the British Isles during the past century as the climate warmed by approximately 0.5°C.[490]

Although species tend to shift in the same general direction, existing biological communities do not necessarily move in synchrony.[1192] Conversely, species are likely to respond individually to climatic change, forming stable, but, by present-day standards, unusual, assemblages of plants and animals. The range shifts are the sum of many local processes of extinction and colonization that occur in response to climate-caused changes in habitat suitability, determined by both direct climatic effects on physiology, including temperature and precipitation, and effects caused by other species. Species with fragmented populations and reduced ranges are especially vulnerable to climate change.

A species range may change both altitudinally and latitudinally. When the climate warms, species shift upward. In general, a short increase in altitude corresponds to a major shift in latitude: the 3°C cooling of 500 m in elevation equals roughly 250 km in latitude. Because mountain peaks are smaller than their bases, species shifting upwards in response to warming occupy smaller and smaller areas, have smaller populations, and may thus become more susceptible to genetic and environmental pressures.[1102] Species originally situated near mountaintops might have no habitat to move up to. Some species lived in Europe during the cold periods, but could not survive conditions in postglacial forests. One previously widespread dung beetle, *Aphodius hodereri*, now occurs only in the high Tibetan plateau where conditions remain cold enough for its survival.[318] Other species, such as the springtail *Tetracanthella arctica*, now live primarily in the boreal zone, but also survive in a few cold, mountaintop refuges in temperate Europe.[318]

The speed of the temperature rise will far exceed the regeneration time of many woody species in the world and their migration to new habitats.[362] This rapid change would likely result in the death of many individual plants and their replacement with early successional species that, in general, are adapted to live in an environment with initially high resource levels.[88] Modeling results based on changes in temperature caused by the increase in CO_2 and other greenhouse gases have suggested a significant change in the patterns of regional plant productivity,[435,436,1364] in the distribution ranges of some plant species,[363,364] and in the species composition on a regional scale,[1183] which may affect the ranges of insect species. For example, the range of American beech (*Fagus grandifolia*) could drastically change, and its distribution could be several hundred miles north of its current position.[363] Additionally, based on the direct response to increased CO_2 alone and the resultant decrease in water consumption, it was also predicted that the ranges of species can expand into drier habitats.[234]

There is little evidence that annual differences in the nutritive quality of host-plant tissue may arise as a result of changes in climate. Campbell[228] demonstrated that annual climatic variation can affect the nutritive value of host-plant tissue for a herbivore. Larvae of a single generation of the western spruce budworm (*Choristoneura occidentalis*) were fed young vegetative shoots that had been collected in five consecutive springs from a stand of balsam fir (*Abies balsamea*). Both sexes were about 20% heavier at pupation when fed foliage that had been produced in early (warm, dry) spring seasons rather than later (cool, wet) springs (Figure 12). This may cause an estimated 25% difference in the reproductive rate of females.

These results correlated with the climatological observations of Wellington.[1583] He noticed that the outbreaks of the eastern spruce budworm (*Choristoneura fumiferana*) in spruce-balsam fir forests of northern Ontario were initiated only after the occurrence of more than two consecutive springs that were earlier than normal. In early spring the days are unusually warm, and the nights unusually cold, while both day and night temperatures are moderate in late springs. Campbell[228] found that the trees produced the least nutritious foliage for insect growth and reproduction during late springs. The most nutritious foliage was produced under those early spring conditions that prevailed during successive spring seasons just prior to the initiation of the spruce budworm outbreaks.

Earlier-than-normal leaf formation by the trees could increase survival of the over-wintering budworm population by decreasing prefeeding stage starvation.[226,227,1481] Laboratory and field studies[226,227] indicated that the earliest emerging moth larvae had the inherently greatest growth and reproductive potential, so that early leaf formation of the host trees selected for genotypes with the greatest capacity for foliage consumption and generation of offspring.

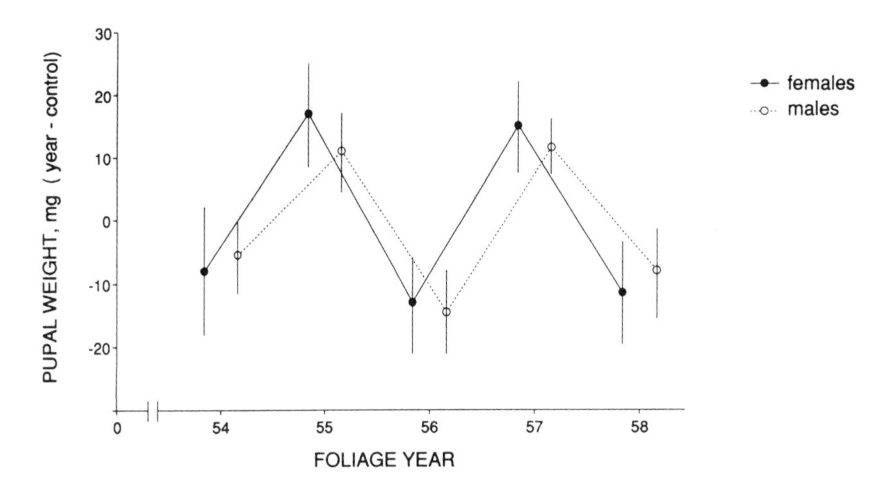

FIGURE 12. The difference between the mean pupal weight of western spruce budworm *Choristoneura occidentalis* (Lepidoptera, Tortricidae) that had been fed a foliage collected on one of 5 consecutive years from the same stand of balsam fir (*Abies balsamea*), and that of their sibs that had been fed a common control foliage (a single collection made from another stand of fir) is plotted against the year on which each "year foliage" was collected. The error bars (2 SE) represent the combined within-group (experimental plus control) variances for each "year foliage." (From Campbell, I. M., *Oecologia,* 81, 341, 1989. With permission.)

Kemp[824] reported that during an infestation of the western spruce budworm, defoliation following an early spring was severe, whereas that following a late spring was only light to moderate. Warm, dry weather of early springs may promote larval feeding and reduce the incidence of diseases.[1109]

IX. METALS

A. BACKGROUND

The distinction between metals and nonmetals is not sharp.[706] Chemists define a metal as an element that conducts heat and electricity, has a lustrous appearance, and participates in chemical reactions as cations (positive ions). Some elements (e.g., antimony, arsenic) are typical metals in terms of physics, but they have the chemical properties of nonmetals. The term "heavy metal" is widely used in the literature concerning pollution. However, the border between a metal that is heavy and one that is not heavy is not clearcut. The position of aluminium, especially, an element that plays a major role in aquatic ecosystems and acidification processes, is questionable in this context. Thus, we mainly follow Hopkin[706] in speaking simply about metals (not "heavy metals" or "trace elements"). Eighty-seven elements out of 105 in the periodic table are considered metals or metalloids. It is clear that we cannot consider all of these metals here, and the review is hence restricted to those metals that are potential or actual pollutants.

Metals are released into the atmosphere and soil by volcanoes, the weathering of rocks, and a wide range of human activities. Due to the massive dilution of particles in the atmosphere, soil is only rarely contaminated by metals as a consequence of a volcanic eruption. In some cases the natural weathering of metal-rich rock may contaminate the surroundings, for example, lead.[584]

The major sources of anthropogenic metal pollutants (in terrestrial ecosystems) are mining, smelting, the combustion of fossil fuels, and agriculture.[706,1341] Usually metal concentrations increase with dept soils over ore bodies or on spoil tips in mining areas, but in aerially polluted regions, concentrations are much higher at the surface where most soil insects live. Metals emitted accumulate in the soil because of their affinity to clay minerals and humus. Fluxes of metals in the environment have increased greatly since the industrial revolution. Nowadays, metal contaminants of soil, vegetation, and waters are derived from a number of sources in heavily industrialized areas. In Scandinavia the pattern of metal deposition shows a decreasing gradient from relatively high values in the south to low values towards the north. The gradients are steep for arsenic, cadmium, lead, and vanadium, whereas those for chromium, copper, iron, nickel, and, to some extent, zinc are less distinct.[1309]

The common weakness of many field studies is that several metals are generally present simultaneously, and information about the biological effects of the individual metals is scant.[584] In polluted areas organisms are exposed to a mixture of metals, and studying the effects of only one metal at a time can give spurious results.[123,386,1046]

B. EFFECTS ON INSECTS

A large number of laboratory experiments on metal toxicity have been conducted using the banana fly *Drosophila melanogaster*. Williams et al.[1617] calculated 4-day LD_{50} values for 1-day-old adult flies fed on a medium containing different concentrations of metal chlorides. The values (per liter of fresh diet) were: cadmium 3.6, mercury 5.7, nickel 12, silver 13, copper 16, cobalt 16, chromium 23, and zinc 34 mmol. The differences between the toxicity of metals were even more pronounced when the effects on larval viability were assessed. The extreme toxicity of cadmium and mercury to adult and larval *Drosophila* was confirmed in several other studies. Damage to chromosomes may occur at levels of cadmium, arsenic, or mercury far below those that disrupt growth and reproduction.[706]

Reports on alterations in the chemical composition of insects caused by metal contamination are rather rare. The mode of action seems to depend on the metal, its concentration, and the species itself.[236] Metals may affect the total protein content of the individual by interfering with the main metabolic pathways, as decribed for *Mysidopsis bahia* (Crustacea).[236] Cadmium stress caused a shift from lipid/CH-catabolism to protein-catabolism, which may result in a reduced total protein content. Another mechanism affected could be the synthesis of metal-induced, metallothionein-like proteins.[46,706,980,981] This would not be revealed by a total body analysis because the mechanism is restricted mostly to specific tissues. The total protein content of aquatic *Chironomus tentans* larvae dropped to 70% of the control after a 1-h exposure to 10 μg/g Cd.[1258] The protein content of the parasitoid wasp *Pimpla turionellae* declined in individuals treated with cadmium or did not change significantly.[1168]

There are some reports dealing with effects of metals on the fat content, fatty acid oxidation, and fatbody, which is the energy source in insects.[988] Fatbody cells failed to differentiate if 2μg Cd per gram fresh weight were injected into *Locusta migratoria*, and egg production was impaired.[988] The pupal parasitoid *Pimpla turionellae* showed a striking decrease in fat content when exposed to cadmium, while the carbohydrate content was hardly affected by metal contamination.[1168] Increasing fat content was observed in some spider species from industrial polluted areas.[1672] Cadmium affected lipid levels in *Chrysochloris stolli* (Heteroptera) in both ways, depending on the tissue.[747]

Contamination of terrestrial ecosystems with metals may upset normal ecological processes by altering the relative abundance and diversity of species of microorganisms, plants, and animals. A reduction in the abundance and diversity of terrestrial invertebrates is well documented on sites that have been heavily contaminated with metals emitted by smelting industries[107,141,654,655,710,755,1444] and steelworks,[383] and soils to which metal-containing pesticides have been applied.[1265]

The most obvious effect of metal contamination on forests near to a smelting works is the presence of a considerable layer of undecomposed leaf material on the surface of the soil.[312] In Hallen Wood, an area subjected to substantial aerial metal deposition from smelting works in southwestern England,[985] this effect was thought to be due to the almost complete absence of earthworms and millipedes that had been poisoned by the synergistic effects of low pH and high levels of metals in their diet.[710] These invertebrates normally stimulate decomposition by converting dead leaves into smaller fragments that are more easily decomposed by microorganisms. Nevertheless, the persistence of this thick layer of leaf material provides a stable habitat for some other groups of invertebrates such as Collembola and mites, which reached higher population densities per unit area in Hallen Wood than in uncontaminated forests. For instance, the number of Collembola was 20,800, Coleoptera larvae 120, and Diptera larvae 4590 per square meter in metal-contaminated leaf litter and soil, while the numbers on uncontaminated sites were 8690, 4, and 291, respectively.[706,710]

A detailed study was performed near a brass mill in Sweden where the soil was contaminated by copper, lead, and zinc. The abundance and species number of Enchytraeidae near the mill were low.[106] The density and biomass of earthworms was also reduced.[115] Collembola were numerous at high metal levels, but the relative dominance between species was changed.[108] Bengtsson and Rundgren[107] examined the abundance and species diversity of ground-living invertebrates, such as spiders, harvestmen, slugs, beetles, and ants, by pitfall trapping in mature conifer forests at various distances from the brass mill. Metal pollution was suggested to be the most important factor adversely affecting the community structure of ground-living invertebrates. The number of species and the number of specimens per species were reduced significantly within 650 m of the mill where the litter layer contained 2500 μg/g copper and 3600 μg/g zinc. Peak numbers and species diversity were restored at the sites where litter concentrations were 600 μg/g copper and 1300 μg/g zinc. The species abundance distributions of spider and beetle communities showed an impoverished zone in the vicinity of the brass mill, and a boundary zone with high relative abundances some distance away.[107] Concentrations of copper and lead in these animals were high close to the mill, especially in spiders.

Tranvik and Eijsackers[1496] examined the food preference, metal-avoidance behavior, and desiccation tolerance of two springtail species, *Folsomia fimetarioides* and *Isotomiella minor*, around the same brass mill at Gusum. A survey of these species along the metal gradient revealed that the distribution of *F. fimetarioides* was inversely related to that of *I. minor*. *F. fimetarioides* is an early colonizer, highly adaptive to extreme environments, and able to utilize the polluted soil close to the brass mill, while *I. minor* is a slowly adapting species, more sensitive to environmental changes induced by metal contamination.[108] *F. fimetarioides* showed a higher preference for metal-tolerant fungi than *I. minor* when they were offered fungi of different metal tolerance. When they were given a choice between polluted and unpolluted fungi and substrate, *F. fimetarioides* significantly avoided the polluted source, while *I. minor* did not.[1496] The capability of *F. fimetarioides* to avoid an artificially polluted substrate may be of significance in a soil that is heterogeneous and where metal concentrations vary between microhabitats. However, *I. minor* also tended to avoid polluted soil, as observed when both species were given a choice.[1496] Both species were negatively affected by drought. *I. minor* was more tolerant when exposed to drought in combination with metal pollution. An abundant supply of preferred fungal species and a better ability to avoid heavy metals favored *F. fimetarioides* and determined its dominance over *I. minor* in polluted soils.[1496]

Migula et al.[1046] examined the separate and combined effects of cadmium, lead, and zinc on energy budget parameters of the last larval stage of the house cricket *Acheta domestica* (Gryllidae). Their results suggested that the mechanisms of lead elimination and detoxification were less effective when it acted alone. Decreased energy concentration in the tissues followed the increasing toxicity of the food, especially under exposure to cadmium.[1046] This means that under such conditions, the insects could not accumulate energy-rich reserves needed at the beginning of the adult stage for the growth and maturation of the reproductory organs.[1647] Such negative effects of environmental pollutants have also been reported for some species of Lepidoptera and Homoptera.[1044]

Joosse and Verhoef[801] found that Collembola are able to discriminate between "clean" and contaminated food and to regulate their consumption rate (if the food contained high concentrations of lead), and thus they may avoid the uptake of higher doses of lead. In crickets such avoidance mechanisms were found only in the case of lead.[1046] The assimilation rate of cadmium was higher in crickets fed with excessive amounts of lead and cadmium. Only zinc, when added with cadmium, reduced cadmium toxicity in crickets.

The mechanism involved in metal detoxification in insects is poorly known, but in general, two detoxification systems appear to be involved. One system participating in the detoxification of metals in insects could involve hemocytes because insect blood cells also play an active role in physiological processes, leading, for example, to a detoxification of insecticides.[576]

Maroni and Watson[980] demonstrated that *Drosophila melanogaster* larvae respond to the presence of toxic metals by producing nonenzymatic low-molecular-weight proteins of high metal and sulfur content, similar to mammalian metallothioneins. They showed that the synthesis of a cadmium-binding protein was induced by the metal. On the other hand, zinc was not bound and did not induce the synthesis of a similar protein. Metallothionein-like proteins have been detected in Plecoptera,[448] Diptera,[46,48,980] and several other invertebrates.[706] Some insect species may not possess such proteins.[1455] Martoja et al.[988] were not able to confirm the occurrence of metallothioneins in *Locusta migratoria* injected with cadmium and mercury solutions. They stated, however, that cadmium induced the production of cadmium-binding glycoprotein.

In terrestrial arthropods exposed to metal pollution, certain metals may be transported and incorporated into intracellular granules connected with the midgut epithelium. These granules are membrane-bound accumulations of homogeneous material. Metals may be transported to granules by metallothionein proteins.[705,706]

Metal concentrations differ greatly between the various species of soil arthropods (Figure 13).[800] Lead, zinc, and cadmium concentrations in forest-floor arthropods inhabiting an area polluted by zinc factory emissions in the Netherlands were determined by van Straalen and van Wensem.[1437] There was a striking difference in metal concentrations between the carabid beetle *Notiophilus biguttatus* and the linyphiid spider *Centromerus sylvaticus*, although both species were reported to specialize on collembolan prey. Their cadmium concentrations differed by a factor of one (18.2 nmol/g dry weight in *N. biguttatus*; 176.9 in *C. sylvaticus*) possibly due to the fact that spiders suck the contents of their prey and defecate only little, while carabids swallow their prey whole and produce relatively large amounts of feces. Van Hook and Yates[704] obtained a rather long biological half-life for cadmium in lycosid spiders compared to crickets. On the other hand, Hopkin and Martin[707,708] reported low assimilation of lead and cadmium in the spider *Dysdera crocata* and in the centipede *Lithobius variegatus*, fed on wood lice. The results of van Straalen and van Wensem[1437] demonstrated that forest-litter arthropods contain variable amounts of zinc and cadmium that cannot be explained on the basis of their body size nor on the basis of the trophic level at which they are operating. Although body-size considerations may be relevant as regards to lead, both views are inadequate with respect to zinc and cadmium. The concentrations of these metals seem to be connected to the physiological

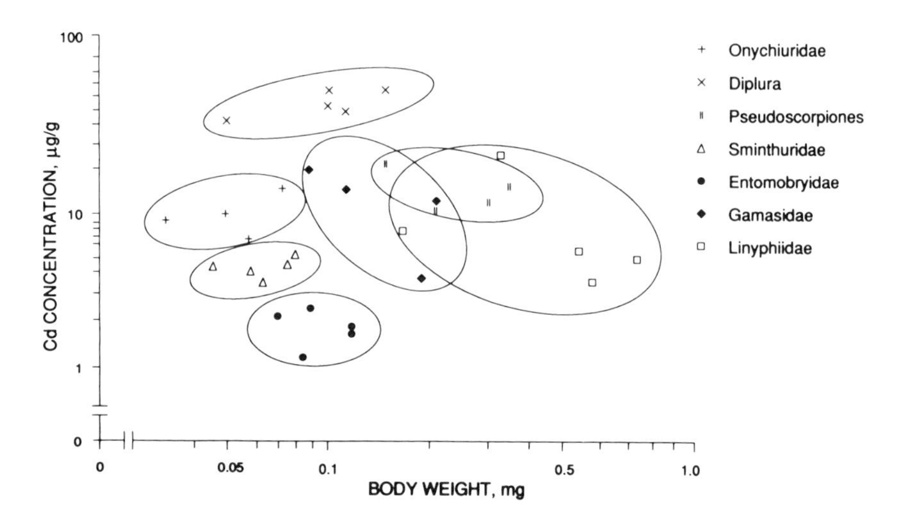

FIGURE 13. Cadmium levels of several microarthropods in a contaminated pine forest soil at Stolberg-Binsfeldhammer, Germany. Cadmium concentrations of individual animals are plotted as a function of body weight. The average soil concentration was 189 µg/g. (From Joosse, E. N. G. and van Straalen N. M., in *Ecological Responses to Environmental Stresses,* Rozema, J. and Verkleij, J. A. C., Eds., Kluwer, The Netherlands, 1991. With permission.)

equipment of the species, rather than to body size or trophic level.[800,1437,1439] This was also shown by a comparison of cadmium, copper, nickel, and lead levels in a sap-feeding aradid bug (*Aradus cinnamomeus*) and a gall-forming tortricid moth (*Retinia resinella*), both feeding on Scots pine. In an area heavily contaminated by metals, metal levels (highest values for Cd 17 µg/g, Cu 1900 µg/g, Ni 220 µg/g, and Pb 32 µg/g) in the bug were ten times those in the moth. This pattern could be explained by the discontinuous alimentary system of the bug in which secretions accumulated in the posterior bulb of the midgut.[665]

The concentrations of metals in moths (*Bupalus piniarius*, Geometridae, and *Panolis flammea*, Noctuidae) and diprionid sawflies (*Gilpinia virens*, *G. frutetorum*, *Microdiprion pallipes*, *Neodiprion sertifer*) and their effects on insect size were examined around a copper smelter in southwest Finland.[656,658] All species were pine defoliators. The highest metal concentrations in each species were recorded in those pupae/cocoons reared on needles from the vicinity of the pollutant source. The decreasing metal concentration gradient followed a linear regression model with increasing distance from the industrial source. The results showed considerable between-species differences in metal concentration. Of the moths, *P. flammea* clearly contained more nickel and cadmium than did *B. piniarius*, but no such difference was found for iron and copper. Cadmium biomagnified in *P. flammea*, but no corresponding biomagnification was observed in *B. piniarius*. In the case of sawflies even fourfold differences in metal concentrations were observed between closely related species reared on the same food. Cadmium also accumulated in the sawflies, the relative increase in cadmium being greater as the distance from the emission source increased. In contrast to cadmium, the concentrations of iron and nickel in cocoons never exceeded those found in the needles.[656] Metal concentrations in food, feces, empty cocoons, and adults of *Neodiprion sertifer* (Hymenoptera, Diprionidae) were analyzed in more detail.[659] The metal levels were higher in the needles than in the feces or insects, except in the case of cadmium. The cadmium concentration in the insects was about twice that in their food intake and much higher than in their feces. The levels in the empty cocoons were relatively low.

C. LEAD

Lead (Pb) is emitted into the atmosphere and deposited on the soil surface due to human activities. The main source of lead in the terrestrial environment is from petrol to which the metal has been added as an "antiknock" agent. The distribution of lead within ecosystems and its transport along food chains have been studied on mine waste,[1289] and in soil and biota near major ways[548,1245] and industrial complexes.[983,1532,1656]

Field studies on polluted soils have shown that soil animals are sensitive to lead. Watson et al.[1568] reported lower total arthropod density and biomass near a lead-mining-smelting complex with high concentrations of metals in the litter. Hågvar and Abrahamsen[584] examined the microarthropod and enchytraeid fauna of a naturally lead-contaminated soil in a spruce forest along a gradient running from unpolluted soil to the middle of a vegetation-free area. Owing to the dissolution of lead from the bedrock, about 50 m^2 of the forest floor lacked vegetation, and another 50 m^2 had poorly developed vegetation. The lead content in the raw humus was up to 10 to 15% dry weight, which is much higher than that reported for areas polluted by man. The ability of microarthropod species to live in the lead-polluted soil varied, and increasing lead levels resulted in a gradual reduction in species numbers among the plants and microarthropods (Figure 14). The results showed that some microarthropods, including the springtail *Isotoma olivacea* and a mite (*Nanorchestes* sp.), were favored by heavy lead pollution. Because a few species were abundant in the vegetation-free area, the total abundance of microarthropods remained almost the same along the gradient.[584]

Controlled laboratory studies on the effects of lead confirm that lead alone can rapidly impoverish the soil fauna. Joosse and Verhoef[801] showed that the springtail *Orchesella cincta* had reduced growth, respiration, and egg production when fed lead-polluted algae. A number of laboratory studies on *Onychiurus armatus* showed that high levels of lead or copper ingested via polluted fungi increased mortality and reduced growth, reproduction, and population growth.[112,113,116] Because fungi accumulate lead efficiently,[112] and many microarthropods are known to ingest fungal hyphae, the main trends found in the study of Hågvar and Abrahamsen[584] may well be explained by toxic effects via the uptake of lead-loaded fungi.

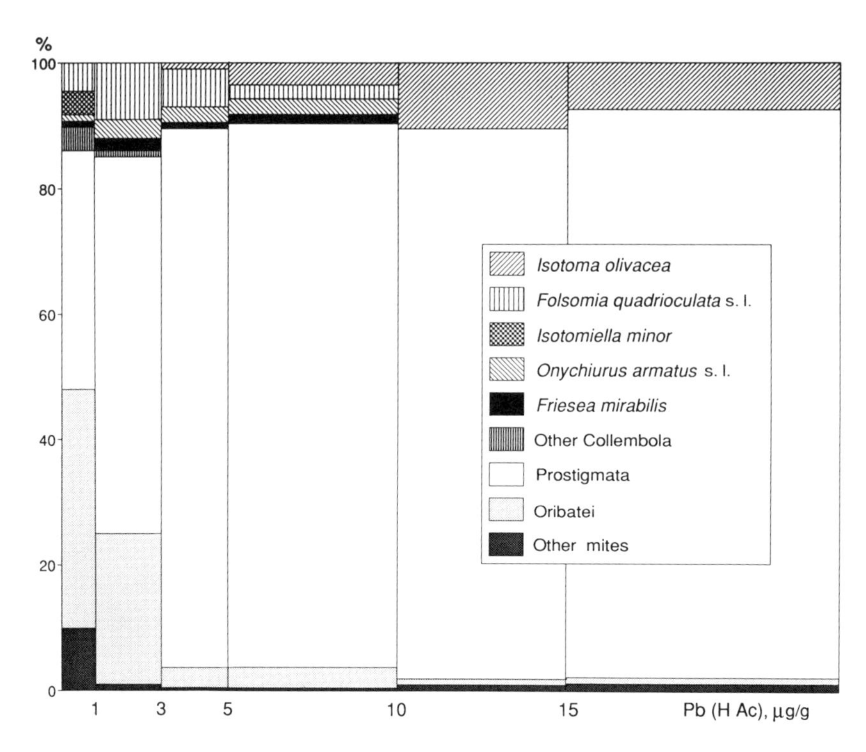

FIGURE 14. The relative composition of the microarthropod community in the 0 to 3-cm layer at different lead levels in naturally lead-contaminated soil in a young spruce (*Picea abies*) forest in Norway. (From Hågvar, S. and Abrahamsen, G., *Environ. Entomol.,* 19, 1263, 1990. With permission.)

Van Straalen and van Meerendonk[1436] suggested that lead in the springtail *Orchesella cincta* is present in three compartments, each with different turnover rates. The most important compartment for lead is the digestive tract, both in size and turnover (Figure 15). Lead in the gut has a half-life of less than a day; the fast body burden only halves within a week, while the slow body burden takes about 3 weeks. The results of van Straalen et al.[1439] showed that excretion of lead and cadmium in *Orchesella cincta* varied between individuals. The average excretion efficiency is, however, remarkably constant: it is not influenced by exposure level, presence of other metals, weight, or sex and is not increased by exposure to metals for more than one molting interval. The differences between populations could not be explained by physiological acclimatization. In later papers genetic differences have been demonstrated.[1222,1223] In other species, metal tolerance may be acquired through exposure during life.[497]

The availability of lead to plants is dependent upon a combination of soil conditions and the response of the roots to lead in the soil water.[43] Particularly important are soil pH and the calcium, organic matter, and clay contents and the amount of hydrous metal oxides. Soil texture, density, and moisture content

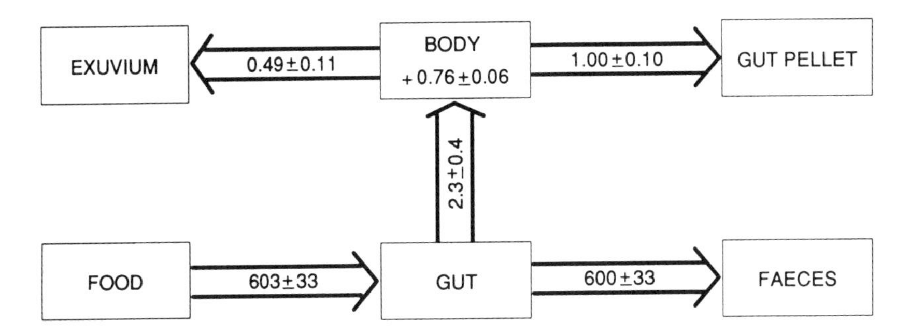

FIGURE 15. Lead budget of *Orchesella cincta* (Collembola, Entomobryidae) exposed to 4.8 μmol/g Pb in the food. Fluxes are in nanograms per molting interval (4 days). Means with their standard errors are indicated. (From van Straalen, N. M. et al., *J. Appl. Ecol.*, 24, 953, 1987. With permission.)

and the abundance of other ions can affect the rate of diffusion to the root system and also absorption.[723] Andrews et al.[43] examined lead and its accumulation within the components of the grassland ecosystem on a revegetated tailings dam at Cavendish Mill, Derbyshire, U.K. The higher soil lead levels at the tailings dam were reflected in the vegetation. Similar concentrations have been reported in vegetation growing on other metalliferous mine wastes.[1288,1289] Lead levels in invertebrates reflect dietary concentrations and feeding strategy, a pattern true also of the indigenous small mammals. The lead concentration in detritivores was higher than that of other categories. Accumulation of lead in herbivores and carnivores showed no consistent pattern. When data from different sampling periods were combined, the lead concentrations in the animal body followed the order: *Lumbricus* spp. > Diptera > *Araneae/Opiliones* > Carabidae. Accumulation of lead in the carabid beetles was considerably lower than that in spiders and harvestmen, even though all three groups are carnivorous. Concentrations of lead in the insectivorous common shrew (*Sorex araneus*) were significantly higher in the contaminated grassland than in an uncontaminated area, but the concentration ratios (body:diet) were less than unity, and there was no evidence of age-dependent accumulation of lead.[43]

Lead is bound tightly by plant tissue, and little translocation occurs.[27] Roadside plants contain more lead than plants from control sites. Price et al.[1235] showed that roadside insects also contained higher lead residues. In areas exposed to high lead emission from the exhaust of vehicles, insects sucking plant juices contained less lead (10.3 μg/g) than chewing (15.5 μg/g) or predatory insects (25 μg/g). This was interpreted as an indication of biomagnification of lead from the herbivore to the carnivore trophic level.[1235] Beyer and Moore[131] demonstrated that lead contamination of the plant *Prunus serotina* adjacent to a busy road was not assimilated, to any great extent, by eastern tent caterpillars (*Malacosoma americanum*). Larvae collected 10 m from the road contained 7 μg/g lead, whereas their food plant contained 10 μg/g lead.

D. MERCURY

Mercury (Hg) is one of the compounds considered most hazardous for the environment.[947] It occurs naturally in low concentrations (<0.02 to 0.05 μg/g) in the soil. However, in agricultural soils mercury levels may exceed 1 μg/g.[854] Mercury has been widely used as a fungicide and for several industrial purposes. Biomagnification of mercury may occur in food chains as shown, for example, by Nuorteva et al.,[1143] who fed *Tenebrio molitor* (Tenebrionidae) larvae on flies containing 40 μg/g mercury. When the beetle larvae died after 4 months, they contained more than 200 μg/g mercury.

Schmidt et al.[1341] used standardized sandy substrate as an oviposition medium in investigating the effects of $HgCl_2$ on the eggs of the acridid *Aiolopus thalassinus* during two successive generations. Although egg-laying females were unable to distinguish between treated and untreated substrate, the number of hatching nymphs decreased with increasing mercury concentration in the soil. Similar results were also shown in another acridid species *Acrotylus patruelis*.[1340] Mercury appeared to be more toxic than cadmium and lead in this respect. The nymphs developing in contaminated soil showed prolonged nymphal duration. In addition, the adult weight was lower, and the lifespan shorter, than in the control. Even low levels of mercury were found to affect successive generations by increasing the mercury concentration in the eggs.[1341] In laboratory experiments accumulation of sublethal amounts of mercury led to reduced growth and reproductive activity in *Tenebrio molitor*.[1349] Females of *Acrotylus patruelis* fed on sublethal doses of $HgCl_2$-contaminated food produced more eggs than their counterparts fed on uncontaminated food. The duration of the larval period was prolonged by 40 to 100% in Hg-contaminated soil. Only 30% of the nymphs reached the adult stage.[1339,1340]

The availability to a primary consumer of mercury associated with plant material at a particular concentration will depend on its chemical form. In the methylated form mercury has been shown to rapidly accumulate in aphids to concentrations that are toxic to their natural enemies.[607] The assimilation of mercury from food by sarcosaprophagous dipteran larvae appears to be much greater when the metal is in the methylated form compared to the inorganic form.[946]

E. CADMIUM

The industrial significance of cadmium (Cd) has increased during the last decades, but it is now declining. Only a small fraction of the processed cadmium is recycled. Most of it is transferred to dumping grounds and further distributed in the environment. Cadmium will finally accumulate in the bottom of oceans. However, it may be bound in the soil for hundreds of years before passing into the oceans. Cadmium is toxic to all organisms. The concentration of cadmium is not the only crucial thing, but also its chemical form. Cadmium occurs partially as free ions and partially as hydroxides adsorbed by particles or in organic matter. The amount of bound cadmium depends on the soil and its properties. Acidification increases the availability of cadmium, and greater amounts of cadmium will leach to deeper layers in the soil.

Free cadmium ions are readily taken up by plant roots, but there are differences between plant species.[956] Since only a small proportion of the cadmium is transported to other parts of the plant, the level of cadmium is usually higher in the roots than in the stem or foliage.[770] In addition, cadmium in the atmosphere can pass into plants directly through the leaf stomata. Up to 20 to 60% of the cadmium in a plant may be absorbed through the cuticula.[717] There is evidence that cadmium concentrations are higher in the leaves of herbs than in those of trees.[985]

Martoja et al.[988] demonstrated several toxic effects of cadmium in the locust (*Locusta migratoria*) and showed that the toxicity was dependent on the age and sex of the insects. The growth rate of larvae of *Tribolium confusum* was reduced severely when 50 μg/g of cadmium were added to a diet containing a minimal level of zinc, but growth was normal when zinc was supplemented to give a concentration of 50 μg/g.[1034] Thus, cadmium in the diet raised the requirements for zinc considerably. The effect of long-term exposure of *Galleria mellonella* (Lepidoptera, Pyralidae) larvae to diets containing 0.1, 1, 10, 50, and 100 μg/g of Cd^{2+} on direct mortality, egg hatchability, fecundity, changes in protein patterns, and phosphatase activity was investigated by Mathova.[992] The effect of cadmium was dose dependent. Thus, exposure to concentrations below 10 μg/g produced no marked change, whereas concentrations above 10 μg/g significantly prolonged larval development and induced "albinotic" larvae. Doses of 50 and 100 μg/g reduced the hatchability of laid eggs, reduced the alkaline phosphatase activity in hemolymph and hemocyte lysate, increased the acid phosphatase activity in hemolymph, and induced a new protein fraction that was detected in the whole-body homogenates of early instars and in the hemocyte lysate of late larval instars. The concentration of 100 μg/g was the only one that directly suppressed female fecundity.[992]

Van Straalen et al.[1440] performed chronic toxicity experiments using the collembolan *Orchesella cincta* and the oribatid mite *Platynothrus peltifer*. The invertebrates were exposed to various levels of cadmium in their food, green algae. In *O. cincta*, cadmium primarily affected female growth, without having a direct effect on reproduction, whereas the effect was primarily on reproduction in *P. peltifer*. Uptake of cadmium was higher in *P. peltifer* than in *O. cincta* and caused a loss of zinc in the former species. As a consequence of their differing physiological responses to cadmium, the population growth rates of mites and collembolans also reacted differently. The population increase capacity of mites appeared to be rather sensitive to cadmium, while collembolans were able, to some extent, to maintain their capacity to increase, in spite of toxic effects at the individual level.[1440]

Matsubara et al.[993] showed that the amounts of cadmium in the tissues and organs decreased when an artificial diet that included both cadmium and zinc was given, in comparison to a diet that included cadmium alone. Different physiological mechanisms may underlie the different effects of cadmium on mites and collembolans. In *P. peltifer*, cadmium disturbed zinc metabolism. Since zinc is often related to reproductive functions, this may explain the reproduction-inhibiting effect of cadmium. In *O. cincta*, cadmium inhibits the growth of females without affecting egg laying directly. Cadmium toxicity in this species must be based on an entirely different mechanism: for example, a disturbance of the hormones promoting female growth. Springtails are able to maintain low body concentrations of metals by excreting them through intestinal exfoliation.[799,1439, 1440]

The toxic effects of cadmium on *O. cincta* at the individual level are not always manifested at the population level. Using detailed life-history data for *O. cincta*, it was shown that the no-effect levels of cadmium for the intrinsic rate of increase and intrinsic biomass turnover were much higher than the no-effect level for individual growth.[1435] This may explain why Collembola populations occurred at sites where cadmium con-centration of the forest floor exceeded the no-effect level of 0.04 μmol/g. At the same time there was continuous selection pressure at these sites for increased resistance, which is realized as increased cadmium excretion efficiency.[800,1439]

There are several reports concerning cadmium uptake by insects. Van Hook and Yates[704] did not observe any bioaccumulation of cadmium in grasshoppers in laboratory experiments. In *Chorthippus brunneus* (Orthoptera), Hunter et al.[734] found that a sevenfold increase in vegetation cadmium concentration resulted in an equivalent tenfold increase in grasshopper total body concentration. Nevertheless, cadmium concentrations were much reduced in comparison with other herbivorous invertebrates at a copper refinery site,[736] and significantly lower than values reported for herbivorous invertebrates collected at an abandoned metal mine site with similar vegetation cadmium concentrations. This may indicate a degree of control over cadmium uptake by *C. brunneus* at the copper refinery site.[736] The high cadmium concentrations found in second instar nymphs may result from the transfer of cadmium to the eggs in gravid females, or from a rapid cadmium accumulation on hatching due to a lack of efficient control mechanisms in the first instar nymphs. The progressive reduction in cadmium concentration in successive nymphal instars may arise through molting losses of cadmium contained in the integument, or through growth dilution.[734] Maroni and Watson[980] found that the larvae of fruit flies contained concentrations of cadmium that were 20% lower than in their food. Cadmium bioaccumulation was observed in *Neodiprion sertifer*, *Gilpinia frutetorum*, and especially in *Microdiprion pallipes* (Hymenoptera, Diprionidae), when the larvae were fed on metal-polluted needles.[656] This was also observed in *Panolis flammea* (Lepidoptera, Noctuidae), but not in *Bupalus piniarius* (Geometridae).[658] Larvae of the small crane fly *Trichocera annulata* (Diptera, Trichoceridae; which feeds on decaying vegetation) collected in the vicinity of a copper refinery contained levels of cadmium and copper that were about ten times higher than those in larvae from the control site.[736]

Nuorteva[1142] reported that there were high cadmium concentrations in the cambium of conifers, in cambium-feeding bark beetles, and in *Formica* ants feeding on honeydew of the phloem-feeding cinarid aphids. Roth-Holzapfel and Funke[1303] reported considerable site-specific differences for metals (such as Pb, Cd, Hg, and Al) in *Ips typographus* and *Trypodendron lineatum* (Scolytidae) in Baden-Württenburg, Germany.

Lindqvist[940] determined cadmium in 14 herb species and in the adults of six insect species from two central Swedish sites with equal cadmium burdens. The cadmium concentrations of the herbivorous insects were about twice as high as those in the herbs. Concentrations of cadmium in the larvae of the sawfly *Dolerus nigratus* were double those in its host plant. The cadmium concentrations in carnivorous insects were even higher. *Cantharis pellucida* (Cantharidae) and *Panorpa communis* (Mecoptera) had twice the cadmium concentrations of herbivorous species. Roberts and Johnson[1288] also found higher concentrations in carnivorous species. In contrast, *Cymindis humeralis*, a carnivorous carabid beetle living on a metal-polluted site on a loamy sand soil in the Netherlands showed no accumulation of cadmium, zinc, or lead.[373] The results of Lindqvist[940] indicated that cadmium was biomagnified in the food chain: plants — herbivorous insects — carnivorous insects. Hughes et al.[729] found that cadmium was concentrated in cabbage loopers (*Trichoplusia ni*) feeding on plants growing on soil amended with sewage sludge. Despite increased cadmium concentrations in the loopers, no obvious toxic effects were observed on their growth or survival.

Vogel[1532] found that the cadmium concentration in the predatory beetle *Thanasimus formicarius* (Cleridae) was almost double that in the bark beetles (e.g., *Ips typographus*, *Pityogenes chalcographus*) used as prey. He also reared[1533] larval stages of the meal beetle *Tenebrio molitor* on artificially contaminated substrate (Cd and Zn, each as acetate and nitrate). Based on low concentrations in the pupae, he concluded that cadmium accumulated in the cuticula of the larvae.

Nuorteva[1142] investigated the effects of spraying birch leaves with a cadmium solution on the occurrence of herbivorous arthropods. Spraying elevated the cadmium concentrations in the treated plants to 7, 10, and 100 times higher than the level in normal birch leaves. Despite this, there were no significant changes in the population densities of herbivorous insects and gall mites.

In addition to the reports of Carter[237] and Hunter et al.,[736] detailed studies on Collembola and cadmium at the species level have been reported on the algal-feeding *Orchesella cincta*[802] and the fungal-feeding *Onychiurus armatus*.[105] Considerable differences exist between species as regards to the extent of metal accumulation.[1437] In addition, terrestrial Isopoda showed cadmium accumulation.[984,1621] This is probably due to their need of calcium, which, to some extent, chemically resembles cadmium.

The cadmium taken up or retained in the larvae of the midge *Chironomus yoshimatsui* was mostly bound nonselectively to high-molecular weight proteins, and only a small amount of the metal was bound to the induced low-molecular weight protein.[1449] In flesh fly (*Sarcophaga peregrina*) larvae approximately 90% of the accumulated cadmium was found in the digestive tract, the fat body and Malpighian tube being less effective at accumulation.[48] Cadmium in the digestive tract was bound mostly to an inducible cadmium-binding protein that consisted of a mixture of five isoproteins with several properties characteristic of metallothionein. Suzuki et al.[1455] found that 84% of the cadmium accumulated by the larvae of the silkworm (*Bombyx mori*) fed a contaminated artificial diet containing cadmium at concentrations of 8 and 80 μg/g wet diet was located in the alimentary canal, and 11% in the Malpighian tubules (up to 1100 and 470 μg/g dry weight, respectively). Cadmium was primarily bound to inducible high-molecular weight cadmium-binding proteins.[1458]

Ortel[1168] studied the effects of cadmium and lead on total lipid, protein, carbohydrate, and water content of the pupal parasitoid *Pimpla turionellae* (Ichneumonidae). The carbohydrate content was hardly or not affected by metal contamination, but the fat content declined drastically. This could indicate a shift in energy metabolism to lipid catabolism, due to cadmium stress. As nearly twice as much water is produced in the lipid catabolism compared to carbohydrate catabolism and, additionally, since *P. turionellae* shows a decreased oxygen consumption that involves lower water loss through respiration, this could explain the increased water content. This detailed interpretation only applies to the cadmium-contaminated group. The somewhat different results for the cadmium plus lead individuals could indicate that lead interferes with the effects of cadmium on metabolism.[1168]

It has long been accepted that pollutant concentrations in organisms are affected by their weight.[1400] Differences in concentration between individuals of the same species, between populations, or between species could merely be attributed to differences in weight. For this reason the use of concentration in comparing the accumulation of metals by different species has been questioned.[1076] Janssen and Bedaux[763] studied the influence of body weight on cadmium accumulation. They used nine species of arthropods sampled for four seasons from a metal-contaminated forest soil. The study showed that the relationship between body size and cadmium level cannot be interpreted as a simple accumulation phenomenon, since there are opposite effects of age, weight, and season. Differences were also found between stages of the same species. The slope coefficients of the life stages of the harvestman *Paroligolophus agrestis* and the pseudoscorpion *Neobisium muscorum* were higher than that for the whole population. The species studied were able to maintain a lower concentration than the value extrapolated from the first stage. This suggested that cadmium was accumulated along with weight rather than with age. The animals concentrated cadmium within each stage, but apparently excreted cadmium when passing to the next stage. As a result there was no increase in concentration during the lifetime in *N. muscorum*, and a relatively small increase in *P. agrestis*. In the snail *Cepaea hortensis*, cadmium accumulated along with age rather than with weight, and accumulation continued even when growth had ceased.[1621]

Within the soil ecosystem, large differences in cadmium concentrations have been observed in arthropod species collected at a single site.[736,1437] These differences may be attributed to differences in food choice and consumption and are thus often explained by differences in trophic level.[733,983] However, no evidence for the biomagnification of cadmium along a soil arthropod food chain was found by Janssen et al.[764]

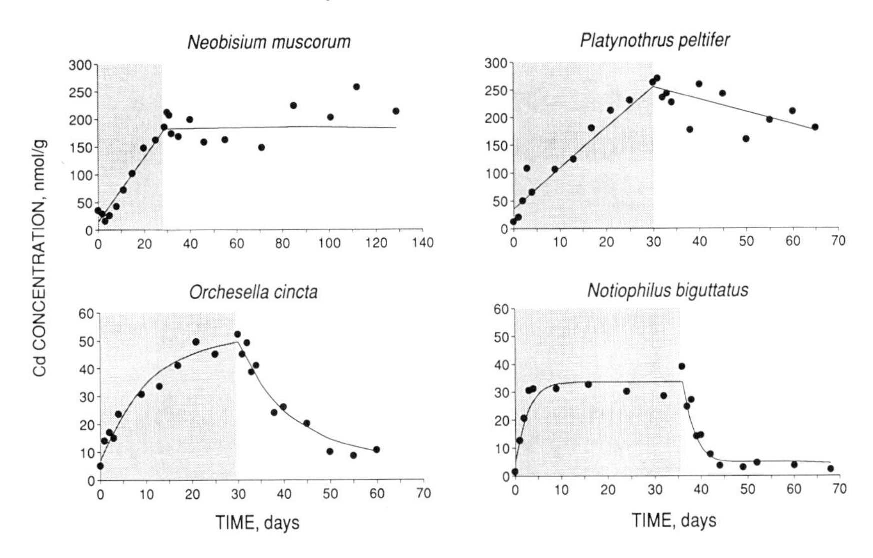

FIGURE 16. Accumulation and elimination of cadmium in four species of soil arthropods: *Neobisium muscorum* (Pseudoscorpionida), *Platynothrus peltifer* (Oribatei), *Orchesella cincta* (Collembola, Entomobryidae), and *Notiophilus biguttatus* (Coleoptera, Carabidae). The smooth line indicates the fitted model. The shaded area indicates the accumulation period. (From Janssen, M. P. M., Bruins, A., De Vries, T. H., and van Straalen, N. M., *Arch. Environ. Contam. Toxicol.*, 20, 305, 1991. With permission.)

As to cadmium accumulation, soil animals can be classified in two or three groups that do not coincide with feeding habits, but follow the gross taxonomic classification.[800] Janssen et al.[764] studied the flow of cadmium through different soil arthropod species by performing laboratory experiments. Four invertebrate species were exposed to cadmium-contaminated food and thereafter to uncontaminated food. Large differences in cadmium excretion and equilibrium concentrations were found (Figure 16). Cadmium assimilation efficiencies were high in the predaceous carabid *Notiophilus biguttatus* and pseudoscorpion *Neobisium muscorum* and were lower in the saprotrophic springtail *Orchesella cincta* and the oribatid mite *Platynothrus peltifer*. Excretion constants were high in the insects *N. biguttatus* and *O. cincta* and low in the arachnids *N. muscorum* and *P. peltifer*. There was no direct relationship between assimilation efficiency and excretion ability. The differences in cadmium assimilation efficiencies reflected differences in trophic level and most probably differences in nutrient demand, which may be determined taxonomically. The influence of excretion ability on the equilibrium concentration was larger than that of assimilation efficiency. Species with a high equilibrium concentration combined a low excretion ability with either low or high assimilation.[764]

The study of uptake, metabolism, and excretion of pollutants by individual species was recommended by Moriarty and Walker[1076] as an approach to predict pollutant concentrations in ecosystems. Together with compartment

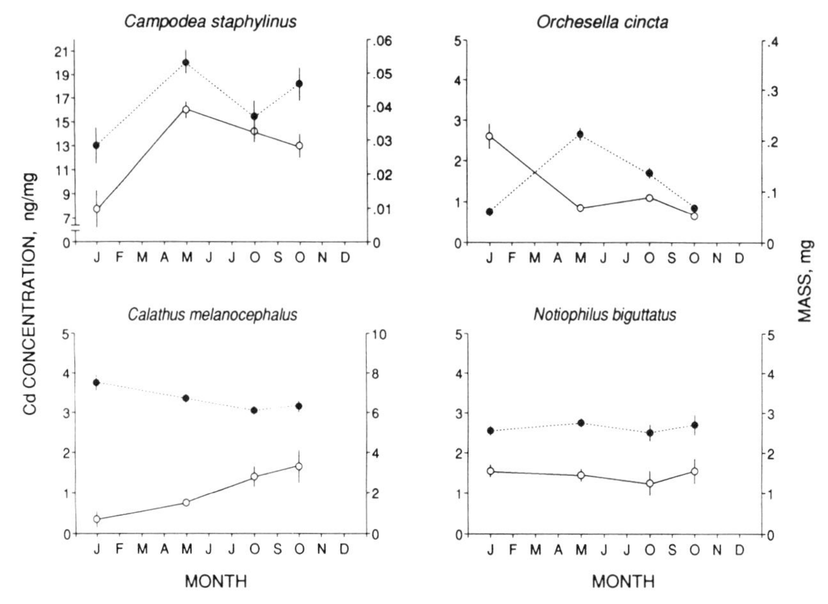

FIGURE 17. Seasonal variation in the mass (dotted line) and cadmium concentration (solid line) of *Campodea staphylinus* (Diplura, Campodeidae), *Orchesella cincta* (Collembola, Entomobryidae), and *Calathus melanocephalus* and *Notiophilus biguttatus* (both Coleoptera, Carabidae). (From Janssen, M. P. M., Joosse, E. N. G., and van Straalen, N. M., *Pedobiologia,* 34, 257, 1990. With permission.)

modeling, these studies may provide a suitable tool for predicting the amount of cadmium in the different soil invertebrates. This approach seems to be more profitable than the common food-chain view in explaining the distribution of chemicals in an ecosystem.[764]

An increase in metal concentrations in spring and summer has been reported for several species and taxonomic groups. Janssen et al.[765] sampled litter arthropods in order to determine the seasonal variation in cadmium concentration in a metal-contaminated site near Budel in the Netherlands. Differences in concentration, as well as different patterns of seasonal variation, were found among species in the same environment (Figure 17).[765] The seasonal variation was species specific, not related to the mean concentration of the species throughout the year, and varied within taxonomic groups. Changes in the cadmium concentration of the food did not play a crucial role in the differential uptake of cadmium during the year. Species such as *Platynothrus* and *Orchesella*, which are known to use algae as an important food component, differed in cadmium level and in seasonal variation. Although the carabids *Calathus* and *Notiophilus* are both confronted with prey species that contain more cadmium in autumn, they showed a different seasonal pattern in concentration. Hunter et al.[736] also noticed seasonal changes in the species abundance, species composition, and age structure of invertebrate communities caused by marked variation in metal contamination levels throughout the year.

Physiologically as well as genetically determined changes have frequently been observed in species characteristics caused by metals.[851] Physiological changes result from individual acclimation; genetically determined changes evolve by adaptation. Adaptation is the process whereby the members of a population become adapted over generations to survive and reproduce in a contaminated environment.[522] Evidence for population differentiation caused by metals has been reported for several species, such as the centipede *Lithobius variegatus*,[707] the isopod *Porcellio scaber*,[231] and the aquatic isopod *Asellus aquaticus*.[497] In none of these examples is it known whether the population differentiation is caused by acclimatization or by adaptation.[394]

Donker and Bogert[394] compared the first-generation laboratory animals of three populations of *Porcellio scaber* collected from a reference wood, a zinc smelter area, and a lead mine site, with respect to the effects of cadmium. The study revealed striking differences between laboratory-raised isopod populations in their response to cadmium. The reduced growth of the reference isopods indicates that these animals were physiologically impaired by cadmium. The mine isopods were not affected by cadmium, and the smelter isopods seemed to be stimulated by some cadmium in their food. This stimulation may be explained by hormesis, described by Stebbing[1423] as an overcorrection of biosynthetic control mechanisms, resulting in increased growth. The fact that smelter isopods were affected by exposure to cadmium and the mine isopods were not suggests that the former are not as well adapted as the latter. This can be explained by the pollution history of the populations: the mine isopods have been exposed for a longer period, a minimum of 2000 vs. 100 years, and to a higher cadmium level than the smelter isopods. Since Cd concentrations in the isopod did not differ between populations, adaptation is probably based on an increased detoxification capacity. The assimilation of Cd did not affect the Cu or Zn content of the isopods, although the adapted isopods regulated their Cu content to a lower level than the reference isopods.[394]

F. ZINC

Zinc (Zn) is an essential component of the enzyme that catalyzes the formation of calcium carbonate. Due to the toxic effects at high concentrations, it has been used as an insecticide to control fly larvae living in the feces of cows.[706] Nuorteva[1142] reported that the adults of five species of moths had lower levels of aluminium, iron, manganese, copper, and cadmium than their food plants, whereas zinc accumulated in the moths to higher levels, possibly in order to eliminate the more toxic elements. Medici and Taylor[1034] reported a protective effect of zinc against cadmium toxicity in the confused flour beetle (*Tribolium confusum*).

Bees are vulnerable in atmospherically polluted sites because pollen becomes heavily contaminated and obtains a surface deposit of metals. In Bulgaria high concentrations of about 700 μg/g of zinc (and copper) were detected in pollen brought back to the hive by bees foraging close to a smelting complex.[1494] The nectar collected by the same bees was not contaminated, and hence the concentrations of zinc in honey were similar in contaminated and uncontaminated areas. This example demonstrates the importance of knowing the feeding behavior of invertebrates, when interpreting the effects of metal pollutants on terrestrial ecosystems. The use of honey for indicating environmental metal contamination has been discussed recently by Jones.[795]

G. COPPER

Copper (Cu) is an essential micronutrient for a wide range of metabolic processes in organisms, but at elevated levels it becomes toxic. Since copper compounds have been used widely in industrial processes (mining, smelting) and agriculture (fertilizers, pesticides), elevated copper concentrations can be found in certain areas of the biosphere.[483]

There are several reports on copper concentrations in insects. For example, copper concentrations in litter arthropods showed high variation in a severely contaminated area in Germany. In general, copper levels in carabid beetles were lower than in spiders, but the values were highly dependent on the species in question.[1634]

Invertebrates from contaminated soil and contaminated grasslands in the vicinity of a major copper refinery with a copper/cadmium alloying plant showed significant elevation of total-body copper and cadmium concentrations relative to control values.[736] Isopods contained the highest concentrations of both metals, which is possibly a consequence of their high calcium requirements,[97] since calcium can be substituted by metals. As a group, the faunal detritivores exhibited an accumulation of copper and obtained 2 to 4 times higher concentrations than found in the organic surface soil and plant litter at the refinery site. Oligochaeta and Isopoda both showed a significant reduction in population size at the refinery site.[736] Seasonal changes in the abundance, species composition, and age structure of insect populations caused variation in metal contamination throughout the year, illustrating the importance of continuous monitoring (Figure 18).

In the grassland ecosystem close to the copper refinery, further biotransfer of metals to carnivorous invertebrates revealed marked differences in metal accumulation by predatory beetles and spiders. The Cu/Cd concentration ratio in predatory beetles was 30 to 35, which may indicate some control over cadmium accumulation in Coleoptera. In contrast, the spiders showed a copper/cadmium concentration ratio of 9 to 11, which may indicate some degrees of control over copper accumulation, but little evidence for control over cadmium accumulation. Regarding the invertebrates with the highest levels of cadmium accumulation, spiders, in particular lycosid spiders, were abundant and frequently encountered in grasslands in the vicinity of the refinery. Spiders therefore represent a potential major pathway for the accumulation of cadmium by predatory small mammals and birds.[736]

Hunter et al.[734] examined metal concentrations in the grasshopper *Chorthippus brunneus* in relation to its precise dietary requirements around a copper refinery. Annual mean copper and cadmium concentrations in

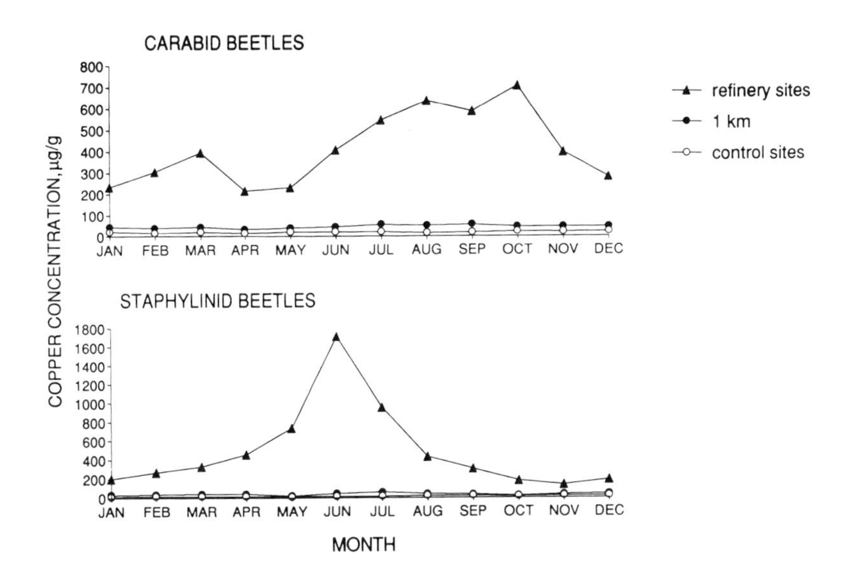

FIGURE 18. Seasonal changes in copper content of carabid and staphylinid beetles (Coleoptera) at refinery, 1 km, and control sites. (From Hunter, B. A., Johnson, M. S., and Thompson, D. J., *J. Appl. Ecol.*, 24, 587, 1987. With permission.)

C. brunneus were an order of magnitude above the control concentrations. Concentrations of copper (1600 µg/g) and cadmium (5.8 µg/g) found in the oldest individuals of *C. brunneus* in November were of an order not reported previously in herbivorous terrestrial insects.[734] The maximum concentration in 2-year-old, sap-sucking bugs (*Aradus cinnamomeus*) feeding on Scots pine was reported to be 1900 µg/g near a copper smelter.[665]

Laboratory experiments conducted by Bengtsson et al.[112] showed that higher rates of growth and average maximum lengths, relative to controls, were attained by the springtail *Onychiurus armatus* subjected to relatively low levels of copper (45 µg/g in the substrate). This effect may be due to the elimination of a copper-sensitive gut parasite;[116] however, copper and lead concentrations 45 µg/g in the substrate (corresponding to levels of more than 1000 µg/g in the fungi), were sufficient to reduce population growth rates. The threshold whole-body concentration in *Onychiurus* at which significant reductions in growth rates could be detected was 200 µg/g (Pb plus Cu), which is similar to levels found in a population of the same species 650 m from a brass mill at Gusum, southeastern Sweden.

H. NICKEL

Though nickel (Ni) smelting activity has resulted in severe environmental damage, e.g., in Sudbury, Canada,[308] its effects on insects are poorly documented. Masui and Matsubara[991] measured the amount of Ni per gram dry weight of certain tissues and organs per individual in silkworms (*Bombyx mori*). The insects were reared aseptically or on an artificial diet containing 250 μg/g nickel given continuously from the day of hatching to the 6th day of the fifth instar. The nickel ingested was excreted by the feces 48 h after feeding. The amount of nickel in the pupa of the aseptically reared silkworms was 4.6 μg on the 2nd day after pupation, while that in the pupae of silkworms fed a nickel-containing diet during their larval stage increased moderately to 6.2 μg.

I. IRON AND MANGANESE

Iron (Fe) and manganese (Mn) appear to be relatively harmless to insects, because they are essential in many physiological processes. However, blast-furnace plants expel large quantities of iron- and manganese-carrying dust, which results in soil pollution that subsequently affects soil-dwelling invertebrates.[803]

Nottrot et al.[1135] studied the uptake and excretion of iron and manganese in *Orchesella cincta* (Collembola) and found that iron decreased the feeding activity and growth, whereas manganese had no obvious effect. Iron caused a decrease in growth, whereas manganese had no effect on growth. A decrease in the feeding activity and growth occurring in the field as a result of iron pollution could have some consequences for the functioning of the soil system, e.g., a decrease of the litter decomposition rate.[1135,1525]

Molting is the most significant pathway for heavy metal excretion by Collembola. Springtails are able to store metals in the gut epithelium,[731] which is expelled as gut pellets at every molt. The manganese content of a gut pellet is much higher than the iron content, whereas in the food and the animals the iron concentration is usually higher.

However, an interesting suggestion for the role of iron in the biology of terrestrial invertebrates has been made in experiments on orientation in Lepidoptera. Deposits of this metal have been found in a subcuticular layer in the abdomen of adult monarch butterflies (*Danaus plexippus*). It is possible that the iron enables the butterflies to orientate on their long migratory flights with respect to the Earth's magnetic field.[792,846]

X. AGRICULTURAL CHEMICALS

A. BACKGROUND

Insecticides, herbicides, and fungicides are all agricultural chemicals. They can be considered pollutants when they have adverse effects on nontarget organisms (Table 6). The testing of 1169 different pesticides on 19 beneficial arthropods showed that a large number of fungicides (71%) and herbicides (60%) were harmless, while most (81%) of the insecticides and acaricides were at least slightly harmful.[628]

Synthetic insecticides have saved millions of human lives, e.g., by killing the mosquito vectors of malaria, killing body lice that transmit typhus, and destroying other insects that transmit killing or crippling diseases to man. They are invaluable in protecting crops from pests and preventing damage during their storage. It has been estimated that without pesticides about one third of the world's food would be lost to pests.[1214] However, pesticides have not solved the pest problems.

Pesticides can have an adverse effect on nontarget animals. There are several reviews on the effects of pesticides on beneficial insects.[207,328,433,434,778] Furthermore, there is extensive literature on their effects on pests,[98,146,645,941,] but some of these effects have been poorly investigated as regards nontarget insects. Pesticides can influence the essential functioning of ecosystems by reducing species diversity and modifying food chains; changing patterns of energy flow and nutritional cycling (including nitrogen); reducing soil, water, and air quality; and changing the stability and resilience of ecosystems.[188,1205] Indirect effects of pesticides may be much more important than direct effects. Since an insecticide is selected for its effectiveness against the pest and the degree of effect on other organisms, it is essential to know these effects before the chemical is applied over large areas.[168]

The effects of pesticides on nontarget insects may be direct or indirect, e.g., via depletion of food resources.[778] Short-term effects are related to the scale and level of a given treatment. Exposure and susceptibility will determine the probability of an insect being affected by a pesticide in the short term. Exposure is affected by the pesticide droplet capture efficiency, the proportion of population in the sprayed area, and the degree of protection by vegetation or soil refuges at the time of spraying. The volume and frequency of application are, of course, crucial. Following application the distribution pattern and dial activity of the species, the availability of contaminated food, as well as the persistence and bioavailability of the contaminant all affect the level of pesticide impact. Susceptibility is associated with the genetic, structural, and physiological factors mediating uptake, metabolism, and toxic effect.[777] The long-term effects cover a less-precise period of time including the phase of short-term population recovery. Important factors mediating long-term effects include mobility, subpopulation structures, reproductive rate, isolation from a reservoir habitat, and the degree of polyphagy.

In North American agroforestry the largest insecticide spray programs are those carried out against several species of spruce budworm (*Chroristoneura* spp.), which defoliate a number of coniferous tree species. Other control programs include those against the gypsy moth (*Lymantria dispar*), the Douglas fir tussock moth (*Hemerocampa pseudotsugata*), the hemlock looper (*Lambdina fiscellaria*), the tent caterpillars (*Malacosoma disstria* and *M. americanum*), and bark beetles.[786] In Europe major chemical control programs are restricted to *Lymantria monacha* and *Zeiraphera diniana*.

The reproductive potential of most invertebrates is high, and arthropod numbers on the treatment site are expected to return rapidly to prespray levels.[168] Because of the small area sprayed, the vacancies left by poisoned individuals would probably be quickly filled by mobile invertebrates from areas surrounding the treatment site. Past studies have shown that invertebrate populations are quite resilient to occasional insecticide applications.[168,1410]

For example, Freitag and Poulter[505] and Carter and Brown[238] investigated the long-term impact of the low-persistence organophosphate insecticides, fenitrothion and phosphamidon, on predacious litter-dwelling arthropods. Both investigations reported a reduction in spider, harvestmen, and beetle numbers in the year of the spray, but the populations recuperated the following year.

Some insecticides, such as DDT and lindane, are very persistent in ecosystems, while easily biodegradable insecticides, such as malathion and other organophosphates, rapidly break down in the environment into allegedly less-harmful substances. Under field conditions the climate and the soil type affect the persistence and toxicity of insecticides. Knowledge of the chemical properties of insecticides and their metabolites, alone, is not sufficient to predict the persistence of biological activity of their soil residues.[4] One alternative is to use a soil animal such as the hemiedaphic collembolan *Folsomia candida* as an indicator of insecticide toxicity and persistence in soil. It has been known for a long time that *F. candida* is sensitive to carbamates, organochlorines, and organophosphorous compounds.[1482]

Because of their recognized importance as pollinating agents and honey producers, the high susceptibility of honeybees to poisoning by regularly used insecticides has fostered extensive research on the problem. Investigations have determined the hazard posed by numerous insecticides to honeybee adults. However, the effects of insecticides, at concentrations that are sublethal to food-provisioning adults, on the development of the immature stages of the honeybee are largely unknown. The progressive feeding of larvae by many different nurse bees increases the probability of the brood being exposed to insecticides.[359] Wittmann,[1636] in an investigation of the distribution of ^{14}C insecticides in nurse bees, concluded that the older larvae of the honeybee are most likely to receive insecticides in food. However, using morphogenic effects as a criterion, the studies by Atkins and Kellum[57] suggested that older larvae are less susceptible. Detrimental effects on the development and viability of these immature stages would have direct consequenses on colony population levels and would subsequently result in a decline in honey production and pollination activities.

B. ORGANOCHLORINE INSECTICIDES

Organochlorine pesticides, widely used in the protection of health and agriculture, may lead to the contamination of ecosystems,[415,1651] including pesticide bioaccumulation in nontarget organisms. For example, DDT (1,1,1,-trichloro-2,2-bis(*p*-chlorophenyl) ethane) was first registered in 1946. Although it was virtually banned in Europe and North America in the 1970s, it continues to be used elsewhere. Residues of organochlorines have been extensively found in most of the biotic and abiotic components of ecosystems.[6,415,502,1075,1651] The problem with DDT is that it is degraded to DDE, which is very persistent and resistant to biodegradation. Both DDT and its breakdown products are fat soluble. They persist in the fatty tissues of animals and become concentrated there.[1233]

TABLE 6
Examples of Effects on Terrestrial Nontarget Insects of Agricultural Chemicals

Chemical	Taxon	Treatment/concentration	Observation	Refs.
Organochlorine insecticides				
pp′-DDT	*Apis indica* (Hymenoptera)	227 ng/l in body	Residue analysis	7
	Danau chrysippus (Lepidoptera)	207 ng/l in body	Residue analysis	7
Lindane	*Formica polyctena* (Hymenoptera)	20 ng/l in body	Long-distance transport	369
Lindane	Collembola	198 μg/l, 15.7 and 78.3 kg/ha	94–100% decrease in abundance[a]	721
Lindane	Collembola	Trees treated with 0.5% solution	Decreased density at tree base[b]	630
Organophosphorous insecticides				
Fenitrothion	*Bombus* spp. (Hymenoptera)	Aerial spray, 0.21 kg/ha	47–100% mortality[c]	1211
Fenitrothion	Collembola	Trees treated with 2% solution	Short-term effect on density	630
Chlorpyrifos	Collembola	479 μg/l, 52.2 and 261 kg/ha	Obliterated	721
Dimethoate	*Apis mellifera* (Hymenoptera)	0.1, 1.0, 10.0 mg/l in syrup	Colonies destructed at 10.0 mg/l	1429
Carbamate insecticides				
Carbaryl	Collembola	479 μg/l, 52.2 and 261 kg/ha	95–100% decrease in abundance	721
Aminocarb	Lepidoptera	175 μg a.i., 2.24 l/ha	Decreased activity	168
Pyrethroid insecticides				
Deltamethrin[d]	Hymenoptera, Diptera	12.5 μg a.i. per hectare, 5 replicates	Reduced density	450
Herbicides and fungicides				
Propiconazole	Wheat field invertebrates	250 μg/l a.i., 0.5 l/ha, 200 μl water per hectare	No response	1405–1407
Pyrazophon	Wheat field invertebrates	300 μg/l a.i., 2 l/ha, 200 μl water per hectare	Decrease in aphid predators Twofold increase in aphids	
Triadimephon	Wheat field invertebrates	125 μg a.i., 500 mg/ha, 200 μl water per hectare	No response	

Table 6 (continued)

Chemical	Taxon	Treatment	Observation	Refs.
Atrazine	*Onychiurus armatus* (Collembola)	2.5–160 µg/g, 30 and 60 days	Dose-dependent mortality, up to 100%	1067
Atrazine	*O. apuanicus* (Collembola)	2.5–160 µg/g, 30 and 60 days	Dose-dependent mortality, up to 97.3%	

[a] Results obtained after 138 days; 88–97% decrease after 10 days.
[b] Collembola did not return to pretreatment numbers for at least 2 years.
[c] Threefold population density in unsprayed areas; population recovery within a few years.
[d] One treatment partially with permethrin, 40 µg a.i. per hectare.

The large-scale use of organochlorines (DDT, lindane, dieldrin, endosulfan) for the control of the tsetse fly in Africa has caused major kills of nontarget insect groups, and their relative abundances have remained markedly depressed for at least a year after insecticide exposure.[409,860,995,1096] The area affected may be large: in Botswana, endosulfan concentrations were high enough to kill tsetse flies more than 12 km downwind of application, and detectable levels were found 30 km downwind.[41]

Reports on the pesticide burden on nontarget beneficial insects are available mostly from the U.S.[130,431,1268,1646] Concentrations of DDT and its metabolites were reported to vary from 10 to 20 ng/g in a predatory mayfly,[1646] 10 to 4440 in a tiger beetle,[1268] and 50 to 5600 in moth species.[130,431] Ahmad et al.[7] investigated the pesticide burden of some insects of economic importance, in Lucknow, India. Chlorinated pesticide residues of DDT, BHC, and aldrin, as well as their metabolites and isomers, were detected in pollinating insects, honeybees (*Apis indica*) and butterflies (*Danais chrysippus* and *Eurema* sp.), and predators, dragonfly (*Platythemis* sp.) and wasps (*Polistes herebreus*). The relative concentrations of organochlorine residues included high levels of DDT and its metabolites (231 to 796 ng/g) followed by BHC (10 to 60 ng/g) and aldrin (0.2 to 6.7 ng/g) in different insect species. The data of Ahmad et al.[7] indicated a sharp difference in the rate of bioaccumulation of persistent pesticides. Such a difference in pesticide residues in moths from natural environments was also noted by Beyer and Kaiser,[130] who considered the feeding habits and the habitats of insects as possible factors contributing to such variation. Taking into account several other factors, e.g., enzyme systems, fat content, age, and weight, would be helpful in explaining such variation in pesticide bioaccumulation in different species.[7]

Debouge and Thome[369] measured the concentration of lindane in five ant species (*Formica polyctena*, *F. rufa*, *Myrmica ruginodis*, *M. laevinodis*, and *Lasius niger*) in Belgium. The mean concentrations were relatively low, around 20 μg/g. The concentrations of insecticide residues were not significantly different in agricultural and forested areas. Debouge and Thome[369] suggested that insect contamination by organochlorine insecticide residues in forested areas would not result from direct intoxication due to the use of these insecticides on the sampling sites. Atmospheric transport of contamined particles could cause environment contamination far away from the sites of insecticide application.

Relatively little information is available about the effects of insecticide applications, such as bark beetle spraying, on the terrestrial saprophagous arthropods in soil and litter. Lindane, used as a seed dressing, changed the community structure of Collembola in beet fields.[342] Studies in a pine forest in California[720,721] indicated a direct relationship between insecticide stability and reductions in nontarget soil arthropod numbers. Plots sprayed with lindane to protect ponderosa pine (*Pinus ponderosa*) from attack by bark beetles had severe changes in the abundances of most arthropod groups, especially in the detritivorous collembolans and oribatid mites. The overall effect of lindane was a broad disruption of the community structure. According to Hastings et al.,[630] spraying 0.5% lindane on white pines to control the southern pine beetle *Dendroctonus frontalis* caused long-term reductions in litter and soil mesofauna populations in the mountains of North Carolina. Mites, collembolans, and other arthropod fauna did not return to pretreatment levels for at least 2 years, and the soil mesofauna remained below initial populations even after 963 days. Joy and Chakravorty[806] found that aldrin and endosulphan, as well as organophosphorous dimethoate, persistently decreased the density of microarthropod groups in the soil of wheat and mustard fields in India. In the laboratory the direct and residual toxicities of insecticides were more prominent for *Cyphoderus* sp., but *Xenophylla* sp. (both Collembola) were resistant to several insecticides.

In general, no differences have been reported between DDT-treated and untreated tree stands in the complex of parasitoid species or in spruce budworm mortality caused by parasitoids. A study conducted in areas of New Brunswick, Canada that were repeatedly treated with DDT between 1952 and 1958 showed an increase in parasitism by *Apanteles* in treated stands compared with control plots. The highest increase occurred in the location sprayed for 6 consecutive years. There was no difference in the incidence of *Glypta* between treated and untreated localities.[954] Most chemical control operations against the spruce budworm are aimed at the larval stage, and larval parasitoids are the ones likely to be affected by commercial spraying operations. However, Blais[147] concluded that the principal hymenopterous parasitoids (*Apanteles fumiferanae*, *Glypta fumiferanae*) were not exposed to the sprays in Quebec because they occur as adults at a somewhat later date.

Dieldrin's parent compound, aldrin, was used extensively in the 1960s and 1970s to control cutworms (larvae of several noctuid moth species) in the U.S. Although use of this compound was prohibited in 1974, dieldrin is highly persistent in soils where it has been applied. Heptachlor was substituted after aldrin was banned, but it was also prohibited in 1978. Pesticide-residue-induced mortality was reported in the endangered gray bats (*Myotis grisescens*) in Missouri, which provides an important summer habitat for the species. The dead gray bats had lethal concentrations of dieldrin in their brains, and some of them also had lethal concentrations of heptachlor epoxide. Insect samples from the feeding range of the gray bats showed widespread pesticide contamination of the food of this endangered species.[272] Twenty-five of 29 insect samples contained measurable dieldrin, heptachlor epoxide, or both. The mean dieldrin concentration for all Coleoptera was 0.41 mg/g wet weight; for Trichoptera, 0.27 μg/g; and for Lepidoptera, 0.07 mg/g. Dieldrin in all the insect samples averaged 0.21 μg/g, and heptachlor epoxide averaged 0.18 μg/g.

C. ORGANOPHOSPHOROUS INSECTICIDES

The organophosphate fenitrothion (applied at an emitted rate of 140 to 280 mg a.i. per hectare) has been the principal insecticide used against the spruce budworm (*Choristoneura fumiferona*) in New Brunswick, Canada since the late 1960s.[1211] Although Atkins et al.[58] reported that fenitrothion was highly toxic to honeybees, reports of its effects on wild-bee populations have been conflicting. Varty and Carter[1523] concluded that the abundance and diversity of wild bees were not drastically influenced by the perennial spray programs in New Brunswick. Buckner[199] obtained variable results; in one study he found no significant mortality among bumblebee queens, but a reduction of about 50% in another.

However, there is evidence of wild-bee mortality in or near forests sprayed with fenitrothion.[834,1210,1643] Kevan[834] stated that the numbers of native pollinators of blueberry, which included bumblebees, were sharply reduced in sprayed areas. Solitary bees may be more susceptible to spray drift than bumblebees.[1643] Markedly reduced activity of native bees (*Bombus* spp., Andrenidae spp., Halictidae spp.) was recorded by Wood[1643] in low-bush blueberry fields in New Brunswick following spraying with fenitrothion over nearby woodland for control of the spruce budworm (Figure 19). The low-bush blueberries (*Vaccinium angustifolium* and *V. myrtilloides*), which are harvested commercially, depend on insects for pollination.

The long-term effect of fenitrothion spraying on native bee populations is not well documented. Kevan[835] estimated that it may take up to 10 years for native bee populations to recover. Wood[1643] reported that the native bee populations had returned to normal levels, and there was satisfactory pollination of blueberry within 3 years after the discontinuance of fenitrothion spraying in the vicinity.

Population recovery of bumblebees appeared to be complete within a few years after discontinuation of fenitrothion spraying.[1211] Plowright et al.[1211] showed that aerially sprayed fenitrothion (0.21 kg/ha) caused mortalities from 100% among experimentally caged bumblebees in exposed habitats, to 47% in cages placed under dense forest canopy. The exposure experiments demonstrated that the ambient insecticide concentrations during the first few hours after spraying are highly toxic to unprotected bumblebees. Bumblebees found foraging in sprayed areas during the days immediately following the spray suffered significantly higher subsequent mortality than those in unsprayed areas. Late-summer population abundances in the unsprayed areas for all *Bombus* species combined averaged three times higher than those in the fenitrothion-treated areas. The results indicated that high toxicity is present at a distance of at least 150 m from the flight path of the spray aircraft. The foraging performance of laboratory-reared bumblebee colonies was significantly higher in sprayed areas with reduced bee populations than in a control area, possibly because of a relaxation of competitive stress. The diversity of plant species used for pollen collection was nearly twice as great in the control, compared to the sprayed areas, suggesting that the effect of fenitrothion spraying on crosspollination may be greatest for plants that are subdominant in the hierarchy of bee preference. Reduced seed-set was demonstrated in one such plant, red clover.[1211]

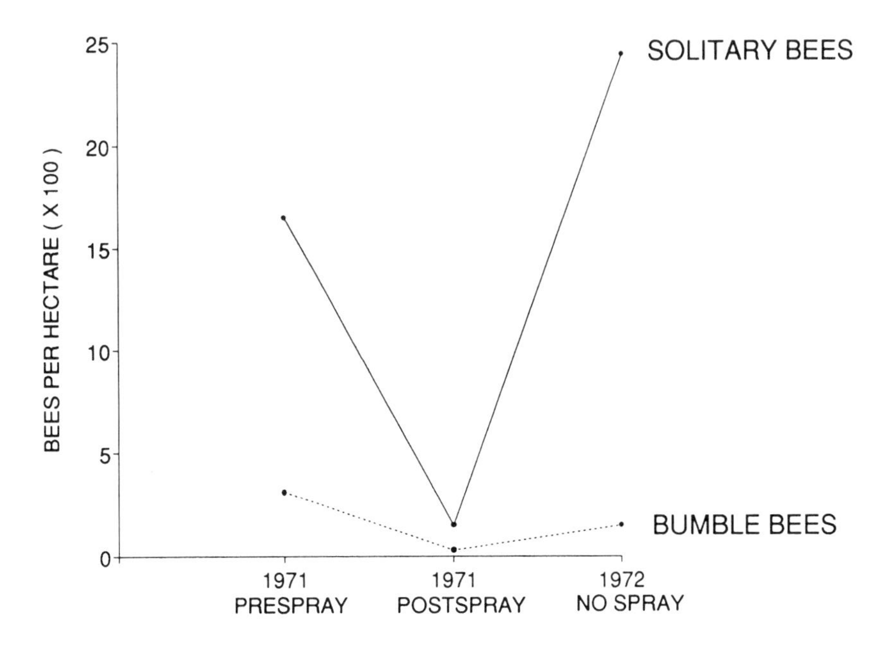

FIGURE 19. Bee densities in a blueberry field near Sussex, New Brunswick, Canada, after spraying with fenitrothion over nearby forest. (From Wood, G. W., *J. Econ. Entomol.*, 72, 36, 1979. With permission.)

Population surveys reinforced the experimental results: a substantial reduction in bumblebee population densities was considered to be associated with fenitrothion spraying operations in New Brunswick.[1211] Those species that emerge earliest from hibernation (*Bombus terricola* and *B. ternarius*) show the greatest reduction in sprayed areas, while those that emerge late in the spring (*B. borealis*, *B. fervidus*, *B. rufocinctus*) may even increase in recently sprayed areas.

Hastings et al.[630] compared the effects of fenitrothion and lindane on the soil fauna. Aqueous 0.5% lindane and 2.0% fenitrothion were applied directly to the forest floor. Although initial fenitrothion residues were 7.5 times greater than those of lindane, the transient nature of these residues and the reduced effects on mites and collembolans indicated that it had a shorter-term effect on soil and litter fauna.[630]

The toxicity of dimethoate, another organophosphorous insecticide, has been extensively studied.[359] Barker et al.[81] suggested that toxicity from dimethoate was cumulative in the honeybee and that the insecticide accumulated to a lethal level without much degradation of dimethoate in the living bee. Adult bees are not repelled by toxic concentrations of dimethoate.[81,359] Dimethoate at a concentration of 10 mg/l in sugar syrup destroyed standard-size field colonies, whereas colonies fed 0.1 or 1.0 mg/l performed better than, or equally as well as, untreated colonies over a period of 8.5 weeks.[1428] However, the literature on dimethoate poisoning in honeybees is conflicting.[359]

Atkins and Kellum[57] reported that approximately 10% of the adult honeybees exposed to dimethoate as larvae had various wing deformations.

Chlorpyrifos has been used extensively in a variety of formulations to control a broad spectrum of agricultural and forest pests.[1147] Hoy and Shea[721] investigated the effects of chlorpyrifos, used against bark beetles, on nontarget soil arthropods. The progressive negative effect of chlorpyrifos on collembolans, severe effects on actinedid mites, and moderate effects on all other elements of the arthropod community made chlorpyrifos equal to lindane in disruption. Perfecto[1188] reported that chlorpyrifos significantly reduced ant foraging activity, as well as densities of the fall armyworm *Scodoptera frugiperda*, in a tropical maize agroecosystem.

D. CARBAMATE INSECTICIDES

Fenoxycarb acts as an insect growth regulator. It does not inhibit cholinesterase and is effective against many insects, including scale, leaf miners, mosquito larvae, ants, stored-product pests, and cockroaches.[1150]

Carbaryl, a carbamate insecticide, varies greatly in its toxicity, depending on the species affected, the formulation, and the pH of the mixture. Furthermore, Green and Dorough[561] reported pronounced differences in the susceptibility of the female house fly to dosages of carbaryl and two other carbamates (Banol®, Baygon®) without corresponding changes in cholinesterase activity or in its sensitivity to the insecticides. They concluded that accumulation of the insecticide at the site of action varied with the age of the flies.

Hoy and Shea[721] reported drastic changes in soil arthropods caused by spraying carbaryl to protect ponderosa pine (*Pinus ponderosa*) from bark beetle attack. Carbaryl had a severe negative effect on collembolans as well as on certain gamasid and actinedid mites, but there was a positive effect on the oribatid mites. Stegeman[1424] reported a dose response by soil mites and collembolans to carbaryl applications of up to 11 kg/ha in both pine and hardwood forest floor in New York State. The response of an arthropod-herbaceous plant community structure to carbaryl is complex.[1453] Investigations involving the carbamate insecticides carbaryl, carbofuran, methomyl, and bufencarb demonstrated that applications of carbamates at dosages ranging from 0.56 to 5.00 kg/ha frequently also caused large reductions in earthworm numbers and biomass.[721]

The nonpersistent insecticide aminocarb (4-dimethylamino-m-tolyl methylcarbamate) is used for the control of the spruce budworm *Choristoneura fumiferana* in Canadian forests. Varty[1522] investigated the impact of aminocarb on litter-dwelling invertebrates and found that the total catch of spiders, beetles, millipedes, and other invertebrates in pitfall traps was 19% lower on an area treated with aminocarb than in a similar forest that had never been sprayed. Bracher and Bider[168] studied the effects of aminocarb on the activity of forest animals in Canada. Changes in the ratios of animal activity between the treatment and control transects, prespray and postspray, were used to identify the possible effects. The decrease in the activity of Arachnida, Lepidoptera,

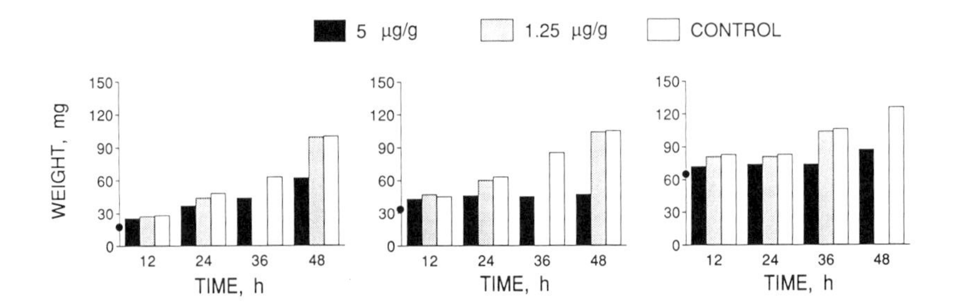

FIGURE 20. Mean weights of surviving larvae of honeybees (*Apis mellifera*, Hymenoptera, Apidae) feeding on royal jelly diets contaminated with carbofuran and on diets of pure royal jelly. Initial weights of the larvae are indicated by a black dot on the weight (Y) axes: (A) 18.5 ± 1.92 mg (72 h), (B) 31.8 ± 4.55 mg (84 h), (C) 64.4 ± 7.02 mg (96 h). Time on the diet is indicated on the X axes. (From Davis, A. R., Solomon, K. R., and Shuel, R. W., *J. Apicult. Res.*, 27, 146, 1988. With permission.)

and Mollusca could be attributed to the direct toxic effects of aminocarb, since this insecticide is particularly lethal to these animal groups. An increase in Diplopoda activity could have resulted from an increase in food availability through a greater availability of carrion or a decrease in competition. Although millipedes feed mainly on plant material, they have been known to eat dead insects, molluscs, and worms. Chilopoda activity possibly increased in response to less competition or predation from arachnids.[168]

The toxicity of carbofuran to honeybees has been reviewed by Davis et al.[357,359] Stoner et al.[1429] reported that the threshold for serious damage in standard-size colonies of honeybees was 1.0 mg carbofuran per liter of syrup. Davis et al.[359] compared the effects of adult-sublethal doses of carbofuran on honeybee larval growth (Figure 20) and development in the laboratory to those of organophosphorous dimethoate. Both carbofuran and dimethoate are systemic insecticides having the capacity to enter, and be translocated within, plants away from the point of application. They might contaminate the honeybee larval diet both fortuitously through spray drift and directly through secretion in nectar. Despite the acute toxicity of carbofuran, the chronic toxicity to adults and larvae caused by the feeding of dimethoate eventually resulted in the mortality of adults and failure at pupation at lower concentrations than with carbofuran.[359] Bai and Reddy[72] found similar results for adults of *Apis cerana indica*. Carbofuran was also found to reduce foraging activity.[1188]

Reduction in acetylcholinesterase (AChE) activity has been found useful as a measure of insecticide poisoning of adult honeybees. The AChE levels of honeybee larvae suggested that assays of this enzyme, which is of widespread importance in nervous transmission, will be less useful for the diagnosis of carbofuran and dimethoate poisoning in larvae than for adults.[359] As the activity of AChE in homogenates of untreated honeybee larvae was highly variable, it does not appear to be as reliable a criterion for detecting insecticide poisoning of larvae as is the reduction in larval weight.

Achik et al.[4] tested the effects of the physical and chemical properties of clay-loam and sandy-loam soils on the LC_{50} values of the carbamates carbofuran, carbosulfan, and thiofanox, and terbufos, an organophosphorous compound, on the springtail *Folsomia candida*. The order of toxicity (terbufos > thiofanox > carbofuran > carbosulfan) was independent of soil type, but the LC_{50} values of carbofuran, in particular, and of the four insecticides, generally, were increased in clay-loam soil compared with sandy-loam soil. This effect was suggested to be due to the greater sequestering capacity of the higher organic matter content in clay-loam soil, which has the effect of reducing the availability of the insecticide for toxic effects. In spite of significant rainfall, the toxicity of carbofuran, carbosulfan, and terbufos continued to cause about 80% mortality in the field for nearly 15 weeks following the incorporation of these insecticides into the sandy-loam soil.[4] Carbosulfan, which had the lowest acute toxicity of the materials tested by Achik et al.,[4] had the lowest water solubility and retained its highly persistent toxicity to Collembola for more than 19 weeks. This result is comparable with the highly toxic terbufos.

E. PYRETHROID INSECTICIDES

Pyrethroid insecticides are characterized by high knockdown and immediate lethal activity, a wide spectrum, and good residual activity, together with repellent and antifeeding activity. Pyrethroids have been widely used for plant protection and in public health and vector pest control.[239,691,743] Deltamethrin has been used for aerial drift spraying against the tsetse fly (*Glossina* spp.) in several African countries. Drift spraying of deltamethrin resulted in a broad-spectrum kill of tree-canopy and grassland insect species, as well as of aquatic Crustacea.[75,1390,1465] However, Inglesfield[743] concluded in his review that application of permethrin and cypermethrin at recommended field rates will have no significant effect on populations of soil microarthropods. Such a lack of effects can be largely ascribed to the limited persistence of these compounds in soil and to their lack of mobility in soil and their strong sorption to organic matter. Pyrethroids can be considered acutely highly toxic to honeybees, but when taking the application rates into account, they are significantly less hazardous. The field observations have mostly indicated the repellency of bees from the crop. In cereal crops any effects of pyrethroids on predatory and parasitic arthropods are limited both in magnitude and duration. In Canadian forests the effects on nontarget species were short lived, although a significant knockdown was observed immediately following the application. The results on the use of pyrethroids against tsetse flies in Africa seem to confirm the minimal effects on terrestrial nontarget insects.[743]

F. OTHER INSECTICIDES

Diflubenzuron, 1-(4-chlorophenyl)-3-(2,6-difluorobenzoyl) urea, is an antimetabolite insecticide that acts primarily as a stomach poison. Since it acts by interfering with the deposition of insect chitin,[1092] it can be used against all stages of insects that are known to form new cuticles. When forest soil was sprayed directly with diflubenzuron, no residues were found after 2 months.[4] Extensive deposition of diflubenzuron on pine foliage was observed when the insecticide was applied by helicopter against the pine sawfly, *Diprion pini*,[4] but it was found not to be very persistent in water. Diflubenzuron has also been used to control the pine looper moth (*Bupalus piniarius*) in a Scots pine stand in Finland.[1105]

Arsenic compounds are serious environmental contaminants and some of them, e.g., calcium arsenate, have been used as insecticides.[188] For example, arsenic compounds are used in the pest control of coca plantations on the slopes of the Andes and cause unpredictable effects in the western Amazonian lowlands.[1318] The global anthropogenic input of arsenic compounds into soils has been estimated at 5.2 to 11.2×10^7 kg/year.[1141]

Goldstein and Babich[551] evaluated a *Drosophila* assay that yields data about acute toxicity to adults and about developmental toxicity to the offspring. After 4 days a reduction in survival of about 10% was noted for flies exposed to 0.25 to 0.75 m*M* arsenite or arsenate and of about 15 and 25% for flies exposed to 1.0 m*M* arsenite and arsenate, respectively. After 7-days exposure, the 7-day LC_{50} values were 0.54 m*M* arsenite and 0.79 m*M* arsenate. Similar studies with white-eyed mutants also supported the greater toxicity of arsenite than of arsenate. The number of adult flies emerging from the control and arsenic-amended media was determined 3 weeks after initial introduction of the parental flies into the vials. Arsenite was more toxic than arsenate, with the LC_{50} values for the developmental exposure assay being 0.21 m*M* arsenite and 0.45 m*M* arsenate. Arsenite was more toxic than arsenate both to the survival of adult flies and to the developmental processes from egg to young adult. The differential toxicities of these arsenic compounds have been attributed to their distinct modes of action. Arsenite is an enzyme inhibitor, specifically interacting with sulfhydryl groups; arsenate is a structural analogue of phosphate and, as such, is an uncoupler of oxidative phosphorylation.[289,926]

G. HERBICIDES

Atrazine, 2-chloro-4-ethylamino-6-isopropylamino-s-triazine, is a herbicide used for selective weed killing in maize and sorghum crops. It is usually applied at a concentration of 2 kg/ha and has limited vertical diffusion and a short active lifespan in the soil.[1067] The effects of atrazine on soil animals have been studied both in the laboratory and in the field, but with contradictory results. Fox[492] reported a decrease in the number of Collembola, Lumbricidae, and larvae of Elateridae, even 14 months after the treatment of fields with 9 kg/ha of atrazine. Liang and Liechtenstein[933] and Liechtenstein et al.[935] suggested that atrazine and insecticides have synergistic effects on the larvae of Diptera. Popovici et al.[1218] obtained no effect on Nematoda and reported a dose-dependent lethal effect on Protozoa, Acarina, Collembola, Enchytracidae, and adult pterygote insect populations in a maize field treated with 5 and 8 kg/ha of atrazine. Moore et al.[1070] and Mallow et al.[971] reported that atrazine affected the numbers of Acarina and Collembola, but they attributed this to a high natural variation of these populations, due to their aggregated distribution. The only clear effect was the increase in number of the springtail *Tullbergia granulata* following atrazine treatment.

Subagja and Snider[1447] reared two parthenogenetic species of Collembola on brewer's yeast treated with commercial atrazine. At a concentration of 5000 mg/g they found a mortality rate of 30% after 4 weeks for *Tullbergia granulata*, but only 5% after 6 weeks for *Folsomia candida*. Fecundity decreased and instar duration lengthened in both species.

Mola et al.[1067] investigated microarthropod populations in maize fields treated with 0, 2, 4, and 6 kg/ha atrazine in Italy. There was a decrease in the number of microarthropods living in the superficial soil layer only in the field that had never been treated before, especially with doses of 6 kg/ha. Atrazine treatment resulted in group-specific and direct effects. Losses were, for the most part, followed by a recovery within 1 month. The different results obtained in the different areas might be due to the specific binding properties of the soil. However, the effects of atrazine were insignificant at the recommended agricultural dose of 2 kg/ha.[1067] Direct tests performed under controlled conditions indicate a rather high mortality rate for the springtail *Onychiurus apuanicus* (18.7% after 30 days and 46.7% after 60 days), even at a dose of 2.5 μg/g. At 80 μg/g atrazine all specimens of *O. apuanicus* were killed, whereas some *O. armatus* even survived a 60-day treatment with 160 μg/g atrazine. The different sensitivity is confirmed by LD_{50} after 30 days: 17.2 μg/g for *O. apuanicus* and 20 μg/g for *O. armatus*. Fecundity in *O. apuanicus* was unaltered at 2.5 and 5 μg/g atrazine, and there were no effects on oviposition and reproduction of newborns. At a dose of 20 μg/g or greater, oviposition did not occur in either species. The results of laboratory tests confirmed that the herbicide may have a direct effect. There is a decrease in the number of Collembola in fields immediately after treatment, which is almost completely restored after 1 month. Doses corresponding to those used in the field caused direct effects, but they did not influence fecundity and thus should not have long-term effects on the populations.

For herbicides, atrazine, simazine, paraquat, and glyphosate had no significant acute or chronic effects on carabid (*Amara* sp., *Agonum* sp., *Pterostichus* sp., *Anisodactylus* sp., and *Harpalus* sp.) longevity or food consumption during one year after exposure to initial field-rate applications.[195] Simazine and atrazine had a repellent effect on carabids, which lasted approximately three days in greenhouse studies. Large carabids did not return to paraquat- and glyphosate-treated fields until about a month after application.

The use of herbicides affects the insect fauna in cereal ecosystems.[258,464,1225] Chiverton and Sotherton[258] investigated the effects on beneficial insects of the exclusion of herbicides from cereal crop edges in Hampshire, U.K. Unsprayed headland plots of a spring wheat crop supported higher densities of both weeds and nontarget arthropods. These plots also contained higher densities of predatory groups, especially the polyphagous species. The carabids *Pterostichus melanarius* and *Agonum dorsale* had taken significantly more meals, and preyed upon a wider variety of arthropod food, in untreated plots. The number of eggs per female was lower in the carabids collected from sprayed plots.

H. FUNGICIDES

The insecticidal properties of foliar fungicides have been reported for a range of pest species and beneficial species.[1405,1526] Much of the work concerning the effects on beneficial species has, until recently, primarily been based on laboratory screening against the predators and parasites of orchard, vineyard, and glasshouse crop pests.

Sotherton and Moreby[1406] investigated the effects of foliar fungicides on beneficial arthropods in wheat fields. These included the natural enemies of cereal aphids; for example, the polyphagous and aphid-specific predators and parasitoids that can reduce the frequency and size of aphid outbreaks. They carried out replicated field trials over a 2-year period in order to quantify the insecticidal properties of three foliar fungicides commonly used in wheat fields. Pyrazophos significantly reduced the numbers of many groups of the natural enemies of cereal aphids in crops of winter wheat and the insects known to be vital in the diet of gamebird chicks, for up to 4 weeks after spraying. On plots sprayed with pyrazophos, 19 species of predatory Coleoptera were observed dead on the soil surface immediately after the spraying, including many larger nocturnal carabid species (*Pterostichus*, *Harpalus*, and *Nebria*). The extent of the activity spectrum of pyrazophos against groups of nontarget arthropods, the extent of the insecticidal action of this compound, and the duration of its effects on small plots closely resembled work carried out in the same seasons and on the same farms, but in crops of barley.[1407] No measurable reductions of beneficial insects were found after the use of either triadimefon or propiconazole. However, the possibility of sublethal effects of either propiconazole or triadimefon was not excluded. Removal of predatory arthropods in the first 2 weeks after spraying was responsible for the subsequent increased numbers of cereal aphids on plots sprayed with pyrazophos 3 or 6 weeks later, despite an observed initial knockdown of aphids immediately after spraying. Aphid resurgence was observed on plots sprayed with pyrazophos on both farms in both years. The high numbers of cereal aphids necessitated the use of additional pesticide applications on plots treated with pyrazophos.[1406] Thus, pesticide use may lead to further pesticide use: the so-called "pesticide treadmill."[163]

The physicochemical properties and relative environmental stability of pentachlorophenol (PCP) have been well studied. In order to understand the transfer of pentachlorophenol residues in the terrestrial food chains, Gruttke et al.[567] constructed a model soil food chain of a ruderal ecosystem consisting of a springtail species (*Folsomia candida*) and a predatory carabid beetle (*Nebria brevicollis*) (Figure 21). The insects were fed continuously with food containing [^{14}C]PCP-Na, and after a steady-state concentration relative to the body burden was reached, the radiochemical was withdrawn. A comparison of PCP residues in the two insect groups showed that there was a difference in the body concentration by a factor of about 100 between the collembolas and the predatory beetles. Accumulation in the beetles was relatively low because of low absorption and rather rapid excretion of the radiocarbon isotope. This process was different in the springtails, and there was a larger accumulation. The residue levels of PCP out of the total radiocarbon present in the body of springtails and beetles were 50 and 41%, respectively. The turnover of PCP was comparable to that in aquatic mosquito larvae (33%) and *Daphnia* (55%). The rapid excretion of [^{14}C]PCP-Na and transformation into the conjugated form indicated a single compartment system in *F. candida* and *N. brevicollis*.[567] A significant amount of the radiocarbon was bound to the body tissues of the soil animals. This related to the quantity of PCP bound to the soil and plant tissues analyzed in the microecosystem studies under outdoor conditions.[614] Gruttke et al.[567] thus concluded that the PCP residues of the animals, which were mainly in the bound and conjugated form, will not endanger other organisms feeding on them or their excrements. This assumption is supported by the fact that PCP residues in the soil food chain decreased with increasing complexity of animals belonging to the decomposers and predators group analyzed in the model ecosystem.[567,614]

XI. INDUSTRIAL CHEMICALS

A. BACKGROUND

A large number of toxic chemicals are emitted from factories, or accidentally released from industrial sources, into terrestrial ecosystems, but less attention has been paid to this than in aquatic ecosystems. The chemicals studied in respect to insects include polychlorinated biphenyl congeners and dioxines. The appearance of polychlorinated biphenyls was first identified in Scandinavia in aquatic ecosystems and in eagles.[775,776]

B. POLYCHLORINATED BIPHENYL CONGENERS (PCBs)

Polychlorinated biphenyls (PCBs) are ubiquitously present in the environment as a result of their use in electrical equipment, paper, plastics, adhesives, paints, waxes, and many other products. Widespread environmental contamination can also result when PCB-containing waste oils are applied along unpaved roads or other areas for dust control. PCBs are readily absorbed by soil constituents. Plants can absorb PCBs in low amounts. PCBs appear to have some effect on photosynthesis and respiration.[1443] Considerable variation among plant species was found in the concentration of PCBs in foliage. Less than 1% Aroclor® 1242, one of the slightly more water-soluble groups of PCB isomers, was found to be taken up through the roots of goldenrod and corn plants.[198]

Anderson and Wojtas[30] reported considerable concentrations of PCBs in honeybees sampled throughout Connecticut. Morse et al.[1081] analyzed PCBs extensively in the U.S. in honeybees, honey, propolis (the tree gums and resins that bees collect to waterproof and varnish the hive interior), and related samples. The highest concentrations were detected in wax (0.74 μg/g). They

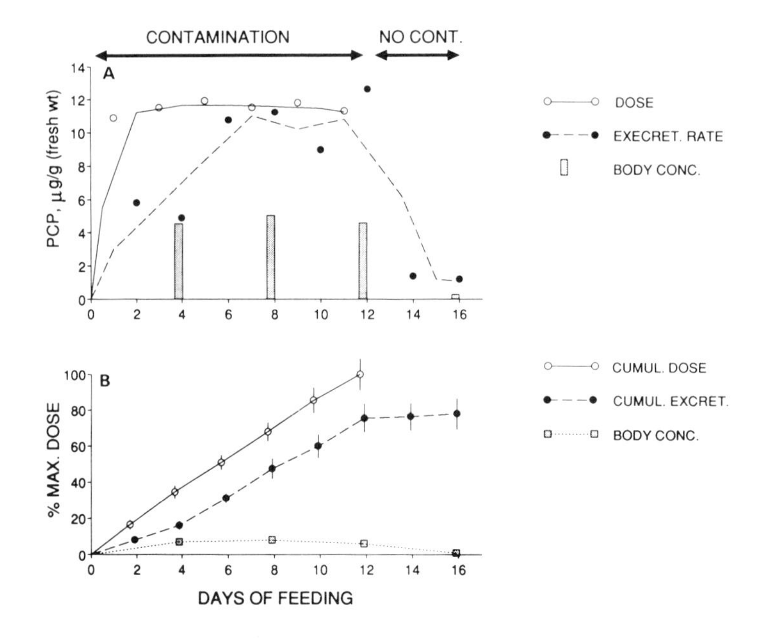

FIGURE 21. Dose and excretion rate every second day, and body concentration (A); and cumulative doses, cumulative excretion, and body concentration (B), of [^{14}C]pentachlorophenate in ground beetles (*Nebria brevicollis,* Coleoptera, Carabidae) after feeding on [^{14}C] pentachlorophenate-contaminated springtails (*Folsomia candida*, Collembola, Isotomidae). (From Gruttke, H., Kratz, W., Weigmann, G., and Haque, A., *Ecotoxicol. Environ. Safety,* 15, 253, 1988. With permission.)

found no relation either between the PCB content of the bees and their manner of collection or between the concentration of PCBs in bees and the time of year sampled. No detectable concentrations of PCBs (less than 0.1 µg/g fresh weight) were found in any of the honey samples from New York State and Vermont. Among other insects analyzed by Morse et al.,[1081] the wasps, which are predatory insects, showed the highest PCB levels, perhaps because they acquire it during consumption of PCB-containing prey insects. PCBs have also been shown to increase the toxicity of organophosphorous insecticides to house flies.[515]

C. DIOXINES

The toxin 2,3,7,8-tetrachlorodibenzo-p-dioxin (TCDD) is known to occur both as a contaminant of products made from trichlorophenol and as a byproduct of the low-temperature incineration of wastes containing chlorinated precursors. The fate of TCDD near Seveso, Italy, following an industrial accident in 1976 resulted in the contamination of 1800 ha of land in an industrial-agrarian community. The magnitude of environmental contamination by the 2,3,7,8-TCDD isomer is the subject of intense debate. Although a number of TCDD sources have been identified, environmental monitoring programs for TCDD have generally been unsuccessful in documenting contamination.[1668] Additional data on the distribution of TCDD in natural ecosystems may soon be forthcoming as a result of the continued development of sophisticated instrumentation for detecting TCDD in picogram (1×10^{-12} g) quantities.

Young et al.[1668] reviewed the results of a long-term field study of an ecosystem contaminated with TCDD. The 15-year study focused on a 3.0-km^2 military test area in Florida that was aerially sprayed with massive quantities of the herbicides 2,4,5-trichlorophenoxyacetic acid (2,4,5-T; 73,000 kg) and 2,4-dichlorophenoxyacetic acid (2,4-D; 77,000 kg) in the course of developing defoliation spray equipment for use in the Vietnam War during 1962 to 1970. Significant concentrations of TCDD have been present in the soils for at least 20 years. The results suggested that less than 1% of the approximately 2.8 kg of TCDD spread over the test area persisted in the soil. Detectable residues (up to 2.9×10^{-9} µg/g TCDD) were found in 32 different animal species, including insects. Biomagnification did not occur. Organisms that came into direct contact with TCDD-contaminated soil generally became contaminated themselves. Crickets, ground spiders, and soil-borne insect grubs were significantly contaminated, while grasshoppers were not. The role of insects in contaminating birds and toads was not clear. The ecological studies demonstrated no significant, adverse, chronic toxic effects of TCDD in animal populations exposed to soil concentrations of TCDD in the range of 0.1 to 1.5 ng/g.[1668]

XII. FUELS

Studies on aquatic arthropods and other organisms[158] indicate that fuels and their components have significant acute and chronic toxicities in ecosystems. However, little research has been conducted on the impact of fuels and their related components on terrestrial ecosystems and/or specific animals within them (see Chapter 3, Section III.C).[11a]

Bombick et al.[158] evaluated the acute toxicity of a variety of Air Force jet fuels (JP-4, JP-8, and JP-9) for several terrestrial insects. JP-4, which consists of both gasoline and kerosene fractions, has been the predominant Air Force aviation fuel for the past 30 years. Both JP-4 and JP-8 are distillation products of crude petroleum and are composed primarily of aliphatic hydrocarbons. JP-9 is a totally synthetic cruise missile fuel. In general, shale fuels were more toxic than their petroleum-derived counterparts. The most toxic fuel was the shale-derived JP-8. The increased toxicity of shale fuels may be due to the presence of additional hydrocarbons or other compounds that are not removed in the shale extraction technique. The order of decreasing susceptibility to the petroleum-derived JP-4 was earwigs (*Forficula auricularia*), rice weevils (*Sitophilus oryzae*), flour beetles (*Tribolium confusum*), lady beetles (*Hippodamia convergens*), tenebrionid beetles (*Tenebrio molitor*), and cockroaches (*Blaberus cranifer*). However, species response varied with different fuel types.[158]

Insects with higher hydrocarbon content (cockroaches) or harder wax layers (tenebrionid beetles) were less affected by jet fuel. Bombick et al.[158] suggested that the chemical composition of the insect cuticle is an important factor determining susceptibility to the toxic effect of a fuel, because it is the initial site of contact in any topical exposure. The rate of penetration of a nonpolar and lipophilic substance into an insect is directly related to the hydrocarbon content of the insect cuticle. Nonpolar compounds do not readily penetrate the insect cuticle[185] because the wax layer on the cuticular surface acts as a primary sink for the toxicant.[1157] Insects with higher concentrations of cuticular hydrocarbons are therefore less susceptible to nonpolar toxicants than those with less. Hard waxes, waxes with high molecular weights, or waxes with high melting points reduce the permeability of insect cuticles.[92]

Evaporation plays a critical role in the relative toxicities of fuel for insects.[158] When insects were exposed to jet fuels that had been evaporated, the toxicities were greatly reduced, with the exception of JP-8. The fuels, except for shale JP-8, were more toxic at higher temperature and low humidity. An increase in temperature results in changes in the cuticle that increases permeability and, possibly in insects, increases susceptibility to the fuel. Higher temperatures increase the metabolic and physical activity, resulting in increased spiracular opening and ventilation. Since the toxicity of the fuel depends on contact with the surface cuticle or trachea, the presence of water molecules might reduce the toxic effect of the fuel.[158]

XIII. LIGHT

Outdoor lighting has drastically increased during the last few decades. This may have contributed to the decline of some nocturnal insects.[493,1243] For example, Gepp[535] reported that an illuminated advertisement comprising of three 2-m-high letters situated at a height of 35 m on the top of a building attracted 350,000 insects in a single year, and a strongly illuminated industrial hall attracted 100,000 insects per night at Graz, Austria.

Frank[493] assessed, on the basis of diverse literature, the impact of lighting on moths. He concluded that outdoor lighting disturbs the flight, navigation, vision, migration, dispersal, oviposition, mating, feeding, and crypsis of some moths, and possibly also circadian rhythms and photoperiodism. Moths attracted and grouped around lights are exposed to increased predation by birds, bats, geckos, spiders, skunks, toads, and other predators. For example, Rydell[1315] reported that during spring and autumn, numbers of insectivorous northern bats (*Eptesicus nilssoni*) were observed along rows of street lights in southern Sweden. He concluded that by attracting insects, artificial lights may provide local patches of food for some species of bats during periods that may be critical for their survival and reproduction. On the other hand, electric lighting may also increase moth populations by suppressing parasitoid populations.[493,1653] The results of several long-term lepidopterological investigations have shown no population changes in spite of the use of light traps killing vast numbers of moths. Elaborate efforts to use light traps for pest control have failed.[493,681]

The effects of lighting vary depending on species, type of lamp, habitats,[493] and latitude. Each insect species obviously has its own mode of spectral orientation, although there are a few characteristics that appear to be typical of insect groups. For example, the attractiveness of yellow-green radiation was weaker than that of near-UV radiation to Trichoptera, Lepidoptera, and Coleoptera, but was strongly attractive to Ephemeroptera and Ichneumonoidea (Hymenoptera).[1047] Although a number of moth species are commonly found in illuminated urban environments, there are no documented cases of extinction due to electric lighting.[493] However, moths living in small, endangered habitats may be severely affected. Low-pressure sodium lamps are less likely than mercury lamps to elicit flight-to-light behavior.[493] There are also great between-species differences in flight to light. The simultaneous use of suction traps measuring aerial moth densities with light traps suggested that, for instance, among British Noctuidae, *Xestia c-nigrum* is 5000 times as likely to fly to light as is *Amphipyra tragopoginis*.[1470] Although there are no records of evolutionary changes in flight-to-light in moths, the reduction of moths flying to urban lamps is consistent with this idea, although other explanations exist.[493]

XIV. TEMPERATURE

One consequence of the burning of fossil fuels is the production of heat. This is probably most important on a local scale, where it can be identified in the form of urban heat islands. Urban surfaces also absorb more solar radiation than rural surfaces, because a higher proportion of the reflected radiation is retained by the high walls and dark-colored roofs of city streets. The heat is stored during the day and released at night. Air temperatures in urban areas are measurably higher than in their surroundings, and the intensity of the difference tends to increase with the size of the urban agglomeration.[495] However, even quite small centers have a heat island. The maximum thermal modification has been reported to be about 12°C.[555,1152]

The importance of two thermal constants, the lower development threshold LDT (temperature when development ceases) and the sum of effective temperatures SET (number of day degrees above LDT for completion of a development stage), for understanding insect life histories has long been recognized. Both reflect the temperature dependence of the rate of ontogenetic development. LDT and SET are good predictors of the timing of life-history events, particularly in eggs and pupae whose development depends only on the rate of intrinsic processes.[703]

XV. RADIATION

A. INSECT TOLERANCE

Terrestrial invertebrates are highly radioresistant in the adult stage, but less so than unicellular organisms.[68] Lethal doses for insects, earthworms, arachnids, and Oniscoidea were 0.5 to 2 Gy, which is much higher than for plants or vertebrates.[884] According to published data, exposure to doses ranging from several to tens of Gy caused male and female sterility in *Lymantria dispar* (Lepidoptera, Lymantriidae) and death of the eggs in a number of invertebrate species, under laboratory conditions. In natural conditions other adverse factors, e.g., low temperatures combined with irradiation, may damage the animals at much lower doses than irradiation alone.[884]

Extensive research has been conducted over the past 35 years on the effects of gamma irradiation on fruits, vegetables, and other food commodities. Much of this research has been done to determine whether radiation would be an effective technology for reducing or eliminating insect and pathogen infestations. Doses of less than 1000 Gy can be used for insect disinfestation, as well as to prevent sprouting and delay ripening of some plant commodities. However, doses as low as 100 Gy or less, when applied to insect eggs or larvae, will prevent adult eclosion or result in the eclosion of abnormal adults unable to reproduce.[204]

In addition to total dose, the dose rate affects insect development.[191,1110] Brown and Davis[191] reported that mortality of 18- or 42-h-old eggs of the red flour beetle was higher when exposed to 25 Gy from a ^{60}Co source at the rate of 26 Gy/min than at the rate of 4 or 11 Gy/min. Burditt et al.[204] demonstrated a differential effect when mature, nondiapausing, cocooned codling moth (*Cydia pomonella*) larvae were exposed to gamma irradiation from a cobalt source at dose rates from 1 to 204 Gy/min. There were significant differences at the high- and low-dose rates in the percentage of larvae that pupated, but were unable to continue their development and emerge as adults (Figure 22). However, organisms throughout the phylogenetic continuum are stimulated by small doses of ionizing radiation. Radiation hormesis is well documented in insects, whose growth rate, life span, and reproductive capacity are increased.[950] For example, males of the larch bud moth *Zeiraphera diniana* (Tortricidae) showed increased longevity following 20 to 40 kR gamma rays.[121] *Acheta domestica* (Orthoptera) and *Tenebrio molitor* (Coleoptera) showed increased longevity when exposed to low doses at 3 kR/min.[1036]

B. FIELD EXPERIMENTS ON SOIL INVERTEBRATES

Soil invertebrates may account for about 80% of the ^{90}Sr and 90% of the Ca involved in the biological turnover through animals. In actual fact, soil dwellers appear to account for 95 to 99% of all the radionuclides accumulated by animals in proportion to their biomass.[885] Experimental laboratory studies have revealed that insects respond to radioactive pollution, although they are more radioresistant than vertebrates.[864,883] In some cases this resistance is explained by the shielding effect of soils.

The effect of an increased level of radiation of 0.5 to 40 Gy/h caused by a natural radionuclide, ^{226}Ra, on soil invertebrates was studied under field conditions in the middle taiga with meadow vegetation during several seasons.[883] The number of dipteran and elaterid larvae was lower on the plots with a high-radiation background. The mean number of mature oribatids was 72 individuals per cubic decimeter in the control against 39 individuals per cubic

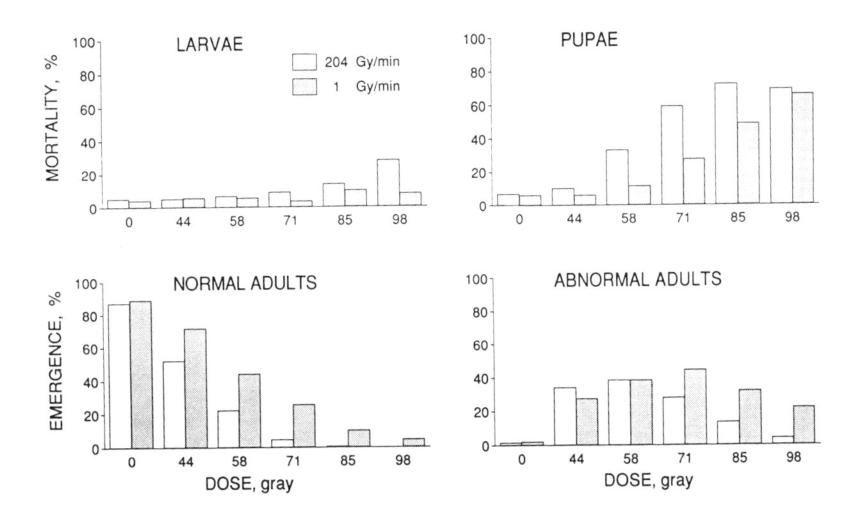

FIGURE 22. Effects of gamma irradiation on subsequent pupation and adult emergence of mature, cocooned, nondiapausing codling moth larvae (*Cydia pomonella*, Lepidoptera, Tortricidae) reared on media and treated at dose rates of 1 and 204 Gy/min. (From Burditt A. K., Jr., Hungate, F. P., and Toba, H. H., *Radiat. Phys. Chem.*, 34, 979, 1989. With permission.)

decimeter on a radium-polluted plot, but no faunistic differences were observed. Earthworms proved to be particularly sensitive to an increased Ra background in their numbers, individual size, and reproduction. Changes in the structure of the midgut epithelium were also reported.[883] A similar experiment using ^{241}Am also resulted in a decrease in the abundance and biomass of Lumbricidae.[886] However, the abundance of predatory beetles increased, possibly due to improved feeding conditions.

Krivolutsky[884] investigated terrestrial invertebrates on experimental plots polluted with ^{90}Sr, ^{137}Cs, ^{144}Ce, ^{106}Ru, ^{95}Zr, ^{65}Zn, ^{125}Sb, and ^{239}Pu. In ^{90}Sr-polluted forest plots (8×10^{10} to 1.2×10^{11} Bq/m^2), mesofauna numbers declined significantly. The greatest reductions in numbers (10 to 100 times) were recorded in earthworms, Diplopoda, and Chilopoda dwelling largely in the upper soil layers where the bulk of the radionuclides are accumulated. Changes in the species composition, in particular a reduction in species diversity, were also found in the microfauna, e.g., in Oribatei. Notably less marked changes were found in predatory beetles, flying insects, or species passively distributed by other animals. When 13 years had elapsed after exposure of the forest litter to ^{90}Sr, some of the ^{90}Sr (up to 5 to 10%) was absorbed by the above-ground parts of woody plants, but the bulk remained in the topsoil, having migrated only down to a depth of 10 cm. The ^{90}Sr levels in invertebrates of the upper forest layers were much lower than in those of soil dwellers. Exposure to these doses resulted in a decline of the populations of several species. This may be due to the fact that several invertebrates overwinter in the soil at the egg stage, which is considered to be more sensitive to radiation than other stages.[884]

C. EFFECTS OF RADIOACTIVE WASTE ON INSECTS

As a part of a study of radionuclide migration through biota associated with mine wastes, Clulow et al.[283] studied ^{226}Ra levels in samples from an inactive uranium tailings site at Elliot Lake, Ontario, Canada. The mean ^{226}Ra levels were 9.1 mBq/g dry weight in the substrate, 62 mBq/g dry weight in rye (*Secale cereale*), less than 3.7 mBq/g dry weight in oats (*Avena sativa*), the dominant species established in revegetation of the tailings, and 117 mBq/g dry weight in washed and unwashed black cutworm (*Agrotis ipsilon*, Noctuidae) larvae. The concentration ratios were vegetation to tailings, 0.001 to 0.007; black cutworms to vegetation, 3.6; and black cutworms to tailings, 0.01. Because the levels of ^{226}Ra in the substrate were at least two orders of magnitude higher compared to those in the plants, ingestion of tailings would contribute to higher levels in the black cutworm larvae. The values were considered too low to be a hazard to herring gulls (*Larus argentatus*), which were observed feeding on cutworms.[283]

D. NUCLEAR ACCIDENTS AND NUCLEAR WEAPONS TESTS

Since the atomic bomb at Hiroshima, the effects of radiation exposure have been of increasing importance for both mankind and ecosystems. It is difficult to precisely predict the ultimate magnitude and extent of global radionuclide contamination following an atmospheric atomic test or a nuclear accident, such as that at Chernobyl or that on Three Mile Island.[489,982,1403] Radionuclides are presumably ejected at the site of such accidents as particles or become attached very early to aerosols and grow by coagulation with other particles during transport. ^{134}Cs and ^{137}Cs are two of the radionuclides released by the Chernobyl reactor.[489] The extent of radioactive contamination in Europe and the meteorological factors that affected it shortly after the Chernobyl accident have been described in detail.[1189]

Rainfall plays a dominant role in the deposition of radioactive particles. It has been reported that radionuclide (e.g., ^{90}Sr) concentration in rainwater shows a marked seasonal variation, with peaks in the late spring and troughs in the late fall. This is now generally believed to be due to a maximum rate of transfer of material from the stratosphere to the troposphere during the spring. Mechanisms for the deposition of submicron particles carrying significant amounts of radioactivity onto foliar surfaces, and physicochemical aspects of the retention, uptake, and desorbability of such particles are not well understood.[489]

There seem to be no papers available about the effects on insects of radioactive pollution originating from atomic bomb or other nuclear tests. However, in the 1961 volume of *Nature*, Kettlewell and Heard[833] reported a migrating moth specimen of *Nomophila noctuella* (Pyralidae) caught in England that contained a strongly radiating particle. It was concluded that the particle originated from the vicinity of the French atomic bomb tests in the Sahara.

E. CHERNOBYL: EFFECTS ON TERRESTRIAL INVERTEBRATES

Assessment of the aftereffects of ionizing radiation on animals and plants and the role of biota in the further redistribution of radionuclides in terrestrial and aquatic ecosystems in the vicinity of the Chernobyl Nuclear Power Station was started in May 1986.[1403]

Genetic effects of ionizing radiation on land invertebrates were studied using three natural *Drosophila* populations located on plots characterized by various levels of contamination.[1403] The flies were caught in July 1986 on the plots exposed to dose rates of 0.8 and 0.002 Gy/h. The highest incidence of dominant lethals was characteristic of those populations from areas with the highest radiation level.

In radioactive fallout the bulk of the fallout products are concentrated in the upper soil layer. Ecological studies on invertebrates showed that radioactive pollution in the nearest zone (litter surface dosage rate of 0.07 to 0.15 Gy/h) had affected the soil fauna. The number of soil mites and the early developmental stages of mesofauna were reduced by a factor of about 30 (as of mid-July, 1986). The reduction in the populations of mesofauna was considerably smaller (two- to threefold) in the fields and in the deep layers of ploughed land, where no single invertebrate group showed any catastrophic drop in numbers. The total exposure dosage of 30 Gy was only 3 to 30% the LD_{50} for the majority of the groups of soil dwellers. However, this dose was sufficient to cause the death of eggs and early larval stages of nearly all invertebrates. Among microfauna the first larval instars of invertebrates in the polluted area were lacking from forest soil, but in ploughed land, the mortality rates were lower. Sokolov et al.[1403] concluded that the absorption of a dose of 29 Gy caused catastrophic changes in microfauna, while doses of about 8 Gy led to minor changes among the fauna of the soil surface. In ploughed soils, even at a total surface dose of about 40 Gy, the deep-soil dwellers were affected only to a small extent. The effects of radiation were most distinct in invertebrates whose breeding or mating periods occurred 1 to 2 months after the accident. The process of recovery of soil fauna proceeded regularly after autumn 1986, and by autumn 1987 both the reproduction and structure of the populations of soil microfauna, even on the most polluted plots, were comparable to the control.

Radioactive pollution has had a considerable impact on other insect fauna as well. For example, Silphidae (Coleoptera) numbers increased sharply, whereas Staphylinidae decreased.[392] The process of ecological shifts associated with the removal of sensitive species and rearrangement of the structure in the irradiated species begins at higher dosage rates of chronic irradiation of 2.6×10^{-4} C/kg/day and over.[1403]

The study of genetic effects caused by ionizing irradiation in terrestrial invertebrates was conducted in natural populations of *Monochamus galloprovincialis* (Cerambycidae) dwelling on fallen pines.[1403] On the experimental plots with dosage rates of 0.04 Gy/h and 0.4 Gy/h, the developmental rates did not differ, and the beetle larvae had no morphological deviations. According to Sokolov et al.,[1403] a mass outbreak of this pest on weakened pine trees may be expected, and as a consequence, the mass death of pine stands in the polluted area may occur within 8 to 10 years.

Teratological effects have been found in 0.3 to 1.5% of the moth larvae of Arctiidae, Noctuidae, and Geometridae near Chernobyl.[392] Increased variability within populations have been reported, for instance, in the ringlet butterfly *Aphantopus hyperantus* (Satyridae), the Colorado beetle *Leptinotarsa decemlineata* (Chrysomelidae), and the seven-spot ladybird *Coccinella septempunctata* (Coccinellidae).[392]

Preliminary analysis of asymmetry in the wings of three species of the suborder Zygoptera (Odonata) revealed a higher level of asymmetry and anomalies of venation in areas with a dosage rate of 0.04 and 0.4 Gy/h compared with the control plots (0.002 Gy/h).[1403] On the other hand, ten genera of Lepidoptera (Pieridae, Nymphalidae, Lycaenidae, Satyridae) showed no morphological deviations, which may be explained by their mobility and migratory activity.

The radioactivity in nocturnal moths was investigated in Finland in 1986 after the Chernobyl accident.[1053] The level of radiation observed was low, of the order of one millionth the radiation that could be acutely toxic for the moths. Local moths that most regularly showed radiation were hairy moths, such as *Achlya flavicornis* (Thyatiridae; mean 25.3 Bq per moth or 240 Bq/g), which might have acquired the radiation from their resting surfaces, i.e., the trunks and branches of trees.

After the Chernobyl accident possible radioactive contamination of honeybees in the U.S. was measured. Honeybees from hives in Oregon, California, Ohio, and New York were obtained during the period April–August, 1986. ^{134}Cs and ^{137}Cs were determined. Of honeybees collected in May and June 1986, only those from Oregon showed detectable levels of ^{134}Cs that could have originated from the Chernobyl incident. In general, the levels of radioactivity were higher in the west coast bee samples compared to those taken in the east. The radio-activity levels detected were considered to be toxicologically of no consequence,[489] while the problems were apparently more serious in Europe. Contam-ination of honeybees by radionuclides presumably relates back to foliar contam-ination. A study of ^{137}Cs contamination of squash and bean plants showed that about 65% of the contamination could be removed by washing, thus indicating surficial residues. The magnitude of contamination sharply diminished on foliar surfaces at heights more than 20 cm above the soil surface.[1594]

XVI. CONCLUSIONS

Exposure to pollution is normally measured as the concentration of pollutant in the ambient air, soil, food, or water, with indication of the duration of exposure. However, the rate of intake depends not only on the amount in the environment, but also on the efficiency of intake.[1075] This information, the mode of uptake and the relative significance of food and skin permeation, is usually missing as to insects.

The effect of a specific treatment on the insect is commonly studied by excluding its interactions with the biotic environment, under laboratory conditions. The results may therefore not be easily extrapolatable to a real field situation where interactions between the treatment and the biotic environment of the target organism may yield results that differ from cases where biotic interactions are excluded.[957] One way of avoiding problems of this type is to include the biotic interactions of the target organism(s) explicitly in the experimental design.[1122]

The impact of pollution on a terrestrial insect population can be negative, positive, or insignificant. A common observation in the vicinity of pollution sources has been a drastic change in the community structure. Depending on the intensity, areal coverage, duration, and frequency of pollution, a various proportion of the species pool is locally, at least temporarily, lost. However, in most cases some insect species have been able to tolerate the pollution and gain from the decline of their natural enemies or from the biochemical changes in their stressed food plants.

The large body of field data indicating that terrestrial herbivores are often positively affected by atmospheric pollution[27,1284] is supported by experimental evidence that has accumulated in the last few years.[1599] Air pollution may affect the quality of the host foliage, they may have a direct effect on the insects, or they may change the efficacy of the natural enemies of the insect species. The importance of these mechanisms varies from case to case and is associated, for example, with the intensity of pollution, the mode of living and feeding of the insect, and the role of different regulating factors in the population dynamics.

The responses of herbivorous insects to sulfur dioxide seem to be positively dose dependent and mainly mediated through host-plant chemistry.[1599] The response of aphids, at least, appears clearly positive to oxides of nitrogen (especially NO), as well. Evidence on the effects of ozone is at yet contradictory, even though there are cases of enhanced performance of herbivores feeding on ozone-fumigated plants.[294,1599,1600] The effects of elevated carbon dioxide concentrations on insects have been positive or insignificant, but responses at the population level are essentially unknown. Research on this area, particularly plant-insect interactions, would be of great importance in understanding the future of biological systems in a high-carbon-dioxide world.[89]

Surprisingly little is known about the effects of acidification on terrestrial insects, compared to the known effects on aquatic species. On some occasions acid mist has been shown to enhance the performance of herbivores, while other studies have given contradictory results. Acidic precipitation can induce significant local and regional changes in soil fauna.

Climate change stands apart from other pollution-related issues in two respects: first, the whole notion is primarily based on uncertain and poorly understood scenarios of future development, not on reports on what has actually happened; second, if the climate change will correspond to expectations, it will be a major factor affecting insect ecology and biogeography in a global scale.

The toxic impact of metal pollution on soil invertebrates, and the consequent interference in litter decomposition, are relatively well known,[706,800] whereas the effects are documented only to a small extent in insects living in the vegetation layer. However, for example, the use of grasshoppers as test animals for the evaluation of soil quality[1341] indicates that considerable changes may take place also in herbivore populations.

Agricultural chemicals are deliberately released in the field, and some of them are intended to kill insect pests. Thus, much of the information on their impact stems from tests on pest animals, while their effects on nontarget animals are less well known. During recent years more attention has been paid to the ecosystem level impacts and effects on nontarget species. Hitherto, assessments of pesticides have focused on individual applications. However, the pattern of local or regional use may influence the effects. In the case of both agricultural and industrial chemicals, there is a need to develop standardized methods of environmental risk assessment also involving insects.

According to the present state of knowledge, radioactive pollution of the environment has the greatest impact on permanent soil fauna, resulting in an appreciable reduction in their population, lower abundance of the population in the deeper-soil horizons, and distinct changes in soil animal communities.[884] The effects on temporary soil invertebrates appear to be determined, to a greater extent, by secondary factors and by the irradiation impacts on predators, parasites, and plants. This may result both in declines and increases in soil invertebrates.[884]

Socially, there is an outspoken demand for generalizations of the effects of pollutants on terrestrial ecosystems, whether or not mediated by insects. Nevertheless, at this state of art most generalizations on the role of insects seem premature because the evidence is simply not convincing. Field patterns of insect distribution and abundance in relation to pollutant levels are very poorly described and documented (cf. natural fluctuations of insect populations). Experimental evidence is largely based on a few herbivorous insects, mainly aphids and a couple of chewing species. The selection of laboratory test species should be widened, and the results should be verified by field experiments and first-rate auditing studies (e.g., on contaminated sites and effects of long-distance pollution). On the other hand, standardization of methods and test species are needed to allow comparisons of effects of different chemicals in terrestrial ecosystems.

Chapter 5

POLLUTION IN AQUATIC ECOSYSTEMS

I. INTRODUCTION

The major water pollutants include organic compounds causing oxygen deficiency, acid precipitation, sediments, agricultural and industrial chemicals, oil, metals, and other harmful elements. However, there are many other pollutants. Hundreds of different pollutants affecting freshwater fish, plants, and invertebrates have been listed.[853,1203,1359] Even chlorinated drinking water and the discharge of chlorinated effluent into waters may cause environmental problems, since the chlorinating of water containing high levels of humic material results in the formation of TCA (trichloroacetic acid) and other chlorinated byproducts. These compounds were found to increase oxygen consumption in dragonfly (*Aeshna umbrosa*) nymphs at concentrations of 100 to 1000 μl/l.[224,393] As pollution control measures have greatly reduced the incidence of chronic discharges into receiving waters of developed countries, the importance of episodic pollution, such as industrial discharges and farm wastes, has increased. Such incidents also highlight the role of peak, compared to mean, pollutant concentration.[1012] Furthermore, thermal and radioactive pollution can occur.

Although only approximately 5% of insects spend all or part of their life cycle in an aquatic environment, waters have been studied at least as much as terrestrial ecosystems in respect to insect ecotoxicology. Several insect orders (Ephemeroptera, Odonata, Plecoptera, Trichoptera) and families (e.g., Blepharoceridae, Culicidae, Chironomidae, and Simuliidae of Diptera; Nepidae, Notonectidae, and Corixidae of Heteroptera; Dytiscidae of Coleoptera) are more or less confined to water in their young stages. Although many insects live in fresh water, and some in brackish water and intertidal zones, there are only a few species living in the open ocean.

Polluted systems have specific benthic macroinvertebrates that are uncommon in unpolluted waters. One reasonable hypothesis is that these species occur in the most enriched subhabitats of unpolluted systems.[785] Chironomid taxa, which typically increase in numbers in eutrophic conditions, probably occupied the least oligotrophic subhabitats in the Georgian Bay. The relative abundance of *Chironomus* and *Procladius* spp. in cool water increased with an increasing proportion of organic matter in the sediments.

Although most aquatic bioassay evaluations have been concerned with acute toxicity, the detrimental effects of many substances may not be evident for weeks, months, or longer. Such chronic toxicity may be related to changes in appetite, metabolism, morphology, growth, reproduction, development of sex products, maturation, hatching, survival of different life stages, deformities, behavior, or other vital functions that do not result in early death.[866]

Observations on the distribution and community structure of benthic macroinvertebrates in response to abiotic gradients in a relatively unpolluted system may assist in interpretating observations made in enriched systems where the influences of natural and cultural factors may often be confounded.[785] The use of aquatic insects as indicators of water quality is dealt with in several studies.[64,508,766–768,901,1065,1074,1216,1658]

Long-term fluctuations make it difficult to identify the impact of increasing human activity on waters.[526] In shallow eutrophic Lake Myvatn, Iceland, Gardarsson et al.[526] studied the variation in the abundance of algae, crustaceans, and chironomids over the past 2300 years in sediment cores. Densities of the benthic alga *Cladophora aegagropila* and associated chironomids *Psectrocladius barbimanus* and *Eurycercus lamellatus* increased with decreasing water depth. The benthic chironomid *Tanytarsus gracilentus* showed an opposite trend. On a shorter time scale population changes, as documented by harvest records (e.g., ducks, fish) and monitoring studies (e.g., ducks, aquatic insects), are considerable and are often associated with food resources.

II. EUTROPHICATION

A. BACKGROUND

Cultural enrichment of aquatic environments is a widespread phenomenon in the world, including several classic cases such as the Great Lakes, the Lakes at Madison, WI, and Lake Washington, WA. The eutrophication process is affected by the input of both organic matter and mineral nutrients.[502,1111,1148] It is most frequently caused by the fertilization of water with excess nutrients (particularly phosphorus) in sewage that contains detergents, human and animal wastes, as well as agricultural runoff contaminated by fertilizers.[502] These waters, known as eutrophic waters, have a good nutrient supply and a high rate of productivity. This contrasts with oligotrophic waters that are unproductive due to the restricted availability of nutrients. Mesotrophic waters are intermediate between these two states. Eutrophication is the process by which an aquatic ecosystem increases in productivity as a result of an increase in the rate of nutrient input.[502] Eutrophication is characterized by an increase in phytoplankton, causing algal blooms. Other symptoms may be alterations in algal flora, and the vigorous growth of vascular plants. These changes are accompanied by secondary changes, such as poor oxygen status of deep water.[502]

The effects of eutrophication on aquatic insects may be profound. For example, in summer 1953 there was a period of 10 days of hot, calm weather that resulted in the stable thermal stratification of the water column in Lake Erie. Because of the large demand for oxygen in the decomposition of organic material in the hypolimnion, widespread anoxia developed in deep water. Prior to this, the benthos was dominated by mayfly larvae, especially *Hexagenia rigida* and *H. limbata*. In 1929 their density was about 397 m^{-2}, and in 1942–1943 they averaged 422 m^{-2}. The density of *Hexagenia* was 300 m^{-2} just before the severe stratification, but it collapsed to only 44 m^{-2} in September. The density remained small, and by 1961 these insects had almost disappeared. The low-oxygen-tolerant benthos was dominated by tubificid worms, chironomids, and molluscs.[99,626,1591]

The invertebrate fauna usually recovers, at least to some degree, as the discharge of waste waters decreases.[784,1519] Physical, chemical, and biological monitoring has indicated that water quality in the Mississippi River at St. Paul has improved over the past five decades. In 1926 the August mean values of dissolved oxygen ranged from less than 1 to 2 mg/l in the river reach near St. Paul, while in 1987 dissolved oxygen values in the same area were 7 mg/l or greater.[784] In addition to several other species, the pollution-intolerant *Hexagenia* mayfly (species not reported) returned to the Twin Cities stretch of the Mississippi River in the 1980s after an absence of 50 years.[507,784] *Hexagenia* spends almost its entire lifespan burrowing in the sediments of lakes and large rivers where it is susceptible to low, dissolved oxygen concentrations and toxins. The response of *Hexagenia* to improved water quality has been so dramatic that the mayflies have become a nuisance: on June 23, 1987, snow-plows were needed to clear *Hexagenia* from the Mississippi River in St. Paul.[784]

Pearson and Penridge[1185] showed that the discharge of organic effluent from a sugar mill into a stream in tropical north Queensland had similar effects on the macroinvertebrate fauna to those found in temperate streams: increased pollution led to decreased diversity. A pronounced impact of the mill discharge on the fauna (reduced diversity and changed composition) was greatest at some distance downstream of the discharge, where maximum depression of dissolved oxygen occurred. Further downstream the fauna recovered owing to the dilution afforded by the confluence of other rivers. When the dissolved oxygen concentration fell to between about 3.5 and 5.0 mg/l, the predominant taxa were those that elsewhere include pollution-tolerant species (especially Chironomidae and Oligochaeta); below 3.5 mg/l the only taxa that were present were Oligochaeta, *Chironomus* sp., and air-breathers. *Chironomus* sp. only occurred where oxygen levels were depressed, and the other chironomids tended to disappear under the same conditions.

The effects of eutrophication on insects can also be mediated via vascular plants (biochemistry or growth form). However, changes in the plant-insect interface caused by eutrophication have received little attention. The protein, amino acid, and sterol concentrations in the leaves of the water hyacinth *Eichornia crassipes* were higher in plants collected from water bodies contaminated by starch and brewery industries. Jamil and Jyothi[761] found that the reproductive potential of the weevil *Neochetina bruchi* was considerably enhanced when fed on water hyacinths from polluted water bodies. The total number of eggs laid was almost doubled.

B. MUNICIPAL WASTE AND URBANIZATION

High human population densities may dramatically alter the watercourses that drain the urbanizing catchments. Serious impacts occur when poorly treated waste is disharged into watercourses from domestic or industrial sources. In most developed countries the direct discharge of waste is controlled through the introduction of sewage collection systems and modern treatment processes. Sewage is generally collected from large areas and treated at a centralized treatment plant whose effluent can be monitored,[796,1317] although storm overflows may happen.[140]

Suspended-sediment loads from urban areas into waters are often considerable. Sediment is an important pollutant of water in the U.S. and elsewhere and represents a pervasive threat to aquatic ecosystems. Sediment reduces primary production by decreasing light penetration, abrading and disrupting cells, smothering respiratory surfaces, and sorbing or binding nutrients or essential elements.[454] Sedimentation reduces the numbers and diversity of invertebrate populations and fish by filling in interstitial gravel spaces and destroying habitats,[196,1145] smothering gill lamellae and ova,[925] and imposing physiological stress that results in reduced growth, higher mortality, and changes in behavior.[454]

Several researchers have studied the impact of watershed urbanization on the benthic fauna of streams. Benke et al.[118] investigated 21 watersheds in the Atlanta area and found a negative relationship between the number of species (and families) of benthic organisms and the degree of urbanization of the watershed. In a California creek urbanization was associated with a decline in pollution-sensitive groups, such as mayflies, caddisflies, and amphipods, and a dramatic increase in oligochaetes.[1208] The number of taxa was 15 to 30 in the upper unurbanized section of the creek and 5 or fewer in the urbanized section. Duda et al.[403] found that the number of benthic macroinvertebrate taxa in a stream in North Carolina decreased drastically after it passed through an urban area. Urbanization was associated with a decline in sensitive mayflies, stoneflies, and caddisflies and a pronounced increase in midges and oligochaetes. The severe alteration of the urban stream fauna was due to leaking sanitary sewers and illegal discharges. DiGiano et al.[379] reported that the benthic insect fauna of a Massachusetts creek showed a dramatic shift from a diverse assemblage of mayflies, stoneflies, caddisflies, and midges to one almost totally dominated by midges, when the stream passed through an urban area where the stream was bordered by separate sanitary sewer pipes and impacted by combined sanitary and storm sewers.

Several studies[118,379,403,796,1208] have demonstrated that the diversity and/or richness of aquatic insects decreases, and chironomid dominance increases, with increasing urbanization. Jones and Clark[796] compared the benthic insects of several watersheds from unurbanized to highly urbanized areas in Virginia. Watershed urbanization caused marked changes in the composition of stream insect communities (Figure 1). Tolerant taxa, such as some chironomids, increased in abundance, while nontolerant taxa decreased or were eliminated. Highly urbanized streams were strongly dominated by Diptera, with Trichoptera being codominant in some cases. Less-urbanized streams were characterized by the presence of a large number of genera belonging to Diptera, Trichoptera,

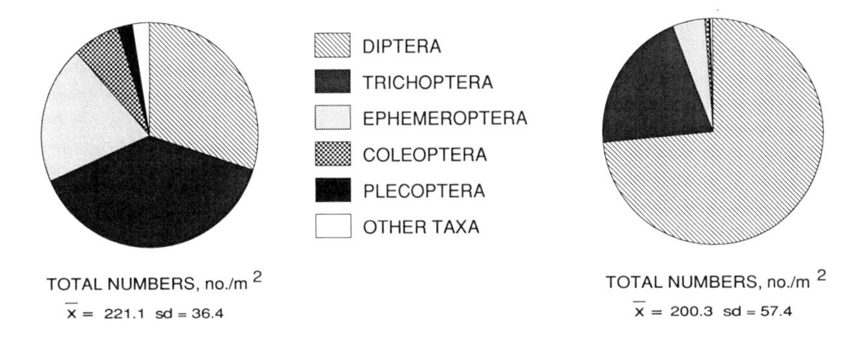

FIGURE 1. Comparison of relative abundances of benthic insect taxa characterizing less urbanized (human population density <10/ha) vs. more urbanized (>10/ha) streams. (Data from Jones and Clark.[796])

Ephemeroptera, Coleoptera, Plecoptera, and Megaloptera. The genera *Cricotopus* and *Orthocladius* were most common in the urbanized streams, while several insects, such as *Constempellina*, *Hydropsyche*, *Stenelmis*, and *Stenonema*, were virtually absent from all moderately-to-highly urbanized streams.[796] In Japan Sasa[1328] reported that the presence of *Chironomus yoshimatsui* larvae indicated that the water was highly polluted with sewage contamination. *C. yoshimatsui* was found in running waters, while *C. samoensis* was found in polluted, stagnant waters.

Dumnicka and Kownacki[405] studied a river ecosystem in a polluted section of the upper Vistula, Poland, where species composition, numbers, and predominance structure of macroinvertebrates indicated heavy pollution. The fauna was represented by only 40 taxa: chiefly Oligochaeta and Chironomidae. The absence of Ephemeroptera, Trichoptera, and Plecoptera, characteristic of pure or slightly polluted rivers, was noted. Janeva[762] investigated the benthic invertebrates in the River Vit in Bulgaria. At the most-polluted sites undiluted organic substances prevented the breathing of the mayfly and stonefly species. Plecoptera and Simuliidae were sensitive to pollution, as well as Trichoptera, except for the genus *Hydropsyche*. The Chironomidae were considerably more tolerant to the increased pollution than were the other benthic groups.

Grzybkowska[568] estimated production and species composition of Chironomidae in cross-sections of the lower course of the rivers Widawka and Grabia in Poland. The discharge of the Widawka had increased by 75% in comparison with its natural discharge, mainly due to inputs of coal mine water. The total macrobenthos density was 18,600 per square meter in the Grabia, and 7400 per square meter in the Widawka. Chironomidae and Oligochaeta were dominant in the Widawka (81%) and Chironomidae in the Grabia (58%). As regards Chironomidae in the Grabia, small Orthocladiinae, inhabiting mainly the middle part of the river bottom, were the most numerous, Tanytarsini and *Polypedilum* being slightly less abundant. The composition of these insects in the Widawka was the opposite of that in the Grabia. The large-sized species,

Chironomus thummi and *Prodiamesa olivacea*, were the most abundant in the zone near the banks, which were covered with a thick layer of silt, while small-sized psammophilous *Polypedilum* inhabited the middle of the river.

Burkhardt[205] studied the effects of pollution on the Trichoptera in running waters in Vogelsberg, Germany. Most species were restricted to clean and slightly polluted sites. More than half of the species that preferred the clean upstream zones could be found in severely polluted sections of the river, too. However, caddisflies living in the upper regions (epirhithral, krenal) were restricted to clean and slightly polluted reaches.

Metals and other toxic compounds are common in urban wastewater.[796] Frenzel[509] investigated whether aquatic communities in the Boise River, ID, were adversely affected by trace elements in the effluents from two wastewater treatment facilities. Trace-element concentrations in the river were less than, or near, analytical detection levels. Chironomidae, Simuliidae, Baetidae, and Hydropsychidae composed more than 90% of the insect population at all study sites. These taxa were also the most abundant insects collected from the Boise River 20 years earlier. Coefficients of community loss indicated benign enriching effects on insect communities, from the effluents. The distributions of trace element-intolerant mayflies upstream and downstream from wastewater treatment facilities indicated that trace-element concentrations in the effluents did not adversely affect these organisms.

The influence of the physical environment in the estuarine portion of a river may mask all but very local effects of pollution. Diaz[377] examined the interaction of pollution and the physical environment with the benthic communities of a large, temperate-zone estuary, the James River, VA. Distribution of benthic communities in the estuarine portion of the river was controlled mainly by salinity. Macrobenthic densities were most severely depressed in tidal freshwater habitats near a major pollutant source. Analysis of only the tidal freshwater portion indicated that the benthic communities reflected the location and concentration of pollution sources along the river. Tidal freshwater communities were dominated by the chironomid *Coelotanypus scapularis*.

C. AGRICULTURE AND FORESTRY

Agriculture (including fish farming) and forestry have recently been recognized as major sources of water pollution, and changes in aquatic insects have been observed. Phillippi[1201] studied aquatic macroinvertebrates as indicators of water quality in the Cache River drainage area in Illinois, where excessive habitat destruction resulting from drainage modification and/or sedimentation from agricultural activity was found at many sites. In general, there was also a relatively high species dominance at disturbed sites, with a high number of individuals and low diversity.

Boreham et al.[162] investigated the benthic macroinvertebrates of a clay stream polluted by livestock slurry on the River Roding in Essex, England. Livestock slurry had adversely affected the species richness of the fauna over a number of years. The pollution-sensitive stonefly *Nemoura cinerea* and the caseless caddisfly larva *Polycentropus* sp. were confined to sites above the outfall. Downstream of the outfall, *Chironomus riparius* and the air-breathing syrphid *Eristalis tenax* were found. At less-polluted sites diverse species assemblages occurred, including the black fly *Simulium* sp., the mayfly *Baetis rhodani*, the cased caddisfly larva *Limnephilus lunatus*, and several beetles. In Greece Koussouris et al.[868] studied invertebrate fauna of the River Louros, which runs through a large area including cultivated plains and livestock units. The dominant insect species were *Chironomus* spp. and *Simulium* sp. (Diptera); *Hydropsyche pellucidula* (Trichoptera); *Ischnura elegans*, *Calopteryx splendens*, and *Platycnemis pennipes* (Odonata); *Dytiscus marginalis* and *Gyrinus* sp. (Coleoptera); *Ephemera* sp. (Ephemeroptera); and *Corixa* sp. (Heteroptera).

A posttreatment analysis was conducted by Phillippi and Coltharp[1202] on the benthic invertebrate community of a small headwater stream in eastern Kentucky that had been affected by forest fertilization. Pairwise comparisons revealed that the nitrate-N concentrations remained significantly higher for 6 years after treatment within the fertilized watershed compared to unaffected streams. There was a lower number of taxa and lower species diversity in the fertilized stream. The decrease in Coleoptera, Diptera, and Ephemeroptera accounted for the difference. The Coleoptera had a significantly higher number of individuals in the unaffected stream. Phillippi and Coltharp[1201] concluded that fertilization did not produce lasting effects upon the benthic community.

Paasivirta[1175] studied macrozoobenthos near a fish farm in Lake Pyhäjärvi, Finland. Eutrophy-indicating species, such as *Chironomus anthracinus* and *C. plumosus*, were found only near the fish farm. The highest organic biomass value for the macrozoobenthos was found at 5-m depth near the fish farm, but the value at 10-m depth in the same area did not differ significantly from those for other areas.

D. DAMS AND RESERVOIRS

Dumnicka et al.[406] studied the effects of stream regulation on the hydrochemistry and zoobenthos in parts of the upper Vistula River with varying pollution loads in Poland. They described the changes in the chemical composition of the water and the animal communities below three reservoirs lying in ecologically different river sectors that differed in the degree of water pollution. The reservoirs transform the chemical composition of the water in various ways, depending on the quality of the inflowing water and the type of reservoir. The insect fauna of one reservoir with elevated concentrations of ammonia-N, phosphates, and organic matter was dominated by detritophagous Simuliidae. Another reservoir was a trap for nitrogen and phosphorus. The effluent water was less enriched than that which flowed in, but in this case its trophic state was not the factor determining the species composition. Owing to the very low discharge, rheophilic forms were absent, while species typical of slow-moving waters, such as Chironomidae, were well represented. Ephemeroptera were the most strongly reduced. One reservoir was built in a heavily polluted sector, and the water leaving the reservoir as overspill became oxygenated as it fell from a height of about 5 m. However, the pollution problem was not eliminated owing to the short retention time. Under such conditions the faunal composition at stations up- and downriver of the dam was similar, though the abundance of animals was four times greater below the dam.[406]

The effects of dams and water pollution on invertebrates in the rivers Innerste and Oker and on the plains of the Harz Mountains, Germany, were investigated by Rehfeldt.[1269] Pollution of the upper courses was low, although the water of the Oker had a low pH. The effects of the dams included a decrease in temperature and oxygen content. At the slope of the mountains the pollution of both rivers, especially with metals, was moderate to excessive. In winter the frequency of Plecoptera increased below the dams, as did the frequency of Trichoptera and Diptera at the slope of the mountains. In summer there were high frequencies of Ephemeroptera in the upper courses of the Innerste, in contrast to the acidic river sections of the Oker. The effects of the dam on the Innerste were greater than on the Oker. Stoneflies, collecting detritus, were most abundant below the dam of the impoundment on the Innerste, where overall diversity was low. However, species diversity along the more variable watercourse of the Oker was high in the respective section.[1269]

E. EXPERIMENTS

The exact mechanism responsible for the observed degradation in benthic insect communities in urban streams has not been demonstrated.[796] Possible causes include increased scouring and erosion, decreased base flow, alterations in trophic relationships, and toxic chemicals such as metals, organics, and road salts. Large amounts of suspended sediments reaching streams will clearly lead to depauperate benthic insect communities.[85,257] However, the recovery of stream communities following the flushing of sediment deposits may occur within a matter of months.[257]

Nebeker et al.[1116] conducted six chronic tests with *Chironomus tentans* to determine its usefulness as a test organism for chronic sediment assays. Larval recoveries ranged from 68 to 92%. In tests starting with second or third instar larvae, adult emergence ranged from 42 to 74%. Sediment from a creek containing high concentrations of copper from mine tailings killed all the larvae; no other sediment was so obviously toxic to *Chironomus*.

Fairchild et al.[454] used three experimental stream ecosystems to determine the effects of sediment and contaminated sediment: the first stream received 1.7 μg/l uncontaminated sediment for 2 h each week for 6 weeks; the second stream received 1.7 μg/l contaminated sediment (50 to 1600 μg/g triphenyl phosphate, increasing weekly doses) for 2 h each week for 6 weeks; and the third stream was maintained as a control. Each stream was monitored for changes in nutrient dynamics, leaf decomposition, primary production, and invertebrate dynamics. Both sediment treatments decreased the percent similarity of benthic invertebrates. Invertebrates in the sediment treatment exhibited delayed nocturnal drift, while those in the sediment/triphenyl phosphate treatment drifted immediately once the toxicity threshold was reached. Both sediment and sediment/triphenyl phosphate decreased the percent similarity of benthic invertebrates, reduced the drift of filamentous algae, increased the production of rooted flora, and increased net nutrient retention. However, neither treatment altered leaf decomposition rates, nor affected the total number of individuals or species, or insect emergence.

Rosiu et al.[1302] determined the effects of sediment from various sediment core depths on the survival and weight gain of larvae of *Chironomus tentans* during 10-d laboratory exposures. Sediment cores were collected from 12 sites in the Trenton Channel of the Detroit River in 1987 and sectioned into 5-cm sections down to a depth of 25 cm. Rosiu et al.[1302] also found that if the reduction in weight gain was more than 30%, naturally reproducing indigenous macrozoobenthos were also restricted in their colonization or were grossly unbalanced in community structure. A threshold toxicity of approximately 25% reduction in weight gain relative to the control was observed in *C. tentans*. Similar results were obtained from bioassays using surfacial sediments.[541]

The presence of mayflies (particularly *Hexagenia* spp.), and sometimes caddisflies, in synoptic surveys of Detroit River sediments is indicative of the quality of the benthic invertebrate habitat.[688,1484] Macrozoobenthic biomass at a less-polluted station was composed of 7.7% mayflies and 5.4% caddisflies, while at other locations where surficial sediments were classified as toxic based on *Chironomus tentans* laboratory assays, no mayflies or caddisflies were observed.[1302]

F. PURIFICATION

Siewert et al.[1372] collected water and macroinvertebrate samples from Finley Creek, IN, both before and 2 years after the cleanup of a nearby hazardous waste site. Below the disposal site, higher concentrations of constituents were found both before and after the cleanup, but 2 years after the cleanup, an improvement of water quality was observed. Total dissolved solids decreased by 52%, and macroinvertebrate taxa increased by 44%. The number of invertebrates, the number of taxa, and the species diversity was lower below the waste site.[1372] *Hyallela azteca* (Amphipoda), *Allocapnia* (Plecoptera), *Baetis* sp., *Ephemera* sp. (Ephemeroptera); *Ischnura* sp. (Odonata); and *Hydropsyche* sp. and *Cheumatopsyche* sp. (Trichoptera) were found only above the site. These invertebrates are sensitive to chemical pollution.[739,925]

III. ACIDIFICATION

A. BACKGROUND

Acidification of freshwaters has become a serious environmental problem in several regions of the world, including northern Europe,[26,819,1126] parts of Canada,[380] and in the northeastern U.S.[74,401,936,1333] Atmospheric deposition of strong acids, such as H_2SO_4 and HNO_3, derived from fossil fuel combustion

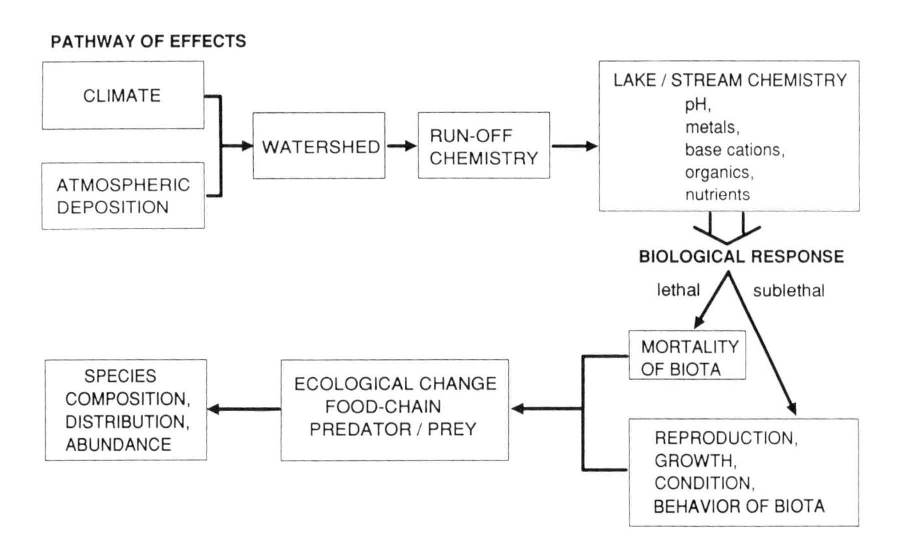

FIGURE 2. Diagram illustrating the probable pathway of effects of, and biological response to, the acidification of surface waters. (Modified from McNicol, D. K., Bendell, B. E., and Ross, N. K., *Can. Wildlife Serv. Occas. Paper*, 62, 1, 1987. With permission.)

has been implicated in surface water acidification in the northeastern U.S.[936] Acid-sensitive watersheds are often underlain by granitic bedrock and have small reserves of readily available basic cations (Ca^{2+}, Mg^{2+}, Na^{+}, K^{+}). These watersheds have a limited capacity to retain strong acidic anions (SO_4^{2-}, NO_3^{-}). Incomplete neutralization is followed by the transport of acidic cations (H^{+}, inorganic monomeric aluminium, Al^{3+}) and metals (Mn^{2+}, Zn^{2+}) from the soil to surface waters.[401,1393] Especially during periods of heavy rainfall, watersheds may not be able to assimilate acidic inputs, and short-term pH depressions of lake and stream waters may have a substantial impact on the aquatic biota well before long-term declines in pH are detectable.[1164] For example, the lakes and streams of the Sierra Nevada are very sensitive to acidic inputs owing to the low buffering capacities of their granitic watersheds[711] (Figure 2).

Some aquatic insects are very tolerant to natural acidity. *Chironomus acerbiphilus* has been found breeding in highly acidic waters in the active volcano area in Japan, even in a lake containing water showing pH 2.0 or less.[1328] However, aquatic communities often show the first and most drastic effects of cultural acidification.[26,317] Numerous studies have been performed in North America and Europe to assess the effects of acidic precipitation on benthic invertebrates.[20,21,210,399,591,597,844,955,1378] Different approaches have included extensive surveys of lakes in regions affected by acid precipitation;[1038,1153] comparisons between acidified and nonacidified lakes[170,326,1248,1606] or rivers;[54,511,600,1060,1171,1393,1451] experimental acidification of rivers[597] or entire lakes;[23,1334] and experimental acidification in artificial stream channels.[210,711,1675]

Increased concentrations of H^+ and trace metals in freshwaters, resulting from acidification, can be detrimental to all trophic levels, including benthic invertebrates.[20-23,170,211,844,1378,1393] However, the studies have often produced conflicting results. For example, Crisman et al.[326] reported only minor changes, while Wiederholm and Eriksson[1606] and Minshall and Minshall[1060] found profound changes in the benthic community composition in acidified environments. These differences may be attributed to several causes. Comparisons were generally made on a pH basis. Metal concentrations were often not monitored, yet the aluminium, whether derived from the sediments and/or the watershed, can be of great importance. Aluminium toxicity can vary at different pH, calcium, and silicon[139] levels, and different conclusions can be made depending on the chosen pH regime. The episodic nature of the acidification process does not facilitate comparisons. The length and strength of the acid pulse, and the resulting consequences, can be widely variable. Most studies present situations without consideration of the state before acidification.[23] Species information also varies depending on the geographic region studied.[1393]

B. CHANGES IN AQUATIC ECOSYSTEMS

Invertebrate community composition in acidified freshwaters is often altered both taxonomically and functionally,[23,104,210,211,326,591,594,597,670,1248] with a reduced species richness and dominance by a few species.[511,844,955,1153,1250,1375,1378,1495,1675] Abundance of benthic invertebrates ies and the total biomass can decrease with increasing acidification,[23,54,471,1248,1296,1606,1675] but some researches reported only slight difference in the abundances of organisms in acidified and circumneutral lakes.[326] Furthermore, species composition[210,442,594,673,1675] and invertebrate life histories can change.[101,211,471,1675] Certain species among the sensitive taxonomic groups can be notably tolerant to low pH.[600]

Mayflies are particularly sensitive: the number of species present decreasing in acidified waters.[511,600,774,844] The disappearance of mayflies from acidified rivers[101,471,623,793,1060] was attributed to higher mortality at emergence,[101,597,1675] to inhibition of oviposition in acidic waters,[1451] or to the avoidance (drifting or mortality) of acidic waters by early instars.[21–23]

Acidic inputs can affect organisms through their effects on other organisms in the trophic web.[711,1163] Fish can be eliminated by acidic deposition, with important repercussions for the invertebrates on which they prey.[442,673] Both food and physicochemistry are important elements of the proximate explanation for community structure in acidic waters.[1312,1452]

Although the benthic invertebrate community is generally less species rich in acidic oligotrophic water than in less-acidic water,[25,301,502,670,671,923,1038,1051,1451,1452] benthic invertebrates can be abundant in an acidic lake, especially if predatory fish have disappeared or have been reduced in numbers. The benthos of acidic lakes is generally dominated by Crustacea and Insecta, especially Notonectidae, Corixidae, Chironomidae, and

Megaloptera; Trichoptera, Ephemeroptera, and Plecoptera may also be present. However, all of the above groups include species that are intolerant of acidity.[502] For example, Raddum[1247] found that while several stonefly species were indifferent to pH in Scandinavian fresh waters, there were also taxa that were sensitive, including *Amphinemura sulcicollis*, *Brachyptera risi*, and *Leuctra hippopus*. The mayfly *Baetis rhodani* tended to dominate its order in circumneutral streams, but disappeared following acidification. In northern Europe, acid-intolerant mayflies include *B. rhodani*, *B. lapponicus*, *B. macani*, and *Centroptilum luteolum*.[600,1249]

Synoptic surveys have indicated that acidification has altered both water chemistry and biotic communities,[399] but only a comparison of present conditions with historical data can determine whether lakes and streams with a previously circumneutral pH have actually become acidified.[594] In Scandinavia both spatial differences and temporal changes in aquatic biota have been documented and attributed to the long-range transport and deposition of anthropogenic mineral acids.[399,1154] In North America documentation of chemical changes resulting from surface water acidification includes studies on a spatial scale either in regions of different acid loadings or at various distances from point sources, such as Sudbury, Ontario.[380]

Models based on chemical survey data and geochemical assumptions have been used to predict the decline in alkalinity and pH of lakes in the eastern and midwestern U.S. Schindler et al.[1335] combined the results for known acid tolerances of different taxonomic groups in order to estimate the extent of damage caused by acidic precipitation to biological assemblages. Over 50% of the species in some taxonomic groups had probably been eliminated from lakes in the Adirondacks, Poconos-Catskills, and southern New England. Moderate damage to biotic communities was predicted for lakes in central New England and northcentral Wisconsin. Damage predicted in Maine, upper Michigan, northeastern Minnesota, and the remainder of the upper Great Lakes region was slight. Crustaceans, molluscs, leeches, and insects were among the most severely affected groups.

The fauna of two lakes in Nova Scotia is an example of the benthos of typical acidic waters.[825] Insects were the most abundant invertebrate group in both the clear-water, pH 5.3, and the brown-water, pH 4.5, lake, comprising 56 and 78% of the total density of benthic invertebrates, respectively. The insects were dominated by Diptera (45 and 68% of total density, respectively), especially Chironomidae (96 and 92% of Diptera). Dominance of Chironomidae is a general characteristic of the benthos of acidic lakes,[640,826,1247,1296] although their species richness decreases with increasing acidity.[1038] Other important insects in the Nova Scotia lakes were Trichoptera (5% of the total invertebrate abundance), Ephemeroptera (1 to 3%), and Odonata (2%). Crustacea comprised 18 to 19% of the benthic invertebrate density. The oligochaete families Naididae and Enchytraeidae accounted for 13% of the invertebrate density in the clear-water lake and 3% in the brown-water lake.

As a result of very acidic precipitation (mean pH 4.17), both biological and chemical changes have occurred at a rather rapid rate in the more sensitive lakes of Pennsylvania, compared to corresponding changes monitored in other parts of North America. Bradt and Berg[170] compared the macrozoobenthos of three lakes in Pennsylvania and attempted to relate the faunal composition to acidification. The total number of organisms, number of taxa, diversity, and evenness did not differ among three lakes with varying sensitivities to acidification. The differences between the macrozoobenthos communities of three Pennsylvania lakes with varying sensitivies to acidification were a higher biomass (wet weight; including Odonata and Mollusca) in the circumneutral lake; more Ephemeroptera (primarily *Caenis*), Gastropoda, and Pelecypoda in the least acidic lake; more predators and Chironomidae (including Tanypodinae and Tanytarsini) in the most acidic lake.

The relationship between lake acidity and water strider (Heteroptera, Gerridae) distribution and abundance was investigated near Sudbury, Ontario, by Bendell.[103] Lake surveys and quadrat sampling showed that the distributions of *Metrobates hesperius* and *Trepobates inermis* and population densities of *Rheumatobates rileyi* were related to lake acidity, but there was no evidence that the distributions or abundances of *Gerris* spp. would be related to lake acidity. Food resources for water striders may be reduced at low pH values. Toxicological effects of acidification on water striders seemed to be limited to the eggs, owing to the limited direct physical contact of adults and immatures with the water.[103]

The causal relationships between acid-related factors and macroinvertebrates are complex and poorly known. Laboratory studies concerning the survival of invertebrates in acidified media do exist, and some of these data are explained by contaminant observations of physiological alterations, especially in Na^+ regulation. Even within the same species (*Corixa punctata*, Corixidae) a very different reaction to Na^+ was found at pH 4 when insects from an acid bog were compared with insects from a circumneutral pond.[1521] Willoughby[1622] suggested that molting is a sensitive stage for pH stress. The same may be true for emergence.[101] During molting and emergence, ion exchange through the integument of the animals increased until the end of the sclerotisation process. Thus, loss of Na^+ and the detrimental effect of increased H^+ concentrations can take place throughout the body, leading to increased stress.

Aquatic insects may actively avoid acidic waters. Sutcliffe and Carrick[1451] noted that female *Baetis rhodani* avoided acidic tributaries when ovipositing. The scarcity of some algae and the impoverished epilithon in acidic streams may provide an inadequate diet for scrapers or surface grazers such as mayflies.[1164,1451,1630] In contrast, shredders, such as some stoneflies and caddisflies, may benefit from the decreased rate of detrital breakdown and increased fungal biomass on leaf litter, at low pH levels.[416,494,683,955] Disturbances in osmoregulation, acid-base balance, and calcium regulation at low pH might explain the scarcity or reduced growth of nymphal mayflies in acidic streams.[471,1535]

Large areas of northern Britain and Wales are subject to episodes or substantial depositions of acidity from the atmosphere.[1535] Many such areas are underlain by base-poor rocks, and afforestation with conifers has represented a further acidifying influence on waters. Wade et al.[1535] described the spatial patterns in invertebrate assemblages in upland Wales. Acidity and aluminium concentration were the variables most closely related to the ordination, classification, and taxon richness of macroinvertebrates. The invertebrate species that were absent from the more acidic sites, e.g., *Baetis rhodani*, *Gammarus pulex*, *Hydropsyche instabilis*, *Hydraena gracilis*, were also absent in other studies involving streams with low or fluctuating pH.[1430,1451,1495]

Baseline monitoring data are important in differentiating between natural and anthropogenic variation in community structure. Hall and Ide[594] compared historical data with current data from the Algonquin Provincial Park, Ontario, to determine whether anthropogenic acidification had had an impact on aquatic invertebrates. They resampled two oligotrophic, low-alkalinity streams in which benthic insects had been thoroughly studied 48 years previously. In one stream not subject to severe acid pulses in the spring (pH 6.4 to 6.19), there were few differences in insect taxa between the two samplings. The other stream currently experienced severe acid pulses (pH 6.4 to 4.9), and there were some notable changes in the benthic insects. Four mayfly and two stonefly species have disappeared since 1942. The occurrence of insect species was a better indicator of anthropogenic stresses than changes in the relative abundance of species.

One of the many external factors possibly modifying the acidification on aquatic insects is humic substances in water. Hämäläinen and Huttunen[600] found different species richness in brown- vs. clearwater streams within the same pH range. Collier et al.[298] studied several brown-water streams with naturally low pH brought about by high concentrations of organic acids under nontoxic aluminium conditions in Westland, New Zealand. Thirty-four of the 37 most widespread aquatic insect taxa were recorded in streams with pH < 5, and 24 were taken from sites with pH < 4.5. Physiological adaptations enabling this tolerance may be the same as those that evolved in response to the physicochemical variability associated with the unpredictable flow regimes of Westland streams.

Smith et al.[1393] compared benthic invertebrate parameters between three streams of differing acidity in the Adirondack Mountains, NY, in an area heavily impacted by atmospheric acid inputs. Although the total invertebrate density was not significantly lower at the acidic site compared with most other sites, Ephemeroptera were absent at the acidic site. *Baetis* and certain heptageniid mayflies (*Cinygmula*) were only found in streams with higher pH values (annual minimum > 5.4). *Baetis* is generally reported to be intolerant of low pH values.[471,1196,1378,1451] *Ephemerella funeralis* was the mayfly most tolerant of lower pH values. It has been collected at acidic sites (pH 4.7 to 4.9[471,1393]) and other locations where the pH ranged from 5.3 to 6.2.[955]

Plecoptera were generally well represented at all the sampling sites of Smith et al.[1393] *Leuctra* were abundant at several acidic sites.[1378,1393,1495] *Peltoperla arcuata*, *Strophopteryx*, and *Acroneuria* were the stoneflies least tolerant of low pH.[844,1393] Several caddisfly genera were only present at high pH sites (e.g., *Glossosoma*, *Apatania*, *Pycnopsyche*, *Micrasema*, *Triaenodes*).[1393]

Hydropsyche sp. and *Diplectrona modesta*, both hydropsychid caddisflies, were present at all sites except the acidic site. Many Trichoptera were also present in the Hubbard Brook Experimental Forest in New Hampshire where low-order acidic streams are numerous.[1017] Elmid beetles were only present in neutral pH waters in the Adirondack Mountains.[1393] Townsend et al.[1495] could find no correlation between elmid density and stream pH in streams in the U.K. However, Simpson et al.[1378] collected *Oulimnius latiusculus* from a site where pH values were always above 5.5 and labile (inorganic) monomeric aluminium concentrations never exceeded 0.04 mg Al per liter.

The biological effects in acidified waters include direct pollution effects as well as the effects of altered predator-prey interaction after the decline or disappearance of fish. Eriksson et al.[442] compared the biological effects of fish removal in a nonacidified lake, with the changes reported from acidified waters. After the fish removal in the experimental lake, a development similar to that in acidified waters occurred. Large-size selective fish predators were replaced by smaller selective predators, such as corixids, *Chaoborus* (Diptera, Chaoboridae) larvae, and other aquatic insects susceptible to fish predation and kept at low densities when fish were present. The absence of fish implied structural changes in both species composition and size composition among zooplankton and phytoplankton. Among the grazers there was a change towards larger species. Species diversity in the plankton as a whole was reduced. The development took place in acidified, as well as nonacidified, lakes devoid of fish, and many structural changes reported from acidified lakes may be ascribed to altered predator-prey interaction after the decline of fish populations, and not regarded as direct pollution effects.[442]

C. EXPERIMENTAL ACIDIFICATION

A variety of approaches have been used to assess the effects of increased acidity on stream invertebrate assemblages.[711] Many studies on the effects of stream acidification on benthic invertebrates have primarily been surveys[1451] or laboratory bioassays.[100,101] The survey approach involves comparison of the distributions and abundances of aquatic organisms in relation to the pH of the waters. However, relationships can be confounded or obscured by the effects of other, perhaps unmeasured, variables. Laboratory bioassays, on the other hand, cannot adequately represent the complex biological and chemical interactions that occur in an organism's natural habitat. Laboratory microcosms and mesocosms, which attempt to mimic natural assemblages, sediments, and water, have been used as compromises between single-species bioassay and field experiments.[211] However, it is difficult to transport and maintain natural assemblages in the laboratory, and in many cases, mesocosms diverge from larger natural systems.[711] Whole-stream acidification experiments have been performed as an alternative approach.[597,1162] Whole-stream experiments have often utilized a downstream manipulative, vs. upstream control, design that does not allow for the replication of treatments. Furthermore, not only are upstream and downstream areas often different,[1164] but they are not independent because upstream processes may affect downstream results.[711]

Experiments can help to pinpoint the early-warning signals of acid stress in order to determine which aspects or processes should be concentrated upon during monitoring programs. Herricks and Cairns[675] reported a 42% reduction in the density of benthic invertebrates in a stream acidified to the pH range 4 to 5 for merely 15 min. In several studies[23,211,1675] significant reductions in benthic density occurred only after prolonged acidification, when changes attributable to long-term stress or to changes in the invertebrates' food supply may have been implicated.

Allard and Moreau[23] investigated the effects of experimental acidification on a lotic macroinvertebrate community. The 3-month experimental acidification (pH 4, with or without aluminium) conducted under seminatural conditions in plasticized wooden channels showed decreased total abundance of benthic macroinvertebrates, except for *Microtendipes* sp. (Chironomini). Most Chironomini are red, and their resistance to acidity is thought to be linked to the presence of the pigment hemoglobin, which can buffer the hemolymph against pH changes.[638] However, Tanytarsini, as well as Chironomini, possess hemoglobin. The reaction of Tanytarsini to acidity was very different from that of Chironomini. This suggested that the presence of the pigment was not the only important factor involved in the resistance of invertebrates to acidification.

The microdistribution of macroinvertebrates illustrated the different habits of invertebrates and was representative of their relative vulnerability in the experiment carried out by Allard and Moreau.[23] The most sensitive groups, Ephemeroptera and Orthocladiinae (Chironomidae), were usually found in surface and pebble samples. These organisms were usually adapted to remain at the surface of the sediment. This habit could make them more vulnerable to acidity than larvae, such as Chironomini, that live buried in sediments. Allard and Moreau[23] deduced from their invertebrate results that the degree of exposure to acidity was certainly stronger at the water-sediment interface than more deeply in the substrate. Biological and chemical activities in the sediments may raise the pH, thus providing a more favorable habitat than in the water. For example, the activity of benthic organisms in sediments has been shown to increase the sediment pH.[874]

Trophic web modifications were assumed to be responsible for the modifications in the community.[23] The main groups that were affected by modifications of the algal communities were the Ephemeroptera and the Orthocladiinae. The rapid desertion of these two groups, and the high mortality rates calculated for the drift after acidification,[21] suggested that a factor other than food limitation was responsible for their response, at least in the short term. Experimental acidification by Allard and Moreau[23] showed differences in sensitivity between the life stages of the mayfly genus *Ephemerella*. Large mayfly larvae were able to survive at low pH, while early instars did not colonize the acidified channels.

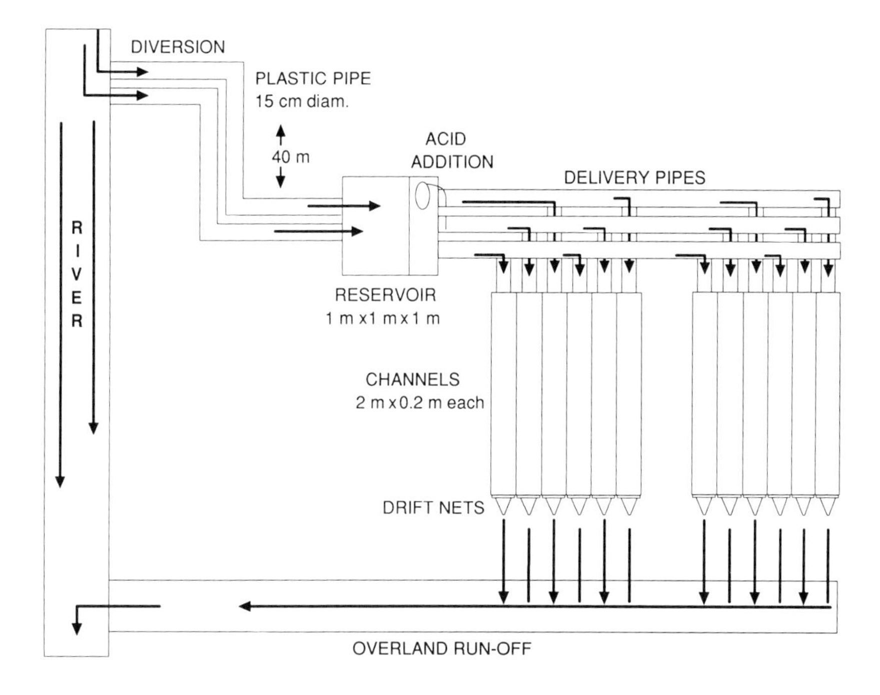

FIGURE 3. Design of the replicate experimental stream channels set up along the Marble Fork of the Kaweah River, a high altitude Sierra Nevada stream in California. Arrows indicate the direction of water flow. (Modified from Hopkins, P.S., Kratz, K. W., and Cooper, S. D., *Hydrobiologia,* 171, 45, 1989. With permission.)

Stream channels have been suggested as a useful compromise between the naturalness of whole-stream manipulations and the control and replication of laboratory experiments (Figure 3). They approximate natural stream conditions while providing the replication needed for statistical analyses.[711] Stream channels have been used to examine the effects of increased acidity on stream systems.[21-23,711,1675] Most stream or stream-channel experiments have involved the continuous additions of acid to stream sections or channels for long periods of time.[21-23,211,597,1675] Ormerod et al.[1162] investigated hydrogen ions and aluminium using short-term experimental acidification. Hopkins et al.[711] performed a simulation of episodic acid rain events (containing sulfuric and nitric acids) in channels constructed alongside a stream in the Sierra Nevada mountains, of similar acidity to those known to occur in the convective storms occurring in the dry season. Except during these episodic events, the pH of freshwater habitats in the western Sierra Nevada is close to 6.2 to 6.5. The replicated stream channels allowed evaluation of the effects of acidic deposition on the stream biota. The experimental design eliminated pseudoreplication, because the treatments were spatially randomized and had independent supply systems.

A significant increase in *Baetis (Ephemeroptera)* drift rates occurred following acidification of experimental stream channels. Increased drift of benthic invertebrates has been reported to follow acidification of streams.[597,1675] This acceleration in macroinvertebrate drift activity is often an animal reaction to lowered pH and levels off to preacid rates over time.[597]

The increased drift of stream invertebrates during acidification may be due to an increased number of killed individuals in the drift.[21] Increases in drift in response to insecticide spraying have been shown to result from an increase in the number of dead invertebrates in the drift.[422] The increased drift may also be due to an active, behavioral response of the drifting organisms[597] or to an increase in the insects' susceptibility to being accidentally dislodged. The high percentage of dead *Baetis* (46%) in the drift immediately following acidification of stream channels indicated that some "toxic" drift occurred.[711]

McNicol and Scherer[1033] examined the effects of acid exposure on *Acroneuria lycorias* (Plecoptera) larvae in laboratory streams. When subjected to a reduction in pH from 8.1 to 2.5 over an 8-h period, larvae showed little behavioral response down to pH 4.2. As the pH fell to 3.0, "head-rubbing" activity appeared and increased in frequency. At pH 3.0 and below, the larvae showed increased gill-ventilatory movements and locomotor activity. Most larvae died within 14 h of exposure to pH 2.5; however, they did not abandon their preferred refuges before death. Larvae exposed to five different pH levels between 4.5 and 8.2 for 30 to 50 days displayed no significant changes in locomotor activity, drift behavior, or microdistribution, when compared to the control animals.

D. COMBINED EFFECTS OF ACIDITY AND ALUMINIUM

The acidification of surface freshwaters and subsequent mobilization of aluminium (Al) from base-poor soils have lead to a decrease in invertebrate diversity and influenced fish stocks and riverine birds in northern Europe, Canada, and North America.[26,93,591,623,1164] Other sources of aluminium include industrial processing, alum treatment of drinking water, sewage treatment, and pulp and paper processing.[1654]

Elevated levels of aluminium may have deleterious effects on fish and invertebrates.[639,640,676-678,1014,1162-1164,1249,1430,1535] Low concentrations of aluminium at pH values below 4.4 do not have a toxic effect and can even enhance the survival of aquatic biota. The toxicity and bioaccumulation of aluminium is species specific in aquatic insects and depends on the speciation of aluminium, which in turn is associated with water pH, temperature, and the concentration of organic ligands.[1654]

While the biological impact of acid precipitation is not contested, it is not clear to what extent the stress is due to the H^+ concentration and the pH-mediated release of aluminium from the watershed.[381,402] Several investigators have attempted to separate the effects of low pH alone from those of aluminium at low pH values, when analyzing invertebrate drift, community modification, or fish physiology.[23,209,1162,1164] The biological effects of low pH and increased aluminium concentrations have been investigated experimentally in artificial or natural stream channels.[23,211,596-598,1675] Attention has been paid to the effects of the elevated concentrations of metals and aluminium that are dissolved and transported to aquatic ecosystems when acid precipitation percolates through poorly buffered soils.

There have been several studies on the toxicity of aluminium and low pH to freshwater invertebrates.[101,102,209,641,678,1014,1162,1164] Ormerod et al.[1162] dosed a soft-water stream in upland Wales for 24 h with sulfuric acid and aluminium sulfate at two points along the stream to create simultaneous episodes of low pH, and low pH with a high aluminium concentration. Both brown trout (*Salmo trutta*) and salmon (*S. salar*) were more sensitive to aluminium at low pH than any of the macroinvertebrates.[1162] Both fish and invertebrates continued to die after being returned to clean water following exposure. Such irreversible effects of brief exposures may occur at concentrations lower than those causing mortality in acute tests. The concentration of aluminium used in the study, although environmentally relevant, may actually be higher than that required to cause mortalities of sensitive taxa (e.g., mayflies) during natural acidic episodes. The mayflies *Baetis rhodani* and *Ecdyonurus venosus* were the most sensitive of several invertebrate species exposed to aluminium at low pH levels. Herrmann and Andersson[678] noted an increased oxygen consumption by mayflies in response to such conditions.

Pretreatment of fish and their eggs to low concentrations of metals may afford the fish some degree of protection in subsequent exposures to higher concentrations. A similar result occurred in the experiment of Ormerod et al.[1162] with *Gammarus pulex* (Crustacea) and *Ecdyonurus venosus* (Ephemeroptera). However, in *Baetis rhodani* the pretreatment with aluminium was accompanied by increased mortality both during and after exposure. Such a pattern may be significant in animals undergoing repeated exposures during a series of acidic episodes.[1162]

Several researchers have shown that the presence of mucus clogging the gills of fish is a stress factor in water of low pH, both in the presence and absence of aluminium.[1015] Baker and Schofield[73] suggested that aluminium toxicity at pH levels of 5.2 to 5.4 involves precipitation of aluminium hydroxide on the gill or other surfaces. The consequent hyperventilation, combined with a loss of plasma salts and lowered blood oxygen tension, produced a toxic response. Herrmann and Andersson[678] proposed a similar toxic mechanism involving mucus production for mayflies. McCahon et al.[1015] examined gills from the mayflies *Baetis rhodani* and *Ecdyonurus venosus*, killed during the experimental dosing of a stream with acid and aluminium,[1162] both chemically and histochemically for the presence of aluminium and mucus. Aluminium was observed over the entire exoskeleton, including the gill plates of both mayfly species as well as in the gut of *E. venosus*, but there was no evidence of mucus production. Herrmann and Andersson[678] reported an increased respiration rate in mayfly nymphs exposed to 0.5 mg/l aluminium at pH 4.0 and pH 4.8 and attributed this to decreased oxygen transport because of impaired osmoregulation and ion transport, and aluminium hydroxide precipitation and mucus production on the gill plates. McCahon et al.[1015] suggested that the gill plates are not the sole target organ, but that aluminium physically occludes the main respiratory surface, the integument, leading to an increased respiration rate and subsequent death.

Rockwood et al.[1290] examined the impact of aluminium on osmoregulation and ionic balance of the dragonfly *Libellula julia* at low pH. The species is abundant in the acidic waters of the Canadian Shield. Aluminium caused highly significant losses of wet and ash weight and of the body burdens of Na^+ and Ca^{2+} in comparison to low pH alone. In contrast, Lechleitner et al.[913] found that the body burden of Na^+ in the stonefly *Pteronarcys proteus* was lower in nymphs after exposure to pH 3.0 for longer than 72 h. They also observed a histological disruption of the osmoregulatory cells of the gills of *P. dorsata* exposed to acutely toxic pH levels. Na^+ loss by *L. julia* was observed only when exposed to Al at low pH, indicating a strong correlation among aquatic invertebrates between sensitivity to low pH and difficulties with ionic regulation.[1290]

Rockwood et al.[1290] also investigated the effects of low pH and elevated aluminium concentrations on oxygen consumption in *Libellula julia*. Significantly depressed oxygen consumption was observed in the nymphs. Low pH (4.0) alone did not inhibit oxygen uptake as much as when used with aluminium in concentrations of 3 mg/l or greater. The reduced oxygen consumption in *L. julia* was hypothesized to be due to mechanical blockage of the gills by denatured mucus and/or Al hydroxide precipitates. In carp (*Cyprinus carpio*), mucus formed during exposure to acid reduced the oxygen uptake by increasing diffusion resistance to oxygen and by forming a nonconvective layer that inhibited water flow between the lamellae of the fish gills.[1512] This response identified as the coagulation film anoxia theory.[1290] The elevated aluminium concentrations may reduce the surface area available for absorption, thus inhibiting ventilation. These observations were in general agreement with those of Correa et al.[306,307] for the dragonfly *Somatochlora cingulata* and the caddisfly *Limnephilus* sp., in which increased concentrations of H^+ and aluminium reduced oxygen consumption. However, Herrmann and Andersson[678] found significantly increased respiration in three species of mayflies exposed to elevated levels of aluminium at low pH. This was suggested to be a homeostatic response to the aluminium, which triggered mucus build- up, precipitation of aluminium hydroxide, and/or impairment of ionic and osmotic regulation. The different responses observed in respiration between mayflies and dragonflies could reflect the latter's inability to ventilate.[1290]

Toxicity of different forms of aluminium to salmonids and the mayflies *Baetis rhodani* and *Ephemerella ignita* at low pH was tested in a 24-h dosing experiment in upland Wales.[1011] Four separate zones were created by the simultaneous addition of sulfuric acid, aluminium sulfate, and citric acid: a control zone, an acid zone (pH 4.9), an aluminium and acid zone (total filterable aluminium 0.27 mg/l, pH 4.9), and a zone of aluminium complexed with citrate (total filterable aluminium 0.23 mg/l, pH 4.9). Elevated aluminium concentrations at low pH caused the highest mortalities of both fish and macroinvertebrates. However, both brown trout and salmon were more sensitive to aluminium at low pH than were any of the invertebrates tested. Mortality data for *B. rhodani* indicated that animals collected from a hard-water site were more resistant than those collected from the softer water of the experimental stream. Nevertheless, any advantage was lost soon after exposure, and a significant mortality was recorded after 24 h.[1011] The survival data of *E. ignita* agreed with the results concerning no mortalities of *Ephemerella* sp. (>1 mm) within 13 days during acidification of artificial channels (total Al 0.37 mg/l, pH 4.0[23]).

Ormerod et al.[1162] took the enhanced drift densities that accompanied treatment as indications of adverse effects in the experiment. Although there were no strong differences between the acid and aluminium zones in the toxic response of *Baetis rhodani*, clear differences were apparent in drift densities. Therefore, sublethal factors could explain the absence of this species from streams with high aluminium concentrations.[623,1430] Laboratory and field studies have indicated that mayflies may be subject to osmotic or respiratory stress at low pH and high aluminium levels.[471,678] The drift responses of most other taxa studied by Ormerod et al.[1162] (e.g., *Leuctra* spp., *Ephemerella ignita*) were consistent with their sensivity to low pH and high aluminium concentrations in the field.[623,1430]

Hall et al.[596] attributed an increase in the drift of the family Dixidae (Diptera), which adhere to the water surface for respiration, in response to aluminium addition to a reduction in surface tension. Such a hypothesis is not consistent with the data of Ormerod et al.,[1162] since the drift of *Dixa puberula* showed a time lag after aluminium addition. Laboratory experiments[1162] with water from the treatment stream (mean pH 7.1) indicated that the surface tension remained above 71 dyn/cm, despite aluminium addition of up to 1.0 $\mu g/m^3$. Therefore, rapid changes in surface tension were probably not the primary reason for the drift.

It has been suggested that aluminium, in one form or another, can be transferred from one trophic level to another and perhaps even occur at a higher concentration at the higher trophic level.[639] Frick and Herrmann[512] studied the occurrence of aluminium accumulation in nymphs of the mayfly *Heptagenia sulphurea* at low pH (4.5). Nymphs were exposed to two aluminium concentrations (0.2 and 2 mg inorganic aluminium per liter) and two exposure times (2 and 4 weeks), the longer time period also including a molting phase. Most of the aluminium was deposited on/in the exuviae, as there was a 70% decrease in the aluminium concentration of nymphs after molting. When the nymphs were exposed for two instar periods, the aluminium content almost doubled (2.34 mg Al per gram dry weight) compared with that of a one-instar treatment (1.24 mg Al per gram dry weight). Internally accumulated aluminium may be transferred to terrestrial predators by mayflies and other aquatic insects that leave their final exuvium in the water. However, aquatic insects that make their final molt in the terrestrial environment, and thereby bring adsorbed aluminium out of the water, are more likely vectors.[512]

According to Krantzberg,[875] who considered the effects of age and weight on metal body burdens, aluminium concentrations were higher in small, than in larger, *Chironomus* larvae. As was the case with Fe and Ca, aluminium concentrations in older larvae were higher than in young larvae of similar size. No essential function has been attributed to aluminium, and it is unlikely that changes in aluminium concentrations with age represent changes in metabolic demand for this element. The ability to eliminate aluminium, or the exposure to bioavailable aluminium, may change with time. Alternately, older larvae could be exposed to higher concentrations of bioavailable aluminium.

E. RECOVERY AND RESTORATION

Rapid chemical and biological recovery of industrially acidified lakes can be accomplished simply by reducing the emission of acidifying substances into the atmosphere. Gunn and Keller[572] surveyed the biological recovery of the acidified Whitepine Lake, near Sudbury, Canada, after reductions in industrial emissions of sulfur. In 1980 the lake was acidic, and its fish population showed signs of acid stress. Water quality improved significantly from 1980 (pH 5.4) to 1988 (pH 5.9). Specific conductance and concentrations of SO_4, Ca, and aluminium all declined with reductions in industrial emissions of SO_2. The densities of benthic invertebrates in the deep zones of the lake (8 to 21 m) that are occupied by acid-sensitive lake trout (*Salvelinus namaycush*), declined from 1140 per square meter in 1982 to 650 per square meter in 1988. In contrast, the density of invertebrates in the shallow areas (0 to 8 m), the principal foraging area of perch (*Perca flavescens*), increased with the decline in the abundance of perch. Although changes in fish predation may explain many of the changes in invertebrate populations, the direct effects of water quality improvement were also detected. The number of taxa of benthic invertebrates increased from 39 in 1982/83 to 72 in 1988, and the relative abundance of many of the original species changed dramatically. Three acid-sensitive species of mayfly (*Hexagenia* sp., *Ephemerella* sp., *Caenis* sp.) appeared in the most recent survey. Taxa scarce in the early survey increased in numbers, coincident with improvements in water quality.

Acidic mine drainage is known to cause reductions in invertebrate densities and in taxa richness.[1383] Mine drainage is usually a point-source pollutant and has a greater potential for reclamation than nonpoint-source pollutants. Addition of a strong base to the acid source will raise the pH and lower the solubility of toxic metals. Skinner and Arnold[1383] evaluated the short-term (1 week) colonization response of invertebrates to the application of base (sodium carbonate) to an acid tributary. Although sodium carbonate did not raise the pH of the river throughout the treated study section, there were apparent biotic responses to its application: namely an increase in brook trout and *Baetis* (Ephemeroptera) colonizers.

Variations in calcium concentration have long been recognized as having important influences on both the quantity and quality of secondary producers and ecosystem processes such as decomposition.[416,683] Some of these effects will be beneficial, but others will be adverse, depending on the target organism and desired endpoint. Present knowledge on the response of aquatic insect populations to liming is still limited.[445,480a1014,1160,1161,1165] For example, experimental limestone application in Welsh lakes was accompanied by a decline among some invertebrate types, such as the chironomid *Zalutschia humphresiae*.[1161]

McCahon et al.[1014] subjected a chronically acidic stream, mean pH 5.2, in upland Wales to an induced episode of acidity during which acid, aluminium, and limestone were added at different points along the stream length for a 24-h period. They showed that the overall mortality was low among the invertebrate species found in acidic waters. Ormerod et al.[1160,1165] also examined the effects of liming acidified streams in Wales by computer modeling. The results indicated that liming and a 90% reduction in sulfate deposition reduce concentrations of toxic aluminium to similar levels. Calcium concentrations and pH were increased by liming to values that were high by comparison with conditions simulated by computer modeling to represent low levels of acidic deposition, either in the past or future. Trout density increased following liming. After liming, the streams acquired aquatic invertebrate species typical of higher calcium concentrations than those simulated under low acidic deposition. Species characteristic of "soft-water" communities were apparently lost. The "soft-water" community declined in the model as a result of acidification, indicating that both liming and acidic deposition resulted in a different faunal community from that prior to acidification. The results supported the idea that liming is suitable for the restoration or protection of a fishery, but indicated that there may be other ramifications, for example, to conservation, which must be considered when liming is implemented.[1165] This conclusion is supported also by empirical data from Norway.[480a]

IV. METALS

A. BACKGROUND

The production of metals and their discharge into waters have increased greatly in recent years.[1072] In Amazonia, for instance, metallic mercury is used in gold mining, and it has been estimated that tens or even hundreds of tons end up in nature.[1318] The effects of these effluents on aquatic invertebrates have been documented in the laboratory.[189,866,867,1297,1414,1554] Although the results have shown considerable variation among species, most aquatic species were sensitive to metals.[694] However, laboratory bioassays have limitations, and these tests should be supplemented with more environmentally realistic procedures, including field sampling and experimentation.[221,843] Most field research on the impact of metals has been descriptive,[35,51,241,276,903,953,1255] and few attempts have been made to predict changes in macroinvertebrate communities exposed to these effluents.[276,1628] Metals are readily taken up by aquatic organisms, where they are bound by the sulfhydryl groups of proteins whose structure and enzymatic activities are changed, resulting in toxic effects at the whole-organism level. The toxicity of metals to the organisms in aquatic ecosystems can vary according to concentration, pH, temperature, organic matter content, hardness, and other properties of the water.[269,491,593,1072,1654] Although metals are reportedly most toxic as free ions, organic complexing agents such as humic acids can increase metal toxicity.[128] Benthic communities in environments subjected to metal stress usually have fewer species, lower diversity, and lower biomasses than unstressed communities.[51,1071,1489] There is also a distinct shift in community composition from sensitive to tolerant taxa.[274]

The occurrence of age- and size-dependent metal accumulation by aquatic biota has been documented by several authors, particularly for marine and estuarine molluscs.[166,1445] The metal burden of an organism can be defined as the total metal content of an individual. It is the sum of surface-adsorbed and internally incorporated metal. If the total metal content remains constant while the body weight changes, measurements of concentration alone may mask the dynamics of metal uptake, storage, and elimination. Metal concentrations in tissues may change while metal burdens remain constant, or concentrations can remain constant while burdens change.[875]

To understand the effects of metals on biota, it is important to know the relationships between metal concentrations in the environment and those in biological tissues. Aquatic insects can concentrate a number of metals.[208,291,1118] Lynch et al.[953] showed that aquatic insects at sites downstream from a mine/mill complex accumulated more molybdenum and copper than upstream insects. Tolerance of insects to metals has been studied biochemically and histochemically in aquatic midges[1449,1659] and mayflies,[1448] as well as in terrestrial flesh flies[48] and silkworms.[1455] In cases where organisms are able to regulate metals, correlations between environmental concentrations and the concentrations in these organisms would be poor. Tolerance to elevated concentrations of trace elements can involve the exclusion, active excretion, or intracellular storage of metals.[877] Histochemical and ultrastructural studies have revealed that metals can be concentrated in insects into specific storage units that are generally associated with digestion, excretion, or ionic regulation.[190,877] For some organisms, metallothionein or other low-molecular-weight, high-S proteins sequester metals. Many invertebrate species can produce metallothioneins to package metals in membrane-bound vesicles and to localize metals in specific tissues.[474] Intracellular granular inclusions may contain several metals. Granule structure and content not only differ among taxa, but also within taxa collected from contaminated or uncontaminated environments. Metal-containing granules have been reported for all the major invertebrate phyla.[190] The function of these granules is often uncertain, although in some species they may serve to immobilize excess concentrations of a metal. The occurrence of metal-containing granules in freshwater insects has received little attention, although Ballan-Dufrancais et al.[76] observed mineral concretions in the midgut cells of 12 species, including the anisopteran *Gomphus* sp. and the chironomid *Chironomus plumosus*. Lhonore[932] observed metal-containing granules containing a number of metals (Ca, Mg, Mn, Zn, and Fe) in the Malpighian tubules, midgut, and adipose tissue of the larvae of *Phryganea varia* (Trichoptera). Darlington and Gower[351] found copper-containing granules in the larvae of *Plectrocnemia conspersa* (Trichoptera). Sumi et al.[1448] found that copper was present in the midgut epithelial cells. The copper accumulated in the epithelial cells could scarcely diffuse into the outer parts of the gut through the epithelial basement membrane. The copper that accumulated in the cells appeared to be granular, indicating that the copper was bound to some proteins and that the formalin used for fixation had resulted in granulation of the Cu-bound proteins.

Larvae of *Baetis thermicus* from a metal-contaminated river (River Mazawa, Yamagata, Japan) accumulated cadmium, copper, iron, magnesium, and zinc markedly. Specific metal-binding proteins induced by copper and cadmium in the tolerant mayfly sequestered these potentially toxic metals.[47,1456,1457] When the larvae of a metal-resistant mayfly species (*Baetis thermicus*) and two metal-susceptible species (*B. yoshinoensis* and *B. sahoensis*) were exposed experimentally to cadmium, a cadmium-binding protein was induced only in the larvae of the resistant species (*B. thermicus*), indicating species-specific differences.

Sumi et al.[1448] collected larvae of the mayfly *Baetis thermicus* from a metal-contaminated river (River Mazawa, Yamagata, Japan) and a noncontaminated river (in Yokohama, Japan), and the histochemical localization of copper, iron, and zinc was examined using staining agents. The results showed that copper absorbed into the midgut epithelial cells induced metal-binding proteins and that the proteins bound tightly to copper were accumulated in the luminal cytoplasm of the epithelial cells. Therefore, the protein-bound copper scarcely diffused into outer parts of the midgut of the larvae. In contrast, zinc is thought to have been bound loosely to native proteins and/or related compounds, since it did not induce any specific metal-binding proteins in the larval tissues.[980,1456] Zinc can diffuse easily into the outer parts of the midgut through the epithelial basement membrane. The high tolerance to metals of mayfly larvae inhabiting a polluted river may be associated with selective induction of metal-binding proteins in their gut.[1448]

Hare et al.[617] determined the concentrations of metals (Cd, Cu, Zn, and As) both for the gut contents and for the body of the nymphs of the burrowing mayfly *Hexagenia limbata*. Trace elements in the gut contents represented up to 22% of the trace elements in the whole animal. *H. limbata* individuals varied substantially in the gut clearance time. Radioisotopes of cadmium, lead, and zinc, added in trace amounts to lake sediments, were used by Hare et al.[617] to measure the uptake and efflux of these metals from various body parts of nymphs of the burrowing mayfly *Hexagenia rigida*. There was no accumulation of radioisotopes in gill tissues, suggesting that the gills were not the main organ of metal uptake in *Hexagenia*. Net uptake of ^{109}Cd and ^{65}Zn by the gut exceeded that by all other body parts in both quantity and concentration, suggesting that the primary source of these metals to *Hexagenia* is sediment consumed as food. ^{210}Pb was not detected in the gut, but was found mainly on the body surface.

Krantzberg and Stokes[877] used atomic absorption, scanning electron microscopy, and X-ray microprobe analysis to investigate metal accumulation and localization in chironomid larvae. The gut was an important site for metal accumulation in chironomid larvae, particularly for chironomids from an acid- and metal-contaminated environment. The internal surfaces of the anal papillae also had detectable metal concentrations. For some insects the midgut pH is slightly acidic (pH < 7.0), and the hindgut somewhat more acidic (pH < 6.0) due, in part, to the secretions of the Malpighian tubules.[245,1608] This acidity may render metals more available through the hydrolysis of metal-ligand bonds. These regions may be sites for metal deposition and thus could be detected by X-ray microanalysis. The midgut, rectum, and Malpighian tubules are sites of extensive ion transport and are structures that could potentially be involved in metal metabolism.[877] In terms of storage capacity, the chironomid midgut epithelium, hemolymph, and fat body are analogous to other invertebrate organs that often have metal inclusions.[877] Nutrient storage occurs in the insect fat body and in the hemolymph itself, and metals have been found in the Malpighian epithelia of *Musca*,[1402] the midgut epithelia of *Drosophila*,[1467] the digestive epithelium and fat body of *Chironomus*,[1449] in the epithelium and lumen of the Malpighian tubules of house flies,[1402] and in the alimentary canal and Malpighian tubules of silkworm larvae.[1455] Krantzberg and Stokes[877] found iron, lead, and occasionally manganese in this gill-like structure of chironomids. Other metals (Ni, Cu, Zn, Cd, and Al) were also sometimes detected. The anal papillae of chironomid larvae may act in a manner similar to that of other invertebrate gills.[877]

Krantzberg and Stokes[877] also compared chironomids collected from two sites that differed in their degree of contamination. Chironomids from two populations differed in their metal storage and distribution. Larvae from the more contaminated site had an increased incidence of metal granule formation. For many invertebrates, specific tissues or organs are involved in the metabolism of essential elements. In polluted environments, however, nonessential metals can compete for binding sites and be incorporated within storage tissues as intracellular granules. Larvae from a contaminated environment may have an enhanced ability to detoxify metals when exposed to elevated concentrations of available metals.[877]

Earlier developmental stages of insects may be more sensitive than later ones,[281,963,1320] although toxic effects on the egg itself have been largely ignored. Gauss et al.[530] found that fourth instar larvae of *Chironomus tentans* were 12 to 27 times more resistant to copper stress than first instars, and that the eggs were much more resistant than either of these instars. Kosalwat and Knight[876] also found *C. decorus* eggs to be more tolerant to copper than other developmental stages.

Several researchers have reported a general pattern of increased tolerance to metals, from mayflies to caddisflies to chironomids.[274,1605,1627] The Chironomidae, particularly Orthocladiini, seem to be especially tolerant to metal pollution.[51,241,1417,1450,1567,1627,1628] Surber[1450] reported increased abundance of *Cricotopus bicinctus* at sites receiving copper and chromium. Winner et al.[1627] found that two species of Orthocladiini, *C. bicinctus* and *C. infuscatus*, dominated stations impacted by heavy metals. In the study of Clements et al.,[276] Orthocladiini were the dominant organisms at all effluent stations in the field and in treated experimental streams. The increased abundance of Orthocladiini may have resulted from reduced numbers of potential competitors that were eliminated from treated streams and effluent sites. Alternatively, these organisms may have responded to increased abundance of a particular resource. Surber[1450] noted that Orthocladiini feed on resistant blue-green algae that often dominate metal-polluted areas. In contrast, Tanytarsini chironomids appear to be quite sensitive to metals,[276] urging caution when making generalizations among taxa such as Chironomidae.

B. LEAD

Lead (Pb) is a worldwide pollutant originating from atmospheric emissions (e.g., combustion of leaded petrol), liquid effluents, and solid waste material.[1654] It occurs in trace amounts in rocks, soil, water, air, plants, and animals. The toxicity of Pb to aquatic organisms and its accumulation in water and the aquatic biota[739,754] are well documented, but records on insects are few.[34,128,876,877,1155,1654]

Oladimeji and Offem[1155] investigated the accumulation, depuration, and toxicity of lead to *Chironomus tentans* larvae. The mortality data indicated that *Chironomus* larvae were the most tolerant to lead poisoning. The "safe" concentrations of lead for chironomid larva would be 0.27 mg/l, and concentrations higher than 0.33 mg/l would damage both the fish population and the organisms that make up their food.[1155]

Erosion and leaching of abandoned deposits of lead-mine tailings in southeast Missouri have resulted in the heavy-metal contamination of surface waters, sediments, and aquatic biota. Besser and Rabeni[128] examined bioaccumulation and toxic effects of mine tailings on the crayfish *Orconectes nais*, *Hexagenia limbata* (Ephemeroptera), and *Chironomus riparius* (Diptera). The bioaccumulation of lead and cadmium may reflect differences in the physicochemical behavior of these metals that affected their availability to invertebrates. Lead concentrations in mayflies were much higher than those in crayfish, while cadmium concentrations were generally higher in crayfish, suggesting that different routes of uptake exist for these two metals. The relatively high concentration of dissolved lead in pool water and the formation of lead/organic complexes apparently favored lead uptake from the aqueous phase. In contrast, the weak correlations of cadmium bioaccumulation with dissolved cadmium levels suggested that the uptake of dissolved cadmium was less important at the low cadmium concentrations in pool water. Ingestion of cadmium-enriched solids, more accessible to free-ranging crayfish than to

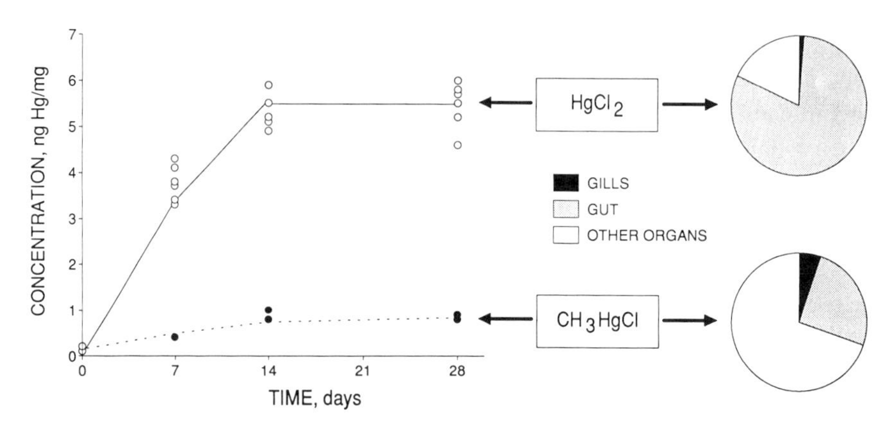

FIGURE 4. Total mercury concentration in *Hexagenia rigida* (Ephemeroptera, Ephemeridae), function of exposure duration and chemical form of the metal initially added to the sediment ($HgCl_2$, 10 μg Hg per gram and CH_3HgCl, 1 μg Hg per gram); and mercury distribution in the gills, the gut, and the rest of the body of *H. rigida*, expressed in relative burdens (%), after contamination via the sediment with $HgCl_2$ (10 μg Hg per gram) and CH_3HgCl (0.5 μg Hg per gram). (Modified from Saouter, E., Ribeyre, F., Boudou, A., and Maury-Brachet, R., *Environ. Pollut.*, 69, 51, 1991. With permission.)

mayflies confined to enclosures, may have contributed to the high cadmium levels in crayfish, from some treatments.[128] Dissolved and complexed metals in leachates apparently contributed to the toxic effects in invertebrates. Toxic effects were strongly correlated with dissolved concentrations of several metals in effluent and pool water. Effects on survival, growth, and development suggested that leachate exposure caused chronic toxicity in midge larvae, although they were resistant to the acute toxicity of metals. Reductions in red pigment suggested that heme synthesis was inhibited by exposure to tailing leachates, as has been reported in fish and invertebrates exposed to lead and zinc in the laboratory. Such inhibition of important biochemical processes may have contributed to the chronic effects observed during leachate exposures.[128]

C. MERCURY

Mercury (Hg) pollution comes from industrial processes associated with mining and smelting; pulp and paper production; manufacturing; the combustion of coal, oil, natural gas, and wood; and, in the past, agricultural applications of mercurial compounds to control fungal pathogens. Methyl mercury is the most hazardous form of mercury to aquatic organisms. It is readily synthesized by bacteria from inorganic mercury.[1654]

Barak and Mason[78] studied the concentrations of mercury, cadmium, and lead in water, sediment, and invertebrates from the Rivers Brett and Chelmer in eastern England. They observed elevated levels of mercury in Trichoptera, cadmium in Ephemeroptera, and lead in Trichoptera, Ephemeroptera, and Gastropoda. Hammer et al.[602] recorded high mercury concentrations in corixids

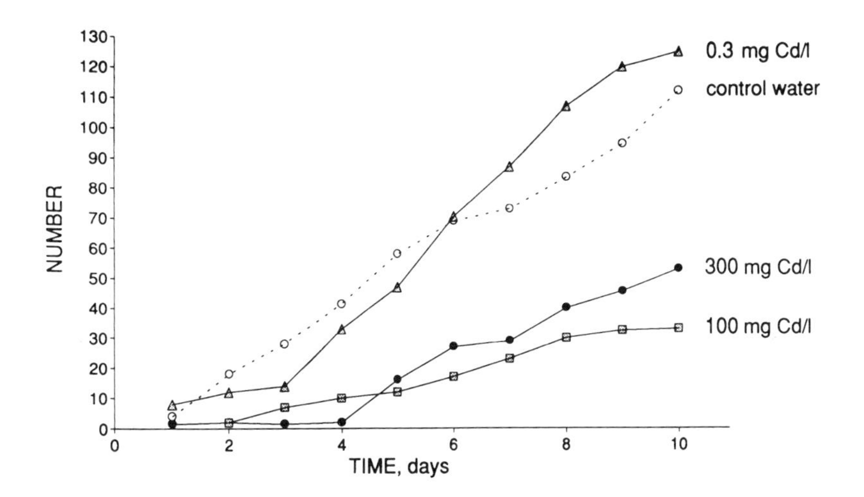

FIGURE 5. The cumulative number of egg masses laid by *Chironomus riparius* (Diptera, Chironomidae) at various concentrations of cadmium and in control water. (Redrawn from Williams, K. A., Green, D. W. J., Pascoe, D., and Gower, D. E., *Bull. Environ. Contam. Toxicol.*, 38, 86, 1987. With permission.)

and gastropods. Elevated mercury concentrations have also been investigated in Chironomidae, Notonectidae, and Odonata in various freshwaters.[1654]

Saouter et al.[1324,1325] investigated the accumulation of two mercury compounds ($HgCl_2$ and CH_3HgCl) by the mayfly *Hexagenia rigida* from artificially contaminated sediments. During the first 2 weeks, mercury accumulation in the nymphs was much more rapid for organic mercury (Figure 4). More than 80% of the mercury body burden for the inorganic form was localized in the gut, and only 1% was accumulated in the gills. The relative burdens in the gut and gills were 25 and 5% for the organic form, respectively.

The concentrations of mercury acutely (LC_{50}) toxic to aquatic insects range from 2.0 mg/l for the mayfly *Ephemerella subvaria* to 1200 mg/l for some damselflies and caddisflies.[1270,1654]

D. CADMIUM

Anthropogenic sources of cadmium (Cd) include mining, smelting, and also agriculture if contaminated sewage sludge is repeatedly used as a fertilizer. Cadmium from water is known to be concentrated in the tissues of aquatic biota through respiration and surface adsorption.[542,1641,1654] However, the role of food in the uptake of cadmium is not well understood, and current evidence is contradictory.[986] The toxicity of cadmium has been investigated in several aquatic insects.[1414,1654] There are considerable differences in cadmium tolerance among insects. Concentrations of less than 10 µg/l were lethal to the mayfly *Ephemerella* sp., while 238 µg cadmium per liter had little effect on the caddisfly *Hydropsyche betteni* and the stonefly *Pteronarcys dorsata*.[1654]

Williams et al.[1615] investigated the extent of selection by *Chironomus riparius* females between a range of cadmium solutions as sites for oviposition and the effect of cadmium on egg hatching. Female *C. riparius* discriminated between different concentrations of cadmium or some related physical/chemical parameter. Significantly higher numbers of egg masses were laid in the control and lower concentrations of cadmium (0.3 and 30 mg Cd per liter) than in solutions of 100 and 300 mg Cd per liter (Figure 5). The egg stage of *C. riparius* was resistant to cadmium toxicity. Exposure of egg masses to cadmium, subsequent to their complete formation in control water, had no effect on the development of the embryos, with 80 to 100% of the masses hatching at each concentration. The percentage viability was lower when egg masses were laid directly into cadmium solutions. Exposure during oviposition disturbed and prevented the complete development of the embryos. Similarly, Wegner and Hamilton[1576] found that the majority of *C. riparius* eggs failed to develop and hatch at 56 and 100 mg calcium sulfide per liter, but at relatively low concentrations (up to 3.2 mg sulfide per liter), little or no toxic effect was observed.

Yamamura et al.[1659] showed that a large proportion of the cadmium uptake in the larvae of *Chironomus yoshimatsui* was bound to high-molecular-weight proteins. The cadmium was subsequently gradually transferred to low-molecular-weight proteins showing characteristic properties of metallothionein during the 5-day experiment. Moreover, Sumi et al.[1449] used the histochemical method to demonstrate that body fat was important for cadmium storage in the organs of this midge larva. Seidman et al.[1356] demonstrated that 63 to 81% of the cadmium accumulated by *Chironomus thummi* was associated with midgut tissues, presumably reflecting the importance of the gut in metal bioaccumulation. Cadmium-binding proteins have also been detected in other cadmium-tolerant insects, including flesh flies[46] and silkworms.[1455] The mechanisms responsible for the tolerance and including induction processes and induced cadmium-binding proteins, were different among those insects. In addition, metal-binding proteins have also been studied in stoneflies[282,448] and terrestrial insects, such as locusts,[988] cockroaches,[159] and *Drosophila* flies.[980]

Cadmium-binding protein was suggested to detoxify cadmium by sequestering this metal in mayfly larvae. Aoki et al.[47] compared the ability of cadmium accumulation and the inducibility of cadmium-binding protein among the larvae of three mayfly species: *Baetis thermicus*, a metal resistant species,[635,1456] and metal susceptible species *B. sahoensis* and *B. yoshinensis*. The larvae were exposed to cadmium (10 mg/l), and the relative induction of cadmium-binding protein was correlated to their susceptibility to the metal. Amounts of cadmium in the larvae increased linearly with the duration (10 days) of exposure. The cadmium concentration differed little among the species. Cadmium-binding protein was induced only in the larvae of resistant species.

The susceptibility to toxic substance of animals subjected to stress may increase.[1180] Pascoe et al.[1181] examined the effects of the provision of food and

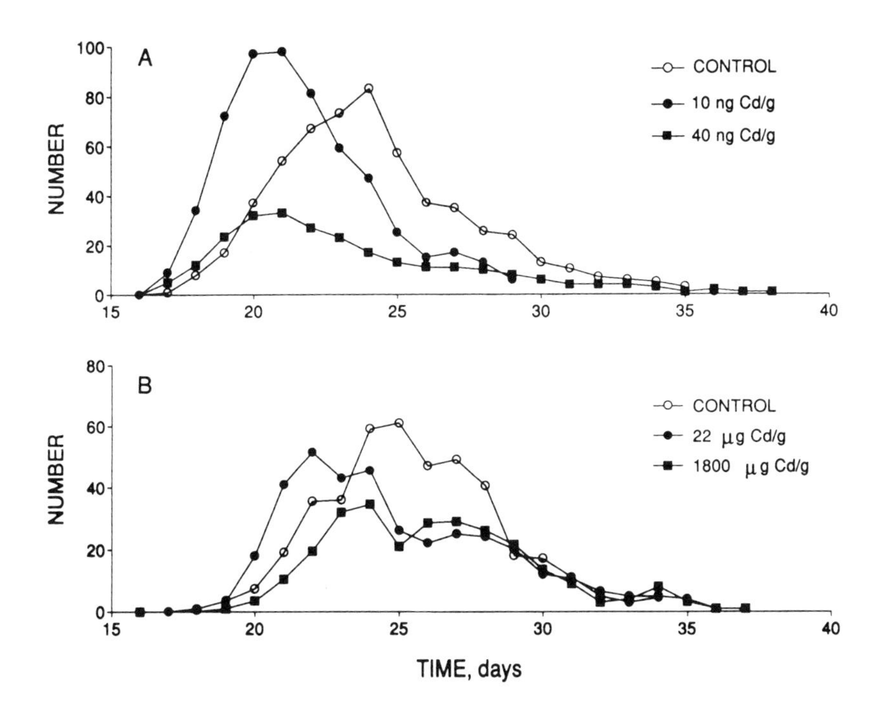

FIGURE 6. Numbers of *Polypedilum nubifer* (Diptera, Chironomidae) females daily emerging from the flow-through (cleanwater) aquarium. (A) Aquarium received a continuous flow of cadmium-contaminated water; (B) larvae were fed cadmium-contaminated food from immediately after the hatch. (Modified from Hatakeyama, S., *Environ. Pollut.*, 48, 249, 1987. With permission.)

artificial sediment on the response of the benthic detritivorous larvae of *Chironomus riparius* to cadmium. Acute toxicity tests demonstrated that the presence of food within the system increased the observed toxicity of cadmium to *C. riparius* fourth instar larvae, while the artificial sediment alone decreased toxicity. Since larvae lived within, and fed upon, the sediment, those animals supplied with food were exposed to higher levels of cadmium than those without food. Cadmium entered via the gut, as well as through the body surface. Because animals provided with sediment are able to construct burrows and thereby adopt normal behavior patterns (e.g., ventilation, search activity), they are likely to be subjected to less physiological stress than those without sediment. This could account for their greater tolerance to cadmium.

Hatakeyama[631] studied the effects of cadmium on the emergence rate, oviposition rate and hatchability of the eggs produced by a chironomid *Polypedilum nubifer* in a flow-through aquarium. No significant impact of cadmium on reproduction was observed in the midge larvae that had been exposed to 10 or 20 mg Cd per liter. Emergence of the cadmium-exposed larvae peaked several days before that of the control, although growth was impaired in first and/or second instars (Figure 6). The percentage emergence success decreased to about 46% of the control at 40 μg Cd per liter. Adult sex ratio, oviposition success, and egg hatchability were not impaired. The emergence success decreased to less than 3% of the control at 80 μg Cd per liter. The emergence success of midge larvae that had been fed contaminated food containing 220 or 1800 μg Cd per gram decreased to nearly 60% of the control, while that of larvae exposed to food containing 22 μg Cd per gram was unaffected.

Studies on the effects of age and weight on metal body burdens of freshwater larvae of *Chironomus* showed that cadmium concentrations were greater in small and young larvae than in older and larger individuals.[875] There is a decrease in the surface area volume ratio with increasing body size, and the relative contribution of surface-adsorbed metal to the total body content should become less important. Only 25% of the total cadmium burden was adsorbed on the exoskeleton of chironomid larvae,[876] and it was apparent that physiological mechanisms exerted some control on cadmium dynamics.[875]

McCahon et al.[1016] studied the acute toxicity of cadmium to the first, third, and fourth larval instars of the caddisfly *Agapetus fuscipes*. First-instar larvae were more sensitive than older instars, and for all age classes, cased animals were more resistant than uncased ones. Case-building activity and the aggregation response were reduced. The ratio of 1.5 between the sensitivity of the youngest and oldest larvae was comparable to differences found between juveniles and adults of several species exposed to a range of compounds.[963,1270,1579] This value was much lower than the ratio reported between the first and fourth instars of *Chironomus riparius* (1:950) during cadmium exposure.[1614]

Acute toxicity of cadmium to 12 species of aquatic macroinvertebrates was examined by Brown and Pascoe.[187] The median lethal time (LT_{50}) for each of 12 predator species decreased with increasing cadmium concentration. In general, the insects were more tolerant of cadmium than other taxa. The most sensitive predator, *Glossiphonia complanata* (Hirudinea), was 10,000 times more sensitive than *Aphelocheirus aestivalis* (Heteroptera), the most tolerant species. Predators were relatively resistant to cadmium poisoning.

Martin et al.[986] monitored the flux of dietary cadmium, using the mass balance technique with the dragonfly *Aeshna canadensis* nymph. The nymphs were first fed food with a cadmium concentration typical of prey items found in relatively unpolluted waters and were then exposed to a cadmium-enriched diet in order to determine whether a change in metal flux and body accumulation occurred at elevated levels of dietary intake. The results showed that the cadmium concentration of fecal pellets was higher than that of the food source. Only 10% of the dry weight of the food intake was excreted in the fecal pellets, whereas all the cadmium was excreted, resulting in the high fecal concentrations. Not only was all the ingested cadmium excreted, but the fecal pellets actually accumulated cadmium from some other source because total cadmium egested was greater than that ingested. It was hypothesized that the excess cadmium in the fecal pellets is probably due to the nymphs' ability to scavenge metal ions out of the water column. Odonate fecal pellets are wrapped in a peritrophic membrane, which is a semipermeable structure lining the midgut region of invertebrates consuming solid food. This membrane is continually replaced because it passes with the excreta into the hindgut and is egested with the feces. It resembles the cuticle, or outermost layer of the arthropod exoskeleton,[1199] and, like the exoskeleton, has a strong affinity for metals.[282] However, the increase in cadmium in the fecal pellets may also be due to the reduction in weight of the fecal material compared to the food ingested, because of the uptake of nutrients by the gut.

Hatakeyama and Yasuno[634] used midge larvae (*Chironomus yoshimatsui*) as prey in a food-chain model to assess the chronic effect of cadmium through food on the growth and reproduction of the guppy fish *Poecilia reticulata*. The transfer rate of cadmium from the midge (270 μg/g dry weight) to the fish was between 0.5 and 1% during the 30-day experiment. The growth rate of the fish was not impaired, but the cumulative numbers of fry produced decreased to 80% of the control. The cadmium concentrations of the digestive tract, liver, and kidney of the fish increased strongly, indicating that cadmium accumulation was mainly derived from the cadmium-accumulated midges. Mortality of the female fishes increased abruptly 6 months after the start of the experiment, whereas no males died.

E. ZINC

The effects of zinc (Zn) on aquatic insects have often been investigated in connection with copper and other metals. Hatakeyama[632] studied the effects of zinc on mayflies. The growth of mayfly larvae exposed to 100 or 300 μg/l Zn ceased after the second week, and all died before emergence. At 30 μg/l Zn the growth rate decreased gradually, and many larvae died before emergence. The molting interval also nearly doubled at these concentrations. Zinc was very toxic to *Epeorus latifolium* (Ephemeroptera), although less so than copper. Zinc concentrations have usually reported to be much higher than those of copper, in river water receiving effluents from abandoned mines.[189,1662] Therefore, it is difficult to assess which is more hazardous to aquatic insects in rivers where the Zn concentrations are several times higher than copper, although the effects of copper on the growth and emergence of mayfly larva was higher than zinc.[632]

Armitage[51] found a negative correlation between the zinc level and the number of taxa in rivers, although the situation was complicated by localized inflows of calcium-rich water. Wilson[1625] surveyed the zinc-polluted River Nent and the East and West Allen, England, using chironomid pupal exuviae. The area of the North Pennine orefield is notable for the extensive mining operations that were carried out principally during the 18th century. At the time of the survey there was very little mining activity in the area, but previous workings and accumulations of mine spoil had left a legacy of acid- and metal-enriched streams. The rivers were affected by acidic mine drainage and were particularly rich in zinc. Chironomid pupal exuviae from zinc-enriched sites showed consistently lower diversities than those from low-zinc sites. The dominant species found at zinc-enriched sites were *Krenosmittia camptophleps* in the Nent and *Eukiefferiella clypeata* and *Tvetenia calvescens* in the West Allen.

F. COPPER

Copper (Cu) is one of the elements essential to the metabolism of many plants and animals. Copper is important because it participates in many enzymatic reactions and is a constituent of the active sites of a number of enzymes. In larger quantities copper can be toxic and have detrimental effects, especially in the lower trophic levels of food chains. Copper is one of the metals mobilized in the environment by man at rates exceeding those of natural geological processes. Since it is toxic and readily accessible to aquatic organisms, copper has received considerable attention.[275-277,351,556,866,867,924,1413,1448,]

Copper has a great affinity for organisms and organic matter. A significant portion of the copper present in waters is complexed with organic matter.[167,915] Copper is also removed from aqueous solution by both coprecipitation and sorption during manganese and iron oxide precipitation. Most of the copper entering waters is strongly bound on inorganic and organic exchange sites or complexed with organic matter in the sediment.[866,1666]

Leland et al.[924] described changes in the population densities of benthic insects and secondary production of an oligotrophic stream (Convict Creek, CA) in response to continuous, low-level experimental additions of copper (2.5 to 15 μg/l total Cu, equivalent to 12 to 75 ng/l Cu^{2+}). The threshold for tolerance of most benthic insects in Convict Creek was between 2.5 and 10 μg/l total Cu. Declines in the population density of species belonging to Plecoptera, Coleoptera, Trichoptera, and Diptera occurred at 5 and/or 10 μg/l total Cu. Differences in sensitivity among families were apparently related to the trophic levels of taxa, with herbivores and detritivores being more sensitive than predators. Few increases in population density of benthic insects were attributed to prolonged Cu exposure. The detritivorous stoneflies *Malenka* sp. and *Pteronarcys princeps*, the herbivorous-detritivorous mayfly *Ironodes lepidus*, and three predatory taxa (caddisflies *Rhyacophila vaccua* and *R. angelita*, and the fly family Empididae) were more abundant at 2.5 μg/l total Cu than in the control. Only the caddisflies were more abundant at 5 or 10 μg/l total Cu than in the control. The benthic communities in the 5 and 10 μg/l total Cu treatments were different from the control. The difference in percentage similarity 11 months after dosing was due to the lack of recolonization by several taxa, such as *Calineuria californica*, *Doroneuria baumanni* (Plecoptera, Perlidae), and *Optioservus divergens* (Coleoptera, Elmidae), in the medium- and high-dosed sections. In contrast, several univoltine taxa that hatch in autumn, e.g., *Brachycentrus americanus*, *Epeorus dulciana*, *Micrasema* sp., *Paraleptophlebia pallipes*, and *Rhyacophila vaccua*, were more abundant in the medium- and high-dosed sections than in the control.[924]

Winner et al.[1627,1628] monitored changes in macroinvertebrate communities exposed to copper in experimentally polluted streams and concluded that these communities showed predictable changes in structure along a metal stress gradient. Clements et al.[275,276] exposed natural assemblages of aquatic insects to copper in laboratory and outdoor streams and showed that the number of taxa and individuals were highly sensitive to this toxic element. Clements et al.[275] examined the effects of copper on aquatic insect communities, using rock-filled trays colonized in the field for 30 days, transferred to laboratory streams, and dosed with $CuSO_4$. Each stream was assigned to one of three treatments: control, low dose (15 to 32 μg/l), and high dose (135 to 178 μg/l). Experiments were repeated during winter, spring, and summer. Simple community-level parameters, such as number of taxa and number of individuals, were highly sensitive to copper exposure. Within 96 h the number of taxa per tray in the low-dose streams (15 to 32.0 μg/l) was reduced by 24 to 36% and the number of individuals by 35 to 52%, relative to the controls. Species diversity was significantly reduced in low-dose streams only during the summer. The effects of copper on Ephemeroptera, the dominant aquatic insects in the winter and summer experiments, showed some seasonal variation. Mayflies, particularly *Baetis* sp., were greatly reduced in the treated streams during each season, but were especially sensitive during summer when water temperatures were higher. Exposure of 15 to 32 μg/l Cu reduced the total number of mayflies by 64 to 82%, suggesting that these communities were considerably more sensitive to metals than indicated by single-species tests. The effect of copper exposure on the abundance of Diptera and Plecoptera varied seasonally, but was less severe than for mayflies.[275]

Geckler et al.[531] reported reduced abundance of most macroinvertebrate groups, but increased numbers of chironomids, at sites affected by copper. In the study of Clements et al.,[275] the chironomids also increased in treated streams during experiments in winter and spring. The results demonstrated significant variation in copper toxicity among groups of aquatic insects. Tanytarsini (Chironomidae) and the mayflies *Baetis brunneicolor* and *Isonychia bicolor* were especially sensitive to copper, whereas Orthocladiini chironomids and *Cheumatopsyche* sp. were quite tolerant. Sensitivity to copper was unrelated to trophic relationships: Tanytarsini and Orthocladiini, which were among the most- and least-sensitive groups, are both classified as collector-gatherers, and *Cheumatopsyche* sp. and *I. bicolor*, which differed greatly in their sensitivity to copper, are collector-filterers.

Morphological characteristics of aquatic insects may affect their susceptibility to metals. Hodson et al.[694] noted that variation in acute toxicity among invertebrate taxa is related to the nature of the body covering. Stoneflies and beetles, which have heavily sclerotized plates, are highly tolerant of metals.[1117,1414,1554] Clements et al.[278] suggested that the external, plate-like gills of mayflies increased their sensitivity to copper. The highly active species, such as *Baetis brunneicolor* and *Isonychia bicolor*, which presumably require more oxygen than the less active taxa, were more sensitive. Copper toxicity within taxa appears to be greater among small individuals.[278,282,530,694]

Kosalwat and Knight[866] investigated the acute toxicity of copper in water and substrate to the midge *Chironomus decorus*. They also studied the ability of *C. decorus* to bioconcentrate copper. This chironomid species was more sensitive to copper than many other aquatic insect larvae. *C. decorus* larvae have a rather thin body covering that may facilitate copper penetration. Dodge and Theis[384] concluded from their experiments on the uptake of copper by dead and living larvae of *Chironomus tentans* that the uptake process is largely passive and involves chemical interactions between Cu^{2+} and sorption sites at the surface or in the interior of the organism. *C. decorus* could conceivably bioconcentrate copper from the water (no food present), but not biomagnify it from the food substrate.[866] Food-chain biomagnification is unlikely to occur, since most of the copper ingested is excreted by the body, and very little is retained. According to acute toxicity tests, copper was more toxic to midges when present in water (aqueous forms) than when present in the food substrate (organically complexed forms). In nature, unless the bodies of water are heavily polluted by copper or polluted with acidic wastes or acidic precipitation, it is unlikely that immediate toxic effects would occur, since most of the copper would be tied up in the sediment. Over an extended period, however, low levels of copper in the sediment may produce chronic toxic effects in aquatic organisms in terms of growth reduction, life cycle alteration, reduced resistance to diseases, and skeletal deformities. These chronic effects of copper may be more important than its acute effects because these subtle changes can proceed unnoticed in nature.[867]

The results of field investigations on metal-polluted rivers in Japan showed that the density of *Epeorus latifolium* (Ephemeroptera) was very low in places where the copper concentration in water exceeded 25 μg/l.[635] Copper in water at a level of 25 μg/l seemed to be so toxic that the low density of *E. latifolium* in these rivers could be attributed solely to copper in the water. Hatakeyama[632] investigated the effects of copper and zinc, via food as well as via the water, on the growth and emergence of the young larvae of *E. latifolium*, using an indoor model stream. A test designed to assess the effects of metals via the water in which the aquatic organisms live is inevitably accompanied by metal contamination of their food, since metals are easily adsorbed on, and/or accumulated in, their food. However, the copper concentration of algae exposed to 100 μg/l Cu for 1 week was about 450 μg Cu per gram. The main cause of the high mortality of mayfly larvae exposed to 100 μg/l Cu was thus attributed to copper in the water because at this algal copper level, the mayfly larvae did not die at all.

According to Kosalwat and Knight,[867] the embryonic development and hatchability of *Chironomus decorus* eggs were not affected by 0.1 to 5 mg/l Cu in water. The embryos developed normally and hatched at about the same time (after 55 h of incubation). All the larvae survived to the end of the test (72 h), except those subjected to 5 mg/l Cu in water, which died after only partial emergence from the egg. The eggs were apparently protected by their shell from copper. The growth of *C. decorus* larvae was reduced significantly when

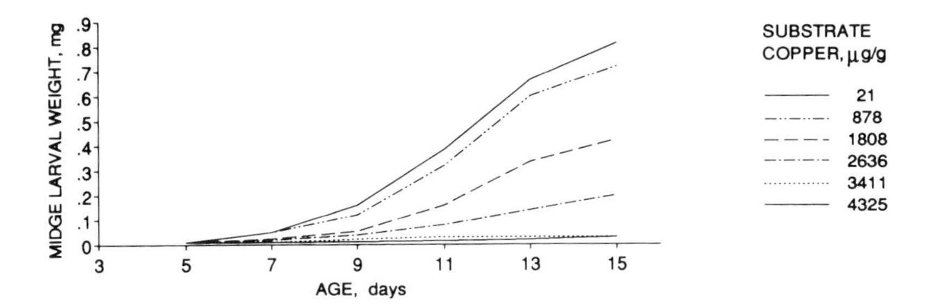

FIGURE 7. Growth curves of *Chironomus decorus* (Diptera, Chironomidae) larvae reared in food substrate with different levels of substrate copper. (Modified from Kosalwat, P. and Knight, A. W., *Arch. Environ. Contam. Toxicol.*, 16, 283, 1987. With permission.)

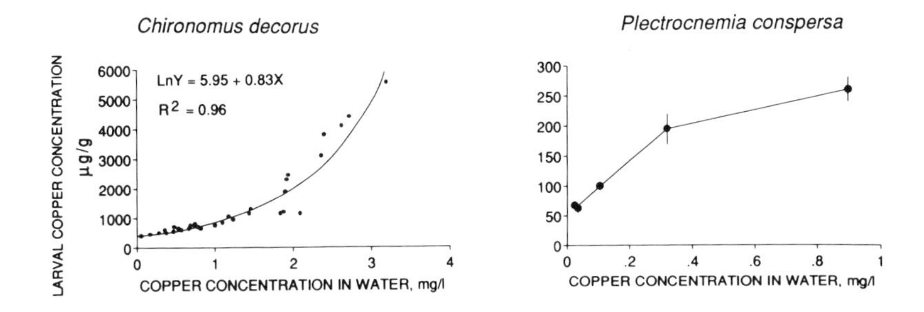

FIGURE 8. Relationship between copper concentration in larvae of two aquatic insects, *Chironomus decorus* (Diptera, Chironomidae) and *Plectrocnemia conspersa* (Trichoptera, Polycentropodidae), and in water. *P. conspersa* were collected in a copper-rich stream in England. *C. decorus* larvae were experimentally exposed to copper concentrations for 48 h. (Redrawn from Kosalwat, P. and Knight, A. W., *Arch. Environ. Contam. Toxicol.*, 16, 275, 1987; and Gower, A. M. and Darlington, S. T., *Environ. Pollut.*, 65, 155, 1990. With permissions.)

they were reared in copper-spiked food substrate (bound Cu) from the age of 1 to 15 days (900 to 4500 µg/g Cu) (Figure 7). The substrate copper concentration at which larval growth was reduced to 50% was 1600 µg/g. Substrate copper caused deformities in the epipharyngeal plate of the larval mouth parts, and a copper concentration higher than 1800 µg/g delayed adult emergence. The copper concentration in pupal exuviae and adults was positively correlated with the copper concentration in the substrate. The larval stage was the most sensitive to copper toxicity, while eggs were the least sensitive. Larval growth could be used to detect copper at relatively low concentrations. Larval deformities offer a quick tool for evaluating copper pollution.

The relationship between copper in water and copper in *C. decorus* larvae could be described by an exponential equation on the basis of acute toxicity tests.[866] However, different relationships between lower concentrations in water and another aquatic insect, *Plectrocnemia conspersa* (Trichoptera), was found in a main drainage streams[56] (Figure 8). Larvae of *Plectrocnemia* from copper-rich sites contained elevated concentrations of this metal.[189,352,353]

Brown[189] found an almost proportional increase in the copper concentration of free-living Trichoptera in the River Hayle, Cornwall, U.K., with increasing copper levels. In *P. conspersa* the similar life cycles of control and contaminated stream populations[353] and the similar survival and growth of introduced and captive resident larvae in the transfer experiment indicated that copper tolerance is a characteristic of the species. *P. conspersa* may also be widely tolerant of metal-contaminated waters, since its prominence within a depleted community is a feature of streams in southwest England affected by copper, lead, zinc, and aluminium.[556]

Gower and Darlington[556] investigated the relationships between the copper content of final instar larvae of *P. conspersa* and their environment. Material in the gut was apparently the single most important copper-containing component, since clearing the larval guts reduced copper concentrations by 61, 57, and 51% in instars V, IV, and III, respectively. However, the percentage reductions possibly included copper excreted during the starvation period. Using histological and electron microscopy techniques together with X-ray microanalysis, Darlington and Gower[351] determined the location of copper stored in the larvae of *P. conspersa* collected from the field. Copper-containing granules were observed in the cells of the Malpighian tubules and in the subcuticular region of larvae. The granules in both these regions may be primarily pigment granules, which provide a mechanism for removing potentially toxic concentrations of copper, and possibly other metals, from circulation. This mechanism of metal tolerance may, in part, account for the reported occurrence of larvae of *P. conspersa* in waters with elevated metal concentrations. This mechanism of metal tolerance may also increase the potential for transfer of metals to higher trophic levels.[353]

Certain electrophoretically defined genotypes and levels of heterozygosity have been associated with differential resistance to the effects of environmental contaminants or have been found in higher frequencies than other genotypes in impacted areas. Benton and Guttman[120] designed a study to test the relative sensitivity to copper toxicity of allozyme variants at several electrophoretically detectable loci in the heptageniid mayfly *Stenonema femoratum*. Mayflies were exposed to 1.6 mg/l copper sulfate for 126 h. All the individuals were then analyzed at three polymorphic loci: glucose phosphate isomerase (GPI), phosphoglucomutase, and malate dehydrogenase. The study produced evidence for differential survivorship among allozyme genotypes of *S. femoratum*. Individual times to death were significantly shorter in two GPI genotypes (one of them was very rare). A reduction in genetic variability caused by differential elimination of sensitive genotypes may reduce the ability of impacted populations to recover from additional impacts or adapt to slowly changing environmental conditions.

G. IRON

The effects of aquatic pollution by iron (Fe) compounds have been studied in connection with coal mining where, in addition to iron, the effects of low pH and high concentrations of sulfate and certain metals may affect the ecosystem. The effects of iron on invertebrates have mainly been investigated in acidic streams in coal-mining areas,[865,1178,1287,1489] but some studies have also considered the influence of ferric hydroxide (ochre) precipitation in neutral streams.[562,1170,1352] Only the concentrations of total iron were determined in these studies.

Laboratory tests on the toxicity of ferrous iron to aquatic insects have given varying results. Walter[1546] found that concentrations below 1.2 mg Fe^{2+} per liter were not harmful in ecological systems, while Warnick and Bell[1554] showed that a concentration of 0.32 mg Fe^{2+} per liter had a toxic effect on the mayfly *Ephemerella subvaria.*

Rasmussen and Lindegaard[1257] studied the effects of iron compounds on macroinvertebrate communities in the River Vidaa in a Danish lowland river system. The area contains considerable amounts of pyrite (FeS_2) and siderite ($FeCO_2$). The groundwater had been lowered in many localities by the activities of lignite mining and intensive farming. As a result of lower water levels, ferrous iron had leached into the rivers where the iron precipitated as ferric hydroxide when the pH exceeded 3. The River Vidaa, which receives iron-rich drainage from farming land, had neutral or only slightly acidic water and low sulfate and metal concentrations. Thus, it was possible to isolate the effects of iron and to determine which of the two parameters, dissolved iron (Fe^{2+}) or total iron (Fe^{2+}, Fe^{3+}, and ferric hydroxide), best describes the observed decrease of invertebrates in iron-enriched portions of the stream.[1257] No correlation was found between the relative number of individuals and dissolved-iron concentrations below 1 mg Fe^{2+} per liter. A distinct decline in the number of individuals was observed at higher concentrations.[1257] The same trend was observed by Skriver[1384] in a number of iron-polluted streams in Denmark. Rasmussen and Lindegaard[1257] found a significant correlation between the numbers of taxa and dissolved-iron (Fe^{2+}) concentrations. An abrupt drop from 67 to 53 in the number of taxa occurred when the dissolved-iron concentration increased from 0.2 to 0.3 mg Fe^{2+} per liter. The precipitation of ferric hydroxide occurred at this concentration and probably initiated a decrease in both the diversity and production of the periphyton. The eliminated invertebrates were mainly grazers feeding on the biofilm. Their study suggested that those invertebrates that are able to live in streams polluted by organic matter were also present at high iron concentrations (Tubificidae, some Chironomidae, and Tipulidae). Streams polluted by organic matter are normally characterized by few species occurring in high numbers, while streams affected by iron compounds showed few species in small numbers. Torup and Lake[1493] drew the same conclusion from a study of a Tasmanian river polluted by a number of metal ions.

Skriver[1384] and Dannisöe et al.[350] found that the number of taxa recorded in the summer in a number of iron-effected streams in Denmark were greater than in the winter. This was due to the immigration of insects with short-term life cycles (especially Diptera) during periods of low dissolved-iron concentrations. Their larvae were eliminated in winter when the dissolved iron concentrations increased.

Krantzberg[875] investigated the effects of age and weight on metal body burdens in Chironomidae. Iron concentrations were higher in older and larger individuals. The chironomids produced hemoglobin and were tolerant of low oxygen conditions. Early instars are typically planktonic and become progressively more benthic with age. The likelihood that larvae will encounter conditions of low oxygen availability increases with time, and it is possible that chironomids can increase their rate of hemoglobin synthesis with age to enhance their ability to tolerate short periods of anoxia. Similarly, as body size increases, the surface area-volume ratio decreases, and oxygen absorption across surfaces also decreases. If increased iron accumulation is, in part, related to increased hemoglobin production, then larger individuals would be expected to produce disproportionately more hemoglobin to compensate for allometric changes in size.

H. COMBINED EFFECTS OF METALS AND ACIDITY

In contrast to the effects of pH and aluminium on invertebrates (see Section III.D), our knowledge of the combined effects of acidification and the toxicity of other metals on invertebrates is limited.[229] Cadmium toxicity is reduced by lower pH values, while the effect of pH on mercury and lead toxicity to invertebrates may be largely species specific.[1654]

Wickham et al.[1604] studied metal accumulation in organisms inhabiting various pH-metal systems represented by three ponds in the Rouyn-Noranda mining region in Canada. Accumulation of copper and aluminium occurred mainly in the organisms of the tailing ponds, with higher values in the acidic, than in the alkaline, pond. The results showed that the effect of a combination of acidity and metals was more severe on the structure of the benthic community than alkalinity and metals. As in other mine pollution studies,[189,208] insects dominated the benthic populations of mine-contaminated ponds, while in the control pond they trailed behind the crustaceans, with less than 25% of the population. There was also a correlation between metal stress and the numerical dominance of chironomids.[872,1627] The large chironomid community of the acidic pond was represented solely by *Chironomus* species, while the communities of the alkaline and control ponds were much richer (22 and 23 genera).

Gerhardt[538] studied cadmium uptake, emergence, survival, and locomotory activity at two pH levels in experiments with the mayflies *Leptophlebia marginata* and *Baetis rhodani*, using model ecosystems containing recirculating stream water, patches of sediment, and leaves, simulating a natural stream. Cadmium was taken up by the species at a rate that indicated linear uptake kinetics. The larvae of *B. rhodani* contained significantly more cadmium than did the adults. Uptake of cadmium by *B. rhodani* was higher at pH 7 than at pH 5. The survival of *L. marginata* was not influenced by pH and cadmium stress, but its emergence was significantly reduced. The survival and emergence of *B. rhodani* were reduced by low pH as well as by cadmium addition.

At pH 7, cadmium had no effect on survival. The locomotory activity of *B. rhodani* was reduced by low pH and additionally by cadmium stress. The different bioconcentration factors of cadmium in the two species at pH 5 may be due to species-specific accumulation rates and tolerances, as well as to the cadmium concentration in the water and the variation in insect size. Surface adsorption of cadmium on the animals may be faster than uptake into the organism. While the concentration of cadmium in the water was twice as high at pH 5 as at pH 7, the cadmium uptake was higher at pH 7 than at pH 5.

Three explanations for the elevated cadmium concentrations in mayflies at pH 7 have been proposed.[538] First, precipitation and adsorption of the metal onto the exoskeleton may be the major uptake mode at pH 7. This was supported by the fact that larvae contained more cadmium than did the adults, and the initial uptake of cadmium was faster at pH 7 than at pH 5. Emergence and molting may be mechanisms that decrease the metal content in the insect.[876] Second, because of precipitation processes, more cadmium may accumulate in the sediment at pH 7 than at pH 5, which will increase exposure to those animals feeding on sediment particles. Third, although direct uptake from the water may be more important at low pH, because of increased concentrations of Cd^{2+} ions in the water and the great amount of water passing respiratory organs, metal ions probably compete with H^+ ions at the binding sites, which may counteract metal uptake from the water. An ameliorating effect of H^+ ions (pH 4.5) on aluminium toxicity (mortality) was also observed for *Baetis rhodani*.[678]

V. AGRICULTURAL CHEMICALS

A. BACKGROUND

Pesticides in aquatic ecosystems often originate from terrestrial spraying operations directed against pests like the spruce budworm (*Choristoneura* spp.) in North America or the tsetse fly in Africa. Pesticides in waters may also originate from the direct application of pesticide to water bodies. Examples include the control of harmful aquatic insects, such as the biting black fly (*Simulium* spp.), fish, and the removal of aquatic weeds by herbicide application.[1091] Rivers and lakes are often contaminated by drainage, run-off, and leaching from agricultural or forestry land treated with pesticides in the recent past. In addition, pesticide contamination of waters is possible as a result of leaf fall in the autumn or through aerial transport by wind and rain.[1091] Muirhead-Thomson[1091] considered that the contamination occurring through spillage, careless disposal of surpluses, or accident is liable to take place in any country where pesticides are regularly used. Industrial effluents also often contain pesticides. In the manufacture of textiles, for example, dieldrin has been used for moth proofing, and dieldrin-containing discharges have resulted in increased residues in aquatic insects.[192,1539]

Chironomid and *Simulium* larvae, stonefly nymphs (e.g., *Pteronarcys californica*, *Acroneuria pacifica*), and caddisfly larvae (e.g., *Hydropsyche* spp.) have been used extensively in laboratory tests on the effects of pesticides on insect fauna. Today the tolerance levels of many aquatic insects to a wide variety of insecticides and other pesticides are known as a result of several laboratory and field investigations.[1089,1091] The toxicity of pesticides can depend on pH, temperature, hardness, and the amount of dissolved organic material in waters.[476-478,889]

Another topic that is largely beyond the scope of this book is pesticide residues in various organisms, which has received much attention since the 1960s.[1089,1091] Insects often play important roles in the food chains along which the biomagnification process takes place. However, predators do not necessarily contain higher pesticide concentrations than their prey. One explanation is that aquatic species can accumulate chemicals from both the water and the diet. Many pesticides are also rapidly metabolized by organisms at higher trophic levels.

Various behavioral responses to pesticide applications have been recorded. They include drift reactions and the immobilizing of various insects, vigorous escape reactions of dragonfly nymphs, case leaving by caddisfly larvae, and burrow leaving by chironomids.[1091] It has been reported that species present on the surface of stones, such as *Baetis* spp. and *Nemoura* sp., are more prone to drift and respond rapidly to toxicant effects, whereas species living below stones or in the substratum are less prone to drift and require greater toxicant exposure to elicit drift response.[1663,1664] Differences in drift reactions may also affect predator-prey relationships.

B. ORGANOCHLORINE INSECTICIDES

Organochlorine residues are persistent and continue to be found in waters, even though their use in agriculture has decreased or ceased in many developed countries. Some organochlorines (e.g., lindane) are still in domestic use. Sewage sludge often contains organochlorines and other harmful chemicals (e.g., PCBs). The application of sewage sludge to agricultural land may contaminate waters at low, but continuous, levels.[1091]

The widespread use of DDT and other organochlorine compounds caused mass destruction of aquatic fauna (e.g., mayflies[181]) in the early years of their application.[740,827,1089,1091] Increases in invertebrate drift following the application of pesticides to running waters have been recognized since the early days of DDT usage.[1091] Even when the levels of DDT in waters are low, aquatic insects and other invertebrates may contain higher levels, and fish feeding on them even higher levels. DDT has had a catastrophic effect on several birds at the top of food chains. On the other hand, several insect species have become resistant to DDT and other organochlorines.[502,722,1041,1075] Furthermore, Webber et al.[1574] concluded that there was an upper limit to the amount of DDT and its metabolites in benthic macroinvertebrates in a heavily contaminated (concentrations of bottom muds ranging from 12 to 2730 μg/g dry weight) sections of two backwater streams in Alabama. The pollutant concentrations in detritivores averaged 125 and 158 μg/g when sediment concentrations averaged 2730 and 96 μg/g, respectively.

Maund et al.[1000] studied the acute and chronic toxicities of lindane to *Chironomus riparius* (Diptera), *Chaoborus flavicans* (Diptera), and *Sigara striata* (Heteroptera) in mesocosm compartments of an experimental pond. The median lethal concentrations (LC_{50}s) of lindane were 240-h LC_{50} of 2.0 μg/l for second instar, 72-h LC_{50} of 6.5 μg/l for fourth instar *C. riparius*, and 96-h LC_{50}s of 4.0 and 3.9 μg/l for fourth instar *C. flavicans*, and fourth and fifth instar *S. striata*, respectively. Lindane significantly reduced the growth over 10 days of second instar *C. riparius* at concentrations where larvae survived (1.0 to 7.0 μg/l). A significant increase in the median emergence time for *C. riparius* was also observed (0.8 to 2.0 μg lindane per liter). The accumulation of lindane has been observed to be pH dependent in *C. riparius*[476] and *Heptagenia fuscogrisea* (Ephemeroptera).[889]

Lohner and Collins[948] determined the uptake rate constants of six organochlorines under nonequilibrium conditions, using *Chironomus riparius*. Uptake-rate constants were calculated from uptake data, using a first-order kinetic expression. Significant correlations were found between the uptake constants of the six chlorinated hydrocarbons and their water solubility and octanol/water partition coefficient values. The results indicated that molecular size may affect the uptake of a chemical in an aquatic system. Those compounds with high partition coefficients and high molecular weights did not accumulate in organisms to the extent that their partition coefficient alone would indicate.

Methoxychlor (2,2-bis-(p-methoxyphenyl)-1,1,1-trichloroethane) has been extensively used in North America for the control of black flies, e.g., *Simulium arcticum* and *S. luggeri* (Simuliidae). The numerous conflicting studies regarding the effects of methoxychlor treatments on nontarget macro- invertebrates[374,482,499,500,637,1541] are usually difficult to interpret due to the variation in experimental design, e.g., the lack of suitable control stations, inconsistency among habitats sampled, little or no collection of pretreatment data, improper statistical analyses, and inadequate taxonomic identification.[1355]

The injection of methoxychlor into lotic systems usually results in the increased drift of benthic macroinvertebrates in the path of the pesticide (Figure 9).[335,482,1355,1541,1543] Methoxychlor-induced drift can be described as catastrophic because of the large, rapid, short-term increases resulting from such treatments.[482,1541] Flannagan et al.[481] concluded that live drifting invertebrates would be "ecologically dead" because suitable habitats for recolonization would not exist for such great numbers of organisms even if some were physiologically capable of recolonization. Muirhead-Thomson[1089] proposed that drifting, injured organisms would face increased predation downstream, while Mohsen and Mulla[1066] predicted that many dislodged organisms would drift into unfavorable habitats where they could not survive. However, Fredeen[500] hypothesized that some drifting organisms were only temporarily incapacitated by methoxychlor exposure and could recolonize downstream.

Species may respond to differences in methoxychlor concentration and exposure time. The concentration of methoxychlor becomes progressively lower with increasing distance travelled downstream, but the exposure time

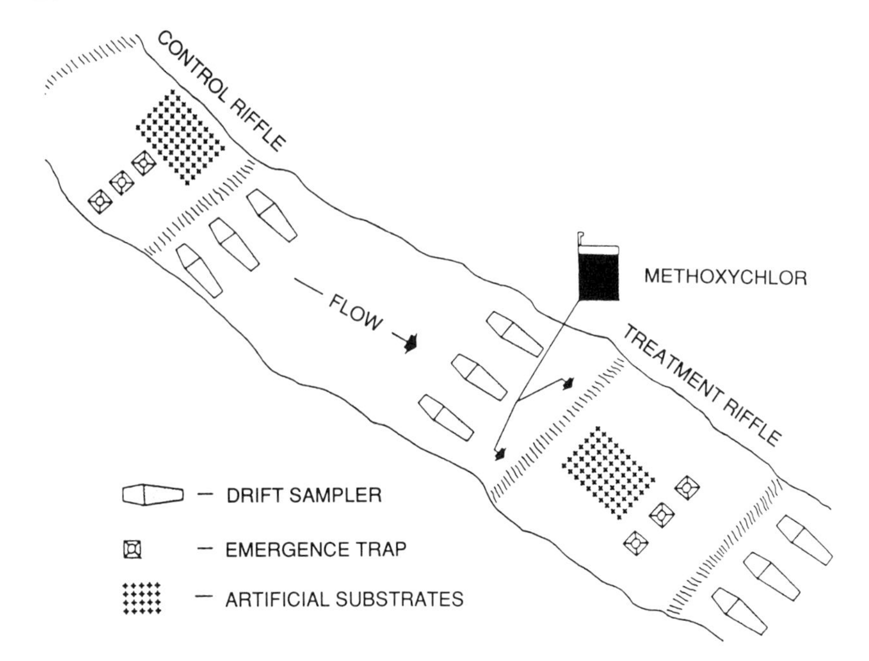

FIGURE 9. Arrangement of sampling devices used at the control and treatment sites by Sebastien et al.[1355] to study the impact of methoxychlor on selected nontarget organisms in a riffle of the Souris Stream in Manitoba, Canada. (Redrawn from Sebastien, R. J., Brust, R. A., and Rosenberg, D. M., *Can. J. Fish. Aquat. Sci.*, 46, 1047, 1989. With permission.)

lengthens.[1089,1543] Some species may be sensitive to prolonged exposure to low concentrations rather than to high concentrations for short periods.[1089]

Downstream drift of benthic invertebrates has been observed to increase more than 1000-fold following exposure to methoxychlor.[335,482,1355,1541,1543] Catastrophic drift and high mortality rates among organisms affected by the pesticide characterize the first phase of environmental impact following pesticide exposure.[1542] Although this catastrophic phase lasts only a few hours, the following reversion phase may last days or weeks. During the reversion phase, drift and mortality levels decline while competition apparently occurs among the survivors, and sublethal and chronic toxic effects become manifest. The recovery phase is characterized by recolonization.[1542] Insects recolonizing an area of denuded stream substrate come from four main sources: downstream drift, upstream migration, upward migration from the hyporheic zone, and egg laying by aerial adults.[1612]

Over relatively short downstream distances (<5 km), the effect of methoxychlor on the drift of aquatic insects was shown to be relatively uniform with regard to the time of drift initiation among species and the time to peak drift responses. For example, Cuffney et al.[335] reported that the onset of drift density increases and the times of peak drift occurred at similar times for different species. Similarly, Sebastien et al.[1355] observed that regardless of trophic level, invertebrate taxa were affected simultaneously by methoxychlor injection.

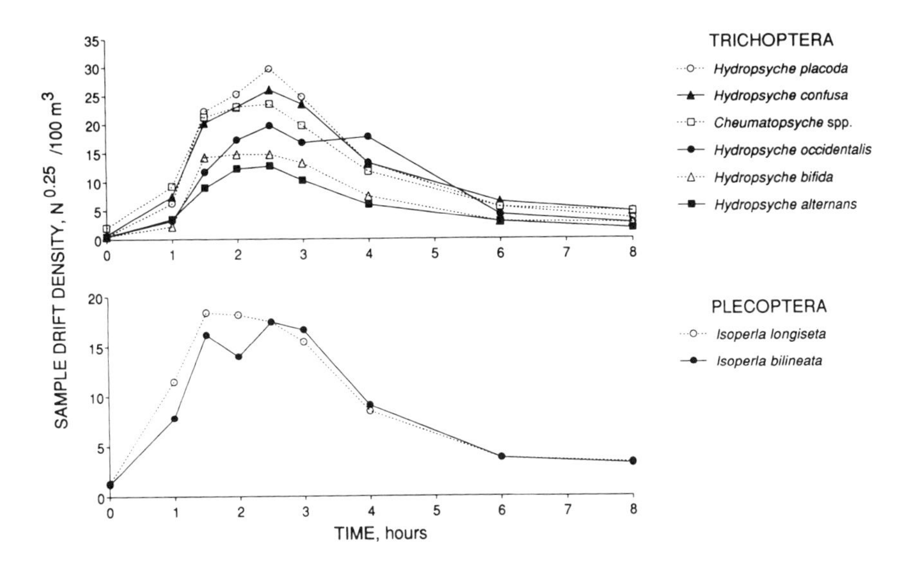

FIGURE 10. Mean sample drift densities of closely related taxa compared during the catastrophic phase following methoxychlor injection on the Saskatchewan River system, Canada. (A) Drift of Hydropsychidae (Trichoptera) larvae; (B) drift of *Isoperla* spp. (Plecoptera) nymphs. (Modified from Dosdall, L. M. and Lehmkuhl, D. M., *Can. Entomol.*, 121, 1077, 1989. With permission.)

Over longer downstream distances (up to 107 km), however, Dosdall and Lehmukuhl[396] found that taxa occupying microhabitats directly exposed to methoxychlor, and taxa having morphological attributes that enhanced absorption of the pesticide, entered the water column more rapidly than did species occupying sheltered microhabitats or having morphological characteristics that ameliorated exposure to the pesticide.

Dosdall and Lehmkuhl[395-397] compared the drift of aquatic insects at three sites downstream from a methoxychlor treatment point on the North Saskatchewan River, relative to an upstream untreated site. The drift responses of the insects differed depending on species (Figure 10), distance from the injection point, and time after methoxychlor injection. Exposure to methoxychlor (300 μg/l for 15 min) initiated a catastrophic drift of insects: of 22 species compared before treatment and following methoxychlor injection, posttreatment drift of 17, 21, and 13 species significantly exceeded pretreatment drift at the 21, 38, and 107 km sites, respectively. The treatment caused drift of several normally nondrifting species.

Sebastien et al.[1355] applied methoxychlor (300 μg/l for 15 min) at a shallow point on the Souris River, Manitoba. Insect taxa were monitored using drift nets, emergence traps, and artificial substrate samplers. All taxa drifted catastrophically for 4 to 24 h immediately following methoxychlor addition. Some taxa recolonized within days, whereas populations of others required months to recover relative to the control site. Species having a high propensity to drift naturally (e.g., *Baetis* spp.) recolonized most rapidly. Taxa that required a long period to recolonize following methoxychlor treatment were usually univoltine, had a low propensity to drift, and had a limited ability to disperse as adults. The impact of methoxychlor was influenced by the prevalent life-cycle stage of some species at the time of treatment. Juvenile crayfish (*Orconectes virilis*) were more sensitive to methoxychlor than were mature individuals. Species richness and total numbers of drifting individuals were significantly reduced for at least 33 days following treatment, whereas richness and numbers on artificial substrates were significantly lower for only 4 and 8 days, respectively.

Wallace et al.[1540] found that the dragonfly *Lanthus vernalis* (Gomphidae) was among the insect taxa least affected by methoxychlor treatment (four applications from February to November). Larval diets of this predator (gut analyses) reflected changes in community structure within the treatment stream. Insects represented only 13% of the prey during the treatment year and more than 82% during the second year of recovery. In contrast, *L. vernalis* in the reference stream consumed primarily insects (73 to 78%) in both years. This indicated that when prey communities undergo massive changes, generalist predators may shift to alternative prey. The disturbance reduced the abundance of prey to a level where less-suitable prey were included in the diet.[1540]

Methoxychlor is usually added directly to the water as a point-source application. Sites closest to pesticide application, although exposed to the highest pesticide concentrations, should have the greatest potential for rapid recolonization because of their proximity to sources of colonization by drift of invertebrates from uncontaminated upstream reaches. The time required to recolonize should progressively increase with increasing distance downstream from the treatment site. One year after applying methoxychlor to the Athabasca River, Canada invertebrate standing crops at the two sampling sites furthest downstream were still significantly reduced, whereas the upstream sampling sites had completely recolonized.[482] The greatest long-term effect on nontarget invertebrates was observed at the sampling site furthest downstream, 400 km from the treatment site. In rivers the effect of distance diminishes with time because sources of colonization shift progressively downstream.[1355] Since the toxicity of methoxychlor increases with increasing exposure time for most organisms,[396,397] species at sites further downstream could be at higher risk, although they would be exposed to the pesticide at lower concentrations. Downstream recovery can be prolonged because of the greater distance from upstream sources of colonization. Multiple methoxychlor treatments, as carried out in some black fly larviciding operations in western Canada,[501] may be more harmful to benthic invertebrates than a single application.[1355]

Recent studies have indicated a link between structural deformities in chironomid larvae and high levels of organochlorine compounds[1565] and other pollutants (see Chapter 6, Section II.B).

C. ORGANOPHOSPHOROUS INSECTICIDES

Organophosphorous insecticides, such as fenitrothion and chlorpyrifos, have largely replaced the organochlorines. Fenitrothion has been used extensively to control the spruce budworm (*Choristoneura fumiferana*) in New Brunswick, Canada, since 1968,[421,1461] replacing DDT, which had been used since 1952.[827] In Japan insecticides such as fenitrothion have usually been sprayed twice a year for more than 10 years to control a forest pest, *Monochamus alternatus* (Cerambycidae), that carries a nematode, *Bursaphelenchus xylophilus*, into the Japanese red pine (*Pinus densiflora*).[636]

Chlorpyrifos is used in agriculture as well as to control mosquitoes in wetlands. It is known to affect nontarget species in aquatic ecosystems at exceedingly low concentrations.[738,1147] Chlorpyrifos binds to organic material,[1331] thus greatly increasing its residual life. Wallace et al.[1544] described the effects of a chlorpyrifos spillage in a temperate river. The catastrophic phase of the spillage lasted for 2 to 3 days, during which chlorpyrifos concentrations in river water exceeded the 48-h LC_{50} levels of most arthropods by 10 to 1000-fold. The effects included catastrophic drift and mortality of larval Plecoptera, Ephemeroptera, Trichoptera, Simuliidae, and Chironomidae after 15-min exposure to a concentration of 100 mg/l. The mortality of the drifting organisms has usually not been investigated.

The impact of a major spillage of the insecticide "Dursban" on riffle macroinvertebrates was studied along 19 km of the River Roding, Essex, U.K.[161,1263] Initial concentrations of the active ingredient, chlorpyrifos, in river water (up to 2.5 mg/l) exceeded the level lethal to all the aquatic arthropods present, by at least tenfold. Arthropods were eliminated, but annelids (Oligochaeta, Hirudinea) and molluscs survived. Chlorpyrifos residues in the water declined below 1 μg/l within 11 weeks, but sediment within 5 km of the spillage site remained contaminated for considerably longer. Of arthropod taxa previously common to all sites, chironomid larvae were the first to recolonize the affected reaches, 13 weeks after the spill. Although other arthropods had recolonized most sites within 79 weeks, the coleopteran *Oulimnius tuberculatus* and the ephemeropteran *Caenis moesta* had failed to return to the lowermost reaches after 108 weeks.

Following various applications of fenitrothion to Canadian forests, invertebrate drift rates ranging from zero[420,1522] to greater than eightfold[421] have been reported. When fenitrothion concentration in streams was up to 6.4 mg/l after an aerial spraying in New Brunswick,[421] mortality of Ephemeroptera (*Baetis* spp.) and Plecoptera (*Leuctra* spp., *Amphinemoura* spp.) was high. The dead component of the drift increased from 26 to 90%. Eidt[419] showed that the increased drift could mean a reduction of standing crop, even though direct sampling may not show it.

Hatakeyama et al.[636] used a drift net to investigate the effects of aerial spraying of fenitrothion in a Japanese mountain stream. The insecticide concentration in the river water increased to 20 μg/l 3 h after the spraying and decreased rapidly to half the peak value after 2 h. The drift of a large number of aquatic insect species started immediately after the spraying and was almost coincident with the peak of the insecticide concentration in the water. The number of species (e.g., *Baetis*, heptageniid, and chironomid spp.) which drifted during the 24-h period following the spray increased from 17 on the previous day to 43. Natural night drift disappeared almost completely following insecticide spraying. The results suggested that some species of Trichoptera (three *Hydropsyche* spp. and *Cheumatopsyche brevilineat* were tolerant of the fenitrothion application. After the second insecticide spraying (20 days after the first), the number of individuals that drifted during the 24-h period following spraying decreased to only 0.85% of that in the first spraying, although insecticide concentrations were similar. The number of species found in the drift was less than one half of that after the first application. The density of Plecoptera was very low even 11 months after the insecticide spraying, but a relatively rapid recovery in Ephemeroptera and several species of Trichoptera and Diptera occurred. Similarly, after the application of temephos to a mountain stream, chironomids recovered earlier than other invertebrates and reached a higher density level than before the application.[1661]

D. CARBAMATE INSECTICIDES

In recent years carbamate insecticides, especially carbofuran (2,3-dihydro-2-2 dimethyl-7-benzofuranyl methylcarbamate), have been among the most widely used insecticides on the Canadian prairies, especially for grasshopper control.[1572,1573] Laboratory and field studies have been conducted on the acute toxicity of carbofuran to *Daphnia*,[625,783,1669] chironomid larvae,[783,816,1093] and mosquito larvae.[899]

Wayland and Boag[1573] assessed carbofuran-induced mortality in selected macroinvertebrates known for their importance as waterfowl food, by confining the animals in small cages in prairie ponds and subsequently treating the ponds with carbofuran. The effect of carbofuran on both *Gammarus lacustris* (Amphipoda) and *Chironomus tentans* (Chironomidae) varied with depth, but survival in the shallow zone was significantly reduced in treatment ponds compared to control ponds. The survival of *Enallagma* damselflies was lower in the treated ponds where the initial carbofuran concentrations were 9 and 32 μg/l, respectively.

The impact of carbofuran applications (Furadan 3G, dose rate 1 kg a.i. per hectare) on aquatic macroinvertebrates has also been studied in irrigated rice fields in Senegal.[1097] The total biomass and the numbers of Odonata, Ephemeroptera, Nematocera (Diptera), and Hydrocorisae (Heteroptera) were severely reduced after treatment and remained low for at least 24 days after treatment. Ephemeroptera was the most sensitive taxon.

The effects of carbaryl on aquatic insects have been extensively studied during gypsy moth and spruce budworm control programs in North America. For example, Courtemanch and Gibbs[314] monitored three streams in areas sprayed with carbaryl (Sevin-4-oil®, 840 g a.i. per hectare) for the first time, three streams in areas sprayed in two consecutive years, and three streams in unsprayed areas in Maine. Initial postspray response was an up-to-170-fold increase in drift in the treated streams. Significant declines were observed among Plecoptera, Ephemeroptera, and Trichoptera. The long-term effect of carbaryl was most apparent on Plecoptera, which did not repopulate the treated streams by 60 days after the treatment and had very low population densities in the next year.

Hanazato and Yasuno[606] applied carbaryl (1-naphthyl-N-methylcarbamate) at 0.1 or 0.5 mg/l to pond ecosystems with or without *Chaoborus flavicans* (Diptera, Chaoboridae) larvae to determine the interactive effects of *C. flavicans* larvae and insecticide on the zooplankton communities. Hanazato and Yasuno[604,605] had earlier shown that *C. flavicans* larvae interfered with the recovery of the zooplankton community after chemical application to plankton communities in outdoor concrete ponds. The lower concentration of the chemical was harmful only to Cladocera. The higher concentration reduced *C. flavicans*, Copepoda, and some rotifer species, as well as Cladocera. In the ponds with a low density of *C. flavicans*, chemical application altered the cladoceran community from dominance by *Daphnia* to that by *Bosmina* and *Moina*. In the ponds with a high density of *C. flavicans*, *C. flavicans* excluded cladocerans from the zooplankton community, presumably through predation and supported the dominance of rotifers. Cladocera did not recover after application of the chemical, even when *C. flavicans* was eliminated by the higher concentration of the chemical. The relatively rapid recovery of *C. flavicans* seemed to interrupt the recovery of Cladocera.[606]

Fenoxycarb treatment induced varying degrees of morphogenetic aberrations in nontarget insects.[1063] Treatment (34 g a.i. per hectare) of recently oviposited eggs of the back swimmer *Notonecta unifasciata* (Heteroptera) resulted in dead embryos. Apparently the insecticide did not interfere with the development of embryos, but the treated eggs were unable to hatch. Treatment of fourth and fifth nymphal stages of *N. unifasciata* produced various degrees of morphogenetic abnormalities, which occurred primarily at the nymph-adult ecdysis. Treated fourth-stage nymphs usually molted to the fifth stage without any visible abnormalities, but affected fifth-stage nymphs were unable to emerge to the adult stage. Morphogenetic aberrations were also observed in treated dragonflies *Anax junius*, *Pantala hymenaea*, and *Enallagma civile* at the nymph-adult molting.

E. ANTIMETABOLITE INSECTICIDES

Insect growth regulators are useful in pest control because of their low fish toxicity, low persistence, and high elimination rate from fish tissues.[1330] Dimilin® or diflubenzuron [1-(4-chlorphenyl)-3-(2,6-difluorobenzonyl)urea] has been used for mosquito control,[1094,1295] nuisance midge control,[13-16,1463] and black fly control.[1066] Diflubenzuron also affects nontarget invertebrates during hatching, molting, pupation, and emergence.[896] Continuous exposure of diflubenzuron in indoor laboratory stream channels had direct toxic effects on the insect fauna at concentrations of 1 ng/g and greater.[610] Numbers of *Simulium* (Diptera), *Baetis* (Ephemeroptera), and *Hydropsyche* (Trichoptera) larvae declined moderately at 0.1 μg/g when exposed for 15 min in field experiments.[1066] Concentrations of 1 and 10 μg/g for half an hour in an artificial stream had direct toxic effects on invertebrates.[1330]

Diflubenzuron was applied to the Kokawa River, Japan, at a concentration of 1.25 μg/g for 1 h to control simuliid larvae.[1330] Most invertebrates were eliminated within 2 weeks. Most of the *Baetis* nymphs died within a few days after exposure to the chemical, due to the short duration of each larval instar. Recolonization by newly hatched larvae of *Baetis* spp., Chironomidae, *Antocha* sp., and Simuliidae was prominent 3 to 4 weeks later in all the treated areas. The application had long-lasting negative effects on Trichoptera and Ephemeroptera, other than *Baetis* spp., but induced, after a rapid recovery, an enormous increase in target dipteran larvae, including simuliids. Frequent applications are subsequently required once this chemical has been used for simuliid control in a river. The recovery of dipteran larvae, which was prominent 3 weeks after the application, was likely attributed to oviposition.

Ali et al.[19] assessed the effects on invertebrate populations in a pond that received off-target diflubenzuron (Dimilin®, 560 g a.i. per hectare) from surrounding citrus plantations commercially treated for the control of citrus rust mite. Diflubenzuron was found to have no adverse effects on the zooplankton and benthic invertebrates in the exposed pond after diflubenzuron treatment in the surrounding area. The main reason for the lack of any ill effects from diflubenzuron might be the low concentrations of the insect growth regulator present in the pond.[19]

A growth regulator, benzonylphenylurea (BPU), UC-84572, was recently tested in the laboratory against seven species of mosquitoes and two species of midges. It proved to be highly toxic against mosquitoes, with LC_{90} values ranging from 0.53 ng/g (*Culex nigripalpus*) to 11.64 ng/g (*Anopheles albimanus*).[17] This BPU analogue was 4 to 13 times more active against five species of mosquitoes, and 2 to 4 times against midges, than diflubenzuron.[17] The insecticide was then applied at different rates to assess the control of natural populations of chironomid midges in outdoor ponds. Ali and Kok-Yokomi[12] reported the impact of UC-84572 on nontarget invertebrates coexisting with the midges. Field applications of UC-84572 at rates up to 10 ng/g a.i. or higher have simultaneous adverse effects on some nontarget aquatic invertebrates, such as nymphal Ephemeroptera and larval and adult Coleoptera, Cladocera, and Copepoda. No displacement or replacement of any animal group was noted in the ponds, indicating that UC-84572 did not cause any long-term or permanent alterations in the aquatic food chain. Similarly, diflubenzuron, Bay SIR-8514, and UC-62644 affected nontarget organisms when used at comparable field rates,[12-14,18,49,1062] but the adverse effects were minimal and usually short lived.

F. PYRETHROID INSECTICIDES

Pyrethroid insecticides were developed for extensive use in agriculture and human disease control.[239,691] Current commercial products were evolved from the natural pyrethrins, which possess high insecticidal potency, low mammalian toxicity, and very short persistence. The modern synthetic pyrethroids retain some of the attributes of the natural products, but have been tailored to provide enhanced residual activity through greater photostability. They are also more resistant to chemical and biological degradation by virtue of changes at several sites in the molecule.[284]

Aquatic vertebrates seem to be much more sensitive to pyrethroids than are terrestrial vertebrates. Pyrethroids are also very toxic to aquatic insects and crustaceans, the LC_{50} values in most cases being well below 1 µg/l.[284] When a variety of mosquito and midge larvae and pupae were tested, 24-h LC_{50} values for deltamethrin, cypermethrin, fenvalerate, and permethrin ranged from 0.02 to 13 µg/l.[1115] In addition to acute toxicity, many pyrethroids may have deleterious effects at sublethal levels. Anderson[32] observed behavioral alterations within hours of exposure in several aquatic invertebrates (at concentrations as low as 0.022 µg/l for fenvalerate and 0.030 µg/l for permethrin), resulting in a cessation of feeding and insect drift. When three natural ponds in the British Isles were treated with cypermethrin, the invertebrate species assemblage was considerably affected, although there was variation in the rate at which different insects were affected. Aquatic species (*Notonecta, Dytiscus*), which frequently come to the surface for air, were affected most rapidly.[331] Many deleterious effects occurred at concentrations below 1 µg/l, including effects on reproduction, growth, and behavior.[33,169,367,1021]

Pyrethroid insecticides are more toxic to insects at lower temperatures.[193,1286,1412] This phenomenon is termed the negative temperature coefficient and is relatively uncommon. Nerves may be more sensitive to the effects of pyrethroid-induced toxicity, although toxicokinetic factors, such as uptake, distribution, and detoxification, may contribute to increases in toxicity at lower temperatures. Other environmental stresses, such as pH, oxygen concentration, sunlight, nutrient input, and turbidity, can also affect organisms at lower temperatures.[284]

Aquatic mosquito larvae are exposed to lipophilic xenobiotics by two routes of entry. The pyrethroid can enter the body of the larva by penetration through the cuticle after contact with the chemical in the water, or by ingestion after adsorption to food particles in the water.[284] *Chironomus tentans* larvae accumulated significantly more permethrin when allowed to enter sediment than when held in the water above the sediment.[1088] Dissolved material such as humic acids can affect toxicity. When mosquito larvae (*Culex pipiens pipiens*) were exposed to the pyrethroid fenvalerate in the presence of humic acid, significant differences in toxicity were noted.[284] Fenvalerate was six times less toxic to mosquito larvae in water containing 50 mg/l humic acid than in clean water. Feeding tests with mosquito larvae showed that after the first hour, when a high concentration of fenvalerate was reached in their bodies, the larvae became restless and stopped feeding. The same irritant or feeding-deterrent action of synthetic pyrethroids has been documented for other species of insect.[53,1281,1329] After 16 h, the larvae had recovered and resumed feeding, which resulted in the high residue levels in their bodies. Fenvalerate was approximately six times more toxic to mosquito larvae through cuticular exposure than through oral exposure. This lipophilic compound was rapidly adsorbed on the aquatic larvae and readily penetrated the cuticle, while much of the ingested insecticide possibly remained biologically unavailable in the digestive tract of the immature mosquitoes.[284]

The pyrethroids are of very low water solubility/high lipophility and therefore are rapidly adsorbed onto particulate material. In the adsorbed state their bioavailability to aquatic organisms is greatly reduced. Thus, under field conditions the aquatic impact of these insecticides is likely to be much less than might be predicted by laboratory acute or chronic toxicity test data. The effects are mostly transient and unlikely to cause adverse changes in the populations or productivity of aquatic ecosystems. Surface-living Heteroptera, Coleoptera, and also Ephemeroptera are often the taxa most affected.[684]

Pyrethroid deltamethrin has been widely used against grasshoppers in Canada and has also been recommended for use against armyworms, cutworms, flea beetles, pea aphids, and plant bugs, on forage crops. However, deltamethrin may pose a significant threat to prairie ponds because pyrethroids are much more toxic to many aquatic invertebrates than most nonpyrethroid insecticides.[1360] Tooby et al.[1492] noted that chironomid larvae, baetid mayfly nymphs, and corixid waterbugs completely disappeared following the application of deltamethrin to ponds. Morrill and Neal[1077] applied deltamethrin insecticide by air, at recommended field application rates (7.5 μg/ha), to two prairie ponds. Following treatment the chironomid larvae rapidly declined to approximately 1% of their pretreatment densities in the treated ponds, whereas there was little change in larval densities in the two untreated ponds. The two treated ponds recovered at different rates: the community in one pond appeared to have recovered 2 months after treatment, whereas that of the other treated pond showed little recovery until 1 year following treatment.

Kreutzweiser and Kingsbury[880] investigated the impact of aerially applied permethrin on stream invertebrates in Canadian forests. The permethrin treatments at rates of 8.8 to 70 μg/ha to cold-water streams resulted in catastrophic drift and usually produced a measurable depletion of bottom fauna. In double-application experiments, the second treatment further decreased benthic populations and substantially extended the time required for recovery of benthos numbers. Recovery of benthic fauna following the various permethrin applications was apparent from 1 to 18 months postspray.

Sibley et al.[1365] evaluated the impact of a concentrated pulse (16 μg/l) of permethrin on macroinvertebrate community of Icewater Creek, a northern Ontario headwater stream. Permethrin induced an immediate catastrophic drift response characterized by an increase as high as 20,000-fold within the Leuctridae (Plecoptera) and as low as 21-fold among the Hydropsychidae (Trichoptera). In most cases peak densities were observed within 30 min of the arrival of the insecticide at the test station. Drift densities declined rapidly up to 4 h after treatment. The greatest impact was observed in the mayflies *Baetis flavistriga*, *Heptagenia flavescens*, and *Epeorus* sp., the stonefly *Leuctra tenuis*, and the caddisfly *Dolophilodes distinctus*.

In another study in the same stream, Kreutzweiser and Sibley[881] also found that a permethrin injection (peak concentration 8.6 μg/l in water) resulted in catastrophic (from 100- to 5600-fold), but transient, increases in the drift density of macroinvertebrates. Their results suggested that the initial concentration of permethrin was more important than exposure duration, in eliciting a drift response. Significant reductions in benthos occurred at sites with drift densities of 4000 and 5600 times pretreatment levels, but not at sites with 100- or 300-fold drift increases.

During tsetse control operations near Bouaflé, Ivory Coast, Africa, Everts et al.[450] conducted a study of the environmental side effects of helicopter spraying with permethrin and deltamethrin on a riverine forest. Deltamethrin applications caused a considerable impoverishment of the aquatic insect fauna. Mayfly larvae and small shrimps (*Caridina africana*) were virtually eliminated after the spraying cycle. The chironomid populations recovered rapidly. An increase in chironomid larvae has also been observed after temephos (Abate®) treatments against Simuliidae.[372] Even the most sensitive insect taxa in the samples, Baetidae and Tricorythidae (Ephemeroptera), tended to recover within a few weeks after their depletion. The reversibility of damage caused by the insecticides was probably promoted by the fact that at the time of application, many species were in diapause or otherwise protected against unfavorable weather conditions.[450]

G. COMPARISON OF DIFFERENT INSECTICIDES

Laboratory bioassays provide a rapid, appropriate technique for assessing the toxicity of various insecticides to organisms, but whether the results can be generalized to field situations is debatable.[1212] The toxicity of an insecticide varies among the organisms used,[1611] individuals of different size,[1024] experimental procedures,[32,1090] and insecticide formulations.[385,420] Pyrethroids such as permethrin may be detoxified quickly by the mixed-function oxydase enzymes,[1018] and insect populations that are severely depleted, knocked down by a pyrethroid application, may recover.[32] Conversely, the mixed-function oxydase enzymes initially increase the toxicity of the organophosphorous and carbamate insecticide.[1020]

Poirier and Surgeoner[1212] used a flow-through laboratory bioassay system that incorporated the important environmental parameters encountered during a spruce budworm control program (e.g., water temperature, pH, exposure period). They examined the toxicities of two registered and two candidate forest insecticides to aquatic invertebrates. The toxicities of fenitrothion, aminocarb, mexacarbate, and permethrin depended on the species, with no single species being more or less susceptible to all of the chemicals, although *Simulium venustum* (Simuliidae), *Isonychia* sp. (Ephemeroptera), and *Acroneuria* sp. (Plecoptera) generally responded more rapidly to the toxicants and were most sensitive. Permethrin was between 10- and 636-fold more toxic than the other compounds tested. However, the researchers estimated that a 100-fold increase in the concentration of permethrin is required to increase the mortality of dragonflies from 5 to 100%, whereas only a tenfold increase in the

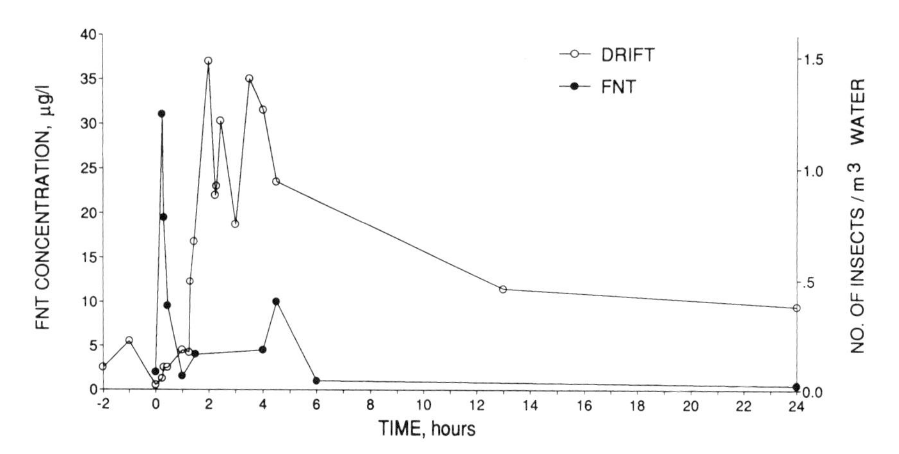

FIGURE 11. Concentration of fenitrothion (FNT) and a number of drifting aquatic invertebrates (insects per cubic meter) sampled from Icewater Creek, Canada, after experimental aerial application of fenitrothion at 280 µg a.i. per hectare. (Modified from Poirier, D. G. and Surgeoner, G. A., *Can. Entomol.*, 120, 627, 1988. With permission.)

concentration of fenitrothion would create the same effect. Fenitrothion was more toxic than mexacarbate to dragonflies, but was equally toxic to black flies and caddisflies. Aminocarb was least toxic to black flies and caddisflies, but was similar to mexacarbate in toxicity to dragonflies. This indicated that several representative species are required to assess the overall risk of an insecticide.

Sublethal behavioral responses of aquatic invertebrates also indicate greatest sensitivity to permethrin.[1212] The case-leaving response of *Pycnopsyche* sp. (Trichoptera) and the drift response of the other invertebrates were observed at concentrations of permethrin >0.25 µg/l, whereas it took 40-fold more fenitrothion and 100-fold more aminocarb or mexacarbate to induce a similar response. *Brachycentrus americanus* (Trichoptera) larvae left their cases at 0.064 µg/l permethrin,[32] and *B. numerosus* larvae left their cases after a 24-h exposure to between 55 and 76 µg/l fenitrothion.[1462]

Poirier and Surgeoner[1213] used field bioassays to assess the toxicities of four formulated insecticides to representative stream invertebrates. Toxicities (48-h LC_{50}) after a 1-h application period ranged from 2.0 to 7.1 µg/l for permethrin, 82 to 284 µg/l for fenitrothion, 344 to 1276 µg/l for aminocarb, and 251 to 1504 µg/l for mexacarbate. Invertebrates drifted at concentrations of permethrin greater than 0.5 µg/l, and at fenitrothion, aminocarb, and mexacarbate concentrations greater than 10 µg/l. An aerial application of 280 µg active ingredient per hectare fenitrothion was made to compare bioassay results with the impact on aquatic invertebrates under operational spray programs. Concentrations of fenitrothion peaked 30 min after spraying at 31.0 µg/l and declined to less than 1.0 µg/l within 14 h. Numbers of drifting invertebrates increased 20-fold 3 h after spraying and declined to before-spray numbers within 24 h (Figure 11).

Mortalities of caged invertebrates in the stream ranged from 0% for *Pycnopsyche* sp. to 16% for *Simulium venustum.*

H. *BACILLUS THURINGIENSIS*

Bacillus thuringiensis serovar. *israelensis* (*B. t. i.*) is being used more frequently to mitigate nontarget effects of pesticides. In several regions of North America, *B. t. i.* has replaced methoxychlor and temephos as the larvicide of choice for reducing populations of black flies and mosquitoes. *B. t. i.* is now the main dry-season larvicide against simuliids in the onchocerciasis control program in Africa.[255]

The safety of *B. t. i.* was currently reviewed by Lacey and Mulla.[897] Colbo and Undeen[290] found that *B. t. i.* applications (10^5 cells per milliliter) had no significant effects on nontarget insects in a small stream in Newfoundland. Similarly, no significant changes of posttreatment population levels of nontarget insects were observed in two streams in New Zealand.[255] However, temporary increases in the drift of the mayflies *Epeorus fragilis* and *Baetis brunneicolor* and in the caddisflies *Parapsyche apicalis* and *Pycnopsyche divergens* were recorded following *B. t. i.* treatment (10 μg/l for 1 min) in a small undisturbed stream in New Hampshire.[1207] Increase in rate of drift was inversely related to increase in distance from the treatment point. Direct mortality caused by the larvicide was observed only in black fly larvae and some Chironomidae. Back et al.[67] recorded reduced densities of *Eukiefferella* and *Polypedium*, but found no significant effect on *Rheotanytarsus* (Chironomidae), after a high-dosage *B. t. i.* treatment of a Quebec stream. Gibbs et al.[540] found no adverse effects of *B. t. i.* on Blepharoceridae (Diptera) at operational dosages in the field, but mortality was reported at high dosages by Back et al.[67]

Merritt et al.[1039] studied the effects of *B. t. i.* (Teknar® HP-D), applied in the Betsie River, Michigan, on black flies and nontarget aquatic invertebrates. Although black fly mortality was high, there were no detectable effects of *B. t. i.* application on nontarget invertebrate drift, numbers of invertebrates in benthic Surber samples, survival or feeding of drifting and nondrifting insects, growth or survival of caged *Stenomena* sp. (Ephemeroptera) larvae, invertebrate functional group composition, survival or weight change of caged rock bass, or fish numbers, species composition, length-weight relationships, or diet. Sampling of *Rheotanytarsus* sp. (Chironomidae) on natural substrates indicated low (27%) mortality caused by *B. t. i.* only 100 m downstream from the application site, with negligible mortality at all other downstream and upstream sites. These results indicated that midge populations were not adversely affected by *B. t. i.*

I. PISCICIDES AND MOLLUSCICIDES

Piscicides or fish toxicants are widely used to eradicate some or all fish from a body of water in order to permit stocking with desirable fish species and freeing them from predation, competition, or other interference from undesirable fish species.[1091] Studies on the reactions of invertebrates to the piscicides

TFM (trifluoromethyl nitrophenol) and antimycin have been conducted in North America.[1114] TFM is a lamprey larvicide, and antimycin is an antibiotic piscicide. TFM was less toxic to invertebrates than to larval lampreys. Treatment with antimycin caused disorientation of caddisfly larvae, many of which left their cases and died.

Molluscicides, e.g., Bayluscide or niclosamide, have been used to control the snails that harbor the parasitic worms causing human bilharziasis in developing countries such as Egypt.[1091] Its impact on insect fauna is poorly known.

J. RODENTICIDES

Compound 1080 (sodium monofluoroacetate) is a white, odorless, nonvolatile powder that is hygroscopic and very soluble in water.[60,249] In New Zealand 1080 is used extensively for rabbit and possum control and, to a limited extent, against wallabies, deer, goats, thar, and wild pigs. It disrupts the Krebs tricarboxylic acid cycle, which is a complex metabolic pathway common to most organisms,[1254] and is consequently toxic to both vertebrates and invertebrates. The compound has been used experimentally as an insecticide in the U.S.,[1306] and the insects studied so far have been found to be very susceptible. Notman[1134] reviewed the toxic effects of compound 1080 on invertebrates: invertebrates from at least nine orders are prone to poisoning by 1080.

K. HERBICIDES AND FUNGICIDES

Muirhead-Thomson[1091] reviewed the impact of herbicides, such as diquat, dichlobenil, paraquat, endothal, cyanatryn, and phenoxy herbicides 2,4-D and 2,4,5-T, on aquatic invertebrates. The use of herbicides on a large scale started after World War II and has been expanding rapidly since then. The direct impact of herbicides on fish is relatively well investigated, but there are few studies on the effects of herbicides on aquatic insects. These studies largely deal with static-water fauna.

Although herbicides in waters may originate from surrounding agricultural land, the most drastic effects on insects are caused by direct application to water for the destruction of undesirable aquatic plants and the clearance of vegetation from waterways.[1091] The indirect effects of herbicide application may be severe on insects. For example, the control of water milfoil in Tennessee by 2,4-D application lead to the disappearance of epiphytic insects, including elmid beetles, dragonflies, and mayflies.[1391] After the cyanatryn application on drainage channels, Corixidae diappeared, but reappeared with the return of macrophytes 1 year after treatment.[1351] On the other hand, the control of weeds in small lakes by applying paraquat was reported to cause only small and transient alterations in the aquatic insect populations.[1571]

Comparative studies showed that insects (especially dragonfly [*Libellula*] nymphs) were relatively resistant to diquat and dichlobenil, whereas amphipods were very sensitive.[1624] Dichlobenil had a narcotizing effect on all invertebrates. Many of the caddisfly larvae had left their cases during the treatment.

Organic tin compounds have been used as pesticides, but now residual tin is primarily a problem of shallow waters and harbors. Such compounds are released to the environment, for example, from antifouling paints used in the bottoms of boats and yachts. For example, tributyltin (TBT) has come under recent regulation because of several environmental issues, including high toxicity at very low concentrations to marine fauna, its growing use, and concern about its persistence in marine waters and sediments.[823,1538] Kelly et al.[823] introduced ^{14}C-labeled tributyltin-chloride (TBT-Cl) into the water column of seagrass microcosms held in the laboratory under flow-through conditions. Benthic macroinvertebrate abundances across a three-treatment, logarithmic-dose gradient were compared to untreated control microcosms. Within 3 to 6 weeks significant mortality appeared in the high treatment. Sensitive species included surface-deposit feeders of several phyla. The results suggest that TBT effects can arise because it is rapidly accumulated in surface sediments.[823]

The herbicide (fungicide) stannic triphenylacetate (Brestan), an organic tin compound with the empirical formula $C_{20}H_{18}O_2Sn$, is used in rice cultivation to suppress algae and molds harmful to rice, and in agriculture against potato and sugar beet blight.[254] Cotta-Ramusino and Doci[311] examined the acute toxicity of the compound to freshwater organisms. They showed that the larvae of *Chironomus riparius*, which is described as living in water polluted by organic matter and poor in oxygen, suffered seriously from this new pollutant that selectively attacks key points in its energy metabolism.

Tecnazene, 2,3,5,6,-tetrachloronitrobenzene, a sprouting inhibitor and fungicide used on stored potatoes, has been found to enter the freshwater environment at specific sites in the U.K., due to the practice of bulk washing potatoes for supermarket presentation. Whale et al.[1592] tested the sensitivity of the mayfly *Caenis moesta* and some other invertebrate and fish species to tecnazene to provide extra information for hazard assessment. Tecnazene was highly toxic to the fish species tested (96-h LC_{50} values from 270 to 2340 μg/l). *C. moesta* and the other invertebrates showed a similar degree of sensitivity to the fish, with 96-h LC_{50}s ranging from 270 to 2340 μg/l and 96-h no-observed-effect concentrations between 50 and 100 μg/l.

The toxicity of the herbicide glyphosate to aquatic invertebrates[487,488] and the effects of a glyphosate application on lentic daphnids[682] have been studied. Glyphosate (Roundup®) was applied aerially for conifer release to part of the lower valley of Carnarion Creek, Vancouver Island, British Columbia.[882] Although the drift patterns of most invertebrates were not measurably affected by the glyphosate applications, the drift response of two particular organisms, *Gammarus* sp. (Crustacea) and *Paraleptophlebia* sp. (Ephemeroptera), suggested a slight and ephemeral herbicide-induced disturbance in and downstream of the treatment areas. Postspray drift of these taxa was not significantly different from prespray levels, but a measurable alteration in the drift patterns of these two genera, especially of *Paraleptophlebia*, was demonstrated. This disruption of drift patterns may have resulted from natural causes (high water levels), but was coincident with glyphosate contamination and therefore cannot be dismissed as being unrelated to the herbicide applications.[882]

VI. INDUSTRIAL CHEMICALS

A. BACKGROUND

Hundreds of chemicals, some of which are very toxic to aquatic insects, are released from industrial sources into waters in the form of normal effluents or accidental spillages. Furthermore, warfare and military training exercises may cause considerable damage to aquatic life. In this respect, halogenated hydrocarbons, including PCBs, have received relatively much attention. However, the number of industrial chemicals is enormous, and their effects on aquatic organisms and ecosystems are poorly known.

Chlorinated aromatic compounds have been identified in sediments and biota from many watersheds and often occur in industrial effluents.[822,1280] Many of the monoaromatic chlorinated compounds (e.g., chlorobenzenes) are highly volatile and are not often detected in the water column. Nevertheless, many of these compounds have been identified in sediments and biota from impacted ecosystems.[468,1156]

PCB contamination of water may result from PCB-containing industrial waste effluents, from sewage wastewater discharged into receiving waters, or from aerial sources.[1100] PCBs can be emitted from hazardous waste landfills,[931] municipal landfills, and incinerators.[1103]

B. POLYCHLORINATED BIPHENYL CONGENERS

Polychlorinated biphenyls (PCBs) have been produced for a multitude of industrial purposes, e.g., electrical transformers, refrigerating briners, lubricants, and glues. Today PCBs occur almost everywhere in nature, but their concentrations are usually low.[1466] Chlorinated aromatic compounds (e.g., PCBs, hexachlorobenzene) are environmentally persistent and highly lipophilic. Subsequently, these compounds tend to become associated with suspended particles, accumulate in sediments, and become a chronic source of contamination to biota. Dredge-spoil removal and encapsulation of contaminated sediments in a dredge-spoil landfill may be ineffective, since rain or snow percolating through the contaminated sediments may cause the leaching of PCBs from the disposal site. The environmental risk posed by such sediment-sorbed compounds depends on their persistence and availability for bioaccumulation. Both benthic and pelagic organisms can accumulate chemicals derived from contaminated sediments and may subsequently exhibit increased mortality, reduced growth and fecundity, or morphological anomalies.[244,856,972] Furthermore, sediment-derived chemical residues may be transferred throughout the food web and eventually to man.[856] Inhibition of plant growth due to PCBs has been documented mainly for algae.[1443]

Benthic invertebrates appear to play an important role in the sediment dynamics of contaminants.[1644] PCBs have been reported in water and caddisfly larvae in the upper Hudson River,[214] in adult mayflies in the upper Mississippi River,[273] and in the macrobenthos of the River Adige, Italy.[1184] Ciborowski and Corkum[263] collected night-flying Trichoptera and Ephemeroptera, using long-wave ultraviolet light from eight sites adjacent to the St. Clair and Detroit Rivers, Canada. Concentrations of highly chlorinated PCBs (hexa-, hepta-, and octachlorobiphenyls) were significantly greater in Trichoptera emerging from Detroit River sites than in samples from St. Clair River sites. Levels of non-PCB organochlorine compounds, pentachlorobenzene (QCB), hexachlorobenzene (HCB), and octachlorostyrene (OCS), were higher in Trichoptera collected from the St. Clair River than from the Detroit River. In both rivers the highest concentrations of less highly chlorinated PCBs (tri-, tetra-, and pentachlorobiphenyls) were found in samples from upstream sites. Similar spatial trends were evident among sites for the mayfly genus *Caenis*. However, concentrations of HCB were typically higher in the mayflies than in caddisflies.

Kovats and Ciborowski[869] captured adult Trichoptera and Ephemeroptera at night at four contaminated sites on the Detroit and St. Clair Rivers and at several uncontaminated central Ontario locations. Gas chromatographic analyses of central Ontario samples revealed low, but detectable (<1 μg/kg), levels of individual PCB congeners (e.g., PCB 180, PCB 66, HCB), and relatively high concentrations of some pesticides (dieldrin, p,p′-DDE). Insects collected near the Detroit and St. Clair Rivers had significantly higher body burdens of nearly all the contaminants considered than insects from central Ontario sites. The spatial pattern of contaminants among insect samples corresponded to that in the sediments. Considerable year-to-year variation in contaminant body burdens may occur. Comparison of these data[869] with those of Ciborowski and Corkum[263] revealed a twofold to 12-fold increase in concentrations of most PCBs in Trichoptera collected from the Detroit River near the mouth of the River Canard, although HCB, OCS, and QCB body burdens remained unchanged. At Windsor concentrations of QCB, HCB, and OCS remained constant in *Hexagenia*, while those of PCBs increased slightly during the same time period (1 year).[869]

The pathways through which pelagic organisms accumulate chlorinated contaminants are not clear, but may be partially related to the food-chain transfer of compounds by benthic organisms.[855] Bottom sediments are a depository for many hydrophobic xenobiotics. These compounds are released to the water column by sediment resuspension when sudden hydrographic events occur.[1644] Chironomid larvae, which are major food sources for many fish species, accumulate many xenobiotic chemicals from sediments.[1088,1274] Wood et al.[1644] investigated the sediment desorption of 71 PCB congeners and their bio-uptake by *Chironomus tentans* in a continuous flow system. Bioconcentration of the PCB congeners by *C. tentans* was selective with the bio-uptake factor, calculated as ng/g (organism) per ng/g (water in the organism chamber), being highest for those with 2 to 4 chlorines.[1644] Larsson[904] reported that midge larvae not only accumulated sediment-sorbed PCBs, but

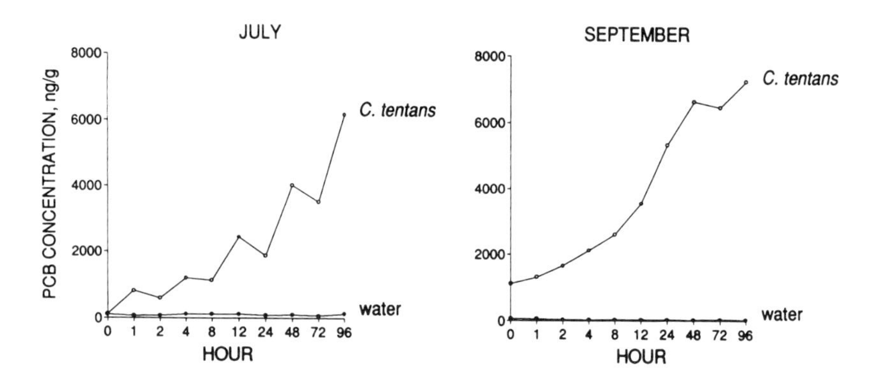

FIGURE 12. Total PCB concentration in *Chironomus tentans* (Diptera, Chironomidae) (ng/g dry weight) and water (ng/l), Hudson River, NY, July and September 1985. Reported values are the means of three samples. (Data from Novak, M. A. et al.[1137])

also retained significant tissue residues after the larvae had metamorphosed to terrestrial adults. Studies on caddisfly larvae indicated selective bioaccumulation of different PCB congeners and differences between the Hydropsychidae and Limnephilidae.[214]

Novak et al.[1138] monitored the concentrations of polychlorinated biphenyls (PCBs) in the Hudson River from 1978 to 1985. PCB contamination of the river occurred over a period of nearly 30 years via discharges by capacitor manufacturing plants in Hudson Falls and Fort Edward, NY. Concentrations of PCBs in caddisflies were higher than in the multiple-plate artificial substrate samplers by approximately a factor of 5 (means: caddisflies 27.9 μg/g, multiplates 5.02 μg/g). This may be associated with the proportion of hexane-extractable compounds (lipids). Caddisflies had lipid concentrations ranging from 5 to 66% of the dry sample, while multiplate samples averaged 1%. Inorganic materials such as silt may affect total PCB concentrations, and PCBs adsorbed to inorganic particulate matter may contribute to the total concentration. The caddisfly results reflected, in part, the results of the multiplate samples, but, on the whole, provided less information. For the caddisfly samples, distance from the original PCB discharge point was not an important predictor of PCB concentration.[1138]

Novak et al.[1137] measured the uptake of PCB congeners in the larvae of laboratory-reared *Chironomus tentans* placed in the upper Hudson River during 2 months in 1985 (Figure 12). Uptake differed with varying degrees of chlorination relative to water concentrations. At the end of the period, concentration factors ranged from 4,000 to over 300,000 times the water concentrations.

The organic carbon content of sediments is likely to be the main determinant of chemical bioavailability for neutral organic compounds. Knezovich and Harrison[855] found that midge larvae *Chironomus decorus* accumulated higher tissue-to-sediment ratios of chlorobenzenes from a low-organic-content sediment than from a high-organic-content sediment. These results were consistent with those reported by Rubinstein et al.,[1305] who found a similar relationship for the accumulation of PCBs by marine worms.

C. OTHER INDUSTRIAL CHEMICALS

Phthalate esters are one of the most produced chemical groups in the world and are used mainly as plasticizers.[1486] Of the phthalates, DEHP [di(2-ethylhexyl)phthalate] seems to exhibit properties typical of organic micropollutants (the ability to accumulate in organisms, persistent to a certain degree, and a low acute toxicity). The compound reduced reproduction in *Daphnia magna*,[1321] and bioaccumulation occurred in invertebrates.[1006] Since phthalate esters are lipophilic they tend to become attached to particles in the aquatic environments and are found in high levels in the sediments.[970] Benthic organisms are therefore more exposed to this substance than those living in the water column. An aquatic laboratory system was constructed to study the behavior of dragonfly larvae (*Aeshna*) exposed to sediment-bound DEHP.[1638] The predation efficiency of dragonfly larvae exposed to DEHP was lower than that for nonexposed larvae. This may be explained by an effect on vision or on the nervous system. In late instars sight is important for the larvae to achieve correct judgement of the prey location before extending the labium. The level of DEHP in dragonfly larvae was much lower than that in the sediment. Although the low level of DEHP in the larvae might indicate metabolism of DEHP in the insects, DEHP does not seem to be easily metabolized in invertebrates.[1042,1401]

As a consequence of a fire in a depot of Sandoz Ltd., Basel, Switzerland, large quantities of chemicals were accidentally discharged into the Rhine on the November 1, 1986.[1516] These chemicals included a fluorescent dye, Rhodamine, two organophosphorous pesticides, disulphoton and thiometon, and a fungicide, ethoxyethylmercuryhydroxide. Initial concentrations of disulphoton and thiometon exceeded 100 μg/l and 20 μg/l, respectively, near the point of discharge, resulting in mass fish kills.[1516] At the German-Dutch border, concentrations of disulphoton and thiometon decreased to about 5.3 μg/l and 2.0 μg/l, respectively. Chironomids living between filamentous algae and mosses on the stones of the jetties showed high mortality rates, which were attributed to the substances discharged into the river at Sandoz, more than 800 km upstream. The damage was not yet visible on the first day after the passing of peak concentrations, but 2 weeks later the majority of the larvae on the stones were dead. High percentages of deformed larvae were also found. The caddisflies *Hydropsyche contubernalis* and *Ecnomus tenellus* did not show such a clear response as chironomids. One month after the accident, strikingly few caddisfly larvae were recorded from the stones, but only a few dead animals were found.[1516]

Spillage from sawmills may have adverse effects on aquatic insect fauna. The spillage of timber preservative containing dieldrin, pentachlorophenol, and bis(tributyltin)oxide contaminated an area of ground around a sawmill in 1983, and an unknown volume entered the Tyock Burn, a small river in

Scotland. The spillages completely destroyed the invertebrate community in the Tyock Burn downstream from the events and brought about a reduction in diversity in the River Lossie below the confluence with the Tyock Burn.[1027] Stoneflies, mayflies, and caddisflies were lost from the river, and recolonization by some of these groups did not take place for 4 years following the second preservative spillage. Continual leaching from the subsoil maintained dieldrin concentrations in the stream at levels that are harmful to fish and lethal to stoneflies. The concentrations of dieldrin in the Lossie almost 2 years after the spillages were still sufficient to cause mortality in stoneflies.[1027]

The acute toxicity of a synthetic hexachloroethane (HC) smoke combustion products mixture to nine freshwater aquatic organisms was determined by Fisher et al.[473] Synthetic HC smoke combustion products are a complex mixture containing Zn, Cd, As, Pb, Al, CC_{14}, C_2Cl_4, C_2Cl_6, C_6Cl_6, and HCl. When the above devices are used in military training exercises, these components are released into the environment where they have the potential to enter the aquatic ecosystem and subsequently damage aquatic life. The U.S. Army indicated that the usage of this munition in training exercises was significant. The synthetic HC smoke combustion products mixture was toxic to all organisms tested. Forty-eight-h LC_{50} values for the insects ranged from 54.1% for the mayfly larvae to 89% of the 100% stock solution for the midge larvae. However, there were no data on the actual amounts of these components that could be released into the aquatic environment after use of the grenades and smoke pots or after disposal of the stockpiled munitions.[473]

VII. OIL

A. BACKGROUND

Oil is a mixture of aliphatic, cycloaliphatic, olefinic, cyclic, and polycyclic aromatic hydrocarbons (PAHs); their oxygen, nitrogen, and sulfur derivatives; and trace amounts of chemical elements. Representative toxic compounds present in crude oil include the organic compounds benzene, toluene, naphthalene, fluorene, phenanthrene, fluoranthrene, pristane, phytane, and pyrene and the metals nickel and vanadium.[333] Oil and its products are frequent contaminants in the aquatic environment ranging from the Arctic to tropical areas such as Amazonia. Approximately 6 million metric tons of oil enter the aquatic environment each year worldwide from natural and anthropogenic sources.[1112] These materials accounted for the third leading cause of reported fish kills in the U.S. between 1961 and 1975.[1517] Incidents like the Persian Gulf War may greatly increase the amount of oil released into aquatic ecosystems.[11a,82]

Oil behavior in the aquatic environment and its effects on freshwater organisms are diverse and complex.[212,333,424,689,1056] Virtually all macroinvertebrates in the aquatic food chain are affected by oil.[148] Surface oil may limit oxygen exchange, coat the gills of aquatic organisms, interfere with respiration, cause pathological lesions on respiratory surfaces, and result in bioaccumulation of hydrocarbons.[1650] Oil may blanket the bottom substrate and prevent invertebrate colonization or directly kill organisms by toxic action.[695]

Oil also has secondary impacts on ecosystems. These include reductions in the decomposition rate of vascular plant litter, nutrient release from sediments, and algal primary production.[333] Decomposition of oil may produce additional toxic effects through the consumption of oxygen.[621] Recovery of aquatic life in an oil-polluted environment may be slow, and the effects of a large oil spill can be devastating and extend over a long period of time.[149,333,378,571,621,944,975] The extent and duration of damage to an aquatic community by an oil spill depends on the volume and chemical characteristics of the oil involved, the initial extent of oil penetration into the stream substrate, the hydrobiological characteristics of the stream, the degree of cleanup, the time of year, and the inherent tolerance of stream organisms.[333]

Oil toxicity is associated with the degree of dispersion of these components throughout the water column.[1056,1468] When oil is spilled on water, volatile substances evaporate rapidly, and the soluble fractions slowly enter the water column.[624] The remaining insoluble fraction floats or combines with the silt and particulate organic matter usually present in aquatic environments and sinks to the bottom where chemical and bacterial degradation occur slowly.[148,621] The toxicity of oil can be attributed to direct contact, exposure to the water-soluble components, and to the products of oil degradation.[695] Toxicity is generally increased by turbulent stream flow, wind, or wave action or the addition of chemical emulsifiers.[1408]

Many field studies measuring the response of the benthic community to accidental contamination by oil or petroleum products have lacked quantitative data on the amounts of contaminants involved.[1300] An estimation of the absolute quantity of material involved per unit area or volume of water is needed to assess the potential for environmental damage.[1174]

B. ACCIDENTS

Most oil spills (3372 incidents in 1981) in and around U.S. waters occur in lotic ecosystems.[1215] Damage to the aquatic environment may be extensive, but information is limited on the ecological consequences of accidental oil spills on freshwater lotic ecosystems.[1056] Recent publications on the effects of oil spillage on benthic macroinvertebrates have produced some contradictory findings concerning the extent of damage to the aquatic community.[200,333,571,621,989,1300,1314]

Stoneflies, mayflies, and caddisflies have been shown to be adversely affected by oil, in several studies.[200,333,1300,1314,1648] *Baetis* and *Isoperla* species were among the most sensitive.[1649] These effects may be due to interference with respiration in organisms with external gills. Oil exposure may also cause abnormalities in the tracheal gills of Plecoptera and Trichoptera.[1374] In contrast, Diptera have generally been tolerant to many pollutants, and some species may show a positive response to oil.[333,1314,1648] Rosenberg and Wiens[1300] found in a study of 29 Chironomidae species on oiled substrates an equal number of species that responded positively, negatively, or not at all. Cushman and Goyert[341] reported an increase in Chironomidae at intermediate levels of oil contamination in experimental ponds. Their study attributed the increase indirectly to the oil-induced loss of macrophytes, coupled with a corresponding increase in benthic algae and detritus, and to decreased predation.

On April 27, 1986, at least 8 million liters of medium-weight crude oil spilled from a ruptured storage tank into the sea on the Caribbean coast of Panama. This was the greatest amount of oil spilled into a sheltered coastal habitat in the tropical Americas.[756] Jackson et al.[756] described the types and extent of damage to coastal populations and communities in the first 1.5 years after the 1986 spill. Intertidal mangroves, seagrasses, algae, and associated invertebrates were covered by oil and died soon after. There was also extensive mortality of shallow subtidal reef corals and the fauna of seagrass beds. After 1.5 years only some organisms in the areas exposed to the open sea had recovered.

Crunkilton and Duchrow[333] monitored the impact of a massive crude-oil spill on the invertebrate fauna of a Missouri Ozark stream. The benthic macroinvertebrate fauna of Asher Creek, a fourth-order stream with a base flow of 0.03 m^3/s, was monitored for 532 days following a 1.5-million-liter domestic-crude-oil spill in August 1979. Oil was visually present in the stream riffle substrate for 453 days following the spill. Aquatic insects and other macroinvertebrates in the affected area were reduced to less than 0.1% of expected numbers at the first sampling period 25 days after the spill. Plecoptera, Trichoptera, and Ephemeroptera were most severely affected, being reduced in numbers for 9 months (Figure 13). Species diversity was less than the minimum values established for unpolluted Missouri streams for 11 months. The initial postspill community was dominated by Chironomidae, Simuliidae, and Oligochaeta. As a group the Chironomidae responded positively after the initial die-off during the early phases of recovery. The positive effect may be due to a tolerance to toxicants present in oil, a response to an increase in appropriate food resources, or to the absence of predators.

The delay in recovery of the benthic community in Asher Creek was partially attributed to seasonal biotic factors.[333] The potential for reproduction of many aquatic species with terrestrial adults is limited by air temperature following a late-summer or autumn spill. Substantial improvement in the community did not occur until 9 months after the August spill and was not complete for at least 17 months. By March, spring flows had scoured the stream channel and warmer air temperatures promoted additional recruitment of species growing most rapidly during the summer. Recolonization of benthic invertebrates following the oil spill was believed to be accomplished by downstream migration of immature individuals within the stream and by egg deposition from adults. Although only 0.6 km of unaffected stream was present upstream of the spill to its origin at Cave Spring, it was sufficient to account for most of the species found in the recovery zone. In contrast, Guiney et al.[571]

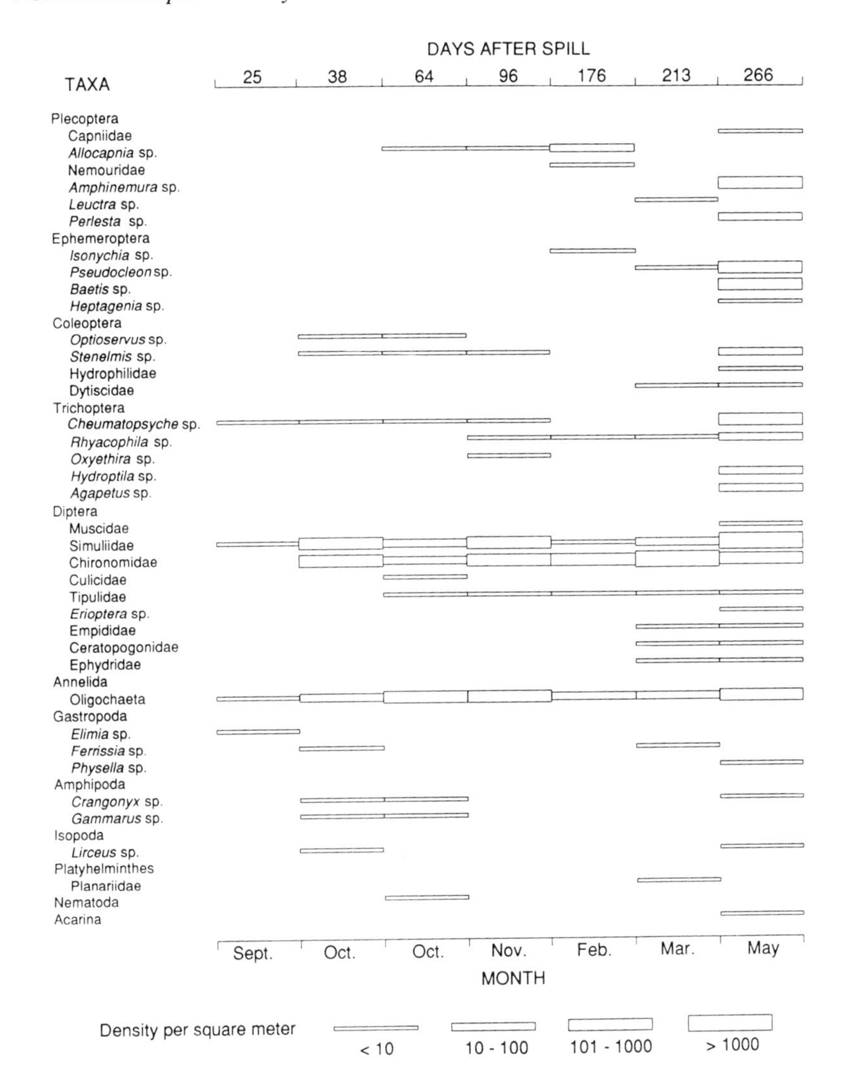

FIGURE 13. Benthic macroinvertebrate fauna of Asher Creek, MO (Site 2) during the first 9 months of recovery after a massive crude-oil spill. (Data from Crunkilton, R. L. and Duchrow, R. M., *Environ. Pollut.*, 63, 13, 1990. With permission.)

attributed delayed benthic recovery following an oil spill in a similar-sized stream in Pennsylvania to the lack of organisms available for recolonization.

Pontasch and Brusven[1215] examined the success of cleanup efforts following the June 4, 1983, spill of over 94 m^3 of unleaded gasoline into Wolf Lodge Creek, ID. The spill occurred as a result of the accidental rupturing of a high-pressure pipeline. Twelve days after stream cleaning, total insect densities and insect biomass in the raked areas were significantly lower than those in nonraked areas. However, 1 month after stream cleaning, and for the remainder of the study, no significant difference was found between the raked and nonraked areas.

Chironomidae and the mayflies *Baetis hageni*, *B. tricaudatus*, *Drunella grandis*, and *Pseudocloeon* sp. and the caddisflies *Rhyacophila vaccua*, *Glossosoma* sp., and *Lepidostoma* sp. had begun to recolonize prior to stream cleaning. Most of the early colonizers were lost from raked areas during the cleaning process. Chironomid, mayfly, and caddisfly densities were lower in the raked section 12 days after stream cleaning, but 1 month after stream cleaning, there was no significant difference in their densities between the raked and nonraked sections. The elmid beetles (*Optioservus quadrimaculatus*, *Heterlimnius corpulentis*, *Zaitzevia parvula*, and *Narpus concolor*) were apparently most adversely affected by stream cleaning. The elmid densities were lower in raked sections through January 1984.[1215] Elmids were found to be adversely affected by physical disturbances of the substrate in an Alberta stream, by Clifford.[280]

Recovery of mayflies, stoneflies, and caddisflies in impacted areas of Wolf Lodge Creek was well underway less than 5 months after the spill.[1215] In comparison, Bugbee and Walter[200] found no stoneflies present in areas containing substrate-trapped hydrocarbons 6 months after a much smaller gasoline spill (19 m^3), and mayflies and caddisflies were found only at the site immediately below the spill. Thus, Pontasch and Brusven[1215] concluded that the stream-cleaning process was beneficial, in spite of similar recovery times in raked and nonraked areas, because it minimized the possibility of long-term hydrocarbon toxicity without causing substantial additional impact.

C. OIL PRODUCTION

Oil production activities create the possibility of environmental pollution. West and Snyder-Conn[1590] studied the effects of pit fluid discharges on water quality and the macroinvertebrate community of tundra ponds in Alaska. Aquatic macroinvertebrate samples were collected from reserve pits, from ponds initially receiving reserve pit fluid discharges, from more distant connected ponds, and from three remote-control ponds. The reserve pits were devoid of invertebrates, the receiving ponds contained 2 to 5 taxa, distant ponds 4 to 10, and the control ponds 10 to 13. Chironomids predominated in all the ponds adjacent to reserve pits where other taxa were rare. The dominant taxa in distant and control ponds included chironomids, calanoid copepods, daphnids, nemourids, and physids. The abundance of organisms was related to the treatment. This pattern was largely attributable to the abundance pattern of crustaceans, primarily daphnids. Diptera, especially chironomids, tended to increase in abundance in ponds adjacent to reserve pits, whereas other insects showed a corresponding decrease. Control ponds had a significantly greater number of invertebrate taxa, a higher diversity, and more organisms. Total taxa was negatively correlated with water hardness, alkalinity, barium, and arsenic.[1590]

D. EXPERIMENTS

The chemical composition and toxicity of three shale crude oils, a hydrotreated oil, and a refined shale oil were assessed to determine the potential hazards to fish species and aquatic invertebrates posed by accidental oil spills.[1649] Shale oils have much higher concentrations of olefins, aromatic substances, and nitrogen heterocyclic compounds than do petroleum crudes. Colorado squawfish (*Ptychocheilus lucius*), fathead minnow (*Pimephales promelas*), cutthroat trout (*Salmo clarki*), and colonies of aquatic invertebrates were exposed to the water-soluble fractions of the shale oils for 96 h in order to determine concentrations lethal to 50% of the exposed organisms. After 96 h in the control solution, invertebrate communities were composed of 18 to 34 taxa and 151 to 211 total organisms. As the exposure concentration increased, the total number of organisms, and both the number and diversity of the taxa, generally decreased. Response of the colonized community was a sensitive index of toxicity when assessed on a taxon basis by percentage composition of total organisms. The most common groups were *Baetis* (Ephemeroptera), *Isoperla* (Plecoptera), *Brachycentrus* (Trichoptera), and Chironomidae, which altogether accounted for more than 50% of the total organisms collected. The most sensitive of these genera were *Baetis* and *Isoperla*. One or both showed a reduction in percentage composition in the lowest concentration tested for all shale oil products. On the other hand, the percentage composition of *Brachycentrus* out of the total number increased as exposure concentration increased.[1649]

VIII. TEMPERATURE

The pollution caused by increasing temperature is called thermal pollution. One source of thermal pollution is from the condenser-cooling water released from electricity-generating stations and nuclear plants. In England, for instance, a temperature increase of 8°C was reported in the River Severn at times of low flow. Thermal pollution also follows from a range of sources during urbanization. For example, on Long Island, NY, the summer temperature of the river water has increased as much as 5 to 8°C.[555] Reservoir construction can also affect water temperatures.[1269]

Temperature, acting either independently or simultaneously with other environmental factors, is recognized as one of the major physical factors controlling the survival, rate of development, and growth of aquatic insect larvae and adults. Consequently, thermal pollution has some significance to aquatic fauna.[859,902] The effects on insects by several pollutants are temperature dependent. Research has been concentrated on the responses of organisms to constant temperature, although constant temperatures are virtually nonexistent in freshwaters and estuarine or coastal waters. Little is known about how fluctuating temperatures affect organisms or specific physiological processes within, for example, estuarine and marine ecosystems.[309,310] Obviously, changes in algal blooms and fish behavior induced by thermal pollution[555] indirectly affect aquatic insects.

IX. RADIATION

Aquatic organisms have always been exposed to radiation from natural sources, both in the external environment (sediment and/or water) and from radionuclides incorporated into the organism. While most aquatic organisms are reported to be relatively resistant to ionizing radiation, there is much interspecific variation in sensitivity. Organisms inhabiting an environment contaminated by radionuclides also receive radiation internally as a result of assimilation from food and absorption from water. In addition, some aquatic biota receive significant external exposure from radionuclides accumulated in sediments. The dose rate from radionuclides in the water is generally low in comparison with that from incorporated radionuclides, which in turn is equal to, or lower than, that from radionuclides in sediments.[1194]

Studies on the effects of irradiation at the ecosystem level have mainly dealt with terrestrial ecosystems. Aquatic microcosm studies have shown that acute doses of 1 to 10 kGy of gamma radiation are required to influence such parameters as the chlorophyll content, biomass, and community development.[467] Studies with two different types of marine microcosms exposed to chronic gamma irradiation for 10 to 12 months failed to demonstrate any effects of radiation at dose rates up to 33 mGy/h.[1194]

The gradual senescence of present-day operating nuclear facilities and the resultant contamination of aquatic and terrestrial ecosystems emphasize the importance of understanding the behavior of radionuclides in the environment. Some few studies deal with levels of radionuclides in insect life stages in waters or soils affected by U operations.[56,91,1459] Whicker et al.[1593] studied the distribution of ^{137}Cs, ^{90}Sr, ^{238}Pu, ^{239}Pu, ^{240}Pu, ^{241}Am, and ^{244}Cm in the biotic and abiotic components of an abandoned reactor cooling impoundment after a 20-year period of equilibration period at Savannah River Site, SC. It received radioactive contaminants via the cooling-water discharges from the reactor from September 1961 to June 1964. Sampling was carried out in 1983 and 1984. Pooled, dried samples representing immature stages of several abundant taxa (Odonata, Diptera, Ephemeroptera) were assayed for radionuclides. Concentrations of ^{137}Cs and ^{90}Sr were one to three orders of magnitude higher than the transuranic radionuclides. Detectable concentrations of ^{241}Am, ^{244}Cm, and ^{239}Pu were found in damselflies and midges, and ^{238}Pu only in midges.[1593]

Over 97% of the total biotic inventory of the radionuclides studied by Whicker et al.[1593] was contained in the aquatic macrophytes. This was primarily due to their large biomass. Concerning the radionuclides in animals, zooplankton accounted for 50% of the ^{137}Cs and 6% of the ^{90}Sr, and fish for

41% of the ^{137}Cs and 49% of the ^{90}Sr; benthic macroinvertebrates contained 36% of the ^{90}Sr. The benthic macroinvertebrates accounted for most of the transuranic radionuclides in animals. No apparent increases with trophic level were evident. The concentrations of ^{137}Cs measured in macroinvertebrates ranged from 1 to 14 Bq/g dry weight, whereas macrophyte and surface sediment samples contained roughly 7 to 30 and 25 to 35 Bq/g dry weight, respectively. In the case of ^{90}Sr, insects contained 40 to 120 mBq/g dry weight, while macrophytes and surface sediments contained 500 to 1500 and 200 to 400 mBq/g dry weight, respectively. Ratios of ^{137}Cs, ^{90}Sr, ^{241}Am, and ^{244}Cm to $^{239,240}Pu$ in the benthic insects were 4, 51, 6, and 28 times higher, respectively, than the same ratios for sediments, indicating accumulation mechanisms other than sediment associations for these radionuclides. Zooplankton and benthic insect larvae have high surface-to-volume ratios, and thus their surfaces, as well as their guts, may carry sediment and sestonic particulates, which in turn could carry measurable ^{137}Cs and transuranic activity. This possibly explained their relatively high ^{137}Cs values and transuranic ratios. The results by Whicker et al.[1593] showed that ponds that have been contaminated with long-lived radionuclides were capable of supporting a diverse and productive flora and fauna.

On Saturday, April 26, 1986, one of four graphite-moderated, water-cooled nuclear reactors at Chernobyl, Ukraine, ignited following an explosion. About 3.5% of the radioactive material of the reactor, corresponding to some 2×10^{18} Bq, were ejected into the atmosphere.[1194] Radioactive substances were transported by weather systems and released as fallout over certain areas of northern Europe. Radioactive substances in the watershed area washed out into waterbodies where the redistribution and accumulation of radionuclides occurred in bottom deposits, aquatic plants, and aquatic animals. According to the preliminary estimates, the irradiation doses of the majority of aquatic organisms in the Kiev Reservoir did not exceed the range of doses at which radiation injuries occur in the populations.[1403]

After the Chernobyl accident, Dahl and Grimås[345] studied radionuclides in *Aedes communis* pupae (Culicidae) in central Sweden. Samples were taken in May, one month after the accident, and in October, six months after the accident. Three radionuclides, ^{131}I, ^{110m}Ag, and ^{140}Ba, were present in higher concentrations in the mosquito pupae than in the substrate. The high concentrations of ^{131}I in the pupae partly indicated accumulation via inhalation by the larvae during the fallout period. Later on during the year, no ^{131}I could be found. The concentration of ^{134}Cs did not differ markedly from the values in the substrate. All the other radionuclides studied (^{137}Cs, ^{103}Ru, ^{141}Ce, ^{144}Ce, ^{95}Zr, ^{95}Nb) had a lower concentration in the pupae than in the substrate. Rearing of the spring population showed no general negative effects on eclosion ability. The substrate had low ^{134}Cs and ^{137}Cs values, the most long-lived radionuclides, as well as the presence of ^{110m}Ag and ^{60}Co. Dahl and Grimås[345] suggested that the main uptake from fallout occurred via feeding, as well as direct contamination through respiration and accumulation in the water.

Measurements of total ecosystem parameters did not appear to be useful for detecting effects at the dose rates that occurred in Swedish waters following Chernobyl. Thus, the available data on the levels of radiation in Swedish aquatic systems due to the Chernobyl accident suggested that direct damage to aquatic organisms was unlikely. However, long-term biological effects on aquatic organisms are possible. The accumulated doses are clearly well above the background levels, and further bioaccumulation connected with sediment loading will increase the dose.[1194]

Baudin et al.[86] studied fish (*Cyprio carpio*) fed on three different natural labeled food types (^{60}Co): a midge larvae (*Chironomus* sp.), a benthic crustacean (*Gammarus pulex*), and the soft tissues of a mollusc (*Lymnaea stagnalis*). Initially, the individuals were fed contaminated food over a seven- to nine-week period and then, during a second phase, received nonlabeled food in order to follow ^{60}Co depuration. The radionuclide transfer factor was twice as high in chironomid larvae and four times higher in gammarids than in molluscs. Irrespective of the differences related to the food type, the values of the transfer factor and the contamination kinetics suggested that although ^{60}Co transfer and accumulation occurred, these processes could not lead to biomagnification of the radionuclide, even over the long term.

X. CONCLUSIONS

The study of aquatic insects in relation to environmental pollution has a long tradition, and there is a huge body of field data available. Compared to terrestrial investigations, the documentation of spatiotemporal changes in insect distribution and abundance appears to be more reliable, although some studies seriously suffer from the lumping of several species into higher taxa or some uncertain ecological groups. Furthermore, there is also an increasing body of information on some insect species from laboratory toxicity testing. What seems to be still largely missing is field experimentation that could reveal mechanisms instead of patterns and link the laboratory results to field situations. The results from studies using microcosms and artificial streams have been promising. However, responses observed in test systems are often scale dependent and may not reflect real changes in natural ecosystems. Distinctly more information is needed on biotic interactions, such as the role of predation under pollution stress. Other major gaps include chronic and sublethal effects of most chemicals on insects.

Eutrophication and associated oxygen deficiency is known to have considerable effects on the benthic community in deep water. This has lead to the disappearance of some intolerant species, while others (chironomids in particular) have increased in numbers. Little attention has been paid to the changes related to the alterations in littoral vegetation. It seems likely that the effects of changes in macrophyte flora have a profound impact on associated insect fauna. The interface of aquatic and terrestrial ecosystems deserves more attention.

Justifiably, acidification of freshwaters has become one of the most investigated fields of environmental entomology. There is a real risk of loss of biodiversity in chronically acidified waters. Alterations of insect communities in various waters have been described, and there is also an increasing amount of information on experimental acidification. In several studies the mechanism causing changes in insect species composition has been partially attributed to mobilization of toxic concentrations of aluminium and other metals, subsequent to acidification. However, the real relative effects on insect populations of episodes vs. chronic acidification have remained obscure. The question on the relative effects of trophic limitations vs. physiological tolerances has also remained unanswered.

Existing studies on the effects on aquatic insects of pesticides applied against terrestrial pests illustrate the connection between terrestrial and aquatic ecosystems. Occasionally, the effects on aquatic fauna seem to be more profound (though usually reversible) than on terrestrial nontarget species.

The relative proportion of active and passive uptake of pollutants by aquatic insects differs drastically from that by terrestrial insects. Biomagnification is apparently too simple a concept for aquatic habitats: the relative importance of intake from food and directly from the abiotic environment is a key factor, and the approach implied by compartmental models seems more appropriate.[1075] The available experimental evidence suggests that intake from food is unlikely to be the major source of residues, e.g., for persistent organic pollutants in aquatic species. However, the rates of active uptake are not known for most insects and chemicals.

Chapter 6

RESPONSE MECHANISMS IN INSECTS

I. INTRODUCTION

Pollutants can affect the physiology, ecology, and evolution of insects. Physiological changes include not only increased mortality, but also sublethal alterations in reproduction, growth, and enzyme activity, which are associated with changes in morphology and behavior. The size of effect can be measured as the deviation from the control, and its duration as the speed of recovery. Pollutants can cause ecological perturbations in plant-insect or fungus-insect interactions, predation, competition, and insect-disease or insect-parasite relations. Evolutionary changes include the industrial melanism induced by changes in the insect habitat, and the economically important development of resistance to various contaminants, e.g., insecticides.

In conventional toxicity testing procedures, insects are exposed to a constant concentration of toxicant until mortality or some other definite response occurs. Several variables may affect response of the organism (Figure 1). In nature the situations where pollutants affect insects are much more complicated. For example, in the case of episodic exposure to pollutants, a range of physical and biological parameters that will influence both organism and ultimately ecosystem response have to be taken into account, in addition to the factors considered in standard toxicity testing (Figure 2).[1012]

II. PHYSIOLOGY

A. LETHAL AND SUBLETHAL EFFECTS

Death is a rather crude index of stress.[867] Methods for measuring the lethal toxicity of pollutants, especially pesticides,[1474] are well established and have been applied to a wide range of invertebrates. A large part of the toxicological literature consists of values of 96-h (or shorter) median lethal concentrations of pollutant for a test species. As shown by Abel,[1] much of the published data on acute lethal toxicity, even that of recent origin, is either trivial or misleading due to methodological limitations. Elzen[433] stressed that direct toxicity and selectivity are interfaced with sublethal effects, i.e., it is a question of degrees of toxicity and levels of organism response. For instance, the relative tolerance of *Gammarus pulex* and *Chironomus riparius* depended on both the chemical type and test period, indicating that there is a need for an increase in the range of species and exposure times used in toxicity testing.[1469]

Any toxic effect of a pollutant is significant if it influences the physiology or behavior of the organism, altering its capacity for growth, reproduction, mortality, or pattern of dispersal, since these are the major determinants of the distribution and abundance of species.[1] Although acute toxicity may have at

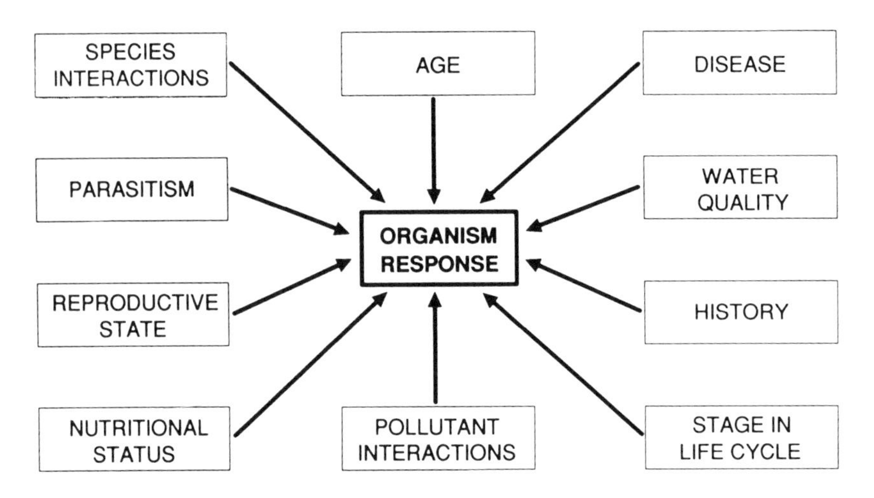

FIGURE 1. Factors affecting organism response during standard toxicity testing. (From McCahon, C. P. and Pascoe, D., *Funct. Ecol.*, 4, 375, 1990. With permission.)

least locally or temporarily significant effects on mortality, sublethal effects may often be more important. Insects are more commonly exposed to relatively low sublethal concentrations of pollutants for long periods than to levels of pollution that will cause rapid mortality. Criteria for these sublethal effects are difficult to define, and their biological significance is often difficult to assess. Examples of sublethal effects are disturbed enzymatic functions, damage to the nervous system, reduced reproduction, etc. These effects may result in altered behavior that may reduce the fitness of the organism.[1638] Figure 3 illustrates documented sublethal side effects among natural enemy orders.

B. GROWTH AND DEVELOPMENT

Growth inhibition is known to be a sublethal response to chronic toxicant exposure.[867] Growth is the change in size of an individual (or part of it), and development is the change in form.[553] Insect development is a sequence of distinct stages: e.g., larval instars, a pupal instar in endopterygotes, and imago. The duration of each stage and changes in form are determined by interactions of internal growth-regulating mechanisms, the current environment, and many cumulative residues (e.g., nutrient reserves) of earlier stages, including the parental generation.[553]

The larval growth of insects may be very rapid. It is not uncommon to have 200-fold size increases in less than 2 weeks, but in some Diptera, for instance, the growth rates may be much higher. An almost 1000-fold increase in size in 4 days has been reported in *Drosophila* larvae.[553] Growth is among the best indicators of organism response to pollutants, since it represents an integration of all the physiological processes. The growth of chironomid larvae has been proposed as a laboratory bioassay to help to define water quality for ecosystem protection.[1469]

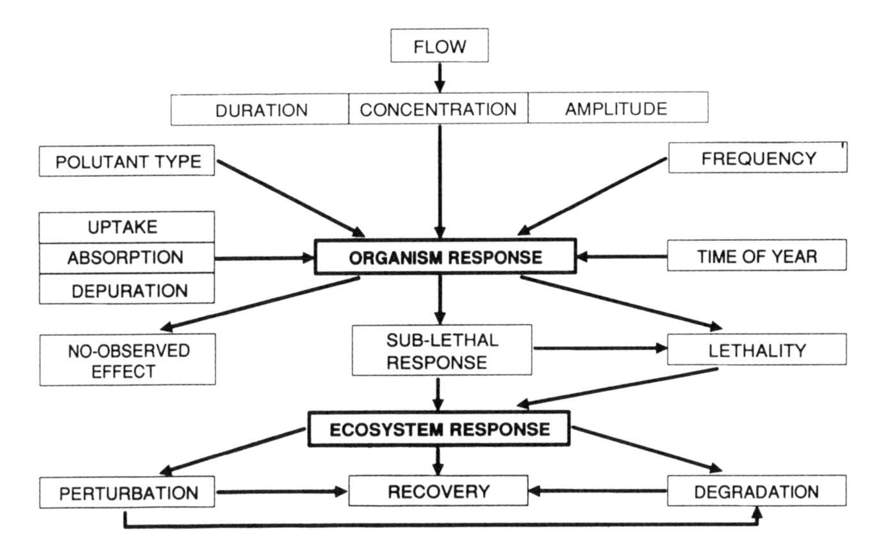

FIGURE 2. Additional factors (see Figure 1) affecting organism response and the interaction between organism and ecosystem responses in the case of episodic pollution. (From McCahon, C. P. and Pascoe, D., *Funct. Ecol.*, 4, 375, 1990. With permission.)

The effects of metals on growth have been studied both in aquatic and terrestrial insects. For example, Wentsel et al.[1585] reared midge larvae of *Chironomus tentans* for 17 days in metal-contaminated sediment collected from Palestine Lake, IN. They reported a significant decrease in midge growth compared to that in the control sediment. Studies on chronic sublethal exposure to boron demonstrated a significant decrease in the growth rate of *Chironomus decorus* at 20 mg B per liter.[961] Chronic effects of cadmium on *Polypedilum nubifer* (Chironomidae) were only observed in terms of the percentage emergence success.[631] The emergence of the cadmium-exposed midge larvae peaked before the control, although the development rate of the larvae was impaired by cadmium in the young stage. Brief exposure of *Chironomus riparius* larvae to equivalent assumed doses (mg/h) of cadmium suggested that exposure to a high concentration for a short time resulted in a reduced adult emergence, in comparison to exposure at a lower concentration for a longer time.[1013] Schmidt et al.[1341] investigated the effects of mercury, cadmium, and lead in the soil on the development of *Aiolopus thalassinus* (Orthoptera, Acrididae). The hatching rate of nymphs developed from eggs laid in treated soil was significantly reduced. The mean durations of the F_1 and F_2 (first- and second-offspring generations) nymphal stages were prolonged in all mercury and cadmium treatments and in treatments with high lead levels (250 to 500 μg/g) (Figure 4).

The entry of fluorine and sulfur into silkworm (*Bombyx mori*) larvae along with their food has an adverse effect on their development. Sulfur retarded the accumulation of larval mass, reduced their activity, and resulted in an increase in mortality and the number of individuals with deformities. Fluorine reduced

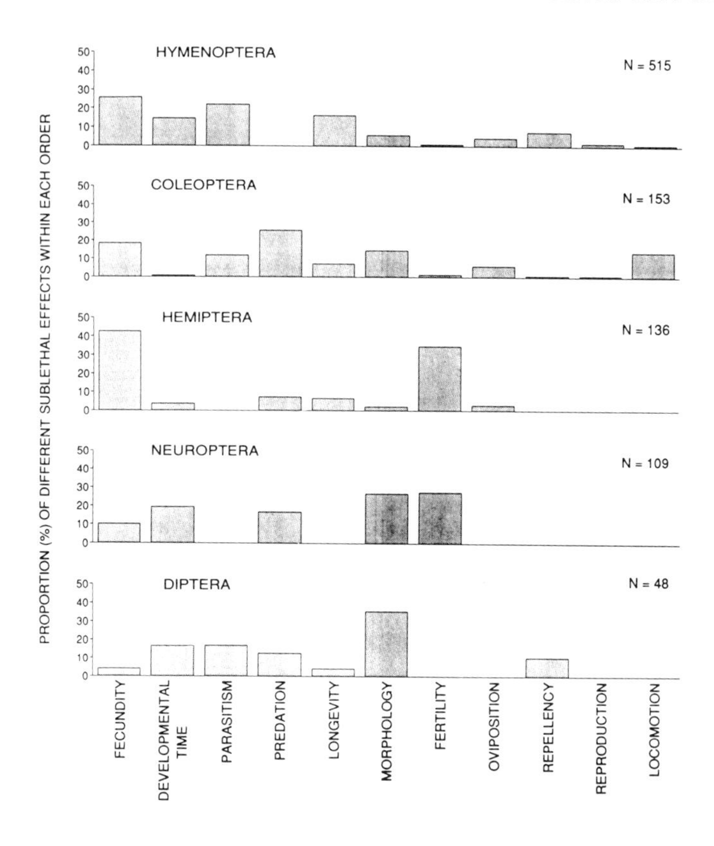

FIGURE 3. Documented (SELCTV database) sublethal pesticide side effects and their distribution within some natural enemy orders of agricultural pests.[1474]

alkaline phosphatase activity in the midgut of the larvae and caused a reduction in their mass and activity.[519]

Katayev et al.[818] reared the moths *Lymantria dispar* and *Orgyia antiqua* on food contaminated by pollutants. The most detrimental developmental effects were caused by hydrogen fluoride: larval mortality increased, larval and pupal mass decreased, the duration of development increased, and the percentage of adult emergence diminished. Comparative fumigation experiments on *L. dispar*, using different food sources, showed that the most significant adverse effects of the pollutants were manifested when the pollutants were ingested with the food, especially in the case of fluorine compounds. Fluorine compounds also had a cumulative effect, which made fluorine dangerous even if present at low concentrations for a long period of time.

Atmospheric pollution may alter the relationships between insects and their host plants. The mean relative growth rate (MRGR) is a measure of aphid performance, which requires no assumption about the shape of the aphid growth curves and thus can be used to compare treatments:[437,1558]

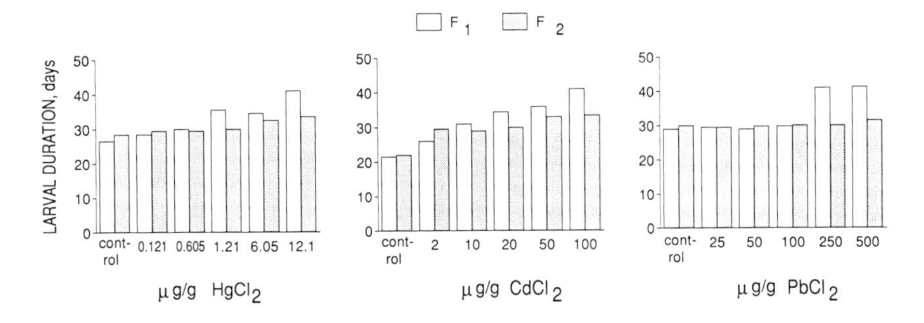

FIGURE 4. Effects of various concentrations of metals in soil on the nymphal duration of F_1 and F_2 generations of *Aiolopus thalassinus* (Orthoptera, Acrididae). The soil was treated only during the embryonic development of the F_1 generation. Statistical differences ($p < 0.001$) from the control indicated by asterisks. (Modified from Schmidt, G. H., Ibrahim, N. M. M., and Abdallah, M. D., *Sci. Total Environ.*, 107, 109, 1991. With permission.)

$$\text{MRGR} = \frac{\ln(\text{final wt}) - \ln(\text{initial wt})}{\text{time end} - \text{time start}}$$

For example, the MRGR of the aphids on Sitka spruce increased when the spruce seedlings were prefumigated with SO_2.[1030] The MRGR of aphids feeding on plants in urban air was also higher than that of conspecifics in a filtered atmosphere.[387,390]

Dohmen et al.[390] suggested that the enhanced growth of the aphid on *Vicia faba* in response to SO_2 fumigation was mediated entirely via the host plant. The physiology of the pea may have changed, and the aphids were probably responding to these changes.[1557] At higher SO_2 concentrations the aphid MRGR may decrease, possibly partly due to the toxic effect of SO_2 on the aphid (Figure 5).[1557] Similar observations on reduced growth have been reported on *Drosophila melanogaster* and honeybees (*Apis mellifera*).[547,686]

Pollutants have been reported to cause structural deformities in several insect orders.[1474] For example, the insect growth regulator Dimilin® caused molting abnormalities in the gypsy moth parasitoid *Apanteles melanoscelus* when hosts fed with pesticide-treated foliage were parasitized.[558] The abnormalities included blackening of the cuticle, twists in the body, and a partially molted cuticle.

High frequencies of structural deformities in chironomid larvae may be associated with high levels of chemicals and radioactive waste (Figure 6).[1566] Although field observations suggested that various types of contaminant may induce deformities in chironomid larvae, experimental data supporting this hypothesis are scarce. Deformities are expected to impair the feeding of the larvae and thus to affect growth and development.[1516]

The chironomid taxa so far recorded as exhibiting structural deformities belong to the subfamilies Chironominae and Tanypodinae.[1198] The frequency of mouth-part deformities in *Chironomus* specimens collected from unpolluted

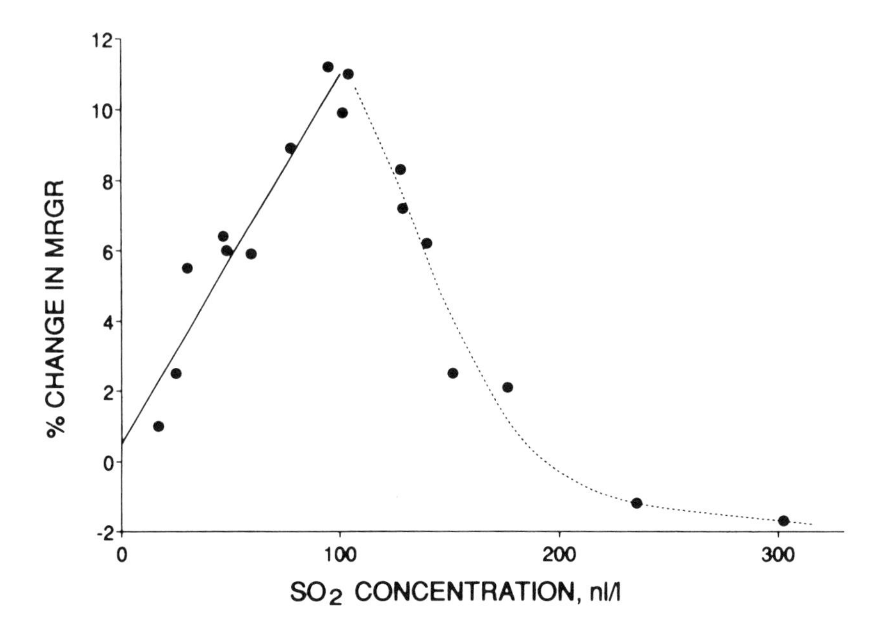

FIGURE 5. Relationship between percentage change, relative to controls, in the mean relative growth rate (MRGR) of pea aphids (*Acyrthosiphon pisum*, Homoptera, Aphididae) feeding on pea plants, and SO_2 concentration. Plants and aphids were fumigated for 4 days. Solid-line shows linear regression relationship for range 0–105 nl/l SO_2 ($r = 0.958$, $p < 0.001$). Dotted curve fitted by eye. (Modified from Warrington, S., *Environ. Pollut.*, 43, 155, 1987. With permission.)

waters was 0.09%.[1563] By comparison, the incidence of such deformities in *Chironomus* from polluted waters reached 25 to 38% in an urban canal in West Berlin,[859] 77% in Parry Sound Harbor, GA,[616] and 83% in the inner harbor area of Port Hope Harbor, Ontario.[1566] The highest recorded incidences of mouth-part deformities in *Procladius* (Tanypodinae) larvae were 8.1% in the Bay of Quinte, Lake Ontario, 2.1% in the Western basin of Pasqua Lake, Saskatchewan, 4.2% in Bultahatchee Lake, and 1.1% in strip-mine ponds, Alabama. Pettigrove[1198] reported a high incidence of mouth-part deformities for the larvae of *Procladius paludicola* collected from sites on the Murray and Darling Rivers, Australia, probably due to organochlorine pesticides (DDT, DDE, TDE, dieldrin, and endrin).

Deformity incidence can be used as an index of toxic stress in aquatic ecosystems.[1516,1564,1565] Industrial effluents containing metals were suspected to be the cause of deformed chironomid larvae collected from Lake Erie and two lakes in British Columbia;[601] Parry Sound, Ontario;[616] Teltokanal, Germany;[859] and Pasqua Lake, Saskatchewan.[1562] The deformities included asymmetrical deformations of larval mouth parts, heavy pigmentation of the head capsule, and thickening of the head capsule and body wall. The mouth-part deformities may affect feeding and growth.[340] Kosalwat and Knight observed deformities in the epipharyngeal plate of the mouth parts at four levels of substrate copper (26, 896, 1741, and 2660 μg/kg).[867]

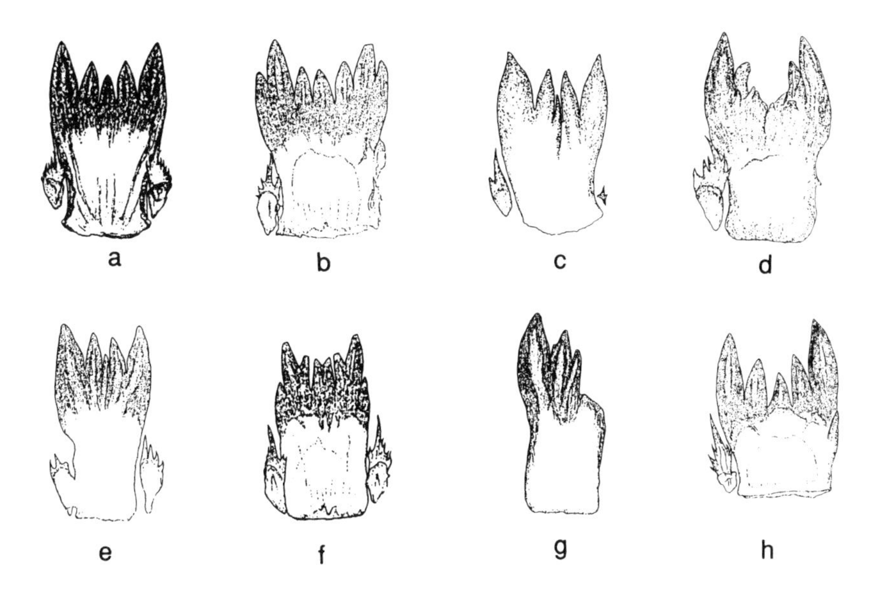

FIGURE 6. Deformities in ligulae of *Procladius* (Diptera, Chironomidae) larvae from Canada: (a) normal and (b–h) deformed ligulae. It is proposed that morphological deformities in larvae can be used in biomonitoring of contaminant concentrations. (Compiled from Warwick, W. F., *Can. J. Fish. Aquat. Sci.*, 46, 1255, 1989. With permission.)

C. INSECT SIZE AND FECUNDITY

Adult body size is a determinant of individual fitness. Smaller individuals complete the larval development with smaller energy costs and sometimes within a shorter time than the large ones. Small females possibly have decreased fecundity and longevity, and small males may be less competitive than the large ones. Size may also affect characteristics such as flight ability, territoriality, or thermoregulation. A species strategy should balance between the advantages (development at less energy cost) and consequences (adult inferiority) of decreasing body size.[702] Organism size also influences toxicity and bioaccumulation of metals, and the high surface-to-volume ratio of aquatic insects makes them especially sensitive. For example, Smock[1400] reported an inverse relationship between body size and metal concentration in mayflies and suggested that surface adsorption was primarily responsible.

Factors basic to egg output include the inherent capacity of the ovaries to produce a given number of eggs, the acquisition of reserves for synthesizing yolk, the hormonal control of vitellogenesis, and the environmental cues that control the timing of hormone synthesis and release. The capacity to produce a given number of offspring resides primarily in the number of ovarioles/ovaries, ovariolar structure, and longevity of the species. The number of eggs produced by a single female ranges from a single egg in the lifetime of the sexual morph of some aphid species to several hundred thousand eggs of the social Hymenoptera and Isoptera.[441]

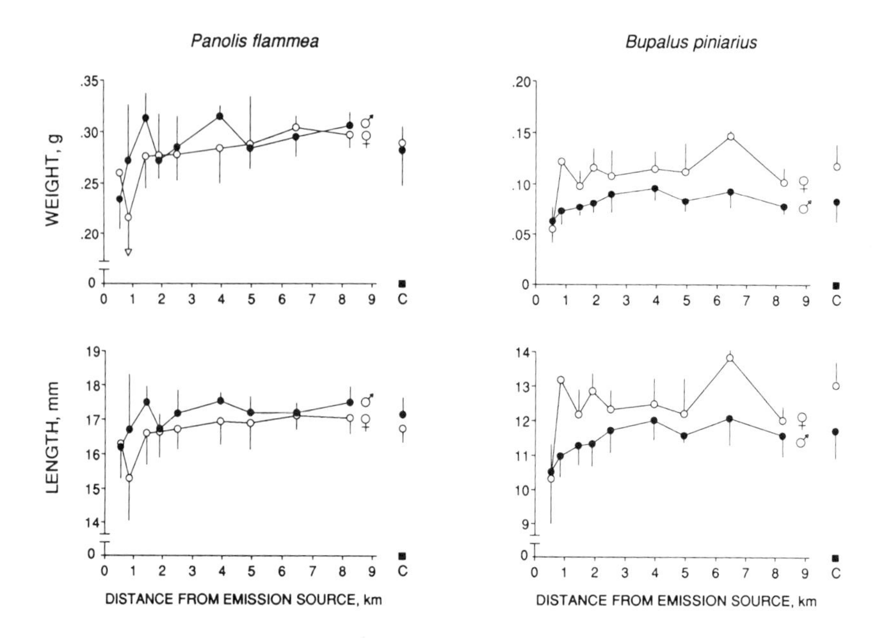

FIGURE 7. Pupal weight and length of *Panolis flammea* (Lepidoptera, Noctuidae) and *Bupalus piniarius* (Lepidoptera, Geometridae) reared on pine needles at increasing distances from the source of pollution in Finland. The values of the reference plots (relatively unpolluted site) are indicated by "c" on the right in the figures. Means and halved standard deviations are given for every distance. (Modified from Heliövaara, K., Väisänen, R., and Kemppi, E., *Oecologia,* 79, 179, 1989. With permission.)

Maternal influences mediated through the variation in egg size were shown to influence the adult size and fitness in Diptera[1425] and pupal weight in Coleoptera.[232] In a satyrid butterfly, variation in egg size did not influence the fitness of the offspring.[815] Postnatal larval development conditions may substantially modify the size of ensuing adults.[702] Starvation of larvae is often the principal determinant of adult size. Slow development, however, does not necessarily increase adult size.[702]

Pollutants in the food plant may decrease the pupal size of herbivorous insects, thus affecting the fecundity of adults. When the moths *Panolis flammea* (Noctuidae) and *Bupalus piniarius* (Geometridae) were reared on pine needles collected at different distances from a point source in Finland, their pupal weight, length, and width were negatively correlated with the distance (Figure 7). This was mainly associated with the concentrations of metals in their food.[666] Similar results were obtained in the laboratory rearing of four species of pine sawflies (Diprionidae).[657] However, the adult size affected fecundity in a complex manner. Detailed analysis of *Neodiprion sertifer* (Diprionidae) showed that lighter females, resulting from rearing with polluted needles, produced fewer, but more viable eggs than heavier females, which, in turn, produced more eggs with lower viability.[667]

Laboratory studies by Katayev et al.[818] showed that fluorine compounds can decrease fecundity in moths. Field investigations indicated that the mean weight of *Hylobius abietis* weevils (Curculionidae) increased as the distance from the source of fluorine emissions became greater. The highest mean weight (50.8 mg) was attained at a distance of more than 13 km from the point source and remained at this stable level.

Schmidt et al.[1341] studied the long-term effects of metals (0.12 to 12 ppm $HgCl_2$, 2 to 100 ppm $CdCl_2$, 25 to 500 ppm $PbCl_2$) in the soil on the development of *Aiolopus thalassinus* (Orthoptera, Acrididae). The adult fresh weight of the F_1 and F_2 generations developed from eggs laid in metal-treated soil was significantly lowered, and the number of egg pods was reduced. Furthermore, the hatchability of the grasshopper nymphs developed from eggs laid by F_1 females was lower than those of the control.

The largest adults of *Culicoides variipennis* (Diptera, Ceratopogonidae) emerged from polluted ponds.[1095] Adult size was correlated positively with longevity and fecundity in *C. variipennis*.[11] Hatakeyama[631] investigated the effects of cadmium on the *Polypedilum nubifer* (Chironomidae). The number of eggs per egg cluster in the cadmium-exposed midges was somewhat larger than that in the control. The latter result suggested that the size of adults that had been exposed to cadmium in the larval stage might be slightly larger than the control. Hatakeyama and Yasuno[633] showed a close correlation between egg number and body size in the midge *Paratanytarsus parthenogeneticus* (Chironomidae).

Pollutants may or may not affect parasitoid fecundity.[433,1186] For example, sublethal doses of carbaryl decreased the number of eggs developing from vitellogenic oocytes in *Bracon hebetor* (Braconidae).[565] Chlorpyrifos, malathion, dimethoate, and methidathion reduced offspring production, and chlorpyrifos shifted the sex ratio of offspring toward fewer females in *Aphytis melinus* (Aphelinidae).[1301] Azinphos methyl and chlordimephon reduced the number of eggs laid on boll weevil larvae by mated *Bracon mellitor* females.[1146]

D. OTHER RESPONSES

The stimulatory effect of low doses of sulfur compounds or slightly increased concentrations of metals on a number of physiological processes and enzyme activity has been documented in various groups of animals, including insects.[727,1044,1282] Such changes in enzyme pattern result from an increased concentration of soluble proteins and intensification of biosynthesis. Increased emission of air pollutants containing metals, SO_2, and NO_2 inhibits the activity of enzymes involved in the glycolysis and amino-acid pathways in homopterans, but most of these enzymes were stimulated in moths.[1044] Moths are protected against these substances by the reflexive regulation of spiracle movement, and their hair cover protects them against the negative effects of dust and acidic precipitation. Consequently, only a slight pH variation was observed in the hemolymph of the satin moth (*Leucoma salicis*).

If mechanisms of detoxication are inefficient, activity of several enzymes with either NAD or flavin nucleotide cofactors is inhibited by addition of sulfites to the active site of the cofactor.[1176] Depressed α-glycerophosphate (α-GPDH) activity in the satin moth was apparently followed by the action of cumulative stressors, and similar inhibition mechanisms might exist. The study on the aphid *Acyrthosiphon pisum* showed that similar environmental stressors may inhibit another shuttle mechanism of NAD via reduced "malic" dehydrogenase activity,[1044] although no α-GPDH activity was detected in this species.

Migula and Karpinska[1045] investigated the effects of combinations of atmospheric pollutants on the activity of α-GPDH in the satin moth (*Leucoma salicis*) originating from differently polluted areas. This enzyme plays several key metabolic roles in the regulation of NAD^+/NADH ratios, the production of ATP for flight, or the production of α-glycerophosphate for subsequent synthesis of glycerolipids or the cryoprotectant glycerol.[1671] Low concentrations of sulfur dioxide caused stimulation of α-GPDH activity through accelerated enzyme biosynthesis or the adaptive potential of allozyme variants. Simultaneous action of SO_2, NO_x, metal-containing dust, and acidic precipitation caused inhibition of α-GPDH activity, probably as the result of sulfite bound with pyridine and flavine nucleotides.[1045] Insignificant changes in the α-GPDH activity of larvae and their fat body, from different experimental groups, indicated that the impact of combinations of pollutants did not affect rapid transport of NADH electrons from the sarcoplasm to mitochondria, or the production of α-GP (glycerophosphate) available for subsequent conversion into di- and triglycerides. The reduced enzyme activity of the pupae exposed to the cumulative action of dust pollutants with additional acidic precipitation could be due to prolonged development of the pupal stage under such conditions. Similar effects have been observed in other lymantriid species as well.[1044] Migula and Karpinska[1045] suggested that the higher activities of α-GPDH found in insects representing a population from heavily polluted Katowice areas might indicate that the increased activity of some metabolic enzymes is a partial adaptive mechanism for living in unfavorable manmade conditions.

Several authors have noted behavioral changes in invertebrates exposed to sublethal concentrations of toxic compounds. For example, Peterson and Peterson[1195] observed anomalies in the net spun by the larvae caddisflies *Hydropsyche angustipennis* in the presence of kraft pulp-mill effluent. Since any effect on an animal's behavior at sublethal concentrations may lead to its eventual death, e.g., through reduced feeding ability or lack of an antipredator device, behavioral experiments should be more widely incorporated into toxicity studies.[1016]

The sublethal effects of pesticides on insect behavior have received relatively little attention.[433,434,645] The behavioral changes may be due either to repellency or to the masking of host odors.[433] Results obtained with the parasitoid *Microplitis croceipes* (Hymenoptera Braconidae) indicated that flight activity and foraging may be altered by sublethal doses of insecticides.

III. ECOLOGY

A. PLANT-HERBIVORE INTERFACE

There are several reports on altered interactions between insects and their host plants. If pollutants are more toxic to the attacking insect than to the plant, the latter will suffer less from the herbivore than under normal conditions.[1367,1472] The response of an insect population to pollution stress will depend greatly on the life history of the species.[646,905] For instance, it has been suggested that species living under bark or in wood are better protected against air pollutants than those living in an exposed position on plants. Pollutants can also induce in plants increased concentrations of compounds harmful to insects, as shown by the study of Trumble et al.[1503] on the phototoxic linear furanocumarins in celery and on the noctuid larvae feeding on it. The degree of insect herbivory can increase, decrease, or not change at all when plants are stressed.[791,905] However, most published studies on plant-insect interactions have dealt with insect outbreaks or improved insect performance on plants stressed by pollutants.

The relationships between plant stress and susceptibility to insects have received much attention.[27,296,589,791,998,1278,1552,1595,1596] It has been hypothesized that plants under abiotic stress undergo physiological changes and become more suitable as food for herbivorous insects. This hypothesis is based on two arguments:[905] first, herbivorous insects have frequently been reported to reach outbreak densities on plants associated with environmental stress;[906,1537] second, certain biochemical changes observed in stressed plants have been suggested to cause improved insect performance.[1277,1595,1596]

The paradigm that plant stress enhances insect herbivore performance and abundance, and the mechanistic explanations that underly this paradigm, are not supported by the evidence.[791] For instance, the results from experimental studies on insect response to stressed trees give little support for the stress hypothesis in its traditional form. Larsson[905] pointed out that there is neither a universally accepted definition of stress in plants, nor an unequivocal measure of stress. For example, xylem water potential, stomatal conductance, or photosynthetic capacity may provide useful measures of plant stress level, but stress should be related to insect performance when plant-insect interactions are to be investigated. In this respect, attention has been paid mainly to single plant traits,[1277,1278,1595,1596] but Larsson[905] stressed that several plant traits are likely to be simultaneously affected by environmental stress. It is also unlikely that insect species with different feeding habits will respond similarly to changes in plant tissues.

It is usually impossible to determine whether the insect outbreaks observed on stressed plants really are due to alterations in host-plant suitability.[1107] Factors having direct effects on insect population density (e.g., weather, natural enemies) are often correlated with those inducing stress in plants.[998] Warm and dry periods in temperate areas often directly enhance the growth, survival, and reproduction of insects, while stressing their host plants at the same time.[905]

Nevertheless, insect outbreaks are often attributed to environmental factors stressing a host plant. One hypothesis assumes that reduced host vigor improves overall insect success.[998] However, the role of pollution may be minor compared to that of other stresses. Drought is one factor that may enhance subsequent insect outbreaks. For instance, the association between water-stressed host trees and outbreaks of pine sawflies has been observed repeatedly.[1537] A plant subjected to moderate water stress may provide conditions for optimum insect survival and reproduction. When stress increases, plant quality may deteriorate to the point where a severely stressed plant is of poorer quality, from the point of view of the herbivore, than either a moderately stressed or even vigorous plant.[998]

Recently, Lechowicz[914] reviewed the alterations that have been observed when plants are exposed to low levels of air pollutants. In general, root biomass is reduced more than shoot biomass in plants exposed to SO_2 or O_3, but NO_2 does not appear to cause such a difference. Quantitative allocation to the leaves increases, and to the stem decreases, after SO_2 exposure. Root carbohydrate concentrations may increase or decrease after SO_2 or O_3 fumigations. Leaf nitrogen concentrations tend to decrease due to air pollutants, whereas leaf carbohydrate concentrations can increase or decrease.[914] Several other changes in leaf quality have been observed as a result of air pollution, including changes in leaf mineral concentrations.[316,1380] Concentrations of different biochemicals, ranging from vitamins to flavonoids, are altered by air pollutants.[1380] Lipid metabolism is variably affected by pollutants.[648,965] Plant reproduction is suppressed by O_3, SO_2, and NO_2, with O_3 having the most marked effects. Seed lipid and protein composition may change following exposure to pollutants.

Plants are frequently protected against grazing by secondary compounds that are toxic or unpleasant to insect herbivores. Coniferous trees produce resin that may prevent insects from attacking bark and wood. If the plant is stressed or weakened for any reason, it may decrease or cease producing secondary compounds. Structural features, such as the layer of wax on coniferous needles, are often destroyed by airborne pollutants.[527] Physiological changes in plants affected by air pollutants change the nutritional suitability of host plants for insect herbivores.[790,914,1500] Substantial modifications in the form and concentration of nitrogen have been reported following plant exposure to pollutants.[242,1085] In tomatoes, ozone fumigations resulted in an increase in free amino acids and soluble proteins,[1502] which may be more readily assimilated by insects than the nitrogen bound in the structural components of cells.

The implications of increased concentrations of assimilable forms of nitrogen for insect development are considerable.[1446,1596] Nitrogen availability affects growth rates, survival, and reproduction.[1158,1228,1229] Because air pollutants often occur over wide geographic areas, the cumulative effects on the herbivore population may be important.[1596] Physiological changes in plants due to stress or direct injury from air pollutants may have more serious consequences than are indicated when the plant system is investigated in the absence of herbivores.[1500]

Several studies have reported an increase in food consumption of, or preference for, foliage exposed to ozone or SO_2.[440,727,771,790,1502] An increased foliage consumption and a weight gain were observed in noctuid larvae on beans exposed to acidic fogs with a pH value of 3.0.[1500] However, there was no significant weight gain by larvae on plants in which soluble protein levels, free amino acid concentrations, or total nitrogen contents were enhanced by acidic fogs with a pH of 2.5 and 2.0.

The feeding preference behavior of insects for plants growing under different environmental conditions has been examined in some species. The Mexican bean beetle preferred bean plants exposed to sulfur dioxide or ozone.[440,728] The gypsy moth showed a variable preference response to ozone-exposed oak seedlings.[771] The feeding preference of insects between different species of food plant may not correlate positively with the suitability of host species preferred for insect development, survivorship, and reproductive fitness.[251,252,1609] The lack of correlation between insect feeding preference and insect performance can also apply to insects choosing between pollutant-stressed and control plants of the same species and genotype.[294]

Tree-dwelling aphids have an intimate relationship with their host plants, influencing plant physiology through feeding and salivary secretions and, in turn, responding to subtle changes in plant chemistry. Such changes can alter aphid growth rates, reproduction, and survival.[839] Pollution damage to plants has been shown to cause similar effects in aphids infesting agricultural crops.[28,390,1557] Acidic mist and gaseous pollutants can enhance the performance of tree-dwelling aphids.[840,1029] Direct effects on the aphids have largely been discounted,[390,1030] indicating that the benefit to the aphids acts through the plant, probably as a result of an improved nutritional quality of sap amino nitrogen.[176,390,966] Braun and Flückiger[176] found substantial changes in the phenolic compounds of hawthorn (*Crataegus* sp.) after exposure to motorway gases. Phenolic and other secondary plant substances are known to influence the well-being of conifer aphids such that changes involving these compounds could have caused some of the described alterations in aphid performance.[840] Moreau[1073] reviewed the effects of air pollutants (insecticides, SO_2, NO_x, ozone, acidic precipitation) on terpene catalysts of corn (maize, *Zea mays*). The reduced activity of the catalysts results in a qualitative and quantitative decrease in terpenes and, consequently, in an increase of aphids.

Several authors have suggested that air pollutants can increase the susceptibility of trees to attack by bark beetles.[164,646,1332,1369,1420] Pollution-damaged conifers may be used for breeding by bark beetles. Most bark beetles can only breed in dead or moribund trees, but a small number of species occasionally attack and kill large numbers of live trees.[261] Living trees possess defense mechanisms against bark beetles, but bark beetles can overcome the defenses when they reach a certain density. The beetles utilize aggregation pheromones by means of which a pioneer beetle can summon conspecifics from the surroundings. Only then will a sufficient number of beetles join the pioneers to exceed the threshold of successful attack. Populations may expand because of a temporary increase in breeding material, or the threshold of successful attack may drop due to reduced tree vigor.[261,262]

However, there are only a few studies on the interactive effect of bark beetles and air pollution. A disease of ponderosa pine was noticed in the San Bernardino Mountains, CA in the early 1950s. The forest was subjected to heavy levels of photochemical air pollutants from the Los Angeles area. The condition was referred to as chlorotic decline or ozone needle mottle. In the 1960s the area was surveyed to determine the association between severity of disease and infestation by two species of bark beetle, *Dendroctonus ponderosae* and *D. brevicomis*. Oxidant injury was positively correlated to the incidence of beetle infestation.[287,1420] Air pollutant injury apparently increased the susceptibility of the trees. In diseased trees, oleoresin pressure, yield, and flow rate were reduced, and the resin crystallization rate increased. Sapwood and phloem moisture contents were decreased. Phloem thickness was less than 60% of that in healthy trees.[285] Concentrations of soluble sugars and reserve polysaccharids decreased.[1058] The methyl chavicol concentration was strikingly lower in the needles of the damaged trees, but no differences were observed in monoterpenoid concentrations.[287] On the other hand, oxidant injury reduced the suitability of the trees as a breeding substrate for bark beetles. Severely diseased trees therefore appeared to act as traps absorbing the resident beetle population and preventing a local buildup. In spite of an abundance of severely diseased trees in the area, no noticeable increase in bark beetle infestations was recorded.[286]

In Europe the spruce bark beetle (*Ips typographus*) has caused repeated large-scale mass attacks over the centuries; the conditions after drought or storms have provided ample breeding substrate. Sometimes *I. typographus* has been mentioned as a serious problem in polluted areas,[1370,1676] but these have been only general observations and not the results of field studies.[1159] In heavily polluted areas of Europe, *I. typographus* clearly prefers healthy trees over declining ones. Furthermore, several outbreaks triggered by drought have not resulted in mass propagation in pollution-damaged forests. Christiansen[261] suggested that this could be due to the lack of attraction to weak trees; a specific chemical composition of the phloem may be needed for the production of essential pheromones.

Stressed trees probably have a reduced production of defensive chemicals, due to inadequate energy stores. However, a more susceptible tree is not necessarily preferred by the bark beetles. The low suitability of damaged trees as a breeding substrate for bark beetles is likely to be caused by the poor nutritional quality of the the phloem.[261] Lowered carbohydrate concentrations have been found in phloem, root tissues, and leaves of forest trees exposed to sulfur dioxide and ozone.[914,1058] Furthermore, damaged trees may have a thinner living bark and hence may be avoided by the larger bark beetles, such as *I. typographus*.[261]

The aggregation pheromones of bark beetles are usually derived from terpenes emanating from trees under attack. Thus, the beetles are able to utilize the defense chemicals of host trees for coordinating the attack. Landing rates of *Dendroctonus ponderosae* have been reported to be higher in trees actively secreting monoterpenes.[1252] The lack of attraction by severely weakened trees may be explained by an inadequate yield of terpenes for the production of crucial pheromone components.[261]

The plant-insect interface also has other dimensions in relation to environmental pollution. There is some concern in the literature[81,358] regarding the secretion of systemic insecticides, applied to control harmful phytophagous insects, in nectar (and possibly in pollen), and the possibility that bees or other nectar-gathering insects may become contaminated in this way. Experimental studies on the distribution of insecticides in flowers and their secretion, using ^{14}C-labeled insecticides, have demonstrated that nectar can become contaminated by insecticides via systemic movement and that the degree of contamination is related to insecticide concentration.[358]

B. PREDATION

Predation appears to be an important factor in structuring biological communities. Arthropod predators and parasitoids may regulate many phytophagous insects, including agricultural pests.[722] Pollution can change predator-prey relations by affecting either the predator or the prey, or both.

In the field, predators and their prey are usually exposed to pollutants simultaneously, and this may result in either similar, increased, or decreased predation intensity. Such variation in predation intensity is most likely to occur if the behavior of either predator or prey is differentially affected by toxicants. Clements et al.[277] examined the influence of chronic copper exposure on predator-prey interactions between the stonefly *Paragnetina media* and its invertebrate prey, two species of net-spinning caddisfly, *Chimarra* sp. and *Hydropsyche morosa*, in replicate artificial streams. Predatory interactions among stoneflies and caddisflies were considerably more sensitive to copper exposure than was caddisfly mortality. Greater vulnerability of *Hydropsyche* to predation in dosed streams possibly resulted from a change in behavior of these organisms. Metals can disrupt silk spinning in Hydropsychidae and result in anomalies in the structure of the capture net.[1195] If these alterations caused *Hydropsyche* to spend more time outside their retreats, maintaining or repairing nets, predation pressure on these organisms would increase.[277] Hershey[679] reported that vulnerability of tube-building chironomids to predation was a function of the amount of time they spent outside their tubes. Thus, benthic communities may be affected through changes in predation at sites moderately polluted by metals.[277]

In order to understand the pollution processes and to find efficient restoration measures, it is necessary to note the difference between the toxic effects and the effects of the altered predator-prey relations. Results from mesocosm experiments have provided predictions about the inlake responses of the invertebrate community to acidification and emphasized the importance of indirect effects of biotic interactions in structuring invertebrate communities.[528] Altered predator-prey relations following the eradication of fish populations are responsible for several changes in acidified lakes.[442] Groups suppressed by fish predation will increase when predation from fish ceases. The decrease of smaller crustacean zooplankton may be explained in terms of predation by the quickly expanding population of invertebrate predators. These animals can crop a substantial proportion of their prey populations.[900] *Chaoborus* (Diptera, Chaoboridae) larvae prey on smaller zooplankton,[460] and corixids (Heteroptera) can cause significant mortalities among cladocerans. Cladocera and mayfly nymphs are important food for larvae of Odonata. The decreases of smaller planktonic crustaceans and of mayfly nymphs in acidified lakes may thus, to a considerable extent, be the result of increased invertebrate predation.[442] The drop in water-mite density in the acidified lakes may also be a result of a change to an ecosystem dominated by invertebrate (e.g., Odonata, Corixidae) predation. *Chaoborus* mandibles have even proved to be useful paleolimnological indicators of the historical status of fish populations in acid-sensitive lakes and acidification trends.[1518]

The predatory larvae of *Chaoborus* seem to have an effect on the recovery process of zooplankton communities after insecticide applications.[603,605,606] Hanazato[603] investigated the effects of repeated application of the insecticide carbaryl on zooplankton in experimental outdoor concrete ponds in Japan, with and without the predator *Chaoborus flavicans*. The results demonstrated that *C. flavicans* altered the response of the zooplankton community to the chemical application by altering community structure (Figure 8). Cladocera dominated in ponds without *C. flavicans*, but the species composition differed among the treatments. In the ponds with *C. flavicans*, rotifers dominated the zooplankton, indicating that they were released from competition with cladocerans and calanoid copepods by the *C. flavicans* predation.

There are several reports dealing with the effects of pollutants on parasitoids and predators of terrestrial insects.[206,207,523,653,686,1221,1366,1367,1472,1530,1587] In industrial areas, reductions in parasitoids have been reported in *Rhyacionia buoliana* (Tortricidae),[1472] *Pristiphora abietina* (Tenthredinidae),[1587] *Exoteleia dodecella* (Gelechiidae),[1367] and aphids.[686] A decreased level of parasitization may partly explain the increased attack of these pests in polluted areas. Parasitization caused by *Spilochalchis* sp. in the swallowtail butterfly *Papilio scamander* was lower in an urban area, than in suburban areas, in Brazil.[1311] Villemant[1530] in aphids and Heliövaara and Väisänen[653] in *Retinia resinella* (Tortricidae) found no significant differences in the proportion of parasitization in areas with different degrees of pollution, while both aphids[655] and the tortricid moth[651] abounded in the zones of high industrial pollution.

Parasitized hosts may be more susceptible to pollutants than are unparasitized hosts, as indicated by some results on pesticides.[433] Examples include the gypsy moth *Lymantria dispar* (Lymantriidae) larvae parasitized by *Apanteles melanoscelus* (Braconidae) in respect to carbaryl,[8] and the tobacco budworm *Heliothis virescens* (Noctuidae) larvae parasitized by *Cardiochiles nigriceps* (Braconidae) treated with methyl parathion and permethrin.[480] However, there are contrasting results on the parasitization of *Scodoptera littoralis* (Noctuidae) larvae.[649,680]

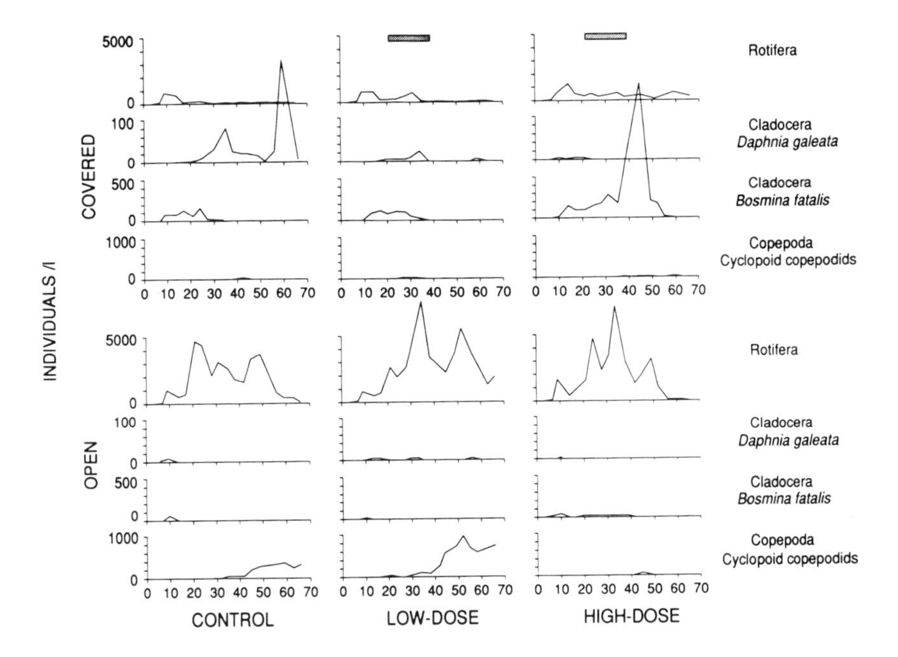

FIGURE 8. Change in density of Rotifera, *Daphnia galeata* (Cladocera), *Bosmina fatalis* (Cladocera), and cyclopoid Copepoda in the covered and open, control, low-dose, and high-dose ponds. The insecticide carbaryl was applied repeatedly at 10 (low dose) or 100 mg/l (high dose) to outdoor experimental ponds in Japan. By covering pods the predation on plankton by *Chaoborus flavicans* (Diptera, Chaoboridae) was prevented. Shaded bars show the insecticide treatment period. (Modified from Hanazato, T., *Environ. Pollut.*, 74, 309, 1991. With permission.)

A factor that may be of great significance for predation is the observed decrease in developmental time of a number of species in polluted areas or on fumigated plants. For instance, larvae of *Diprion pini* (Diprionidae) reared on pine needles polluted by industrial emissions tended to develop at a slower rate, and to hatch later, than those reared on unpolluted needles. The length of the development time affects the probability of being consumed by a predator.[662]

The reason for the improved growth of aphids in polluted air has not been fully established, but there are several possible explanations. Some authors[175,1241,1366] have proposed that reduced numbers of predators would be responsible. However, a lack of predation is not necessary for increased aphid populations,[387] but it may be a contributing factor.[912]

Perfecto[1188] studied the direct and indirect effects of carbofuran and chlorpyrifos in a tropical agroecosystem in Nicaragua. Both insecticides reduced ant foraging activity in the maize field (Figure 9). This was suggested to be the reason for the increase in the fall armyworm *Scodoptera frugiperda* (Noctuidae), while chlorpyrifos also reduced densities of this pest. However, in the latter case the treatment resulted in higher levels of another herbivore, the corn leafhopper *Dalbulus maidis* (Cicadellidae), which appeared to prefer plants that have the least *S. frugiperda* damage.

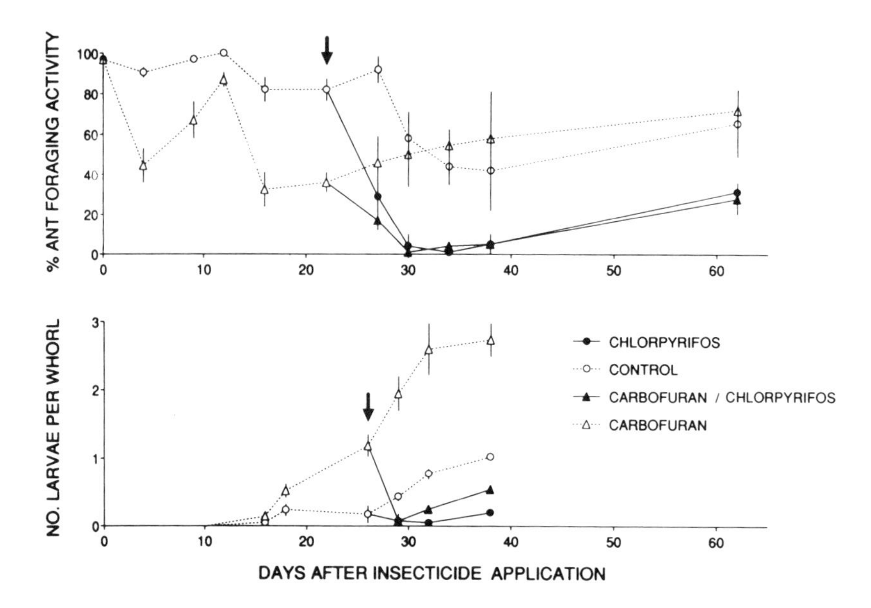

FIGURE 9. Ant foraging activity (Hymenoptera, Formicidae) and the number of larvae of *Spodoptera frugiperda* (Lepidoptera, Noctuidae) in maize-pest-ant system for the carbofuran and the control treatments before and after the application of chlorpyrifos in a tropical agroecosystem in Nicaragua. The arrows represent the last sampling date before the application of chlorpyrifos, after which ant foraging activity for all four treatments is shown. Standard errors are indicated by vertical bars. (Modified from Perfecto, I., *Ecology,* 71, 2125, 1990. With permission.)

Chemicals are known to alter the web-building behavior of spiders.[1633] However, only a few investigations have focused on spiders in relation to air pollution, whereas the effects of pesticides are better known.[449,1478] A negative correlation between the number of spider species and SO_2 concentrations was found by André[39] and Clausen,[270,271] but not by Gilbert.[545] The correlations could be explained by structural differences, for example, the frequency of epiphytes, between the habitats.[271]

Deeleman-Reinhold[371] investigated the spider fauna of a peat bog over 14 years in an industrially polluted area in the Netherlands. The number of larger wandering spider species, such as Lycosidae, decreased, whereas an increase in linyphiids was observed. Luczak[951] also observed a better resistance in linyphiid spiders than in wandering spiders. Such changes in the predators may have considerable effects on the soil and litter fauna.

Needle loss in coniferous trees in southwest Sweden is, in part, induced by air pollution.[36] Branches of Norway spruce (*Picea abies*) loose needles and become sparser, and the proportion of needle-free twigs increases. Gunnarsson[573] studied the possible effects of needle loss on the abundance of spiders living on spruce branches, by comparing two spruce stands with relatively high and low percentages of needle loss, respectively. The abundance of large spiders was considerably lower in the stand with a high needle-loss percentage than in the other stand. One tentative explanation for the different densities between the large spiders in the two stands may be bird predation, which is known to cause high mortality among spruce-living spiders. Large spiders might be easier to detect, and thus more vulnerable to bird predation, on branches with few needles. According to an alternative explanation, the microclimate would have favored large spiders in the sheltered, dense-needled stand.[573] There was a higher density of small spiders (when relating numbers to the needle mass) in the needle-loss stand. A low predation rate on small individuals in the needle-loss stand, due to the scarcity of large spiders, may explain the result.

C. DISEASES AND PARASITES

Environmental pollution may affect insect-parasite and insect-disease relationships. The effect may also be mediated through host plants, since the quality of food may have an impact on the susceptibility of insects to diseases.[80,525,1347] There is some supporting experimental evidence: the foliage of a number of plants contains antibacterial substances;[894,964,1078] the host-plant species of the gypsy moth affects its susceptibility to viral disease;[821] and the concentrations of various chemicals in the food affect the pathogenity of the diseases.[87,354,355,465,845]

Neuvonen et al.[1122] found that about two weeks after the application of a nuclearpolyhedrosis virus, the survival of the sawfly (*Neodiprion sertifer*) larvae was significantly higher in larvae fed with needles from trees treated with artificial acidic precipitation of pH 3 than with needles from control trees. Although the differences between treatments decreased towards the end of the larval period (when mortality was almost 100% in all the larval groups treated with the virus), the difference in the timing of mortality as such may have important consequences for the level of consumption suffered by the host trees. Neuvonen et al.[1122] speculated that the susceptibility of *N. sertifer* larvae to the virus depended on the quality of the host foliage and, furthermore, that simulated acidic deposition could reduce the susceptibility of larvae to the viral disease. The mechanism of the reduced susceptibility to viral disease on the treated foliage was not investigated. However, two possible mechanisms were proposed: first, the larvae fed with foliage from treated trees may have had a better nutritient status (the acidic precipitation contained HNO_3), thus delaying the development of the disease; second, the acid treatment may have changed the pH of the pine needles, thus affecting the progression of the disease. However, field experiments did not support these findings. Heliövaara et al.[667a] observed that the mortality of *N. sertifer* larvae reared at different distances from an industrial point source was the highest at the most polluted site. Although the cocoon mortality caused by parasitoids was lower near the pollutant source, the mortality caused by nuclear polyhedrosis virus was high.

Chemical suppression of antagonistic organisms is considered to be a possible mechanism by which industrial pollutants trigger forest insect outbreaks.[517] Widespread and persistant epidemics of the pamphiliid *Cephalcia abietis* have been reported in spruce stands in the central European uplands during recent decades.[987,1350] Baltensweiler[77] suggested that this was due to the increased impact of air pollution on forests. The hypothesis that *C. abietis* benefits indirectly from air pollution was both confirmed and explained by Fischer and Führer.[472] They studied the effects of soil acidification on an endoparasite (*Steinernema kraussei*), an entomophilic nematode of the body cavity of soil-inhabiting insect larvae, which is associated with the nymphs of *C. abietis*. Resting in the soil for at least one year, the diapausing nymphs are exposed to parasitism by the nematode. The host is killed by the release of a symbiotic bacterium (*Xenorhabdus* sp.). After reproduction new invasive nematodes leave the dead insect, searching for new hosts. During that time they are directly exposed to the conditions prevailing in the soil. Several generations per year and a high reproduction rate make the nematode an efficient mortality factor in the host population. However, being continuously reduced in numbers by extremely acidic soil conditions, the nematode loses its ability to regulate the host density. Fischer and Führer[472] concluded that the ability of nematodes to search for, or to invade, host insects was reduced in some way.

D. COMPETITION

The role of between-species competition varies widely from group to group. It is most important in social insects, e.g., ants and bees, whereas competition is commonly not considered a major determinant of community structure in phytophagous insects.[911] Most examples of interspecific competition in phytophagous insects are found to be highly asymmetrical (i.e., amensal) interactions.[1446]

Competition between pests may occur as a direct result of interactions between exploiters or may be mediated by exploiter-induced changes in the host plant.[769,789,1454] Strong et al.[1446] and Price[1232,1233] considered that competition between insects was a weak force in structuring insect communities because there was too little evidence supporting direct insect-insect interactions. However, evidence for host-mediated interspecific competition has been presented by Karban[813] for insects feeding on *Erigeron glaucus*, and by West[1589] for insects feeding on oak. Raupp et al.[1261] argued that direct inter- and intraspecific competition does occur on willow because the chemical secretions of *Plagiodera versicolora* (Coleoptera) larvae repel other insect herbivores. Direct intraspecific competition between insects was reported by Bultman and Faeth[202] between leaf-mining insects that feed on oak. Karban et al.[814] documented adverse reciprocal host-plant-mediated competition between an insect herbivore and a fungal plant pathogen of cotton.

Prior to 1940s the mosquito *Culex quinquefasciatus*, a vector of bancroftian filariasis, was rare in urban areas in Africa, whereas a bird-biting mosquito, *C. nebulosus*, was common. Larvae of both *Culex* species live in polluted groundwaters. From the mid-1940s such habitats were repeatedly sprayed with DDT, which apparently lead to the virtual disappearance of *C. nebulosus* and its replacement by large populations of *C. quinquefasciatus*. It was suggested that *C. nebulosus* is susceptible to DDT, while *C. quinquefasciatus* is less sensitive and rapidly evolves insecticide-resistant populations.[1214]

In the antimalarial campaigns carried out in east Africa during the late 1950s, the insides of houses were sprayed with residual deposits of dieldrin to kill the malaria vector *Anopheles funestus*. The spraying campaign was successful, but at the same time the population of a closely related mosquito, *A. rivulorum*, increased. The two *Anopheles* species coexist in larval habitats, but adults of *A. funestus* rest indoors, while those of *A. rivulorum* rest outdoors. It has been suggested that the decrease in the *A. funestus* population reduced larval competition between the two species and rendered the increase of *A. rivulorum* possible.[1214]

Plowright et al.[1211] compared the foraging performance of bumblebees in fenitrothion-sprayed and unsprayed areas. Their measurements of foraging performance indicated that the relaxation of competitive stress possibly contributed to population recovery. The extent to which bumblebee populations are regulated by competition for food has long been a matter of dispute. Some researchers[179,592] have concluded that bumblebees could not be regulated by the availability of forage, while Inouye[745] proposed that food competition has played an important part in the evolution of tongue length in bumblebees.

IV. EVOLUTION

A. INDUSTRIAL MELANISM

Correlations between melanic frequency in insects and industrial areas have been observed in many parts of Europe and North America.[279,398,829,1048,1136,1172,1173,1327,1358,1515] The term industrial melanism is used commonly to indicate an association of high frequencies of the genetically controlled melanic forms of normally light-colored animals with areas affected by atmospheric pollution of urban or industrial origin. Many arthropod species exhibit melanic forms. Often the melanic frequencies correlate with the extent of industrial regions.[143,829,918,973] Such associations are best known in insects, especially in Britain, where they have been recorded for more than 200 species of moths,[829] the ladybird *Adalia bipunctata*,[323-325] the psocopteran *Mesopsocus unipunctatus*,[1217] and the spittlebug *Philaenus spumarius*,[920] but also in spiders.[55,958] The northernmost record of industrial melanism so far recorded comes from Kuusamo, northern Finland, and concerned the moth *Xestia gelida* (Noctuidae).[1052] Different contributing factors that influence industrial melanism in insects were described by Brakefield.[171] It seems evident that there is no unique set of selective forces that act to favor the melanic forms. Some general properties of industrial melanism can be recognized,[143,829,918] but the details of the webs of selective influences differ from one polymorphism to another.[171,973]

The increase and spread of the melanic *carbonaria* form of the peppered moth *Biston betularius* over industrial areas in Britain from the mid-19th century onwards has become widely known and quoted as a classical example of microevolutionary change (Figure 10). Berry[125] summed up the history of melanism in this species. Steward[1426] concluded that *carbonaria* was already widely distributed in northern England and the midlands by 1885, but it was still absent from a large area of southern Britain. After 1890 *carbonaria* seems

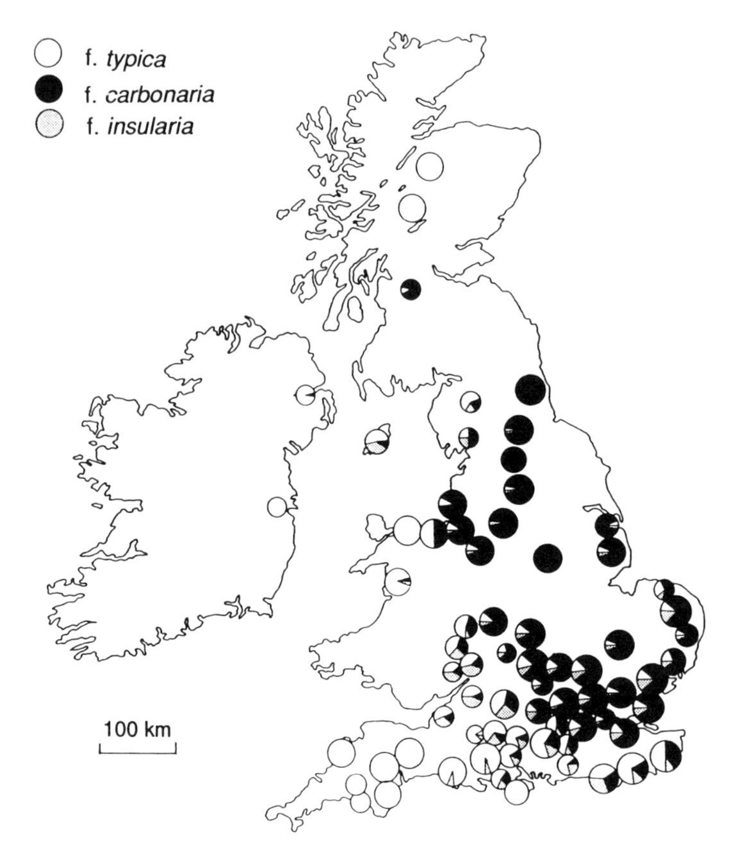

FIGURE 10. The relative frequencies of the normal and two melanic forms of the peppered moth (*Biston betularius*, Lepidoptera, Geometridae) in Great Britain. The results are based on more than 3000 records collected from 1952 to 1970 at 83 sites. (From Kettlewell, B., *The Evolution of Melanism,* Clarendon Press, Oxford, 1973. With permission.)

to have spread rapidly over a large area, including East Anglia and London. For example, James[760] found no melanic forms in 77 *B. betularia* caught in north London in 1894, but had 74% *carbonaria* in a large sample taken in the same place in 1915. Surprisingly, *carbonaria* did not reach a frequency of 1% in London until 1895.[125] Although quantitative information about *carbonaria* is scarce, there are even less data on the other melanic form, *insularia*. It was present in Manchester, U.K., almost at the same time *carbonaria* was first caught, and Kettlewell[832] concluded that it was common in Folkestone on the south coast of England before *carbonaria* appeared there.[125]

The first record of *carbonaria* in continental Europe was made in 1867 at Breda in Holland, and the form was subsequently recorded progressively at places to the east.[125,1136] By the early 1900s it occurred all over northwestern Europe, apart from Scandinavia.[125] *Carbonaria* was not recorded in Denmark and Sweden until the 1940s, and it now occurs at low frequencies in Denmark,

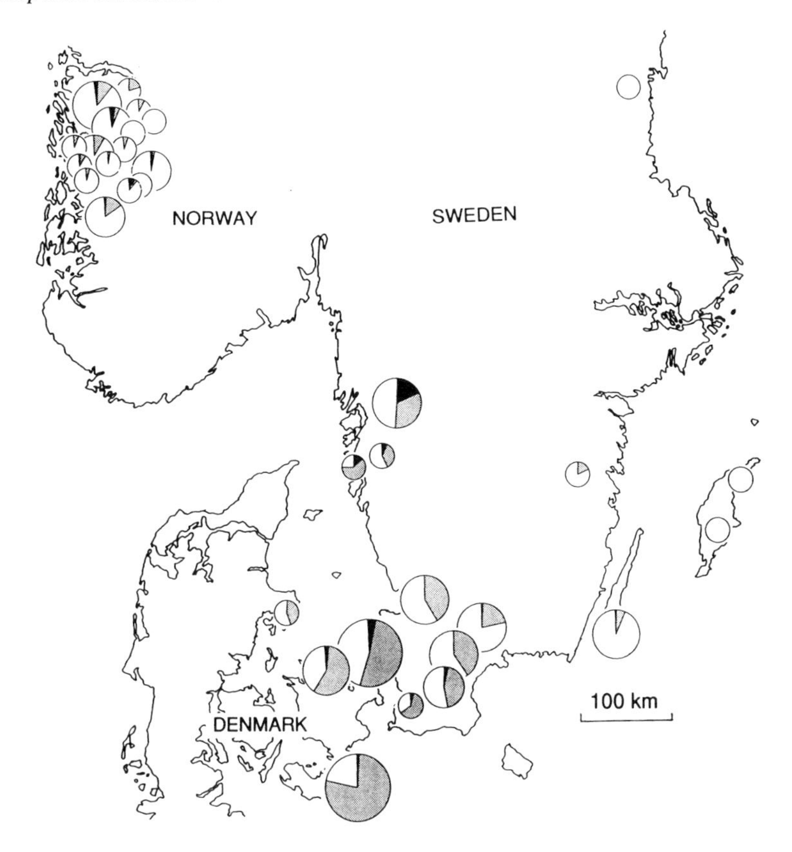

FIGURE 11. The relative frequencies of the normal and two melanic forms of the peppered moth (*Biston betularius*, Lepidoptera, Geometridae) in southern and central Scandinavia. The size of the pies indicates the number of individuals from each locality as follows: <50, 50–99, 100–300, >300 individuals. Colors as in Figure 10.[29,398]

southern Sweden, and western Norway; it was found in Finland in 1982.[29,398,1050] A high proportion of the population in Denmark and southern Sweden now belongs to the dark form *insularia* (Figure 11). The moth population in central Helsinki, Finland, is only slightly darker than the rural populations.[1055]

The polymorphism in *Biston betularius*, as in several other species, is controlled by a single gene locus with melanic alleles dominant to the nonmelanic *typica*.[829,918] In 1896 Tutt[1504] suggested that the *typica* form is well camouflaged when at rest on the pale bark of trees with pale foliose lichens in rural habitats. Tutt described the blackening of the tree surfaces and walls by air pollution in industrial Britain and hypothesized that the light form *typica* then became conspicuous to bird predators, whereas melanic moths were cryptic. The blackening by soot and other particulate pollutants was also accompanied by the killing of epiphytic lichens by gaseous pollutants, particularly sulfur dioxide (Figure 12).[145,171,829,1354] Entirely black resting sites can only favor

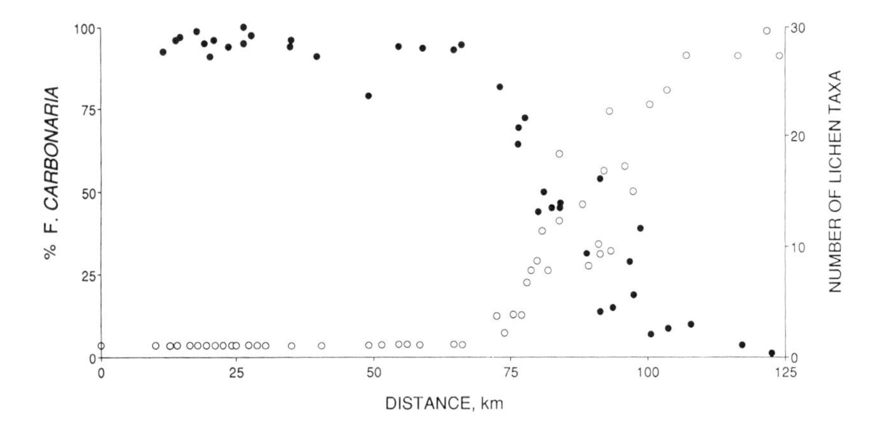

FIGURE 12. The frequency of f. *carbonaria* (black dots) in samples of the peppered moth (*Biston betularius*, Lepidoptera, Geometridae), and the number of different lichens (taxa) on oak trees (*Quercus* spp.) (white circles), along a transect from the Manchester area to central Wales. Distances are measured from the most northeasterly site. (Redrawn from Bishop, J. A., Cook, L. M., Muggleton, J., and Seaward, M. R. D., *J. Appl. Ecol.*, 12, 83, 1975. With permission.)

carbonaria.[934] However, whether the estimates of visual selection are reliable in a quantitative sense is less certain.[171,1051]

Various features of the distribution of the melanic morphs in *B. betularius* have remained obscure:[973] the frequency of *carbonaria*, even in the regions with the most industrial pollution, never exceeded 95%. After air pollution was reduced, the *carbonaria* frequency decreased even in regions where the pale form should have been at a visual disadvantage.[267,268,303,304] A high frequency of *carbonaria* was observed in East Anglia, where the pollution level would suggest a low frequency. Lees and Creed[919] explained this by the nonvisual disadvantage of *typica* compared to *carbonaria*. The cline running from Liverpool to north Wales extended much further than would be expected on the basis of measured values of selective pressure due to avian predation. Bishop[142] proposed that this could be partially attributed to the introduction of heterosis.[919] High *insularia* frequency was observed in the industrial regions of south Wales where the *carbonaria* frequency was low and in the rural areas of the Cotswolds.[973]

The predation experiments carried out since Kettlewell's classic studies[830,831] have employed several methods: direct observation of predation by birds,[829] mark-release-recapture techniques, or studies of the rate of disappearance of dead insects exposed to birds. Sometimes higher densities of moths were used than are found in natural populations.[144] Mikkola[1051] stressed that the experiments have been based largely on moths at rest on tree trunks. Cage experiments with male moths suggested that the normal resting sites are actually beneath narrow horizontal branches. The moth has a specific resting behavior and corresponding coloration: it takes a position with the body perpendicular to the longitudinal axis of the branch and with the wings broadly to the sides

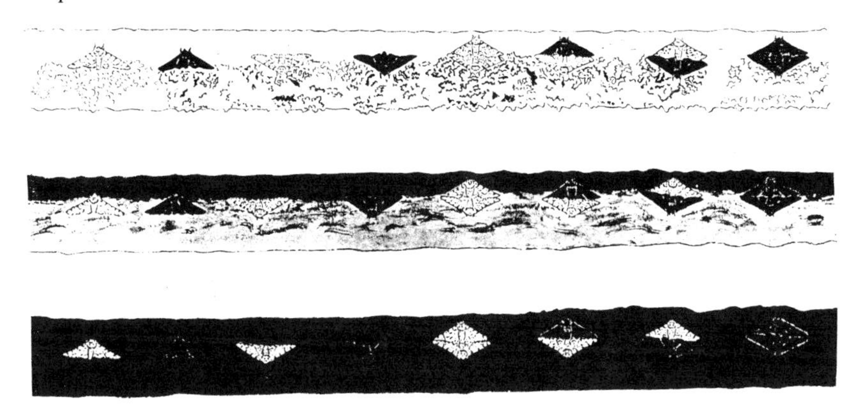

FIGURE 13. Diagram illustrating pairing and single *carbonaria* and *typica* forms of the peppered moth (*Biston betularius,* Lepidoptera, Geometridae) at rest on branches influenced to differing degrees by particulate air pollution. (From Liebert, T. G. and Brakefield, P. M., *Biol. J. Linnean Soc.*, 31, 129, 1987. With permission.)

(exceptional for a night-active moth; also, the abdomen and the hindwings have protective coloration) (Figure 13). An experiment using dead moths suggested that mortality rates differ between moths on trunks and those on trunk-branch joints.[719]

Peppered moths emerge shortly before dusk. Females exhibit a disperal flight before settling. Liebert and Brakefield[934] studied the behavior of virgin females released after flight activity in cages on trees in a rural and an urban wood. Once females had attracted males and paired soon after dark, they did not fly and only walked short distances. Many moths will rest during the day underneath or on the side of branchlets in the tree canopy. However, some moths will rest on nonhorizontal branches, and others on main branches or trunks. Such a range in resting positions is supported by records for the small number of moths found in the wild.[719] The resting behavior brings moths into close proximity with epiphytic lichens if present.[171]

Laboratory experiments indicated that the forms of *B. betularius* may differ in their choice of color and texture of resting background.[143,719,829,918] Mikkola[1051] found that a lower proportion of *typica* males than *carbonaria* males rested on the branches in the experimental cage, but that the forms did not select light and dark backgrounds differentially according to their own color. Any differences in resting site selection could substantially influence patterns of visual selection in the wild,[719,934] but as yet there is little evidence of this from outdoor cage experiments.[171,934,1051]

Kettlewell[829] pointed out the possibility of differential mortality of specific phenotypes in different pairing combinations. Peppered moths remain in copula for nearly 24 h. Copula of *typica* resembles foliose lichens, possibly more than single moths.[934,1050] The survival and relative crypsis of females was monitored on moths released in different parts of the tree, and the data indicated that the phenotype of the pairing partner may influence survival.[171]

The melanism of the ladybird *Adalia bipunctata*[325] and the mottled umber moth *Erannis defoliaria* (Geometridae) was markedly associated with the phurnacite plant at Abercwmboi, Wales.[920] The decline in melanic frequency away from the factory in these species was steep.[918] The other moth species, *Diurnea fagella* (Oecophoridae)[1426] and *Phigalia pilosaria* (Geometridae),[917] had high melanic frequencies at sites immediately adjacent to the plant, but in contrast with *A. bipunctata* and *E. defoliaria*, these species also have high melanic frequencies elsewhere in the region.[920]

The decline of melanic frequency in *A. bipunctata* in Birmingham, U.K., in the 1960s continued until towards the end of the 1970s.[173,324] Creed[324] pointed out that data from the 1920s suggested that there had also been some reduction in melanics prior to the 1960s. The decrease from 1960 to 1978 was associated with a disadvantage to melanics of about 10% compared to the selection regime prior to the decline. It was consistent with the timing and pattern of the decrease in smoke pollution over the same period. The decline in melanic frequency was less similar to the change in sulfur dioxide levels and supported the finding of a substantially weaker association between geographical variation in melanic frequency in *A. bipunctata* and sulfur dioxide than in melanic frequency and smoke.[173]

Description or quantification of the decline in the frequency of melanic *A. bipunctata* cannot alone indicate the nature of the selective influence of smoke. Although it is possible that smoke has some direct selective effect favoring melanics,[323-325] there is, however, only circumstantial evidence in support of this suggestion. Experimental evidence suggested that melanic ladybirds are better able to absorb infrared radiation and, as a result, are more active than the nonmelanics.[117] It was postulated that this could be advantageous in regions with low sunshine levels, including those with high air pollution levels.[1086] Some data supported the hypothesis of a causal association between the effect of smoke in reducing solar radiation and thermal melanism.[1086] This hypothesis predicts that as smoke emissions decline, solar radiation increases at ground level, and the thermal advantage of melanics is reduced via effects on adult fitness, especially reproduction.[171] Such an effect of lowered smoke pollution on adult reproduction should be detectable in the next generation. No corresponding mechanism has been indicated with regard to the direct effects of smoke, although a number could be postulated,[173] for example, one based on the effect of small particles on the permeability of the cuticle.[411] The results on the melanism of *A. bipunctata* around the Gulf of Finland indicate a close association of high melanic frequencies with industrial areas, especially with St. Petersburg (Figure 14).[1054,1358]

The establishment of smokeless zones in Britain after the clean-air legislation in the 1950s led to a fall in smoke and sulfur dioxide pollution in industrial regions. The decline in smoke was rapid, but that in sulfur dioxide was more gradual.[918] Declines in the frequency of the *carbonaria* form of *Biston betularius* were correlated with the reductions in air pollution.[267,268,303] The initial reduction in particulate air pollution and a lightening of tree surfaces clearly played

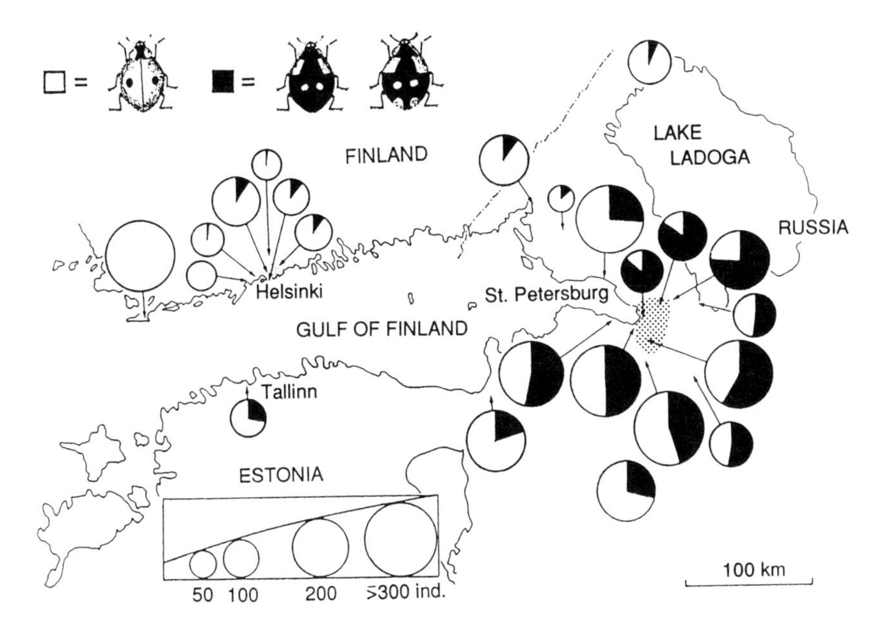

FIGURE 14. The melanism of *Adalia bipunctata* (Coleoptera, Coccinellidae) from selected localities around the Gulf of Finland. The black sector of the pies shows the melanic frequency. The westernmost sample in Finland represents individuals from the 1940s. (Modified from Mikkola, K. and Albrecht, A., *Ann. Zool. Fenn.*, 25, 177, 1988. With permission.)

a part in the decline of *carbonaria* at West Kirby, northwest England. Nevertheless, the best fit to the observed change in frequency was obtained when the fitness of *typica* was linearly correlated with the level of sulfur dioxide. However, Clarke et al.[268] considered that foliose lichens remained virtually absent at West Kirby due to the dominance of aggressive, opportunistic epiphytes and that there was possibly a strong nonvisual component to the selection in favor of *typica*. The period of the decline in frequency of *carbonaria* coincided with reductions of sulfur dioxide to levels below which some grey foliose lichens (e.g., *Hypogymnia physodes*) could be expected to appear on branches in the tree canopy.[934] Initially, colonization of epiphytes may mainly occur on the new growth of branches in the tree canopy with low residual surface pollution and a comparatively high light intensity.[171,934]

A lag between the fall in sulfur dioxide and the dramatic decline in *carbonaria* from the early 1970s is apparent. One explanation for this is that the effect of sulfur dioxide, as distinct from the earlier effect of smoke, is indirect. It is considered to influence the relative crypsis of the phenotypes via reestablishment of epiphytes, especially foliose lichens, on the resting backgrounds of the moths. The frequency of *carbonaria* has declined quite dramatically in the last 20 years or so. At Caldy near Liverpool,[268] *carbonaria* declined rapidly from about 90 to 40% since 1970. Cook et al.[303] showed that this change was consistent with a more or less constant selective disadvantage to *carbonaria* of

FIGURE 15. The forms of *Oligia latruncula* (top row) and *O. strigilis* (lower row) (Lepidoptera, Noctuidae). The "wild" forms on the right, the "dark" form in the middle, and the "black" forms on the left. The border between "wild" and "dark" forms is distinct, but that between the "dark" and "black" forms is less definite. (Photo by K. Mikkola.)

about 12% compared to the earlier period. At Caldy *carbonaria* is being replaced largely by *typica*; *insularia* increasing from below 1% up to only about 4%.[172,973] This suggested that a more complex change in the relative fitness of the various genotypes is occurring in the Netherlands, where both *typica* and *insularia* are increasing at the expense of *carbonaria*.[172] The decline in *carbonaria* at Caldy has been closely correlated with a reduction of sulfur dioxide in the locality.[973] Sulfur dioxide levels have progressively declined since the early 1970s, and lichen species diversity on several tree species has increased. Small colonies of foliose lichens now occur on the trunks and upper surface of tree branches in cities, probably influencing the relative crypsis in the peppered moth together with changes in bark coloration.[172,934]

A sharp increase in pollution by smoke apparently caused industrial melanism in *Oligia latruncula* (Noctuidae) in Helsinki, Finland, not quite 50 years after the commencement of a similar evolutionary trend in *Biston betularius* populations of Manchester.[1048] Museum samples indicated that the industrial melanism of *Oligia strigilis* was of a much later date; the oldest melanic moths from Finland dated from 1961 (Figure 15).[1049] In *O. latruncula* the melanic frequency ranged from 5 to 20% in the rural areas and reached 100% in central Helsinki. The melanism of *O. strigilis* was more strictly limited to the polluted areas, with a maximum of 92% in central Helsinki.[1048] The form *typica*, which was absent in the 1970s, reappeared in the samples from central Helsinki in the late 1980s. This was attributed to improved air quality.[895]

The melanism of *O. strigilis* occurred in Finland as geographical isolates, whereas that of *O. latruncula* had a continuous distribution. The local populations of *O. strigilis* differed from each other in the coloration of the melanic moth and in the relationship between the numerical ratio of the color forms and degree of air pollution. Apparently, the melanism had risen independently in several localities. Melanic color was a semidominant character; i.e., the heterozygotes are evident as intermediate, and the coloration seemed to be controlled by a single gene pair.[1049] Mikkola[1048] even constructed an *Oligia* index on the basis of the probable genetic background to describe the combined melanic gene frequency of these species. The local variations in the index could be explained by the values of the total dust fall. Melanism also occurs at a low level in the *Oligia* species in the unpolluted coastal archipelago.[1049]

B. DEVELOPMENT OF RESISTANCE

Pollutants can change the gene pool of an insect population, and the effects of pollution can be modified as a result of such changes. Exposure to pollutants can lead to genetic selection for individuals with the biochemistry or behavior to nullify their toxic effects. Development of pesticide resistance is the classic and economically most important example of microevolution in which a toxic chemical acts as a selective agent to increase the frequency of the genes responsible for survival.[215,857] Resistance is the development in a strain of the ability to tolerate toxicant levels that are lethal to most individuals in a normal population of the species. Natural tolerance is the preadaptive tolerance shown by some species to some insecticides. Resistance is a preadaptive trait or arises by mutation.[215]

The development of pesticide resistance is an unpredictable and diverse phenomenon.[534,1041,1209,1411] Mechanisms of resistance may vary between populations. The various mechanisms of pesticide resistance can be classified in general terms as activation and detoxification, reduced penetration and transport, larger capacity for storage or faster excretion, and reduced target-site sensitivity, as well as changes in host or habitat preferences (i.e., behavioral resistance) that reduce exposure.[1411] Resistance genes have been located on specific chromosomes in house flies, mosquitoes, and *Drosophila melanogaster*.

Pesticide resistance is the result of random mutation that establishes an R-allele in the natural population of the species. Widespread pesticide application propagates the R-allele through preferential survival, and it becomes dispersed throughout the population. As the R-allele becomes sufficiently common, the effectiveness of the pesticide is reduced. Where the R-allele is partially dominant, completely resistant RR homozygotes are rapidly selected. With recessive alleles or combinations of genes each conferring low-level resistance, selection is much lower.[1041]

A number of generalized mechanisms for insecticide resistance have been identified in terms of specific genetic regulation.[1041] Metabolic resistance involves detoxification of the insecticides by enzymatic processes, including esterases, microbial oxidases, glutathione transferases, and epoxide

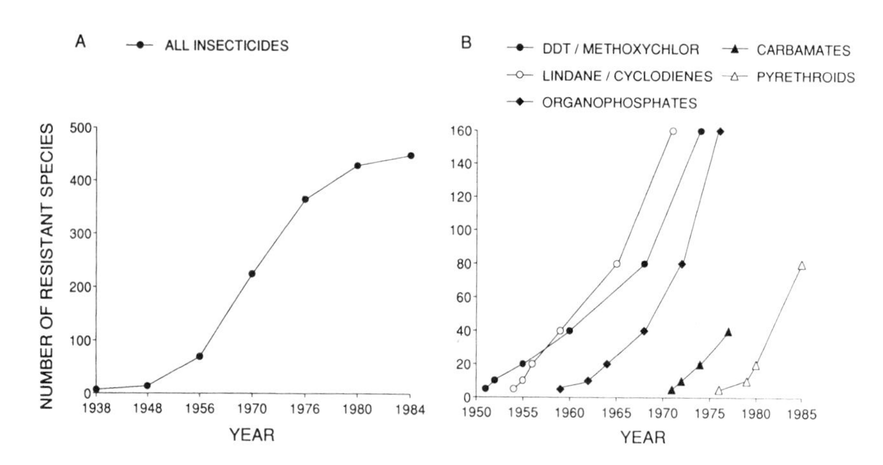

FIGURE 16. Approximate rates of development of insecticide-resistant species worldwide.[1041]

hydrolases. Kdr resistance involves a modification of the target sites for DDT and the pyrethroids at the sodium channels of the nerve axon. AchE-r, or altered acetylcholinesterase, involves a change in the biochemical action of the nerve synaptic enzyme acetylcholinesterase that is the target site for organophosphates and carbamates. Pen is a resistance mechanism involving decreased penetration of insecticides into the insect.

Huffaker et al.[722] reviewed the case of the development of pesticide resistance in the Clear Lake gnat (*Chaoborus astictopus*) in California. The gnat population of the lake was so high that it was a considerable nuisance in this resort area. This lead to the first insecticide application using a low dosage of TDE, in 1949. The treatment at first seemed successful, but the gnats began to increase again already in 1953, and another application with a higher TDE dosage was used. Unfortunately, the effects of this application lasted an even shorter time. The next application in 1957 failed, and the gnat populations were exploding, while the unexpected side effects included dead birds and high insecticide concentrations in fish. Later on, the lake was treated with the less persistent methylparathion, but the best results were gained by introducing a fish species, the Mississippi silverside, into the lake for biological gnat control.

There are now several hundred well-established cases of insects, agricultural pests, or medically important species becoming resistant to insecticides and other chemicals (Figure 16).[533,534,1209] In multiple resistance, resistance is evolved independently after exposure to each particular insecticide. In crossresistance the development of resistance to one insecticide also confers resistance to one or more other insecticides that may or may not be chemically related to the insecticide that stimulates the evolution of resistance within the population.[1075] Recently, Brattsten[174] postulated that all field populations have acquired some degree of resistance, that the genetic constitution has been changed for practically all insects that inhabit areas exposed to insecticides.

Avoidance and retardation of pesticide resistance in insect populations have tried to be achieved by reducing the selection pressure through the use of reduced rates and numbers of pesticide applications, and restricting the applications to a small portion of the population.[215] The likelihood that resistance will develop in a susceptible insect population when exposed repeatedly to insecticide may be influenced by many factors, including the different spread rate of dominant and recessive genes, the interval between successive generations, mobility, and fecundity.[1075]

Pimentel and Levitan[1206] estimated that less than 0.1% of the pesticides applied to crops actually reaches the target pests. Hence, they concluded that most of the pesticides applied enters the environment, contaminating the soil, water, and air, and perhaps poisoning or adversely affecting nontarget organisms. The figure of 0.1% was calculated as the amount of insecticide, for example, that comes into direct contact with the insect. The bulk of the pesticides is probably within the environment of the target pests. In the case of herbicides, a far higher percentage of the chemical reaches the target plant.[1667]

Although hundreds of target species have become resistant to pesticides, there is little information on such microevolution in nontarget species. However, it seems feasible that resistance would have developed in them as well. Croft and Morse[329] and Croft and Strickler[330] reviewed the pesticide resistance in natural enemies of pests. For example, the toxicity tests of methidathion to parasitoids of the red scale *Aonidella aurantii* (Hemiptera, Diaspididae) have shown differences in resistance between parasitoid strains treated in different ways.[1344] The different sensitivities of the two springtail species *Onychiurus armatus* and *O. apuanicus* to atrazine could support the hypothesis of a genetic selection of resistant populations[1316] for a field treated over a period of a few years. Joosse[798] is in agreement with this on the basis of the high genetic variability found in Collembola populations. However, the hypothesis of different binding properties of soil[498] may also explain the contrasting results.

Compared to the rapid development of pesticide resistance, the evolution of metal resistance seems to be a slow process.[800] Resistance to cadmium was demonstrated in laboratory-reared F_1-generations of *Orchesella cincta* (Collembola) populations originating from metal-contaminated sites, providing evidence for genetic differentiation.[1222,1223] This adaptation was associated with increased metal excretion. Although the average metal tolerance was increased in populations subjected to long-term selection, less efficient individuals were not eliminated from adapted populations. In similar experiments on the effects of zinc on *O. cincta*, no significant adaptive changes were found.[1222] Metal resistance in soil animals seems to be of degree rather than of kind.[800] Posthuma et al.[1223] studied population differentiation in *O. cincta* populations, from reference (<5 nmol Cd per gram; 0.5 to 1.2 μmol Zn per gram) and metal contaminated sites (45 to 557 nmol Cd per gram; 14.5 to 23.9 nmol Zn per gram), by comparing cadmium excretion efficiency in F_1-generation laboratory individuals. Significant differences in excretion efficiencies implied that the distribution of toxic metals over body compartments differs, tolerant populations having a higher proportion deposited in the gut. Body concentrations of zinc were consistently higher in individuals from the polluted site during both cadmium and zinc exposure.

V. CONCLUSIONS

Descriptive and correlative field investigations are needed for several purposes, but the bulk of existing information on response mechanisms involved stems from experimental evidence, allowing sound interpretation of field observations. Although there is an extensive literature on insect responses to various pollutants, generalizations are difficult to make, since the information is still fortuitous and incomplete. The present understanding of the response mechanisms lies on all too few well-investigated cases. Crucial research needs include trophic interactions, genetic variation in relation to pollutants, and, to a great extent, patch and population dynamics and associated phenomena (dispersal, population persistance, natural fluctuations, etc.), as well as the chemical speciation of pollutants and the simultaneous effects of several stress factors.

The effects of pollutants on insect populations are mediated via their effects, direct or indirect, on individuals, and the liability of these effects is related to the dose. Concerning the effects of pollutants on individual organisms, Moriarty[1075] suggested that the logical approach is to concentrate on the pollutant and its detailed mode of action, by discovering the primary lesions and by exploring the ramification that result from these lesions. For example, some sublethal effects of pollutants can be elucidated from knowledge of the mode of action. The alternative approach is to investigate the health of the individual and how well it is functioning. With both approaches the effect of other environmental variables deserves more attention, and this could be associated with the studies on pollutant levels in insects.

Several compounds are often simultaneously involved in the ecological effects of pollutants. There may be biologically active impurities, or metabolites or breakdown products that have biological effects. Distribution of a pollutant in the environment is uneven, and some pollutants occur in more than one form. The relationships among exposure, pollutant dose, and effects on the individual organism are complex. Species, populations, sexes, and developmental stages may react in a different way to the same exposure, for both genetic and environmental reasons. The interactions between individuals within a population, and between species within a community, further complicate the picture.

Indirect effects of pollutants on insects have proved to be crucial in an ecosystem context. The plant-insect, and sometimes associated plant-pathogen interactions pose special research challenges in relation to environmental pollution. Plant stress-induced variation in insect herbivory is due to both variation in plant responses to pollution and variation in insect sensitivity to changes in stressed plants.[791] While most pollutant-induced changes in predators have been negative, the changes in experiments on herbivores have often been positive (e.g., increased size, enhanced growth, and decreased mortality) and mediated by the host plant. The success of pest attack is often related to poorly known qualitative changes in certain plant tissues. These changes can ensue from allocational changes in plant organs, resulting from the pathological consequences of the pollutant, or from prophylactic responses to potential pollution damage. Episodic pollution may affect growth rates of herbivores, predation, or competition in a way that the final effects of pollution on insect populations and their host plant become visible only after a time lag. Future work in this area should include the time courses of plant growth and critical metabolite pools in relation to the performance of particular insect species on plants exposed to realistic pollutant regimes.[914] Moreover, studies on indirect effects should be widened to cover also other-than-host plants, since slow, gradual, pollution-induced changes in vegetation (and climate) apparently lead to drastic alterations in insect fauna. Such effects may, perhaps more often than expected, explain the observed patterns of increased insect performance in polluted areas.

The genetic studies on insects and pollution have concentrated on two interesting phenomena, viz. industrial melanism and pesticide (and metal) resistance. The results on industrial melanism illustrate the effects of pollutant-induced changes in insect habitat. The selective advantage of melanic forms must be considerable, as shown by the rapid increase in the proportion of melanic forms in cryptic moths or by the loss of warning coloration in ladybird beetles. However, as shown by the studies on moths, these changes are not irreversible. The development of pesticide resistance is another example of detectable contemporary evolution. Still, very little is known about resistance in nontarget insects and its potential effects on the ecosystem level. Furthermore, it is hard to believe that the genetic effects on insects of pollution would be restricted to these isolated phenomena. It is tempting to speculate, for example, on the possibility of insect herbivores adapting to the biochemistry of pollutant-stressed host plants, or predators adapting to changed species composition of prey. Genetic changes may also reduce the suitability of insects as bioindicators. For example, as a result of altered metabolism, pollutant tolerance or excretion ability may change, causing changes in species abundance and community structure or the concentrations of pollutants in individuals. On the other hand, genetic changes (e.g., industrial melanism) may also be utilized as indicators of environmental perturbation.

Chapter 7

THE COST OF POSSIBLE IMPACTS

I. INTRODUCTION

Insect-mediated effects of pollutants alter both the structure and function of ecosystems. The structural effects include changes in species diversity, biomass, and trophic relationships. The functional effects include changes in the production and decomposition of organic matter. Pollutants can directly or indirectly (e.g., through changes in vegetation) decrease the species richness or genetic variation of insects. Changes in insect number and species assemblages can reduce species diversity in other organisms, including plants, birds, and mammals. Even though changes in species composition would not take place, ecosystem dynamics can be interfered. A prerequisite for the detection of such disturbances is long-term monitoring of the state of the ecosystem, including entomological aspects. The concepts of bioindication and critical loads are associated with environmental monitoring.

Since insects play an essential role in many ecosystems, the effect of pollutants mediated by insects have economic consequences. These consequences may be direct losses of products like honey and silk, or indirect losses of fish, game, agricultural crops, or timber. Water pollution and pesticide applications can cause medical and veterinary problems associated with insect vectors.

II. BIODIVERSITY AND ECOSYSTEM DYNAMICS

A. LOSS OF BIODIVERSITY AMONG INSECTS

Biodiversity is the total variety of genetic strains, species, and ecosystems. Biological diversity should be conserved as a matter of principle because all species deserve respect regardless of their use to humanity.[753] However, biodiversity has also economic, scientific, educational, and aesthetic value. Insects form a crucial component of biodiversity. More than 50% of the terrestrial biodiversity is associated with insects.

Although the use of insecticides, herbicides, fungicides, and other artificial chemicals have caused losses of insect populations, it has not lead to any documented cases of species extinction. However, especially on islands, pesticide use may be a real risk for the native fauna. Small populations are susceptible to aerial application of insecticides. Wiest's sphinx moth (*Euproserpinus wiesti*) was almost exterminated when the area around its last known locality in Colorado was aerially sprayed with malathion for grasshopper control.[1584]

Mainly circumstantial evidence suggests that acidic precipitation is responsible for declines of insect populations in Europe and North America. Some terrestrial species associated with lichens, which have declined due to air pollution, seem to have drastically declined in Northern Europe (e.g., the geometrid moth *Alcis jubatus*).[660] More drastic changes have been reported in acidified waters. For example, Jensen[774] concluded that the number of stonefly (Plecoptera) species had decreased by 25 to 40%, and that of mayfly (Ephemeroptera) by 75%, in catchment areas of 250 to 700 km^2 in Sørlandet, southern Norway.

Water pollution caused by acidic mine drainage, silage pit effluents, and siltation has apparently had detrimental effects on aquatic insect communities, and industrial pollution has had at least transient adverse consequences on insects.[1584] For example, Blepharoceridae (Diptera) are generally unable to tolerate pollution or siltation, and many mayfly and stonefly species are highly sensitive to pollution. It was reported that all but one species of caddisfly in the River Rhine had severely declined as a result of urban and industrial pollution in central Europe, and at least one species (*Hydropsyche tobiasi*) was considered probably extinct. Urban pollution is also regarded as the principal threat to a stonefly (*Capnia lacustris*), which lives at depths of 60 to 80 m in Lake Tahoe on the border between California and Nevada.[1584]

Environmental pollution may result in losses of isolated populations. Environmental changes caused by human activities may increase the extinction rate, or decrease the colonization rate, so much that metapopulation extinction becomes inevitable.[612] A metapopulation may go extinct when the size of suitable habitat patches becomes too small and/or when the number of habitat patches is decreased, thus increasing their degree of isolation.[611] Pollutants affect the habitat suitability. Pollution may, of course, destroy an entire small population, but it can also destroy a whole metapopulation by eliminating some linking local populations. Such consequences are, however, difficult to discriminate from those of land-use changes. The patch fluctuating pattern of agricultural land treated with pesticides may support insect metapopulations. However, as emphasized by Hanski and Gilpin,[613] a species confined to a newly fragmented habitat does not necessarily function as a metapopulation, for it may have so poor a dispersal ability that the local population, once extinct, will remain extinct.

B. LOSS OF BIODIVERSITY AMONG OTHER ORGANISMS

Insects are an important component of diet for many predators, including both vertebrates (e.g., fishes, birds, and bats) and invertebrates (e.g., spiders). Consequently, changes in insects are reflected in these higher trophic levels as well.[258,265,266,1097,1323,1430,1461] Another important point is that several harmful chemicals may concentrate along food chains in which insects play a major role. Terrestrial insects may also mediate harmful residues to aquatic food chains,[1509] and vice versa.

Goriup[554] reviewed the effects of acidic precipitation on insect-feeding birds. Birds that are generalized invertebrate feeders (e.g., dabbling ducks or grey wagtails *Motacilla cinerea*) may remain largely unaffected for some time, especially if competition for food is eased by the absence of fish. In dying forests the dead standing timber can provide additional food and habitat for relatively uncommon or rare species of woodpeckers (e.g., three-toed woodpecker *Picoides tridactylus*)

One bird species that specializes on aquatic invertebrates known to be vulnerable to acidification is the dipper (*Cinclus cinclus*). Dippers rely heavily on caddisfly (Trichoptera) larvae and mayfly (Ephemeroptera) nymphs during the breeding season, and these animals become fewer with increasing water acidity. Local declines of dipper populations have been recorded in parts of Great Britain and Germany.[554] Ormerod et al.[1163] studied the ecology of dippers in relation to stream acidity in upland Wales. Laying dates were later, clutch and brood sizes smaller, and nestling growth slower along acidic streams by comparison with circumneutral streams. Increased acidity has also been linked to reduced food resources for certain waterfowl.[1032]

Pollution may affect vertebrates via unexpected routes through insects and other invertebrates. Air pollution apparently increases needle loss. In Sweden the reported approximate 50% reduction in the density of large spiders in spruces (*Picea abies*) with high needle loss may have severe consequences for birds. Spiders are essential winter food for the overwintering passerine birds, such as the goldcrest (*Regulus regulus*) and tits (*Parus* spp.), and there are few alternative prey items. In October, spiders made up 45% of the prey items for goldcrests foraging in spruce, but later they became more important and comprised about 80% of the prey in January and February. Reduced winter food supply is one possible explanation for some bird declines in coniferous forests.[573]

The large-scale use of insecticides has led to the decline of several nontarget species.[188] In Africa during the tsetse fly (*Glossina* spp.) control campaigns, ground-sprayed dieldrin killed a large number of vertebrate species in Botswana, and DDT and BHC (lindane) spraying caused major kills of several nontarget insect groups in a game reserve in Zululand. Furthermore, it was recorded that the flowering plants of the families Orchidaceae and Asclepidaceae, which are pollinated by a few species of Hymenoptera, had almost vanished from the game reserve 30 years later, possibly due to the lack of pollinators.[995]

In North America most short- and medium-term changes in amphibian, snake, and mammalian activity could be directly or indirectly attributed to the mortality of invertebrates in the area sprayed by insecticides for the control of the spruce budworm (*Choristoneura* spp.).[168] In view of the expected rapid recovery of invertebrate populations, it was unlikely that these changes in vertebrates would persist over the long term. Examination of stomach contents had revealed that spruce budworms could make up approximately 40% of the diet of birds, the majority of the insectivorous birds being facultative feeders preying opportunistically upon a great variety of available prey. The mortality of spruce budworm following an operational aminocarb spray would presumably cause insectivorous birds to switch to other abundant alternate prey species.[168]

The potential for pesticides in aquatic ecosystems to affect fish diet and subsequent growth is high because toxicity degree is often greater for zooplankton and macroinvertebrates than for fish.[788] Elimination or a significant reduction in the density of important food items by a pesticide could reduce growth, increase vulnerability to predation, and lead to starvation, especially in the early life stages, and ultimately reduce year-class strength.[1059] There have been numerous studies on the sensitivity of early life stages of fish to toxicants,[1022] but the link between a pesticide, the invertebrate forage base, and changes in fish diet are not clearly established.[178]

In a recent review Clements[274] summarized that when concentrations of heavy metals in water are high, the contribution of food to total body burdens in fish will be relatively insignificant because of the greater rate and efficiency of transport across the gills. When concentrations in water are low, food-chain transfer of metals may be the primary route of exposure. Consequently, the relative magnitude of food-chain transfer depends on the feeding habits of fish, on metal levels in prey, on the ability of prey species to tolerate and accumulate metals, on the relative abundance of prey, and on the trend of fish to switch to tolerant prey.

C. MONITORING AND BIOINDICATION

Nowadays continuous control over the environment has become important. There is a need to assess the distribution and effects of pollutants: rates of release of pollutants in the environment, degree and changes of environmental contamination, biological effects (Figure 1), etc.[194] There is considerable disagreement about the relative usefulness of monitoring the amounts of pollutants in the environment and of monitoring for biological effects.[1075] Biological monitoring is the systematic use of biological states and effects to gain information about the environment.[1259] In addition to hydrochemistry, biogeochemistry, and ecophysiology, populations and communities are monitored. Natural successional changes, catastrophies, and population fluctuations may mistakenly be interpreted as trends in short-term studies. Furthermore, it may not be worthwhile carrying out long-term monitoring studies or models of organisms that are adapted to a quick evolution.[1259] Biological variables have many advantages over the physicochemical ones, in environmental monitoring. They usually integrate the effects of several environmental variables over a long period of time and also permit the detection of occasional disturbances.[812,1298]

Bioindicators, organisms or biotic processes that respond significantly to changes in an affecting-state variable, can demonstrate or predict the onset or reversal of larger-scale damage (Table 1). The simplest bioindicator is an organism that reveals the presence or absence of some factor by its own presence or absence. Its abundance or frequency may be related to a particular environmental gradient. A response may also manifest as morphological or physiological symptoms. The indicator may also accumulate pollutants or unusual metabolites in altered ratios in all or part of its body. A bioindicator can be used as a monitor when its response is regularly quantifiable

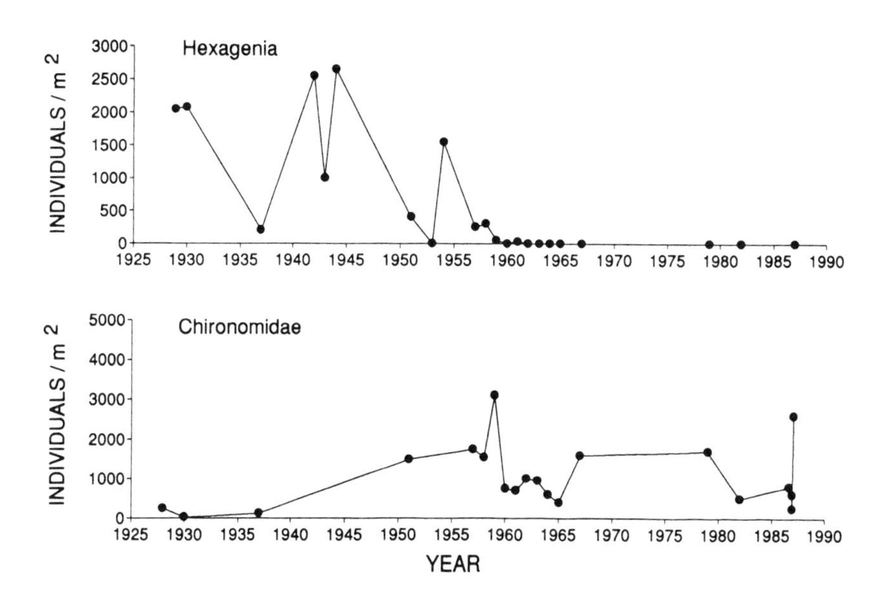

FIGURE 1. Abundance of major groups of benthic invertebrates in the open waters of western Lake Erie, 1928–1987. (From Reynoldson, T. B., Schloesser, D. W., and Manny, B. A., *J. Great Lakes Res.*, 15, 669, 1989. With permission.)

in terms of time and/or space and in relation to contamination level.[983,1259] Bioindicators may also be of use when comparing different pollutants, habitats, or methods.

In order to develop a more comprehensive understanding of pollutant (e.g., metal) contamination and accumulation by the biota, sampling a wide range of developmental stages and size classes will provide substantially more information than will the collection of individuals of uniform age and size. This approach is likely to lead to the development of biomonitoring protocols that are more likely to be successful in tracking spacial and temporal changes in environmental quality.[875]

Honeybees, for example, have been used as indicators of cadmium, arsenic, and radionuclide contamination because they contact innumerable foliar and other surfaces during foraging.[489,1264,1298,1545] Bees are particularly well suited for use as *in situ* sentinels. Hives of various sizes can be moved easily to within or near an area of concern. There are more than 50,000 bees in an average hive. Honeybees have been utilized as an indicator of the "health" of a definite ecosystem.[1264] Even slight contamination of pollutants can be evidenced by either analyzing the bees themselves or their products (honey, pollen, wax, propolis). The activity of the honeybee over any territory is well distributed, and its range can be precisely established. The foraging range can cover more than 2.5 square miles.[1298] Since honeybees have been used as biological indicators of environmental radioactivity, Ravetto et al.[1264] proposed an experimental beehive network to monitor the area on the Po River in Italy where a

TABLE 1
Examples of Insects as Bioindicators

Taxon	Environmental quality	Measure	Refs.
Terrestrial			
Oligia strigilis, O. latrunculus (Lepidoptera)	Air quality, total dust fall	Melanic gene frequency	895,1048,1049
Apis mellifera (Hymenoptera)	Radionuclides	Concentration in bees and honey	544,1264
	Metal pollution	Concentration in honey	795,1491
Adalia bipunctata (Coleoptera)	Air pollution	Melanic frequency	325
Epigeal fauna	Industrial pollution	Catchability	260
Soil invertebrates	Radionuclides	Concentration	884
Aquatic			
Hexagenia bilineata (Ephemeroptera)	Water quality	Abundance and distribution	508
Hexagenia sp. (Ephemeroptera)	Water quality	Bioelectric potential	1074
Ephemera danica (Ephemeroptera)	Sewage pollution	Survival	140
Micronecta spp. (Heteroptera)	Water quality, eutrophication	Relative abundance (aqoustic)	766–768
Chironomus yoshimatsui (Diptera)	Sewage pollution	Abundance	1328
C. acerbiphilus (Diptera)	Acidity	Abundance	1328
C. riparius (Diptera)	Cd, phenol, ammonia, lindane	Emergence, larval survival, concentration	1009
Chironomus spp. (Diptera)	Organic pollution	Relative abundance	1251
Chironomus spp., *Procladius* spp. (Diptera)	Toxic stress	Morphological abnormalities in larvae	1564–1566
Tanytarsus dissimilis (Diptera)[a]	Pesticides and other chemicals	LC_{50}	698
Chironomidae (Diptera)	Lake type concept	Relative abundance, bioindices	812
Aedes aegypti (Diptera)	Organic pollution	Phototaxis	391
Plecoptera, Trichoptera, etc.	Water quality	Index based on relative abundance	901
Ephemeroptera, Trichoptera	Organochlorine pollution	Concentration in adults	869
Macroinvertebrates	Metal pollution	Community structure	1627
	Water quality	Biotic index on artificial substrate	1065
	Water quality	Community and diversity indices	1216
	Water quality	Relative abundance	64,1658
	Acidity	Species tolerance limit	600
		Species optima and tolerances	600

[a] Fish, daphnids, crayfish, snails, tadpoles, and leeches also simultaneously present in the test.

2000-MW nuclear power station was due for installation. The versatility of bees might also be useful at toxic waste sites where unknown chemicals have been dumped.[1298]

The role of soil invertebrates for setting soil and sediment quality criteria so far has been limited, and quantitative dose-response relations are seldom known.[1433] However, studies on species like *Folsomia candida* (Collembola) and *Aiolopus thalassinus* (Orthoptera) have given promising results on the possibilities of testing detrimental effects of metals and pesticides in soil. Soil animals are also suggested to be appropriate biological indicators of an area's radioactive-pollution level, because their population density is generally relatively high.[884] Soil invertebrates can be collected on a mass scale during the entire warm season. Studies on forest-floor arthropods have indicated a negative correlation between the number of animals (in Barber traps) and the degree of industrial pollution (e.g., SO_2). Thus, catchability of the epigeal fauna has also been suggested as a bioindicator of industrial pollution in forests.[260]

Several aquatic invertebrates have been used as indicators of freshwater quality (Figure 2).[52,451,669,767,901,1328,1556,1658] Water quality indices are based on the reduction in the community characteristics of clean water and the progressive disappearance of particular indicator species in polluted waters.[669] Chironomid midges and hydropsychid caddisflies have been used for the biological assessment of aquatic contaminants.[214] These organisms accumulate lipophilic compounds to equilibrium concentrations in excess of those in their environment. Macroinvertebrate assemblages have been used as bioindicators of stream conditions.[1655] The use of artificial substances for collecting invertebrates may represent a means of reducing variability that can be associated with natural substrates.[1065] Courtemanch and Davies[313] proposed a coefficient of community loss to assess adverse effects in aquatic communities. The coefficient used the ratio of numbers of taxa lost between an unaffected reference community and a pollution-affected community, to the total number of taxa. Benthic invertebrates have been used as indicators of pollution because they have relatively long life cycles, they are relatively sedentary, and different taxa have different ranges of tolerance to a variety of pollutants.[552,669] The benthic invertebrate community has been suggested as part of an indicator of the health of mesotrophic aquatic ecosystems.[1276,1597]

Benthic invertebrate densities and drift are considered sensitive indicators of acid stress.[471,597,711,718] Much effort has been devoted to finding early indicators of aquatic acidification. In Europe, mayflies of the genus *Baetis* (e.g., *B. rhodani*, *B. lapponicus*, and *B. macani*),[1060,1249] benthic crustaceans, and snails usually disappear from acidifying ecosystems long before species of sport fish.[1249] The disappearence of acid-sensitive species in streams and lakes has indicated increased surface-water acidification and may predict reductions in fish populations and other biota.[423,594,597,1154,1336]

In spite of the wide use of insects as bioindicators, their reliability and robustness as indicator organisms is often questionable. The term indicator species has been used in ecotoxicology, with a wide variety of meanings.

FIGURE 2. Pollution and eutrophication of Lake Päijänne, Finland, as indicated by (A) distribution of *Micronecta* species (Heteroptera, Corixidae) and (B) summarized data of traditional water quality investigations. Heavier shading indicates increasing pollution. For areas left white, no data available. (From Jansson, A., *Ann. Zool. Fenn.*, 14, 105, 1977. With permission.)

Sometimes it has indicated the idea that knowledge of one species within a community will indicate the biological health of the community. This seems reasonable if the traditional view of community as a supraorganism is accepted, but as Moriarty[1075] argued, it is misleading. The relative degrees of exposure and susceptibility are much more functions of the particular pollutant and the individual species than of the community. There are no apparent grounds to suppose that there are certain indicator populations that will inevitably react when pollutants have affected other species within the community. An indicator species can only be used to assess the impact of pollution on a community if rather a lot is known about both the pollution and the community. The analysis of organisms for pollutant content is to be preferred mainly in cases where the species is of interest in its own right.[1075]

Organisms living in polluted environments may be relatively more tolerant to contaminants, e.g., metals, than organisms from noncontaminated habitats.

Such tolerance may have evolved genetically within a population through natural selection (i.e., adaptation) or may be due to physiological changes within individual organisms during earlier exposure (i.e., acclimation).[563,851] Tolerant individuals may accumulate greater or lower concentrations of pollutants in their tissues than their nontolerant conspecifics, when both are exposed to contaminated diet. For example, if organisms can evolve modified accumulation capabilities in response to the selection pressure of exposure to toxic substances within their habitat, the suitability of using such organisms for biomonitoring purposes becomes a matter of dispute. Studies on various terrestrial invertebrates have shown that between-population differences exist in invertebrates exposed to different levels of pollutant stress.[563]

The whole concept of environmental health can be misleading. A pollutant may cause marked alterations in a community, but it will only become a different community, although it may be a less desirable community for economic, social, scientific, or aesthetic reasons. Furthermore, the concept of environmental health often includes the idea that pollutants produce similar changes in many different ecosystems. Pollutants are usually believed to simplify the structure of both plant and animal communities. Moriarty[1075] also questioned these ideas. Apart from any of the effects pollutants may have, it is unclear whether there are any general laws that govern the structure of communities that contain many species.

D. ECOSYSTEM DYNAMICS AND CRITICAL LOADS

Pollutants affect ecosystem dynamics through changes in insects, and insects through changes in ecosystem dynamics. The use of model ecosystems[1040] has been considered a promising approach for providing essential information on how chemicals will behave in real ecosystems.[1233] Pesticides, polychlorinated biphenyls, and other pollutants (carcinogens, drugs, products from coal, food additives, and metals) have been run through model ecosystems.

Freedman[502] defined bioassay as the quantitative estimation of the intensity or concentration of a biologically active environmental factor, measured via some biological response under standardized conditions. The main goals of the use of bioassays in pollution research are to rank hazards, to set discharge limits, to predict the environmental consequences of discharges, to protect important species, and to protect ecosystem structures and functions.[223] However, Gray[560]considered that the species traditionally used in bioassays are not sufficiently sensitive to detect subtle ecological effects of pollutants.

It is increasingly difficult to attribute any observed faunal changes to a particular chemical, especially at low, but continuous, concentrations.[1091] For example, hundreds of potentially hazardous chemicals have been identified in the rivers Rhine and Meuse in central Europe.[1389] Thus, it is clear that there is a need to develop toxicity tests[1386-1388] that will enable us to predict safe concentrations of these chemicals for aquatic life.

There are several techniques for carrying out acute toxicity tests that are applicable to aquatic insects.[1091] In the static technique the test solution and organisms are placed in test chambers and kept there for the duration of test. The recirculation technique is similar, but the test solution is continuously circulated through an apparatus maintaining water quality. The renewal technique differs from the static technique in that test organisms are periodically exposed to fresh test solution of the same composition. In the flow-through technique, test solutions flow through the test chambers on a once-through basis for the whole duration of the test. Recently, there has been a tendency from single-species to multispecies toxicity testing.[698,1074] Pontasch et al.[1216] evaluated diversity indices, community comparison indices, and canonical discriminant analysis to determine their utility in quantifying macroinvertebrate response to a complex effluent in multispecies toxicity tests.

The effects of pesticides and other harmful substances have also been studied using experimental channels and simulated streams in the laboratory. In artificial community streams and channels *in situ*, the impact of a pollutant on whole-species assemblage can be studied under conditions similar to those in natural conditions, but allowing certain factors to be controlled and studied separately.[1091] This approach has been used, for instance, in studies with both control and experimental channels on the effects of the lamprey larvicide TFM,[962] as well as in Japanese investigations with large open-air artificial streams on the community effects of temephos and chlorpyrifos.[1665] In static waters the use of constructed standard outdoor plots with a miniature ecosystem has allowed the comparison of the effects of different pesticides and dosages under controlled conditions.[1091]

The concept of a critical load or level is based on a dose-response relationship. The concept is gaining international recognition as a practical way of assessing effects and planning and implementing emission control. Various international groups have arrived at definitions that suit their own particular purposes, and it seems difficult to make an agreed single definition. Critical levels mean the pollutant levels above which direct adverse effects on plants, animals, ecosystems, or materials may occur according to present knowledge (Figure 3).[201] Target loads are determined by political agreement, and they may be above critical loads when practical or economic considerations take precedence, or below critical loads when providing a safety margin to ensure critical values are not exceeded.[201]

Critical loads are useful tools in finding areas particularly susceptible to pollution. The uncritical use of the critical load and related concepts may be too wide a generalization to be applied to the complex biological systems.[1075] For instance, the botanical changes in central Europe associated with nitrogen deposition seem to indicate that long before ion inputs into ecosystems develop toxic effects on tissues or individuals, they influence competitive relations between species.[429] Both floristic and vegetational changes have a fundamental impact on herbivorous insects, which may easily pass unnoticed.

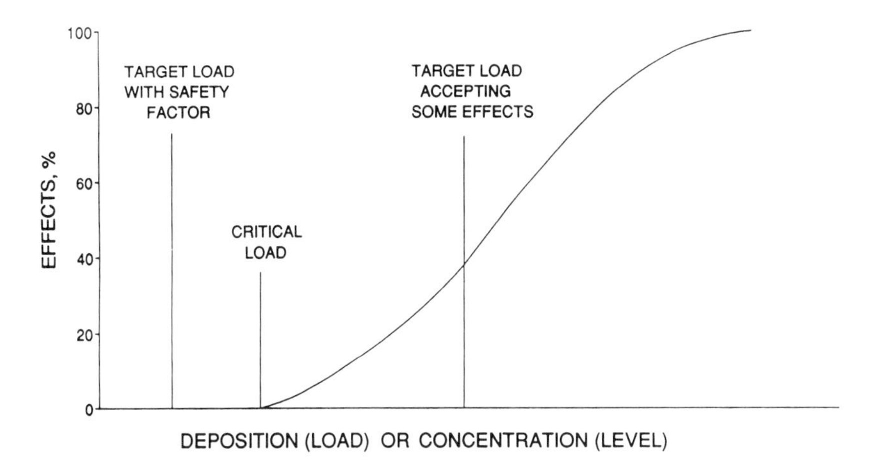

FIGURE 3. Theoretical dose-response curve showing alternative "target loads" compared with "critical load." (Redrawn from Bull, K. R., *Environ. Pollut.*, 69, 105, 1991. With permission.)

Grant[559] discussed the acceptable damage to nontarget invertebrates caused by pesticide applications. The speed of biological recovery from damage will be related to the dose rate of the pesticide or other pollutant, the existence of unaffected areas from which immigration can occur, etc. In the absence of residual toxicity, unacceptable lengthy recovery periods could be measured against the life history of sensitive species. The acceptable damage might be assessed by combining the recovery period and the degree of divergence between treated and control populations.

Metal contamination is a possible threat to the ecological functioning of the soil. In spite of some controversy, van Straalen and Ernst[1434] suggested that metal biomagnification may endanger species in critical pathways, e.g., top carnivores and soil arthropods specializing on soil fungi. According to Joosse and van Straalen,[800] protection of animal life in soil is achieved if there is only a small change that a species from the soil community is exposed to (a level greater than its no-effect concentration). Assuming a statistical distribution of sensitivities, the probability that a no-effect concentration is exceeded can be calculated as a function of the concentration in soil. The parameters of the distribution are estimated from toxicity data on a certain number of test species.

III. ECONOMIC CONSEQUENCES

A. LOSSES OF HONEY AND SILK PRODUCTION

The honeybee (*Apis mellifera*) is economically the most important insect in the world because it provides apiculturists with economic honey yields, and it is the primary pollinator of most crops. The loss of nectar and pollen sources due to the herbicide applications inevitably affects honey production. Many insecticides are regarded as deleterious compounds for bees and other pollinators,[188,578,978] and their application is recommended outside flowering periods.

Violation of these recommendations may lead to the death of individual bees when they intercept spray droplets or contact pesticide-contaminated surfaces, or can lead to contamination of the entire colonies when pesticide-carrying adults return from foraging activities.[1113] Because they are totally reliant on adult workers for their food supply, honeybee larvae are susceptible to pesticides in the colony's stores of pollen and ripening nectar.[356,357] In more acute cases of pesticide poisoning where large numbers of adults are killed outright, the entire social network of the bee colony is disrupted, and the brood stages may die from starvation (larvae) or lack of temperature regulation (eggs, larvae, pupae).[780] Pesticides may reduce the overwintering ability of bee colonies,[1429] as well as the pollinating activity of workers and honey production.[57,578,780] Pollutants in nectar may be concentrated as a result of the evaporation of water that occurs during the elaboration of honey, or may be diluted by uncontaminated nectar.

Fluorine compounds have caused considerable poisoning of bees near industrial plants in parts of France, Germany, and Switzerland,[1001,1003] though low fluoride levels (up to 37 μg/g dry weight in pollen) do not seem to have detrimental effects on bee populations.[1005] The toxicity of fluoride to bees depends on several factors, but generally the LD_{50} varies between 5 and 8 mg per bee.[1002] Other pollutants, such as metals, may also enter the hive with the pollen collected by the workers.[1227]

Pollution also may affect honey production through changes in the dynamics of pest populations. The study of Ortel[1168] showed that *Pimpla turionellae*, a pupal parasitoid of the greater wax moth *Galleria mellonella* (Pyralidae) was sensitive to cadmium. Decline of fat and protein content and increase of water content occurred in *P. turionellae* hatched from contaminated pupae of *G. mellonella* with body concentrations of 1.6 μg Cd per gram dry weight. Such body concentrations of *P. turionellae* were not restricted to individuals from especially polluted areas. This would mean a great disadvantage for species like *P. turionellae* and may account for alterations in their population dynamics and, through this, in those of the pests of beehives.

Air pollution may have detrimental effects on silk production. Widespread damage to the silkworm (*Bombyx mori*) industry due to air pollution has been reported, for instance, from Jiangsu and Zhejiang provinces in China.[1549] When mulberry leaves polluted by fluorine compounds were ingested by the moth larvae, their growth rates decreased and mortality increased.[520,521,741,891,892,1064,1549-1551] Fluoride concentrations above 30 μg/g increased silkworm mortality rates, and those above 80 μg/g severely inhibited cocoon production.[1549]

B. LOSSES OF FISH AND GAME

Fish are often more sensitive to pollution than are aquatic insects. Some aquatic macroinvertebrates provide an important source of food for vertebrates.[1032,1163,1164,1430] Acidity and aluminium concentration in waters are known to have consequences for birds dependent on aquatic macroinvertebrates as a food source. Populations of salmonid fish are also limited in acidic streams, and while the toxic effects of acid-related factors are probably important, their influences on fish through the scarcity of invertebrate food cannot be excluded.[1162,1430]

The indirect effects of a spruce budworm (*Choristoneura* spp.) control campaign on salmonid fish, Atlantic salmon (*Salmo salar*), and brook trout (*Salvelinus fontinalis*) may be considerable. The application of fenitrothion drastically reduces the numbers of stream invertebrates. It has been estimated that at dosage of 280 μg/ha, a reduction of salmon size at the end of the summer of 13 to 15% would be expected in those streams that register an invertebrate biomass reduction in excess of 65%. Longer-term effects on salmon production may be more severe.[1461]

Pollutants have had considerable impact on game through population changes in birds of prey.[1075] There is little information on insect-mediated changes in game populations due to pollution, and there seem to be no estimates on the economic losses. Several agricultural chemicals, such as paraquat[427] and chlordane,[426] have detrimental effects on fish and birds. Consumption of food-chain organisms with excessive selenium concentrations probably caused the impaired reproduction in several aquatic birds at Kesterton, CA. Embryonic deaths and deformities, debilitation and failure of coots to nest, and mortality of adult coots and other birds were observed.[713] Elevated boron loadings also had adverse effects on the growth of waterfowl.[425]

Several nontarget insects, such as Heteroptera, Curculionidae, Chrysomelidae, and Tenthredinidae, feeding on weeds are vital dietary items of variable species.[258,1224,1404] Chiverton and Sotherton[258] found that the exclusion of broadleaved weed herbicides from the outermost 12 m of spring wheat headland resulted in a significant increase in the nonpest insects, especially the Heteroptera, which are important in the diet of insect-eating gamebird chicks. Such differences between herbicide-treated and untreated plots have been found both within a field and in between-field experiments. The increased abundance of the beneficial arthropods has been shown to enhance survival rates of wild game birds chicks both within a season and in the longer term.[258,1256]

C. LOSSES OF CROPS AND TIMBER

In addition to the direct phytotoxic effects of air pollutants, the indirect insect-mediated effects can have a serious impact on crop productivity (Figure 4). The secondary effects may be particularly important, since the pollutant concentrations often do not cause easily detectable effects on plants. Even so, these concentrations may be sufficient to alter the host-insect relationship in such a way that the performance of the insect pest is significantly enhanced.[387] For instance, the tomato pinworm (*Keiferia lycopersicella*) showed greater survival on ozonated plants.[1502] Increased susceptibility of the soybean after sulfur dioxide[728] or ozone[246] exposure to the Mexican bean beetle (*Epilachna varivestis*) has been demonstrated, but the quantity of the loss of yield is more difficult to assess. In a sulfur dioxide fumigation experiment in the field, the

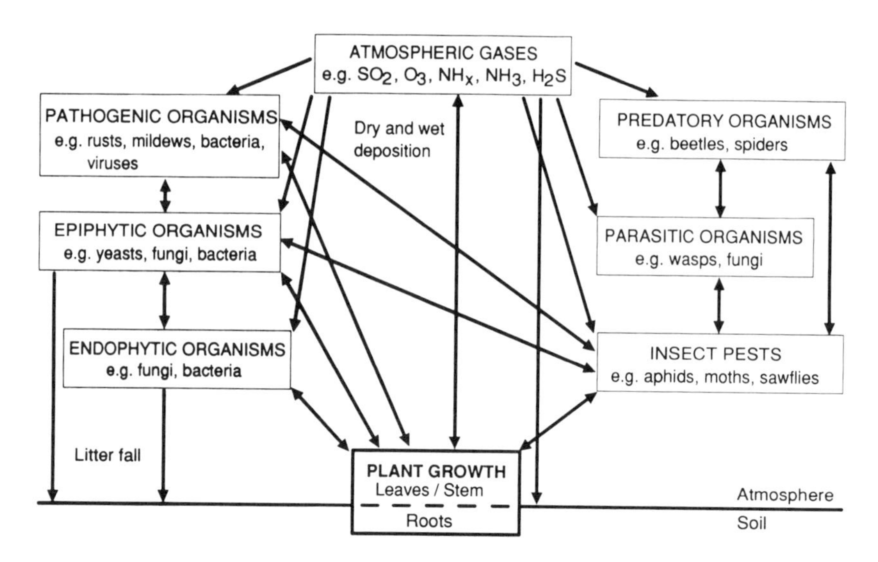

FIGURE 4. Some potential above-ground interactions between air pollutants and a crop. (Redrawn from McLeod, A. R., *Agric. Ecosyst. Environ.*, 38, 11, 1992. With permission.)

proportion of soybean leaf area damaged by the Mexican bean beetle was 45% on control plants and 75% on treated plants.[726] Since the fumigated plants remained smaller than the control plants, the difference in the leaf area consumed by the beetle may have resulted from both the increased number of beetles and the smaller leaf area available. Reduced plant growth may have been due to the sulfur dioxide or to the herbivory or to an interaction between the two.[1599]

Several aphid species damage their hosts not only directly by exploiting the plant nutrients, but also as vectors of diseases, e.g., viruses and mycoplasmas. The growth rate of aphids, e.g., economically important cereal aphids (*Sitobion avenae*, *Metapolophium dirhodum*), increases considerably in polluted air.[28,387,1029,1599] Aphid populations build up much faster, and reach higher numbers, in polluted areas (Figure 5). This is of great economic importance especially in farming and, to some extent, in forests. Dohmen[387] postulated that in polluted areas farmers may be forced to use far more pesticides or may even be prevented from growing certain crops. However, although grain aphids (*Sitobion avenae*) showed enhanced performance in an area polluted by sulfur compounds, the damage to spring wheat was considered to be rather slight.[1242]

Warrington et al.[1561] experimented with the pea aphid (*Acyrthosiphon pisum*) on pea plants (*Pisum sativum*) with and without sulfur dioxide fumigation. After 63 days of exposure, the aphid population was 1.8 times greater on plants in ambient air with sulfur dioxide than on plants in filtered air. Leaf dry weight was reduced 29.4% by sulfur dioxide alone, as mere aphids caused significant loss of total shoot weight by 21.4% and of pea yield by 42.4%. In plants exposed both to aphids and sulfur dioxide, shoot weight decreased further by 12.8%, and total plant weight lessened by 10.4%.

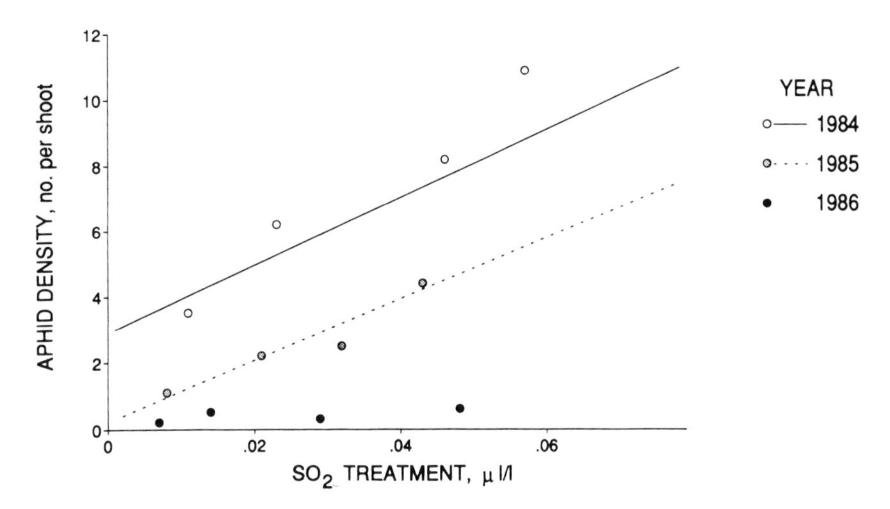

FIGURE 5. Density of the grain aphid (*Sitobion avenae*, Homoptera, Aphididae) on winter wheat in 1984 and 1985 and on winter barley in 1986. Aphid density (number of aphids per 50 shoots at the peak of infestation) was positively correlated ($p < 0.01$) with SO_2 concentration (µl/l). (From Aminu-Kano, M., McNeill, S., and Hails, R. S., *Agric. Ecosyst. Environ.*, 33, 233, 1991. With permission.)

Effects on the host plant may be a direct loss of yield to herbivores or further damage through increased gas uptake via insect-damaged tissues.[1560] Even when insect growth is not increased, the consequences to the plant may still be serious because chewing insects may compensate for the decreased nutrient quality of the plant by consuming more.[1599]

Causal relationships between air pollutants and increased forest insect populations have remained obscure, especially in relation to the effects of silvicultural practices (Figure 6). Despite the high natural variation in insect population size, there are examples of increased pest potential in polluted areas. For instance, negative correlation was observed between the numbers of *Cinara pilicornis* during its population peak, and the final length of the current year's leader shoot of Norway spruce (*Picea abies*) seedlings in the field experiments in the vicinity of a pulp mill.[701] Leader shoot growth of spruce seedlings infested by aphids was positively correlated with the distance from the pollutant source.

Insects can enhance pollutant-induced damage of their host trees. For example, Kidd[840] found that the number of trees (*Pinus sylvestris*) showing symptoms of chlorotic mottling did not increase in treatment with aphids (*Eulachnus agilis*) alone, compared to controls, whereas treatment with acid solution resulted in mottling of 50% of trees, and treatment with both acid and aphids resulted in mottling of all trees (Figure 7). However, the proportion of needles showing damage symptoms did not differ between the aphid treatment with and without aphids.

Ozone-weakened pines in California, Virginia, and Mexico were more susceptible to invasion by bark beetles, but since the infested trees did not

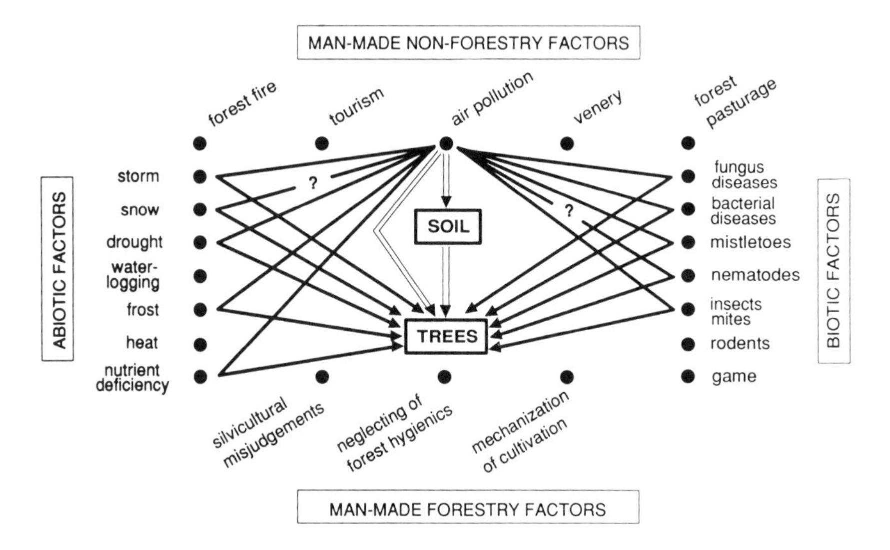

FIGURE 6. Direct and indirect harmful influences of air pollutants on forest trees; interactions between air pollution and natural (abiotic, biotic) stress factors in a forest ecosystem. (Redrawn from Führer, E., *Forest Ecol. Manage.*, 37, 249, 1990. With permission.)

permit good brood production, there was no noticeable increase in the number of infested trees.[285,286,368,1057,1058,1382] The sap-sucking bug *Aradus cinnamomeus* (Heteroptera) of pines showed high population densities around a smelter in Finland, which resulted in poor height growth of pines.[652] However, it is difficult to separate the direct effects of pollutants on pines from the indirect ones through the bugs.

Although the following examples have not been shown to cause forest economic losses, they illustrate the risks involved in insect-pollution relations to timber production. *Tomicus piniperda* (Scolytidae), an expensive shoot-damaging pest of pines, has been shown to abound near industrial sources,[44,663] but the potential loss in timber or forest growth has not been estimated. Field investigations have shown that some bark beetles tolerate high levels of metal pollution. In a heavily polluted area in western Finland, the attack density of *Tomicus piniperda* was very high, often exceeding 300 attacks per square meter, and on average, 1130 new beetles were produced per square meter.[663] Maximum production in *T. piniperda* is reached at rather high densities. Although *T. piniperda* does not usually reproduce on standing vigorous trees, the excess part of the population turns to shoot feeding when the attack shutoff density is reached. Consequently, pollution-damaged forests are a serious risk of *T. piniperda* attack to surrounding forest areas.

It has been proposed that the frequency, extent, and severity of insect outbreaks have increased during the last 200 years.[77] Experimental studies have shown that simulated acid precipitation may decrease egg mortality of the European pine sawfly (*Neodiprion sertifer*).[668] Decreased egg mortality in more acidic conditions inevitably leads to larger larval colonies. In this

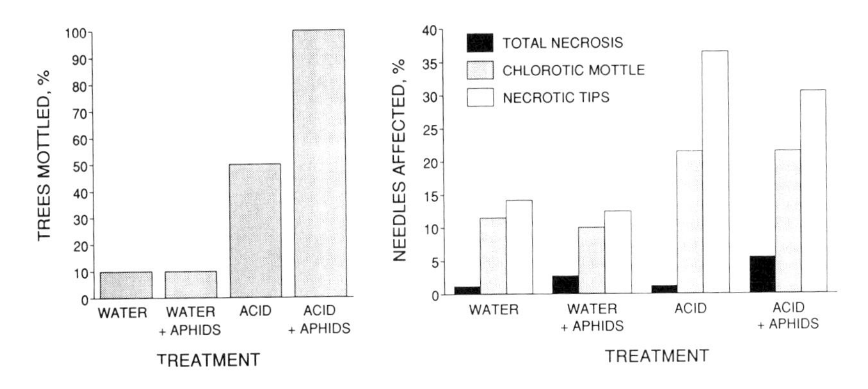

FIGURE 7. The proportion of trees (*Pinus sylvestris*) and needles showing damage symptoms in each of the four treatments using simulated acid mist and pine aphids (*Eulachnus agilis*, Homoptera, Lachnidae). (From Kidd, N. A. C., *J. Appl. Entomol.*, 110, 524, 1990. With permission.)

gregarious species, larger colonies have higher developmental rates and lower larval mortality. Thus, acidic precipitation affecting egg survival may be an additional factor causing outbreaks of pine sawflies.

D. MEDICAL AND VETERINARY PROBLEMS

Organic waste has increased with increasing human population. Some insects that are favored by such eutrophication carry diseases of man and domestic animals. The biting midge, *Culicoides variipennis* (Ceratopogonidae), is the major vector of bluetongue virus in the U.S.[797] and has been reported in greatest numbers in manure-polluted water sources.[852,1167,1342] Schmidtmann et al.[1342] attempted to associate pollution-related factors quantitatively with *C. variipennis* in dairies in New York State. The degree of animal access to a potential breeding area was the most important single predictor of larval occurrence. However, concentrated manure storage systems were mostly negative for *C. variipennis* larvae, and the percentage of organic matter in a sample was actually related negatively to the abundance of larvae.

Mullens and Rodriguez[1095] constructed wastewater ponds on a dairy in southern California to document colonization by *C. variipennis* and to investigate whether adding different amounts of manure affected the density of this species. Manure loading during the summer increased the numbers of immatures and adults in the most-polluted ponds. Chemical oxygen demand (COD) was used to quantify organic pollution. Relative to levels before loading, larval densities in the most-polluted ponds (>0.4 kg COD per cubic meter per week, with average COD levels >516 mg/l) were increased from 21- to 89-fold, and adult emergence was increased from 6- to 61-fold. Increases in the numbers of larvae or emerging adults in cleaner ponds (<0.2 kg COD per cubic meter per week, with average COD levels <306 mg/l) were less than seven times as high as numbers in the interval before loading. Adults emerging from polluted ponds were significantly larger than those emerging from cleaner ponds. Optimum levels for *C. variipennis* were 500 to 1,500 mg COD per liter. Levels below 300 to 400 mg COD per liter limited habitat suitability, whereas levels in excess of 2,000 mg COD per liter appeared to be suboptimal, but not strictly limiting. Mullens and Rodriguez[1095] concluded that controlling the level of manure loading may be promising in the management of this vector.

Pesticide applications may have adverse side effects on the natural enemies of disease vectors. For example, the effects of pesticides on Trichoptera deserve attention, as they belong to the main predators of black fly (Simuliidae) larvae in onchocerciasis infested regions.

Chironomids are often a serious nuisance to people living near eutrophicated lakes or rivers. In Japan, examples of species that have caused problems when occurring in enormous masses include *Chironomus plumosus*, *C. yoshimatsui*, and *Tokunagayusurika akamusi*.[1328] More importantly, chironomids are a causative agent of bronchial asthma near lakes, rivers, sewage ditches, and rice paddies. Bronchial asthma is a serious allergic disease caused by the inhalation of dust containing allergens. It was demonstrated that some 38% of the asthmatic patients in the city of Tokyo gave positive reactions against the chironomid midge antigens. It was also confirmed that chironomids constitute an important allergen, causing various diseases in hypersensitive patients. An early occurrence of bronchial asthma was reported from Sudan, where it was caused by *Cladotanytarsus lewisi*. A kind of hemoglobin similar in structure to that of man was identified as the main allergen. Japanese studies have demonstrated positive reactions in hypersensitivity tests with *Polypedilum kyotoense*, *Chironomus yoshimatsui*, *Tokunagayusurika akamusi*, and the cosmopolitan giant lake midge *Chironomus plumosus*.[1328] Thus, eutrophication can cause medical problems even when insects are not vectors of disease, but the problem arises from the increase of insect numbers per se.

IV. CONCLUSIONS

Insects have been studied in relation to pollution in such a haphazard fashion that there is still only a limited possibility for generalizations, although premature generalizations are by no means uncommon in the literature. We do know bits and pieces about a number of things, but do we understand anything? For example, most environments are patchy. Hassell et al.[629] recently concluded that local movement in a patchy environment can help otherwise unstable host and parasitoid populations to persist together, but that the deterministically generated spatial patterns in population density can be exceedingly complex and sometimes indistinguishable from random environmental fluctuations. By adding various combinations of pollutants to such systems, the result may be chaotic and difficult, or perhaps impossible, to predict.

Pollutants are not the only way mankind affects wildlife. Indeed, it is commonly held that direct destruction of habitat, be it from urbanization and road construction, agriculture, forestry, the control and supply of water, recreation, or other activities, is the main human impact on nature. As stated by Moriarty,[1075] pollutants are but one small part of the total impact man has on wildlife. Furthermore, the direct effects of pollution on agroforestry are more significant than those via insects on a global scale, although severe local problems have been shown to occur.

The concept of sustainable development stresses that the needs of the present generations must be met without destroying the possibilities of the future generations to meet their own needs.[510] *Our Common Future*, commonly known as the Brundtland Report, insists that we attend to the economic needs of the world's people as well as to the quality of the environment. Our current ability to predict ecological effects of pollutants is rudimentary when we take into account the huge diversity of insects and the complex structure of ecosystems. There is no good reason to suppose that there is a constant relationship, for different pollutants or different species, between the dose needed to kill and that needed to impair an organism.[1075]

In environmental policy the concept of critical load has received much attention. However, the concept has serious defects, since species are different, and moreover, communities are complex and diverse assemblages of species. Thus, even if some species were studied and monitored in this context, it is impossible to make predictions about the effects of pollutants and their different doses in nature.[1075] In this context it is also necessary to emphasize that the role of chance in nature may be more important than expected. The result of relatively small pollution-induced changes in ecosystems may result in appreciable secondary changes, possibly appearing only after a time lag.

Environmental pollution is basically due to two uncontrolled processes governed by economic forces. First, the technological developments even in the industrialized countries associated with continuous growth have not been able to solve the problems associated with industrial discharges and environmental risks. Second, the population explosion, especially in developing countries, increases pressures for urbanization, industrialization, and more intensive agriculture with the extensive use of pesticides and fertilizers, all of which cause severe pollution problems. Social questions and the global economic discrepancy between the industrial and developing countries make the solution difficult and indicate that pollution problems will also become more severe, especially in the poorer parts of the world. Pollution is by no means an isolated phenomenon, but is closely associated with these alterations, all contributing to the global change.

Insects play an essential role in many ecological processes, due to their abundance and diverse ecological niches. Although the direct adverse effects of pollutants on insects may be scientifically interesting or sometimes even economically considerable, the indirect effects on the ecosystems mediated through insects are obviously much more profound. Insects are often near the bottom of the food chains and may be important agents of pollutant entry into food chains.[1117,1414] The changes in herbivorous insect populations affect the succession and productivity of vegetation. Many predatory animals, including economically important vertebrates, are dependent on insects. The soil fauna affects the structure and quality of the soil. Pollution tends to induce sudden and unpredictable changes in insect pests, causing considerable damage to agriculture and forestry. Most environmental studies that claim to consider ecosystems hardly mention insects. However, since most pollutant-induced changes in ecosystems are likely to be indirect, studies omitting insects must be regarded inadequate as regards real ecosystem impacts.

REFERENCES

1. **Abel, P. D.,** Pollutant toxicity to aquatic animals — methods of study and their applications, *Rev. Environ. Health,* 8, 119, 1989.
2. **Abrahamsen, G.,** Effects of lime and artificial acid rain on the enchytraeid (Oligochaeta) fauna in coniferous forest, *Holarctic Ecol.,* 6, 247, 1983.
3. **Abrahamsen, G., Hovland, H., and Hågvar, S.,** Effects of artificial acid rain and liming on soil organisms and the decomposition of organic matter, in *Effects of Acid Precipitation on Terrestrial Ecosystems,* Hutchinson, T. C. and Havas, M., Eds., Plenum Press, New York, 1980, 341.
4. **Achik, J., Schiavon, M., and Houpert, G.,** Persistence of biological activity of four insecticides in two soil types under field and laboratory conditions, *J. Econ. Entomol.,* 82, 1572, 1989.
5. **Aebischer, N. J.,** Assessing pest effects on non-target invertebrates using long-term monitoring and time series modelling, *Funct. Ecol.,* 4, 369, 1989.
6. **Agnihothrudu, V. and Mithyantha, M. S.,** *Pesticide Residues and a Review of the Indian Work,* Bangalore Rallis India Ltd., 1978.
7. **Ahmad, I., Siddiqui, M. K. J., and Ray, P. K.,** Pesticide burden on some insects of economic importance in Lucknow (India), *Environ. Monit. Assessment,* 9, 25, 1987.
8. **Ahmad, S. and Forgash, A. J.,** Toxicity of carbanyl to gypsy moth larvae parasitized by *Apanteles melanoscelus, Environ. Entomol.,* 5, 1183, 1976.
9. **Åhman, I.,** Oviposition and larval performance of *Rhabdophaga terminalis* on *Salix* spp. with special consideration to bud size of host plants, *Entomol. Exp. Appl.,* 35, 129, 1984.
10. **Akey, D. H. and Kimball, B. A.,** Growth and development of the beet armyworm on cotton grown in an enriched carbon dioxide atmosphere, *Southwest. Entomol.,* 14, 255, 1989.
11. **Akey, D. H., Potter, H. W., and Jones, R. H.,** Effects of rearing temperature and larval density on longevity, size, and fecundity in the biting gnat *Culicoides variipennis, Ann. Entomol. Soc. Am.,* 71, 411, 1978.

11a. **Al-Houty, W.,** Gulf War impact on desert insects of Kuwait, Proc. XIX Int. Congr. Entomol., Abstracts, Beijing, China, 1992, 181.

12. **Ali, A. and Kok-Yokomi, M. L.,** Field studies on the impact of a new benzoylphenylurea insect growth regulator (UC-84572) on selected aquatic nontarget invertebrates, *Bull. Environ. Contam. Toxicol.,* 42, 134, 1989.
13. **Ali, A. and Lord, J.,** Experimental insect growth regulators against some nuisance chironomid midges of central Florida, *J. Econ. Entomol.,* 73, 243, 1980.
14. **Ali, A. and Lord, J.,** Impact of insect growth regulators on some nontarget aquatic invertebrates, *Mosq. News,* 40, 564, 1980.
15. **Ali, A. and Mulla, M. S.,** The IGR diflubenzuron and organophosphorous insecticides against nuisance midges in man-made residential-recreational lakes, *J. Econ. Entomol.,* 70, 571, 1977.
16. **Ali, A. and Mulla, M. S.,** Effects of chironomids larvicides and diflubenzuron on nontarget invertebrates in residental-recreational lakes, *Environ. Entomol.,* 7, 21, 1978.
17. **Ali, A. and Nayar, J. K.,** Laboratory toxicity of a new benzoylphenylurea IGR (UC-85472) against mosquitoes and chironomid midges, *J. Am. Mosq. Control Assoc.,* 3, 309, 1987.
18. **Ali, A. and Stanley, B. H.,** Effects of a new insect growth regulator, UC-62644, on target Chironomidae and some nontarget aquatic invertebrates, *Mosq. News,* 41, 692, 1981.
19. **Ali, A., Nigg, H. N., Stamper, J. H., Kok-Yokomi, M. L., and Weaver, M.,** Diflubenzuron application to citrus and its impact on invertebrates in an adjacent pond, *Bull. Environ. Contam. Toxicol.,* 41, 781, 1988.

20. **Allan, J. W. and Burton, T. M.,** Size-dependent sensitivity of three species of stream invertebrates to pH depression, in *Impact of Acid Rain and Deposition on Aquatic Biological Systems,* ASTM STP 928, Isom, B. G., Dennis, S. D., and Bates, J. M., Eds., American Society for Testing and Materials, Philadelphia, 1986, 54.
21. **Allard, M. and Moreau, G.,** Influence d'une acidification expérimentale sur la dérive des invertébrés benthiques, *Verh. Int. Ver. Limnol.,* 22, 1793, 1984.
22. **Allard, M. and Moreau, G.,** Influence of acidification and aluminium on the density and biomass of lotic benthic invertebrates, *Water Air Soil Pollut.,* 30, 673, 1986.
23. **Allard, M. and Moreau, G.,** Effects of experimental acidification on a lotic macroinvertebrate community, *Hydrobiologia,* 144, 37, 1987.
24. **Allsopp, R.,** Control of tsetse flies (Diptera: Glossinidae) using insecticides: a review and future prospects, *Bull. Entomol. Res.,* 74, 1, 1984.
25. **Almer, B., Dickson, W., Ekström, C., and Hornström, E.,** Sulfur pollution and the aquatic ecosystem, in *Sulfur in the Environment,* Part II, Nriagu, J. O., Ed., J. Wiley & Sons, New York, 1978, 271.
26. **Almer, B., Dickson, W., Ekström, C., Hornström, E., and Miller, U.,** Effects of acidification on Swedish lakes, *Ambio,* 3, 30, 1974.
27. **Alstad, D. N., Edmunds, G. F., and Weinstein, L. H.,** Effects of air pollutants on insect populations, *Ann. Rev. Entomol.,* 27, 369, 1982.
28. **Aminu-Kano, M., McNeill, S., and Hails, R. S.,** Pollutant, plant and pest interactions: the grain aphid *Sitobion avenae* (F.), *Agric. Ecosyst. Environ.,* 33, 233, 1991.
29. **Andersen, T. and Bengtson, S.-A.,** Melanism in the peppered moth *Biston betularia* (L.) in western Norway (Lepidoptera: Geometridae), *Entomol. Scand.,* 11, 245, 1980.
30. **Anderson, J. F. and Wojtas, M. A.,** Honey bees (Hymenoptera: Apidae) contaminated with pesticides and polychlorinated biphenyls, *J. Econ. Entomol.,* 79, 1200, 1986.
31. **Anderson, J. M., Ineson, P., and Huish, S. A.,** Nitrogen and cation mobilization by soil fauna feeding on leaf litter and soil organic matter from deciduous woodlands, *Soil Biol. Biochem.,* 15, 463, 1983.
32. **Anderson, R. L.,** Toxicity of fenvalerate and permethrin to several non-target aquatic invertebrates, *Environ. Entomol.,* 11, 1251, 1982.
33. **Anderson, R. L.,** A review of the toxicity of synthetic pyrethroids to aquatic invertebrates, *Environ. Toxicol. Chem.,* 8, 403, 1989.
34. **Anderson, R. L., Walbridge, C. T., and Fiandt, J. T.,** Survival and growth of *Tanytarsus dissimilis* (Chironomidae) exposed to copper, cadmium, zinc and lead, *Arch. Environ. Contam. Toxicol.,* 9, 329, 1978.
35. **Anderson, R. V., Vinikour, W. S., and Brower, J. E.,** The distribution of Cd, Cu, Pb and Zn in the biota of two freshwater sites with different trace metal inputs, *Holarctic Ecol.,* 1, 377, 1978.
36. **Andersson, F.,** Acidic deposition and its effects on the forests of Nordic Europe, *Water Air Soil Pollut.,* 30, 17, 1986.
37. **Andersson, M.,** Toxicity and tolerance of aluminium in vascular plants. A literature review, *Water Air Soil Pollut.,* 39, 439, 1988.
38. **Andow, D. A., Kareiva, P. M., Levin, S. A., and Okubo, A.,** Spread of invading organisms, *Landscape Ecol.,* 4, 177, 1990.
39. **André, H.,** Introduction à l'étude écologique des communantés de microarthropodes corticoles soumises à la pollution atmosphérique. II. Recherche de bioindicateurs et d'indices biologiques de pollution, *Ann. Soc. R. Zool. Belg.,* 106, 211, 1977.
40. **André, H. M.,** Associations between corticolous microarthropod communities and epiphytic cover of bark, *Holarctic Ecol.,* 8, 113, 1985.
41. **Andrews, M., Flower, L. S., Johnstone, D. R., and Turner, C. R.,** Spray droplet assessment and insecticide drift studies during the large scale aerial application of endosulphan to control *Glossina morsitans* in Botswana, *Tropic. Pest Manage.,* 29, 239, 1983.

42. **Andrews, S. M., Cooke, J. A., and Johnson, M. S.,** Distribution of trace element pollutants in a contaminated ecosystem established on metalliferous fluorspar tailings. 3: Fluoride, *Environ. Pollut.*, 60, 165, 1989.
43. **Andrews, S. M., Johnson, M. S., and Cooke, J. A.,** Distribution of trace element pollutants in a contaminated grassland ecosystem established on metalliferous fluorspar tailings. I. Lead, *Environ. Pollut.*, 58, 73, 1989.
44. **Anisimova, O. A.,** Characteristics of the ecology of pine beetles (Hylesinae) and weevils in forest impaired by fluorine wastes from aluminium plants, in *Conifers and Dendrophagous Species* (in Russian), Akademiya Nauk SSSR, Irkutsk, 1978, 56.
45. **Anisimova, O. A. and Sokov, M. K.,** The role of insects in tree stands impaired by toxic wastes, in *The Influence of Anthropogenic and Natural Factors on Conifers* (in Russian), Akademiya Nauk SSR, Irkusk, 1975, 61.
46. **Aoki, Y. and Suzuki, K.T.,** Excretion of cadmium and change in the relative ratio of iso-cadmium-binding proteins during metamorphosis of fleshfly (*Sarcophaga peregrina*) larvae, *Comp. Biochem. Physiol.*, 78C, 315, 1984.
47. **Aoki, Y., Hatakeyama, S., Kobayashi, N., Sumi, Y., Suzuki, T., and Suzuki, K. T.,** Comparison of cadmium-binding protein induction among mayfly larvae of heavy metal resistant (*Baetis thermicus*) and susceptible (*B. yoshinoensis* and *B. sahoensis*) species, *Comp. Biochem. Physiol.*, 93C, 345, 1989.
48. **Aoki, Y., Suzuki, K. T., and Kubota, K.,** Accumulation of cadmium and induction of its binding protein in the digestive tract of fleshfly (*Sarcophaga peregrina*), *Comp. Biochem. Physiol.*, 77C, 279, 1984.
49. **Apperson, C. S., Schaefer, C. H., Cowell, A. E., Werner, G. H., Anderson, N. L., Dupras, E. F., Jr., and Longanecker, D. R.,** Effects of diflubenzuron on *Chaoborus astictopus* and nontarget organisms and persistence of diflubenzuron in lentic habitats, *J. Econ. Entomol.*, 71, 521, 1978.
50. **Arakelian, A. G., Elizarova, L. V., and Akramovskaya, E. G.,** On the state of soil fauna of invertebrates in the zone of pollution of surrounding medium, *Biol. Zh. Arm.*, 41, 462, 1988.
51. **Armitage, P. D.,** The effects of mine drainage and organic enrichment on benthos in the River Nent System, Northern Pennines, *Hydrobiologia*, 74, 119, 1980.
52. **Armitage, P. D., Moss, D., Wright, J. F., and Furse, M. T.,** The performance of a new biological water quality score system based on macroinvertebrates over a wide range of unpolluted running-water sites, *Water Res.*, 17, 333, 1983.
53. **Armstrong, K. F. and Bonner, A. B.,** Investigation of a permethrin-induced antifeedant effect in *Drosophila melanogaster*: An ethological approach, *Pest. Sci.*, 16, 641, 1985.
54. **Arnold, D. E., Bender, P. M., Hale, A. B., and Light, R. W.,** Studies on infertile, acidic Pennsylvania streams and their benthic communities, in *Effects of Acidic Precipitation on Benthos*, Proceedings of a symposium, Singer, R., Ed., North American Benthol. Society, Springfield, IL, 1981, 15.
55. **Arnold, G. A. and Crocker, J.,** *Arctosa perita* from colliery heaps in Warwickshire and Leicestershire, *Bull. Br. Spider Study Group*, 35, 7, 1967.
56. **Arthur, J., Grant, J. C., and Markham, O. D.,** Importance of biota in radionuclide transport at the SL-1 radioactive disposal area, in *Idaho National Engineering Laboratory Radioecology and Ecology Programs 1983 Progress Report*, (DOE/ID-12098, June 1983), U.S. Department of Energy, Idaho Operations Office, 1983, 56.
57. **Atkins, L. E. and Kellum, D.,** Comparative morphogenic and toxicity studies on the effect of pesticides on honeybee brood, *J. Apicult. Res.*, 25, 242, 1986.
58. **Atkins, L. E., Anderson, L. D., Nakakihara, H., and Greywood, E. A.,** Toxicity of pesticides and other agricultural chemicals to honey bees, *Univ. Calif. Agric. Exp. Stn. Agric. Ext. Serv.*, July, 1970.
59. **Atkinson, I.,** Introduced animals and extinctions, in *Conservation for the Twenty-First Century*, Western, D. and Pearl, M. C., Eds., Oxford University Press, New York, 1989, 54.

60. **Atzert, S. P.,** A review of sodium monofluoroacetate (compound 1080), its properties, toxicology, and use in predator and rodent control, *U.S. Fish Wildlife Serv., Special Sci. Rep.,* 146, 1, 1971.
61. **Auclair, J. L., Maltais, J. B., and Cartier, J. J.,** Factors in the resistance of peas to the pea aphid, *Acyrthosiphon pisum* (Harr.) (Homoptera: Aphididae). II. Amino acids, *Can. Entomol.,* 89, 457, 1957.
62. **Augustine, M. G., Fisk, F. W., Davidson, R. H., LaPidus, J. B., and Cleary, R. W.,** Host-plant selection by the Mexican bean beetle, *Epilachna varivestis, Ann. Entomol. Soc. Am.,* 57, 127, 1964.
63. **Augustine, S. and Diwan, A. P.,** Population dynamics of midge worms with relation to pollution in river Kshipra, *J. Environ. Biol.,* 12, 255, 1991.
64. **Augustine Vattakeril, S. and Diwan, A. P.,** Community structure of benthic macroinvertebrates and their utility as indicators of pollution in river Kshipra, India, *Pollut. Res.,* 10, 1, 1991.
65. **Ausmus, B. S., Dodson, G. J., and Jackson, D. R.,** Behavior of heavy metals in forest microcosms. III. Effects of litter-soil carbon metabolism, *Water Air Soil Pollut.,* 10, 19, 1978.
66. **Austarå, Ø., and Midtgaard, F.,** A preliminary experiment on the fecundity of *Neodiprion sertifer* reared on *Pinus sylvestris* grown in forest soil of various acidity. *Scand. J. For. Res.,* 2, 365, 1987.
67. **Back, C., Boisvert, J., Lacoursiere, J. O., and Charpentier, G.,** High-dosage treatment of a Quebec stream with *Bacillus thuringiensis* serovar. *israelensis*: efficacy against black fly larvae (Diptera: Simuliidae) and impact on non-target insects, *Can. Entomol.,* 117, 1523, 1985.
68. **Bacq, Z. M. and Alexander, P.,** *Fundamentals of Radiobiology,* Pergamon Press, London, 1966.
69. **Bährmann, R.,** Untersuchungen der Dipterenfauna in natur- und industrienahen Rasenbiotopen Thüringens (DDR) mittels Bodenfallen (Diptera Brachycera), *Dtsch. Entomol. Z. N. F.,* 34, 85, 1987.
70. **Bährmann, R.,** Über den Einfluss von Luftverunreinigungen auf Ökosysteme XIV. Ökofaunistische Untersuchungen an Zweiflüglern (Diptera, Brachycera) industrienaher *Agropyron-* und *Puccinellia*-Rasen bei Jena/Thüringen, *Zool. Jb. Syst.,* 115, 49, 1988.
71. **Bährmann, R. and Weipert, J.,** Die Chloropidenfauna (Diptera, Chloropidae) immissionsgeschädigter Rasenbiotope im Saaletal bei Jena (Thür.) XV. Beitrag über den Einfluss von Luftverunreinigungen auf Ökosysteme, *Beitr. Entomol. Berlin,* 39, 279, 1989.
72. **Bai, A. R. K. and Reddy, C. C.,** Inhibition of acetylcholinesterase as a criterion to determine the degree of insecticide poisoning in *Apis cerana indica, J. Apicult. Res.,* 16, 112, 1977.
73. **Baker, J. P. and Schofield, C. L.,** Aluminium toxicity to fish in acidic waters, *Water Air Soil Pollut.,* 18, 289, 1982.
74. **Baker, J. P., Bernard, D. P., Christensen, S. W., Sale, M. J., Freda, J., Heltcher, K., Marmorek, D., Rowe, L., Scanlon, P., Suter, G., Warren-Hicks, W., and Welbourn, P.,** Biological effects of changes in surface water acid-base chemistry, NAPAP Rep. 13, in *National Acid Precipitation Assessment Program, Acidic Deposition: State of Science and Technology,* Vol. II, Government Printing Office, Washington, D.C., 1990.
75. **Baldry, D.A.T., Everts, J., Roman, B., Boon von Ochssee, G.A., and Laveissiere, C.,** The experimental application of insecticides from a helicopter for the control of riverine populations of *Glossina tachinoides* in West Africa. VII. The effects of two spray-applications of OMS-570 (endosulfan) and OMS-1998 (decamethrin) on *G. tachinoides* and non-target organisms in Upper Volta, *Tropic. Pest Manage.,* 27, 83, 1981.
76. **Ballan-Dufrancais, C., Jeantet, A. Y., and Martoja, M. R.,** Composition ionique et signification physiologique des accumulations minérales de l'intestin moyen des insectes, *C. R. Acad. Sci. (Paris),* 273, 173, 1971.

77. **Baltensweiler, W.,** "Waldsterben": forest pests and air pollution, *Z. Angew. Entomol.,* 99, 77, 1985.
78. **Barak, N. A.-E. and Mason, C. F.,** Heavy metals in water, sediment and invertebrates from rivers in eastern England, *Chemosphere,* 19, 1709, 1989.
79. **Barannik, A. P.,** Ecological faunistic characteristics of dendrophilous entomofauna of green plantings in industrial cities of Kemerovo province, *Soviet J. Ecol.,* 10, 56, 1979.
80. **Barbosa, P. and Saunders, J. A.,** Plant allelochemicals: linkage between herbivores and their natural enemies, *Rec. Adv. Phytochem.,* 19, 107, 1985.
81. **Barker, R. J., Lehner, Y., and Kunzmann, M. R.,** Pesticides and honey bees: nectar and pollen contamination in alfalfa treated with dimethoate, *Arch. Environ. Contam. Toxicol.,* 9, 125, 1980.
82. **Barnaby, F.,** The environmental impact of the Gulf War, *Ecologist,* 21, 166, 1991.
83. **Barnes, R. L.,** Effects of chronic exposure to ozone on soluble sugar and ascorbic acid contents of pine seedlings, *Can. J. Bot.,* 50, 215, 1972.
84. **Barth, A.,** Untersuchungen über Rückstandswirkungen des Häutungshemmstoffes Dimilin im Waldboden mittels eines neuen Biotestverfahrens, *Anz. Schädlingskunde, Pflanzenschutz, Umweltschutz,* 54, 164, 1981.
85. **Barton, B. A.,** Short-term effects of highway construction on the limnology of a small stream in southern Ontario, *Freshwater Biol.,* 7, 99, 1977.
86. **Baudin, J. P., Fritsch, A. F., and Georges, J.,** Influence of labelled food type on the accumulation and retention of ^{60}Co by a freshwater fish, *Cyprinus carpio* L., *Water Air Soil Pollut.,* 51, 261, 1990.
87. **Bauer, L. S. and Nordin, G. L.,** Nutritional physiology of the eastern spruce budworm, *Choristoneura fumiferana,* infected with *Nosema fumiferanae,* and interactions with dietary nitrogen, *Oecologia,* 77, 44, 1988.
88. **Bazzaz, F. A.,** The physiological ecology of plant succession, *Ann. Rev. Ecol. Syst.,* 10, 351, 1979.
89. **Bazzaz, F. A.,** The response of natural ecosystems to the rising global CO_2 levels, *Ann. Rev. Ecol. Syst.,* 21, 167, 1990.
90. **Bazzaz, F. A. and Sipe, T. W.,** Physiological ecology disturbance and ecosystem recovery, in *Potentials and Limitations in Ecosystems Analysis,* Schulze, E. D. and Zwölfer, H., Eds., *Ecol. Stud.,* 61, 203, 1987.
91. **Beak Consultants Ltd.,** Study to assess the distribution of radionuclides in the aquatic ecosystem in the vicinity of the Serpent River mouth, *Rep. Environ. Can.,* 1985.
92. **Beament, J. W. L.,** The ecology of cuticle, in *The Insect Integument,* Hepburn, H. R., Ed., Elsevier, New York, 1976, 359.
93. **Beamish, R. J. and Harvey, H. H.,** Acidification of the La Cloche mountain lakes, Ontario and resulting fish mortalities, *J. Fish. Res. Board Can.,* 29, 1131, 1972.
94. **Beard, R.,** Observations on house flies in high-ozone environments, *Ann. Entomol. Soc. Am.,* 58, 404, 1965.
95. **Becker, H.,** Bodenorganismen — Prüfungskategorien der Forschung, Umweltwissenschaften und Schadstoff-Forschung, *Z. Umweltchem. Ökotoxikol.,* 3, 19, 1991.
96. **Beckerson, D. W. and Hofstra, G.,** Effect of sulphur dioxide and ozone singly or in combination on leaf chlorophyll, RNA, and protein in white bean, *Can. J. Bot.,* 57, 1940, 1979.
97. **Beeby, A.,** Interaction of lead and calcium uptake by the woodlouse *Porcelio scaber* (Isopoda), *Oecologia,* 32, 255, 1978.
98. **Beeman, R. W. and Matsumura, F.,** Basis for hyperexcitation symptom caused by high doses of chlordimeform in the American cockroach, *Comp. Biochem. Physiol.,* 73C, 145, 1982.
99. **Beeton, A. M. and Edmondson, W. T.,** The eutrophication problem, *J. Fish. Res. Board Can.,* 29, 673, 1972.

100. **Bell, H. L.,** Effects of pH on the life cycle of the midge *Tanytarsus dissimilis, Can. Entomol.,* 102, 636, 1970.
101. **Bell, H. L.,** Effect of low pH on the survival and emergence of aquatic insects, *Water Res.,* 5, 313, 1971.
102. **Bell, H. L. and Nebeker, A. V.,** Preliminary studies on the tolerance of aquatic insects to low pH, *J. Kansas Entomol. Soc.,* 42, 230, 1969.
103. **Bendell, B. E.,** Lake acidity and the distribution and abundance of water striders (Hemiptera: Gerridae) near Sudbury, Ontario, *Can. J. Zool.,* 66, 2209, 1988.
104. **Bendell, B. E. and McNicol, D. K.,** Fish predation, lake acidity and the composition of aquatic insect assemblages, *Hydrobiologia,* 150, 193, 1987.
105. **Bengtsson, G.,** The optimal use of life strategies in transitional zones or the optimal use of transition zones to describe life strategies, in *Proc. Third European Congr. Entomology,* Velthuis, H. H. W., Ed., Nederl. Entomol. Ver., 1986, 193.
106. **Bengtsson, G. and Rundgren, S.,** Abundance and species numbers of enchytraeids in a coniferous forest soil near a brass mill, *Pedobiologia,* 24, 211, 1982.
107. **Bengtsson, G. and Rundgren, S.,** Ground-living invertebrates in metal-polluted forest soils, *Ambio,* 13, 29, 1984.
108. **Bengtsson, G. and Rundgren, S.,** The Gusum case: a brass mill and the distribution of soil Collembola, *Can. J. Zool.,* 66, 1518, 1988.
109. **Bengtsson, G. and Tranvik, L.,** Critical metal concentrations for forest soil invertebrates. A review of limitations, *Water Air Soil Pollut.,* 47, 381, 1989.
110. **Bengtsson, G., Berden, M., and Rundgren, S.,** Influence of soil animals and metals on decomposition processes: a microcosm experiment, *J. Environ. Qual.,* 17, 113, 1988.
111. **Bengtsson, B.-E., Elmquist, H., and Nyholm, E.,** Några rön kring apollofjärilen i Sverige samt försök att förklara dess tillbakagång (On the Swedish apollo butterfly with an attempt to explain its decline), *Entomol. Tidskr.,* 110, 31, 1989.
112. **Bengtsson, G., Gunnarsson, T., and Rundgren, S.,** Growth changes caused by metal uptake in a population of *Onychiurus armatus* (Collembola) feeding on metal polluted fungi, *Oikos,* 40, 216, 1983.
113. **Bengtsson, G., Gunnarsson, T., and Rundgren, S.,** Influence of metals on reproduction, mortality and population growth in *Onychiurus armatus* (Collembola), *J. Appl. Ecol.,* 22, 967, 1985.
114. **Bengtsson, G., Hedlund, K., and Rundgren, S.,** Selective odor perception in the soil Collembola *Onychiurus armatus, J. Chem. Ecol.,* 17, 2113, 1991.
115. **Bengtsson, G., Nordström, S., and Rundgren, S.,** Population density and tissue metal concentration of lumbricids in forest soils near a brass mill, *Environ. Pollut.,* 30, 87, 1983.
116. **Bengtsson, G., Ohlsson, L., and Rundgren, S.,** Influence of fungi on growth and survival of *Onychiurus armatus* (Collembola) in a metal polluted soil, *Oecologia,* 68, 63, 1985.
117. **Benham, B. R., Londsdale, D., and Muggleton, J.,** Is polymorphism in two-spot ladybird an example of non-industrial melanism? *Nature,* 249, 179, 1974.
118. **Benke, A. C., Willeke, G. E., Parrish, F. K., and Stites, D. L.,** Effects of urbanization on stream ecosystems, Completion Report Project Number A-055-GA, Office of Water Research and Technology, U.S. Department of the Interior, 1981.
119. **Bennett, W. H.,** The effect of needle structure upon the susceptibility of hosts to the pine needle miner (*Exoteleia pinifoliella* (Chamb.)) (Lepidoptera: Gelechidae), *Can. Entomol.,* 86, 49, 1954.
120. **Benton, M. J. and Guttman, S. I.,** Relationship of allozyme genotype to survivorship of mayflies (*Stenonema femoratum*) exposed to copper, *J. N. Am. Benthol. Soc.,* 9, 271, 1990.
121. **Benz, G.,** Effect of gamma rays to prolong the life of male larch budmoth, *Zeiraphera diniana* (Lepidoptera: Tortricidae), *Experientia,* 26, 1252, 1970.
122. **Berge, H.,** Beziehungen zwischen Baumschädlingen und Immissionen, *Anz. Schädlingskunde Pflanzen-Umweltschutz,* XLVI, 155, 1973.

123. **Berglind, R.,** Combined and separate effects of cadmium, lead and zinc on ALA-D activity, growth and hemoglobin content in *Daphnia magna*, *Environ. Toxicol. Chem.*, 5, 989, 1986.
124. **Berrigan, D.,** The allometry of egg size and number in insects, *Oikos,* 60, 313, 1991.
125. **Berry, R. J.,** Industrial melanism and peppered moths (*Biston betularia* (L.)), *Biol. J. Linnean Soc.,* 39, 301, 1990.
126. **Berryman, A. A.,** *Population Systems: A General Introduction,* Plenum Press, New York, 1981.
127. **Berryman, A. A.,** Biological control, thresholds, and part outbreaks, *Environ. Entomol.,* 11, 544, 1982.
128. **Besser, J. M. and Rabeni, C. F.,** Bioavailability and toxicity of metals leached from lead-mine tailings to aquatic invertebrates, *Environ. Toxicol. Chem.,* 6, 879, 1987.
129. **Beyer, W. N. and Anderson, A.,** Toxicity to woodlice of zinc and lead oxides added to soil litter, *Ambio,* 14, 173, 1985.
130. **Beyer, W. N. and Kaiser, T. E.,** Organochlorine pesticide residues in moths from the Baltimore, M.D.-Washington, D.C. area, *Pest. Monit. J.,* 4, 129, 1984.
131. **Beyer, W. N. and Moore, J.,** Lead residues in eastern tent caterpillars (*Malacosoma americanum*) and their host plant (*Prunus serotina*) close to a major highway, *Environ. Entomol.,* 9, 10, 1980.
132. **Beyer, W. N., Miller, G. W., and Fleming, W. J.,** Populations of trap-nesting wasps near a major source of fluoride emissions in western Tennessee, *Proc. Entomol. Soc. Wash.,* 89, 478, 1987.
133. **Bian, Y.,** Effects of fluoride pollution on mulberry-silkworm production, *Environ. Condit China,* 3, 5, 1985.
134. **Bian, Y. and Wang, J.,** Characteristics of fluoride accumulation in mulberry leaves and pollution forecasting, *J. Environ. Sci. China,* 8, 71, 1987.
135. **Biesiadka, E. and Szczepaniak, U.,** Investigation on water bugs (Heteroptera) in strongly polluted Lake Suskie (northern Poland), *Acta Hydrobiol.,* 29, 453, 1987.
136. **Billings, W. D., Peterson, K. M., Shaver, G. R., and Trent, A. W.,** Root growth, respiration, and carbon dioxide evolution in an arctic tundra soil, *Arctic Alpine Res.,* 9, 129, 1977.
137. **Bink, F.A.,** Acid stress in *Rumex hydrolapathum* (Polygonaceae) and its influence on the phytophage *Lycaena dispar* (Lepidoptera; Lycaenidae), *Oecologia* 70, 447, 1986.
138. **Binns, W. O. and Redfern, D. B.,** Acid rain and the forest decline in West Germany, *For. Comm. Res. Developm. Paper,* 131, 1983.
139. **Birchall, J. D., Exley, C., Chappell, J. S., and Phillips, M. J.,** Acute toxicity of aluminium to fish eliminated in silicon-rich acid waters, *Nature,* 338, 146, 1989.
140. **Bird, S. C., Reynolds, N., and Henderson, R.,** Monitoring effects of a storm sewer overflow upon the Nant Ffrwd, South Wales, *Water Sci. Technol,* 21, 1785, 1989.
141. **Bisessar, S.,** Effect of heavy metals on microorganisms in soils near a secondary lead smelter, *Water Air Soil Pollut.,* 17, 305, 1982.
142. **Bishop, J. A.,** An experimental study of the cline of industrial melanism in *Biston betularia* (L.) (Lepidoptera) between urban Liverpool and rural North Wales, *J. Anim. Ecol.,* 41, 209, 1972.
143. **Bishop, J. A. and Cook, L. M.,** Industrial melanism and the urban environment, *Adv. Ecol. Res.,* 11, 373, 1980.
144. **Bishop, J. A., Cook, L. M., and Muggleton, J.,** The response of two species of moths to industrialisation in northwest England. II. Relative fitness of morphs and population size, *Philos. Trans. R Soc. Lond., [Biol.],* 181, 517, 1978.
145. **Bishop, J. A., Cook, L. M., Muggleton, J., and Seaward, M. R. D.,** Moths, lichens and air pollution along a transect from Manchester to North Wales, *J. Appl. Ecol.,* 12, 83, 1975.

146. **Blackwell, A. and Cox, S.,** The disruption of proleg clasping by sublethal doses of chlordimefom in silk moth larvae: a possible mechanism of crop protection, *Entomol. Exp. Appl.,* 42, 9, 1986.
147. **Blais, J. R.,** Effects of aerial application of chemical insecticides on spruce budworm parasites, *Bi-monthly Res. Notes,* 33, 41, 1977.
148. **Blumer, M. and Sass, J.,** Oil pollution: persistence and degradation of spilled fuel oil, *Science,* 176, 1120, 1972.
149. **Blumer, M., Sanders, H. L., Grassle, J. F., and Hampson, G. R.,** An ocean of oil, *Environment,* 13, 2, 1971.
150. **Bogdanova, D. A.,** Pine trunk pests in a zone of industrial pollution (in Russian), *Ekologija,* 1987, 87, 1987.
151. **Boggs, C. L.,** Nutritional and life-history determinants of resource allocation in holometabolous insects, *Am. Nat.,* 117, 692, 1981.
152. **Boiteau, G., King, R. R., and Levesque, D.,** Lethal and sublethal effects of aldicarb on two potato aphids (Homoptera: Aphididae): *Myzus persicae* (Sulzer) and *Macrosiphum euphorbiae* (Thomas), *J. Econ. Entomol.,* 78, 41, 1985.
153. **Bolin, B.,** How much CO_2 will remain in the atmosphere? In *The Greenhouse Effect, Climatic Change and Ecosystems,* Bolin, B., Doos, B. R. O., Jager, J., and Warrick, R. A., Eds., *Scope* 29, J. Wiley & Sons, Chichester, 1986, 93.
154. **Bolsinger, M. and Flückiger, W.,** Effect of air pollution at a motorway on the infestation of *Viburnum opulus* by *Aphis fabae*, *Eur. J. For. Pathol.,* 14, 256, 1984.
155. **Bolsinger, M. and Flückiger, W.,** Enhanced aphid infestation at motorways: the role of ambient air pollution, *Entomol. Exp. Appl.,* 45, 237, 1987.
156. **Bolsinger, M. and Flückiger, W.,** Ambient air pollution induced changes in amino acid pattern of phloem sap in host plants — relevance to aphid infestation, *Environ. Pollut.,* 56, 209, 1989.
157. **Bolsinger, M., Lier, M. E., Lansky, D. M., and Hughes, P. R.,** Influence of ozone air pollution on plant-herbivore interactions. Part 1: biochemical changes in ornamental milkweed (*Asclepias curassavica* L.; Asclepidiaceae) induced by ozone, *Environ. Pollut.,* 72, 69, 1991.
157a. **Bolsinger, M., Lier, M. E., and Hughes, P. R.,** Influence of ozone air pollution on plant-herbivore interactions. II. Effects of ozone on feeding preference, growth and consumption rates of monarch butterflies (*Danus plexippus*), *Environ. Pollut.,* 77, 31, 1992.
158. **Bombick, D. W., Arlian, L. G., and Livingston, J. M.,** Toxicity of jet fuels to several terrestrial insects, *Arch. Environ. Contam. Toxicol.,* 16, 111, 1987.
159. **Bonquegneau, J. M., Ballan-Dufrancais, C., and Jeantet, A. Y.,** Storage of Hg in the ileum of *Blattella germanica:* biochemical characterization of metallothionein, *Comp. Biochem. Physiol.,* 80C, 95, 1985.
160. **Bordeau, P. and Treshow, M.,** Ecosystems response to pollution, in *Principles of Ecotoxicology,* Butler, G. C., Ed., Wiley, Chichester, 1978, 313.
161. **Boreham, S. and Birch, P.,** Changes in the macro-invertebrate benthos of a rural Essex clay stream folowing pollution by the pesticide Dursban, *London Nat.,* 69, 79, 1990.
162. **Boreham S., Hough R. A., and Birch P.,** The effects of livestock slurry pollution on the benthic macro-invertebrate fauna of a clay stream, *London Nat.,* 68, 77, 1989.
163. **van Den Bosch, R.,** *The Pesticide Conspiracy,* Doubleday, New York, 1978.
164. **Bösener, R.,** Occurrence of bark-breeding forest pests in fume-damaged pine and spruce stands, *Arch. Forstwes.,* 18, 1021, 1969.
165. **Boullard, B.,** Interactions entre les polluants atmosphériques et certains parasites des essences forestières (champignons et insectes), *La Forêt Privée Francaise,* 94, 31, 1973.
166. **Boyden, C. R.,** Effect of size upon metal content of shellfish, *J. Mar. Biol. Assoc. U. K.,* 57, 675, 1977.
167. **Boyle, E. A.,** Copper in natural waters, in *Copper in the Environment, Part I: Ecological Cycling,* Nriagu, J. O., Ed., John Wiley & Sons, New York, 1979, 77.

168. **Bracher, G. A. and Bider, J. R.,** Changes in terrestrial animal activity of a forest community after an application of aminocarb (Matacil®), *Can. J. Zool.,* 60, 1981, 1982.
169. **Bradbury, S. P. and Coats, J. R.,** Comparative toxicology of the pyrethroid insecticides, *Rev. Environ. Toxicol. Contam.,* 108, 143, 1989.
170. **Bradt, P. T. and Berg, M. B.,** Macrozoobenthos of three Pennsylvania lakes: responses to acidification, *Hydrobiologia,* 150, 63, 1987.
171. **Brakefield, P. M.,** Industrial melanism: do we have the answers, *Trends Ecol. Evol.,* 2, 117, 1987.
172. **Brakefield, P. M.,** A decline of melanism in the peppered moth *Biston betularia* in the Netherlands, *Biol. J. Linnean Soc.,* 39, 327, 1990.
173. **Brakefield, P. M. and Lees, D. R.,** Melanism in *Adalia* ladybirds and declining air pollution in Birmingham, *Heredity,* 59, 273, 1987.
174. **Brattsten, L. B.,** Insecticide resistance: research and management, *Pest. Sci.,* 26, 329, 1989.
175. **Braun, S. and Flückiger, W.,** Increased population of the aphid *Aphis pomi* at a motorway. Part 1: field evaluation, *Environ. Pollut. Ser. A,* 33, 107, 1984.
176. **Braun, S. and Flückiger, W.,** Increased population of the aphid *Aphis pomi* at a motorway. Part III: the effect of exhaust gases, *Environ. Pollut. Ser. A,* 39, 183, 1985.
177. **Braun, S. and Flückiger, W.,** Effect of ambient ozone and acid mist on aphid development, *Environ. Pollut.,* 56, 177, 1989.
178. **Brazner, J. C. and Kline, E. R.,** Effects of chlorpyrifos on the diet and growth of larval fathead minnows, *Pimephales promelas,* in littoral enclosures, *Can. J. Fish. Aquat. Sci.,* 47, 1157, 1990.
179. **Brian, M. V.,** *Social Insect Populations,* Academic Press, London, 1965.
180. **Briffa, K. R., Bartholin, T. S., Eckstein, D., Jones, P. D., Karlen, W., Schweingruber, F. H., and Zetterberg, P.,** A 1,400-year tree-ring record of summer temperatures in Fennoscandia, *Nature,* 346, 434, 1990.
181. **Brittain, J. E.,** Biology of mayflies, *Ann. Rev. Entomol.,* 27, 119, 1982.
182. **Brittain, J. E. and Eckeland, T. J.,** Invertebrate drift — a review, *Hydrobiologia,* 166, 77, 1988.
183. **Bromenshenk, J. J. and Gordon, C. C.,** Terrestrial insects sense air pollutants, in *Proc. 4th Joint. Conf. Sens. Environ. Pollutants,* American Chemical Society, Washington, D.C., 1978, 66.
184. **Bromenshenk, J. J., Carlson, S. R., Simpson, J. C., and Thomas, J. M.,** Pollution monitoring of Puget Sound with honey bees, *Science,* 227, 632, 1985.
185. **Brooks, G. T.,** Penetration and distribution of insecticides, in *Insecticide Biochemistry and Physiology,* Wilkinson, C. F., Ed., Plenum Press, New York, 1976.
186. **Brouwer, F. and Falkenmark, M.,** Hydrology: water availability changes, in *Toward Ecological Sustainability in Europe,* Solomon, A. M. and Kauppi, L., Eds., IIASA, Laxenburg, Austria, 1990, 67.
187. **Brown, A. F. and Pascoe, D.,** Studies on the acute toxicity of pollutants to freshwater macroinvertebrates. 5. The acute toxicity of cadmium to twelve species of predatory macroinvertebrates, *Arch. Hydrobiol.,* 114, 311, 1988.
188. **Brown, A. W. A.,** *Ecology of Pesticides,* John Wiley & Sons, New York, 1978.
189. **Brown, B. E.,** Effects of mine drainage on the River Hayle, Cornwall. (A) Factors affecting concentrations of copper, zinc and iron in water, sediments and dominant invertebrate fauna, *Hydrobiologia,* 52, 221, 1977.
190. **Brown, B. E.,** The form and function of metal-containing 'granules' in invertebrate tissues, *Biol. Rev.,* 57, 621, 1982.
191. **Brown, G. A. and Davis, R.,** Sensitivity of red flour beetle eggs to gamma radiation as influenced by treatment age and dose rate, *J. Georgia Entomol. Soc.,* 8, 153, 1973.
192. **Brown, L. G., Bellinger, E. G., and Day, J. P.,** Dieldrin pollution in Holme catchment, Yorkshire, *Environ. Pollut.,* 18, 203, 1979.

193. **Brown, M. A.,** Temperature-dependent pyrethroid resistance in a pyrethroid-selected colony of *Heliothis virescens* (F.) (Lepidoptera, Noctuidae), *J. Econ. Entomol.,* 80, 330, 1987.
194. **Bruns, D. A., Wiersma, G. B., and Rykiel, E. J., Jr.,** Ecosystem monitoring at global baseline sites, *Environ. Monit. Assess.,* 17, 3, 1991.
195. **Brust, G. E.,** Direct and indirect effects of four herbicides on the activity of carabid beetles (Coleoptera: Carabidae), *Pest. Sci.,* 30, 309, 1990.
196. **Brusven, M. A. and Prather, K. V.,** Influence of stream sediments on distribution of macrobenthos, *J. Entomol. Soc. Br. Col.,* 71, 25, 1974.
197. **Bryce, D. and Hobart, A.,** The biology and identification of the larvae of the Chironomidae (Diptera), *Entomol. Gaz.,* 23, 175, 1972.
198. **Buckley, E. H.,** Accumulation of airborne polychlorinated biphenyls in foliage, *Science,* 216, 520, 1982.
199. **Buckner, C. H.,** Supporting document supplied to the Expert Panel on Fenitrothion, in *Publ. NRCC,* 14104, Environ. Secretariat, NRC, Ottawa, 1975.
200. **Bugbee, S. L. and Walter, C. M.,** The response of macroinvertebrates to gasoline pollution in a mountain stream, in *Proc. Joint Conf. on Prevention and Control of Oil Spill,* American Petroleum Institute, Washington, D.C., 1973, 725.
201. **Bull, K. R.,** The critical loads/levels to gaseous pollutant emission control, *Environ. Pollut.,* 69, 105, 1991.
202. **Bultman, T. L. and Faeth, S. H.,** Experimental evidence for intra-specific competition in a lepidopteran leaf-miner, *Ecology,* 67, 442, 1986.
203. **Bultman, T. L. and Faeth, S. H.,** Impact of irrigation and experimental drought stress on leaf-mining insects on Emory oak, *Oikos,* 48, 5, 1987.
204. **Burditt, A. K., Jr., Hungate, F. P., and Toba, H. H.,** Gamma irradiation: effect of dose and dose rate on development of mature codling moth larvae and adult eclosion, *Radiat. Phys. Chem.,* 34, 979, 1989.
205. **Burkhardt, R.,** Untersuchungen über die Trichoptera des Vogelsberges. 2. Auswirkungen anthropogener Verunreinigungen der Fliessgewässer, *Arch. Hydrobiol.,* 111, 107, 1987.
206. **Burn, A. J.,** Assessment of the impact of pesticides on invertebrate predation in cereal crops, *Aspect Appl. Biol.,* 17, 279, 1988.
207. **Burn, A. J.,** Long-term effects of pesticides on natural enemies of cereal crop pests, in *Pesticides and Non-target Invertebrates,* Jepson, P. C., Ed., Intercept, Wimborne, Dorset, England, 1989, 177.
208. **Burrows, I. G. and Whitton, B. A.,** Heavy metals in water, sediments and invertebrates from a metal-contaminated river free from organic pollution, *Hydrobiologia,* 106, 263, 1983.
209. **Burton, T. M. and Allan, J. W.,** Influence of pH, aluminium and organic matter on stream invertebrates, *Can. J. Fish. Aquat. Sci.,* 43, 1285, 1986.
210. **Burton, T. M., Stanford, R. M., and Allan, J. W.,** The effects of acidification on stream ecosystems, in *Acid Precipitation, Effects on Ecological Systems,* D'Itri, F. M., Ed., Ann Arbor Science, Ann Arbor, MI, 1982, 209.
211. **Burton, T. M., Stanford, R. M., and Allan, J. W.,** Acidification effects on stream biota and organic matter processing, *Can. J. Fish. Aquat. Sci.,* 42, 669, 1985.
212. **Bury, R. B.,** The effects of diesel fuel on a stream fauna, *Calif. Fish Game,* 58, 291, 1972.
213. **Buse, A.,** Fluoride accumulation in invertebrates near an aluminium reduction plant in Wales, *Environ. Pollut. Ser. A,* 41, 199, 1986.
214. **Bush, B., Simpson, K. W., and Koblintz, R. R.,** PCB congener analysis of water and caddisfly larvae (Insecta: Trichoptera) in the Upper Hudson River by glass capillary chromatography, *Bull. Environ. Contam. Toxicol.,* 34, 96, 1985.
215. **Bush, G. L. and Hoy, M. A.,** Evolutionary processes in insects, in *Ecological Entomology,* Huffaker, C. B. and Rabb, R. L., Eds., John Wiley & Sons, New York. 1984, 247.
216. **Busvine, J. R.,** *Insects and Hygiene,* Methuen & Co., London, 1951.

217. **Butler, G. D., Jr., Kimball, B. A., and Mauney, J. R.,** Populations of *Bemisia tabaci* (Homoptera: Aleyrodidae) on cotton grown in open-top field chambers enriched with CO_2, *Environ. Entomol.,* 15, 61, 1986.
218. **Butowsky, R. O.,** The pecularities of heavy metal distribution in insects in agroecosystems (in Russian), *Agrochimija,* 2, 84, 1989.
219. **Butowsky, R. O.,** The autotransport pollution and entomofauna (in Russian), *Agrochimija,* 4, 139, 1990.
220. **Butowsky, R. O.,** The influence of autotransport pollution on the distribution of parasitic Hymenoptera in agroecosystems (in Russian), *Agrochimija,* 5, 103, 1990b.
221. **Cairns, J., Jr.,** Are single species toxicity tests alone adequate for estimating environmental hazard? *Hydrobiologia,* 100, 235, 1977.
222. **Cairns, J., Jr.,** The myth of the most sensitive species, *BioScience,* 36, 670, 1986.
223. **Cairns, J., Jr. and Pratt, J. R.,** The scientific basis of bioassays, *Hydrobiologia,* 188/189, 5, 1989.
224. **Calabrese, E. J., Chamberlain, C. C., Coler, R., and Young, M.,** The effects of trichloroacetic acid, a widespread product of chlorine disinfection, on the dragonfly nymph respiration, *J. Environ. Sci. Health,* A22, 343, 1987.
225. **Campbell, B. C., Jones, K. C., and Dreyer, D. L.,** Discriminative behavioral responses by aphids to various plant matrix polysaccharides, *Entomol. Exp. Appl.,* 41, 17, 1986.
226. **Campbell, I. M.,** Reproductive capacity in the genus *Choristoneura* Led. (Lepidoptera; Tortricidae). I. Quantitative inheritance and genes as controllers of rates, *Can. J. Genet. Cytol.,* 4, 272, 1962.
227. **Campbell, I. M.,** Genetic variation related to survival in lepidopteran species, in *Breeding Pest-Resistant Trees,* Gerhold, H. D., Schreiner, E. J., McDermott, R. E., and Winieski, I. A., Eds., Proc. NATO NSF Adv. Study Institute, Pergamon, Oxford, 1966, 137.
228. **Campbell, I. M.,** Does climate affect host-plant quality? Annual variation in the quality of balsam fir as food for spruce budworm, *Oecologia,* 81, 341, 1989.
229. **Campbell, P. G. C. and Stokes, P. M.,** Acidification and toxicity of metals to aquatic biota, *Can. J. Fish. Aquat. Sci.,* 42, 2034, 1985.
230. **Capecki, Z.,** Masowe wystapiene zasnui wysokogorskiej *Cephalcia falleni* (Dalm.), Pamphiliidae, Hymenoptera w Gorcach [Mass occurrence of *Cephalcia falleni* (Dalm.) (Pamphiliidae, Hymenoptera) in the Gorce Mountains], *Sylwan,* 126, 41, 1982.
231. **van Capelleveen, H. E.,** Population differentiation in *Porcellio scaber* Latr. due to zinc and cadmium contamination, *Environ. Pollut.,* in press.
232. **Carbonell, E. A., Frey, J. J., and Bell, A. E.,** Estimation of maternal, sex-linked and additive x additive epistatic gene effects for body size of *Tribolium, Theor. Appl. Genet.,* 70, 133, 1985.
233. **Carlson, C. E. and Dewey, J. E.,** Environmental pollution by fluorides in Flathead National Forest and Glacier National Park, USDA Forest Service, Northern Region, Rep., 1971, 72.
234. **Carlson, R. W. and Bazzaz, F. A.,** The effects of elevated CO_2 concentrations on growth, photosynthesis, transpiration, and water use efficiency of plants, in *Environmental and Climatic Impact of Coal Utilization,* Singh, J. J. and Deepak, A., Eds., Academic Press, New York, 1980, 609.
235. **Carson, R.,** *Silent Spring,* Houghton-Mifflin, Boston, 1962.
236. **Carr, R. S., Williams, J. W., Saska, F. I., Buhl, R. S., and Neff, J. M.,** Bioenergetic alterations correlated with growth, fecundity and body burdens of cadmium in mysids (*Mysidopsis bahia*), *Environ. Toxicol. Chem.,* 4, 181, 1985.
237. **Carter, A.,** Cadmium, copper and zinc in soil animals and their food in a red clover system, *Can. J. Zool.,* 61, 2751, 1983.
238. **Carter, N. E. and Brown, N. R.,** Seasonal abundance of certain soil arthropods in a fenitrothion-treated red spruce stand, *Can. Entomol.,* 105, 1065, 1973.
239. **Carter, S. W.,** A review of the use of synthetic pyrethroids in public health and vector pest control, *Pest. Sci.,* 27, 361, 1989.

240. **Cates, R. G. and Zou, J.,** Douglas-fir (*Pseudotsuga menziesii*) population variation in terpene chemistry and its role in budworm (*Choristoneura occidentalis* Freeman) dynamics, in *Population Dynamics of Forest Insects,* Watt, A. D., Leather, S. R., Hunter, M. D., and Kidd, N. A. C., Eds., Intercept, Andover, Hampshire, U.K., 1990, 169.
241. **Chadwick, J. W., Canton, S. P., and Dent, R. L.,** Recovery of benthic invertebrate communities in Silver Bow Creek, MT, following improved metal mine wastewater treatment, *Water Air Soil Pollut.,* 28, 427, 1986.
242. **Chang, C. W.,** Effect of ozone on ribosomes in pinto bean leaves, *Phytochemistry,* 10, 2863, 1971.
243. **Chantaramongkol, P.,** Light-trapped caddisflies (Trichoptera) as water quality indicators in large rivers: results from the Danube at Veröce, Hungary, *Aquat. Insects,* 5, 33, 1983.
244. **Chapman, P. M. and Fink, R.,** Effects of Puget Sound sediments and their elutriates on the life cycle of *Capitella capitata, Bull. Environ. Contam. Toxicol.,* 33, 451, 1984.
245. **Chapman, R. F.,** *The Insects: Structure and Function,* 2nd ed., Elsevier North Holland, New York, 1979.
246. **Chappelka, A. H., Kraemer, M. E., Mebrahtu, T., Rangappa, M., and Benepal, P. S.,** Effects of ozone on soybean resistance to the Mexican bean beetle (*Epilachna varivestis* Mulsant), *Environ. Exp. Bot.,* 28, 53, 1988.
247. **Charles, P. J. and Villemant, C.,** Modification des niveaux de populations d'insectes dans les jeunes plantations de pins sylvestres de la forêt de Roumare (Seine-Maritime) soumises à la pollution atmosphérique, *Comt. Rend. Acad. d'Agric. de France,* 63, 502, 1977.
248. **Chen, P. S.,** *Biochemical aspects of insect development, Monogr. Dev. Biol.,* 3, 1, 1971.
249. **Chenoweth, M. B.,** Monofluoroacetic acid and related compounds, *Pharmacol. Rev.,* 1, 383, 1949.
250. **Chew, F. S.,** Coevolution of pierid butterflies and their cruciferous food plants. I. The relative quality of available resources, *Oecologia,* 20, 117, 1975.
251. **Chew, F. S.,** Coevolution of pierid butterflies and their cruciferous food plants. II. Distribution of eggs on potential food plants, *Evolution,* 31, 568, 1977.
252. **Chew, F.,** Food plant preferences of *Pieris* butterflies and cruciferous food plants, *Oecologia,* 46, 347, 1980.
253. **Chew, F. S. and Robbins, R. K.,** Egg-laying in butterflies, *Symp. R. Entomol. Soc.,* 11, 65, 1984.
254. **Chiapparini, L., Baldacci, E., and Moglia, C.,** Sull'impiego del trifenil acetato di stagno come alghicida, *Il Riso,* 3, 227, 1964.
255. **Chilcott, C. N., Pillai, J. S., and Kalmakoff, J.,** Efficacy of *Bacillus thuringiensis* var. *israelensis* as a biocontrol agent against larvae of Simuliidae (Diptera) in New Zealand, *N. Z. J. Zool.,* 10, 319, 1983.
256. **Chiment, J. J., Alscher, R., and Hughes, P. R.,** Glutathione as an indicator of SO_2^- induced stress in soybean, *Environ. Exp. Bot.,* 26, 147, 1986.
257. **Chisholm, J. L. and Downs, S. C.,** Stress and recovery of aquatic organisms as related to highway construction along Turtle Creek, Boone County, WV, Water-Supply Paper 2055, U.S. Geological Survey, Washington, D.C., 1978.
258. **Chiverton, P. A. and Sotherton, N. W.,** The effects on beneficial arthropods of the exclusion of herbicides from cereal crop edges, *J. Appl. Ecol.,* 28, 1027, 1991.
259. **Chlodny, J.,** Liczebnosc mszyc Aphididae i fauny towarzyszacej w uprawach brzozy brodawkowatej (*Betula verrucosa* Ehrh.) na obszarze Górnoslaskiego Okregu Przemyslowego. Entomologia a ochrona srodowiska, *Panstwowe Wydawnictwo Naukowe,* 41, 1976.
260. **Chlodny, J., Matuszczyk, I., Styfi-Bartkiewicz, B., and Syrek, D.,** Catchability of the epigeal fauna of pine stands as a bioindicator of industrial pollution of forests, *Ekol. Pol.,* 35, 271, 1987.
261. **Christiansen, E.,** Bark beetles and air pollution, *Medd. Norsk Inst. Skogforsk.,* 42, 101, 1989.

262. **Christiansen, E., Waring, R. H., and Berryman, A. A.,** Resistance of conifers to bark beetle attack: searching for general relationships, *For. Ecol. Manage.*, 22, 89, 1987.
263. **Ciborowski, J. J. H. and Corkum, L. D.,** Organic contaminants in adult aquatic insects of the St. Clair and Detroit Rivers, Ontario, Canada, *J. Great Lakes Res.*, 14, 148, 1988.
264. **Clancy, K. M. and Price, P. W.,** Rapid herbivore growth enhances enemy attack: sublethal plant defences remain a paradox, *Ecology*, 68, 733, 1987.
265. **Clark, D. R., Jr.,** Uptake of dietary PCB by pregnant big brown bats (*Eptesicus fuscus*) and their fetuses, *Bull. Environ. Contam. Toxicol.*, 19, 707, 1978.
266. **Clark, D. R., Jr. and Stafford, C. J.,** Effects of DDE and PCB (Aroclor 1260) on experimentally poisoned female little brown bats (*Myotis lucifugus*): lethal brain concentations, *J. Toxicol. Environ. Health*, 7, 925, 1981.
267. **Clarke, C. A., Clarke, F. M. M., and Dawkins, H. C.,** *Biston betularia* (the peppered moth) in West Kirby, Wirral, 1959–1989: updating the decline in *f. carbonaria*, *Biol. J. Linnean Soc.*, 39, 323, 1990.
268. **Clarke, C. A., Mani, G. S., and Wynne, G.,** Evolution in reverse: clean air and the peppered moth, *Biol. J. Linnean Soc.*, 26, 189, 1985.
269. **Clarke, R.,** The effects of effluents from metal mines on aquatic ecosystems in Canada. A literature review, *Fish. Mar. Serv. Res. Dev. Tech. Rep.*, 488, 1, 1974.
270. **Clausen, I. H. S.,** Notes on the impact of air pollution (SO_2 & Pb) on spider (Araneae) populations in North Zealand, Denmark, *Entomol. Medd.*, 52, 33, 1984.
271. **Clausen, I. H. S.,** The use of spiders (Araneae) as ecological indicators, *Bull. Br. Arachnol. Soc.*, 7, 83, 1986.
272. **Clawson, R. L. and Clark, D. R., Jr.,** Pesticide contamination of endangered gray bats and their food base in Boone County, MO, 1982, *Bull. Environ. Contam. Toxicol.*, 42, 431, 1989.
273. **Clements, J. R. and Kawatski, J. A.,** Occurrence of polychlorinated biphenyls (PCB's) in adult mayflies (*Hexagenia bilineata*) of the upper Mississippi River, *J. Freshwater Ecol.*, 2, 611, 1984.
274. **Clements, W. H.,** Community responses of stream organisms to heavy metals: a review of observational and experimental approaches, in *Ecotoxicology of Metals: Current Concepts and Applications*, Newman, M. C. and McIntosh, A. W., Eds., CRC Press, Boca Raton, FL, 1991, 363.
275. **Clements, W. H., Cherry, D. S., and Cairns, J., Jr.,** Structural alterations in aquatic insect communities exposed to copper in laboratory streams, *Environ. Toxicol. Chem.*, 7, 715, 1988.
276. **Clements, W. H., Cherry, D. S., and Cairns, J., Jr.,** Impact of heavy metals on insect communities in streams: a comparison of observational and experimental results, *Can. J. Fish. Aquat. Sci.*, 45, 2017, 1988.
277. **Clements, W. H., Cherry, D. S., and Cairns, J., Jr.,** The influence of copper exposure on predator-prey interactions in aquatic insect communities, *Freshwater Biol.*, 21, 483, 1989.
278. **Clements, W. H., Farris, J. L., Cherry, D. S., and Cairns, J., Jr.,** The influence of water quality on macroinvertebrate community responses to copper in outdoor experimental streams, *Aquat. Toxicol.*, 14, 249, 1989.
279. **Cleve, K.,** Die Erforschung der Ursachen für das Auftreten melanistischer Schmetterlingsformen im Laufe der letzten hundert Jahre, *Z. Angew. Entomol.*, 65, 371, 1970.
280. **Clifford, H. F.,** Effects of periodically disturbing a small area of substratum in a brown-water stream of Alberta, Canada, *Freshwater Invert. Biol.*, 1, 39, 1982.
281. **Clubb, R. W., Gaufin, A. R., and Lords, J. L.,** Acute cadmium toxicity studies upon nine species of aquatic insects, *Environ. Res.*, 9, 332, 1975.
282. **Clubb, W. R., Lords, J. L., and Gaufin, A. R.,** Isolation and characterization of a glycoprotein from the stonefly, *Pteronarcys californica*, which binds cadmium, *J. Insect Physiol.*, 21, 61, 1975.

283. **Clulow, F. V., Davé, N. K., Lim, T. P., and Cloutier, N. R.,** Uptake of ^{226}Ra by established vegetation and black cutworm larvae, *Agrotis ipsilon* (class Insecta: order Lepidoptera), on U mill tailings at Elliot Lake, Canada, *Health Phys.,* 55, 31, 1988.
284. **Coats, J. R., Symonik, D. M., Bradbury, S. P., Dyer, S. D., Timson, L. K., and Atchison, G. J.,** Toxicology of synthetic pyrethroids in aquatic organisms: an overview, *Environ. Toxicol. Chem.,* 8, 671, 1989.
285. **Cobb, F. W., Jr., Wood, D. L., Stark, R. W., and Miller, P. R.,** Effect of injury upon physical properties of oleoresin, moisture content, and phloem thickness, *Hilgardia,* 39, 127, 1968.
286. **Cobb, F. W., Jr., Wood, D. L., Stark, R. W., and Parmeter, J. R., Jr.,** Theory on the relationship between oxidant injury and bark beetle infestation, *Hilgardia,* 39, 141, 1968.
287. **Cobb, F. W., Jr., Zavarin, E., and Bergot, J.,** Effect of air pollution on the volatile oil from leaves of *Pinus ponderosa, Phytochemistry,* 11, 1815, 1972.
288. **Cockfield, S. D.,** Relative availability of nitrogen in host plants of invertebrate herbivores: three possible nutritional and physiological conditions, *Oecologia,* 77, 91, 1988.
289. **Coddington, K.,** A review of arsenicals in biology, *Toxicol. Environ. Chem.,* 11, 281, 1986.
290. **Colbo, M. H. and Undeen, A. H.,** Effect of *Bacillus thuringiensis* var. *israelensis* on non-target insects in stream trials for control of Simuliidae, *Mosq. News,* 40, 368, 1980.
291. **Colborn, T.,** Measurement of low levels of molybdenum in the environment by using aquatic insects, *Bull. Environ. Contam. Toxicol.,* 29, 422, 1982.
292. **Cole, C. V., Elliot, E. T., Hunt, H. W., and Coleman, D. C.,** Trophic interaction in soils as they affect energy and nutrient dynamics. V. Phosphorus transformations, *Microbiol. Ecol.,* 4, 381, 1978.
293. **Coleman, J. S.,** Leaf development and leaf stress: increased susceptibility associated with sink-source transition, *Tree Physiol.,* 2, 289, 1986.
294. **Coleman, J. S. and Jones, C. G.,** Plant stress and insect performance: cottonwood, ozone and a leaf beetle, *Oecologia,* 76, 57, 1988.
295. **Coleman, J. S. and Jones, C. G.,** Acute ozone stress on eastern cottonwood (*Populus deltoides* Bartr.) and the pest potential of the aphid, *Chaitophorus populicola* Thomas (Homoptera: Aphididae), *Environ. Entomol.,* 17, 207, 1988.
296. **Coleman, J. S. and Jones, C. G.,** A phytocentric perspective of phytochemical induction by herbivores, in *Phytochemical Induction by Herbivores,* John Wiley & Sons, New York, 1991, 3.
297. **Coleman, J. S., Jones, C. G., and Smith, W. H.,** Effects of ozone on cottonwood-leaf rust interactions: independence of abiotic stress, genotype and leaf ontogeny, *Can. J. Bot.,* 65, 949, 1987.
298. **Collier, K. J., Ball, O. J., Graesser, A. K., Main, M. R., and Winterbourn, M. J.,** Do organic and anthropogenic acidity have similar effects on aquatic fauna? *Oikos,* 59, 33, 1990.
299. **Collins, N. M. and Morris, M. G.,** Threatened swallowtail butterflies of the world, in *The IUCN Red Data Book,* IUCN, Gland, Switzerland, 1985.
300. **Compton, O. C., Remmert, L. F., Rudinsky, J. A., McDowell, L. L., Ellertson, F. E., Mellenthin, W. M., and Ritcher, P. O.,** Needle scorch and condition of ponderosa pine trees in the Dallas area, *Oreg. State Univ. Agric. Exp. Stn. Misc. Pap.,* 120, 1, 1961.
301. **Conroy, N., Hawley, K., Keller, W., and LaFrance, C.,** Influences of the atmosphere on lakes in the Sudbury area, *J. Great Lakes Res.,* 2 (Suppl. 1), 146, 1976.
302. **Constantinidou, H. A. and Kozlowski, T. T.,** Effects of sulfur dioxide and ozone on *Ulmus americana* seedlings. II. Carbohydrates, proteins, and lipids, *Can. J. Bot.,* 57, 176, 1979.
303. **Cook, L. M., Mani, G. S., and Varley, M. E.,** Post-industrial melanism in the peppered moth, *Science,* 231, 611, 1986.
304. **Cook, L. M., Rigby, K. D., and Seaward, M. R. D.,** Melanic moths and changes in epiphytic vegetation in north-west England and north Wales, *Biol. J. Linnean Soc.,* 39, 343, 1990.

305. **Cooley, D. R. and Manning, W. J.,** The impact of ozone on assimilate partitioning in plants: a review, *Environ. Pollut.,* 47, 95, 1987.
306. **Correa, M., Coler, R. A., and Yin, C.-M.,** Changes in oxygen consumption and nitrogen metabolism in the dragonfly *Somatochlora cingulata* exposed to aluminium in acid waters, *Hydrobiologia,* 121, 151, 1985.
307. **Correa, M., Coler, R., Yin, C.-M., and Kaufman, E.,** Oxygen consumption and ammonia excretion in the detrivore caddisfly *Limnephilus* sp. exposed to low pH and aluminium, *Hydrobiologia,* 140, 237, 1986.
308. **Costescu, L. M. and Hutchinson, T. C.,** The ecological consequences of soil pollution by metallic dust from the Sudbury smelters, *Proc. Int. Environ. Sci.,* 17, 540, 1972.
309. **Costlow, J. D.,** Effects of cyclic temperature on larval development of marine invertebrates. I. Molting, growth, and survival, in *Physiological Responses of Marine Organisms to Environmental Stressors,* Dorigan, J. V. and Harrison, F.L., Eds., DOE/ER-0317, Research supported by U.S. Department of Energy 1980 to 1986, Washington, D.C., 1987, 83.
310. **Costlow, J. D. and Sanders, B. M.,** Effects of cyclic temperature on larval development of marine invertebrates. II. Regulation of growth as a general indicator of stress, in *Physiological Responses of Marine Organisms to Environmental Stressors,* Dorigan, J. V. and Harrison, F.L., Eds., DOE/ER-0317, Research supported by the Department of Energy 1980 to 1986, Washington, D.C., 1987, 105.
311. **Cotta-Ramusino, M. and Doci, A.,** Acute toxicity of Brestan and fentin acetate on some freshwater organisms, *Bull. Environ. Contam. Toxicol.,* 38, 647, 1987.
312. **Coughtrey, P. J., Jones, C. H., Martin, M. H., and Shales, S. W.,** Litter accumulation in woodlands contaminated by Pb, Zn, Cd and Cu, *Oecologia,* 39, 51, 1979.
313. **Courtemanch, D. L. and Davies, S. P.,** A coefficient of community loss to assess detrimental change in aquatic communities, *Water Res.,* 21, 217, 1987.
314. **Courtemanch, D. L. and Gibbs, K. E.,** Short- and long-term effects of forest spraying of carbaryl (Sevin-4-oil®) on stream invertebrates, *Can. Entomol.,* 112, 271, 1980.
315. **Cowling, D. W. and Bristow, A. W.,** Effects of SO_2 on sulphur and nitrogen fractions and free amino acids in perennial ryegrass, *J. Sci. Food. Agric.,* 30, 354, 1979.
316. **Cowling, D. W. and Koziol, M. J.,** Mineral nutrition and plant response to air pollution, in *Effects of Gaseous Air Pollutants in Agriculture and Horticulture,* Unsworth, M. W. and Ormrod, D. P., Eds., Butterworths, London, 1982, 349.
317. **Cowling, E. B. and Linthurst, R. A.,** The acid precipitation phenomena and its ecological consequences, *BioScience,* 31, 649, 1981.
318. **Cox, B. C. and Moore, P. D.,** *Biogeography: An Ecological and Evolutionary Approach,* Blackwell, Oxford, 1985.
319. **Cracker, L. E. and Bernstein, D.,** Buffering of acid rain by leaf tissue of selected crop plants, *Environ Pollut. Ser. A.,* 36, 375, 1984.
320. **Craft, C. B. and Webb, J. W.,** Effects of acidic and neutral sulfate salt solutions on forest floor arthropods, *J. Environ. Qual.,* 13, 436, 1984.
321. **Craker, L. E. and Starbuck, J. S.,** Metabolic changes associated with ozone injury of bean leaves, *Can. J. Plant Sci.,* 52, 589, 1972.
322. **Cramer, H. H.,** Die geographischen Grundlagen des Massenwechsels von *Epiblema tedella* C., *Forstwiss. Centralbl.,* 70, 42, 1951.
323. **Creed, E. R.,** Geographic variation in the two-spot ladybird in England and Wales, *Heredity,* 21, 57, 1966.
324. **Creed, E. R.,** Industrial melanism in the two-spot ladybird and smoke abatement, *Evolution,* 25, 290, 1971.
325. **Creed, E. R.,** Two-spot ladybirds as indicators on intense local air pollution, *Nature,* 249, 390, 1974.
326. **Crisman, T. L., Schulze, R. L., Brezonik, P. L., and Bloom, S. A.,** Acid precipitation: the biotic response in Florida lakes, in *Ecological Impact of Acid Precipitations,* Drablos, D. and Tollan, A., Eds., Proc. Int. Conf., SNSF Project, Ås-NLH, Norway, 1980, 296.

327. **Croft, B. A.,** *Arthropod Biological Control Agents and Pesticides,* Wiley-Interscience, New York, 1990.
328. **Croft, B. A. and Brown, A. W. A.,** Responses of arthropod natural enemies to insecticides, *Ann. Rev. Entomol.,* 20, 285, 1975.
329. **Croft, B. A. and Morse, J. G.,** Research advances on pesticide resistance in natural enemies, *Entomophaga,* 24, 3, 1979.
330. **Croft, B. A. and Strickler, K.,** Natural enemy resistance to pesticides: documentation, characterization, theory and application, in *Resistance to Pesticides,* Georghiou, G. P. and Saito, T., Eds., Plenum Press, New York, 1983, 669.
331. **Crossland, N. O.,** Aquatic toxicology of cypermethrin. II. Fate and biological effects of pond experiments, *Aquat. Toxicol.,* 2, 205, 1982.
332. **Crowell, K. L.,** Rates of competitive exclusion by the Argentine ants in Bermuda, *Ecology,* 49, 551, 1968.
333. **Crunkilton, R. L. and Duchrow, R. M.,** Impact of a massive crude oil spill on the invertebrate fauna of a Missouri Ozark stream, *Environ. Pollut.,* 63, 13, 1990.
334. **Crutzen, P. J., Heidt, L. E., Krasnec, J. P., Pollock, W. H., and Seiler, W.,** Biomass burning as a source of atmospheric gases CO, H_2, N_2O, NO, CH_3Cl and COS, *Nature,* 282, 253, 1979.
335. **Cuffney, T. F., Wallace, J. B., and Webster, J. R.,** Pesticide manipulation of a headwater stream: invertebrate responses and their significance for ecosystem processes, *Freshwater Invert. Biol.,* 3, 153, 1984.
336. **Culliney, T. W. and Pimentel, D.,** Effects of chemically contaminated sewage sludge on an aphid population, *Ecology,* 67, 1665, 1986.
337. **Culliney, T. W. and Pimentel D.,** Preference of the green peach aphid, *Myzus persicae,* for plants grown in sewage sludges, *Bull. Environ. Contam. Toxicol.,* 39, 257, 1987.
338. **Cummins, K. W.,** Trophic relations of aquatic insects, *Ann. Rev. Entomol.,* 18, 183, 1973.
339. **Cure, J. D. and Acock, B.,** Crop response to carbon dioxide doubling: a literature survey, *Agric. For. Meteorol.,* 38, 127, 1986.
340. **Cushman, R. M.,** Chironomid deformities as indicators of pollution from a synthetic, coal-derived oil, *Freshwater Biol.,* 14, 179, 1984.
341. **Cushman, R. M. and Goyert, J. C.,** Effects of a synthetic crude oil on pond benthic insects, *Environ. Pollut. Ser. A,* 33, 163, 1984.
342. **Czarnecki, A. and Losinski, J.,** The effect of GT seed dressing on the community of Collembola in the soil under sugar beets, *Pedobiologia,* 28, 427, 1985.
343. **Czechowski, W.,** The ants *Lasius niger* L. (Hymenoptera, Formicidae) as indicators of the degree of environmental pollution in a city, *Przegl. Zool.,* 24, 113, 1980.
344. **Dadd, R. H.,** Insect nutrition: current developments and metabolic implications, *Ann. Rev. Entomol.,* 18, 381, 1973.
345. **Dahl, C. and Grimås, U.,** Report of radionuclides in *Aedes communis* pupae from central Sweden, 1986, *J. Am. Mosq. Contr. Assoc.,* 3, 328, 1987.
346. **Dallinger, R., Prosi, F., and Back, H.,** Contaminated food and uptake of heavy metals by fish: a review and a proposal for further research, *Oecologia,* 73, 91, 1987.
347. **Danks, H.V.,** Arctic arthropods. A review of systematics and ecology with particular reference to the North American fauna, Entomol. Society of Canada, Ottawa, 1981.
348. **Danks, H. V.,** Insect plant interactions in arctic regions, *Rev. Entomol. Québ.,* 31(1986), 52, 1987.
349. **Danks, H.V. and Foottit, R.G.,** Insects of the boreal zone of Canada, *Can. Entomol.,* 121, 625, 1989.
350. **Dannisöe, J., Frederiksen, N., Jensen, E. R., Lindegaard, C., and Nilssen, E.,** Födegrundlagets betydning for produktionen af örred (*Salmo trutta L.*) i okkelbelastede vandlöb, Bilag 17 til Okkerredegörelsen, Nat. Agency Environmental Protection, Copenhagen, 1984.
351. **Darlington, S. T. and Gower, A. M.,** Location of copper in larvae of *Plectrocnemia conspersa* (Curtis) (Trichoptera) exposed to elevated metal concentrations in a mine drainage stream, *Hydrobiologia,* 196, 91, 1990.

352. **Darlington, S. T., Gower, A. M., and Ebdon, L.,** The measurement of copper in individual aquatic insect larvae, *Environ. Technol. Lett.,* 7, 141, 1986.

353. **Darlington, S. T., Gower, A. M., and Ebdon, L.,** Studies on *Plectrocnemia conspersa* (Curtis) in copper contaminated streams in southwest England, in *Proc. of the Fifth Int. Symp. on Trichoptera,* Bournand, M. and Tachet, H., Eds., Junk, the Netherlands, 1987, 353.

354. **David, W. A. L. and Taylor, C. E.,** The effect of sucrose content of diets on susceptibilty to granulosis virus disease in *Pieris brassicae, J. Invertebr. Pathol.,* 30, 117, 1977.

355. **David, W. A. L., Ellaby, S., and Taylor, G.,** The effect of reducing the content of certain ingredients in a semisynthetical diet on the incidence of granulosis virus disease in *Pieris brassicae, J. Invertebr. Pathol.,* 20, 332, 1972.

355a. **Davies, M. T., Davison, A. W., and Port, G. R.,** Fluoride loading of larvae of pine sawfly from a polluted site, *J. Appl. Ecol.,* 29,63,1992.

356. **Davis, A. R.,** The study of insecticide poisoning of honeybee brood, *Bee World,* 70, 163, 1989.

357. **Davis, A. R. and Shuel, R. W.,** Distribution of ^{14}C-labelled carbofuran and dimethoate in a royal jelly, queen larvae and nurse honeybees, *Apidologie,* 19, 37, 1988.

358. **Davis, A. R., Shuel, R. W., and Peterson, R. L.,** Distribution of carbofuran and dimethoate in flowers and their secretion in nectar as related to nectary vascular supply, *Can. J. Bot.,* 66, 1248, 1987.

359. **Davis, A. R., Solomon, K. R., and Shuel, R. W.,** Laboratory studies of honeybee larval growth and development as affected by systemic insecticides at adult-sublethal levels, *J. Apicult. Res.,* 27, 146, 1988.

360. **Davis, B. N. K., Lakhani, K. H., and Yates, T. J.,** The hazards of insecticides to butterflies of field margins, *Agric. Ecosyst. Environ.,* 36, 151, 1991.

361. **Davis, B. N. K., Lakhani, K. H., Yates, T. J., and Frost, A. J.,** Bioassays of insecticide spray drift: the effects of wind speed on mortality of *Pieris brassicae* larvae (Lepidoptera) caused by diflubenzuron, *Agric. Ecosyst. Environ.,* 36, 141, 1991.

362. **Davis, M. B.,** Insights from paleoecology on global change, *Bull. Ecol. Soc. Am.,* 70, 222, 1989.

363. **Davis, M. B.,** Lags in vegetation response to greenhouse warming, *Climatic Change,* 15, 79, 1989.

364. **Davis, M. B. and Botkin, D. B.,** Sensitivity of cool-temperature forests and their fossil pollen record to rapid temperature change, *Q. Res.,* 23, 327, 1985.

365. **Davison, A. W.,** Pathways of fluoride, in *Ecosystems,* Coughtrey, P. J., Martin, M. H., and Unsworth, M. H., Eds., Blackwell, Oxford, 1987, 193.

366. **Day, K.,** The growth and decline of a population of the spruce aphid *Elatobium abietinum* during a three year study, and the changing pattern of fecundity, recruitment and alary polymorphism in a Northern Ireland forest, *Oecologia,* 64, 118, 1984.

367. **Day, K. E.,** The acute, chronic and sublethal effects of synthetic pyrethroids on zooplankton in the laboratory and the field: an overview, *Environ. Toxicol. Chem.,* 8, 411, 1989.

368. **DeBauer, M. L., Tejeda, T. H., and Manning, W. J.,** Ozone causes needle injury and tree decline in *Pinus hartwegii* at high altitudes in the mountains around Mexico City, *J. Air Pollut. Contr. Assoc.,* 35, 838, 1985.

369. **Debouge, M. H. and Thome, J. P.,** Dispersion du lindane dans 5 espèces de fourmis en Wallonie et cinètique d'accumulation chez *Formica polyctena.*[Dispersion of lindane on five ant species in Belgium (Wallonie) and accumulation in *Formica polyctena*], *Ann. Sci. Nat. Zool. Biol. Anim.,* 10, 25, 1989.

370. **Decourt, N., Malphettes, C. B., Perrin, R., and Caron, D.,** La pollution soufrée limite-t-elle le développement de la maladie de l'écorce du hetre (*Cryptococcus fagi, Nectria coccinae*), *Ann. Sci. For.,* 37, 135, 1980.

371. **Deeleman-Reinhold, C. L.,** Changes in the spider fauna over 14 years in an industrially polluted area in Holland, *Acta Zool. Fenn.,* 190, 103, 1990.

372. **Dejoux, C.,** Action de l'abate sur les invertébrés aquatiques. III. Effets des premiers traitements de la Bagoué. Rapport O.R.S.T.O.M. No. 14, Lab. d'Hydrobiol., Bouaké Ivory Coast, 1977.
373. **Denneman, W. D.,** Voorkomen van *Cymindis humeralis* op een met zware metalen belaste locatie (Coleoptera: Carabidae), [Occurrence of *Cymindis humeralis* on a heavy metal polluted site (Coleoptera: Carabidae)], *Entomol. Ber. Amst.,* 50, 4, 1990.
374. **Depner, K. R., Charnetski, W. A., and Haufe, W. O.,** Effect of methoxychlor on resident populations of the invertebrates of the Athabasca River, in *Control of Black Flies in the Athabasca River,* Haufe, W. O. and Croome, G. C. R., Eds., Alberta Environment, Edmonton, 1980, 141.
375. **Dercks, W., Trumble, J., and Winter, C.,** Impact of atmospheric pollution on linear furanocoumarin content in celery, *J. Chem. Ecol.,* 16, 443, 1990.
376. **Dewey, J. E.,** Accumulation of fluorides by insects near an emission source in western Montana, *Environ. Entomol.,* 2, 179, 1973.
377. **Diaz, R. J.,** Pollution and tidal benthic communities of the James River Estuary, Virginia, *Hydrobiologia,* 180, 195, 1989.
378. **Dickman, M.,** Preliminary notes on changes in algal primary productivity following exposure to crude oil in the Canadian arctic, *Can. Field Nat.,* 85, 249, 1971.
379. **DiGiano, F. A., Coler, R. A., Dahiya, R. C., and Berger, B. B.,** A projection of pollutional effects of urban runoff in the Green River, Massachusetts, in *Urbanization and Water Quality Control,* Whipple, W., Ed., American Water Resource Association, Minneapolis, 1975, 28.
380. **Dillon, P. J., Jeffries, D. S., Scheider, W. A., and Yan, N. D.,** Some aspects of acidification in southern Ontario, in *Ecological Impact of Acid Precipitation,* Drablos, D. and Tollan, A., Eds., SNSF Project, Ås-NLH, Norway, 1980.
381. **Dillon, P. J., Yan, N. D., and Harvey, H. H.,** Acidic deposition: effects on aquatic ecosystems, *CRC Critic. Rev. Environ. Contam.,* 13, 167, 1984.
382. **Dixon, A. F. G.,** *Aphid Ecology,* Blackie, Glasgow, 1985.
383. **Dmowski, K. and Karolewski, M. A.,** Cumulation of zinc, cadmium and lead in invertebrates and in some vertebrates according to the degree of an area contamination, *Ekol. Pol.,* 27, 333, 1979.
384. **Dodge, E. E. and Theis, T. L.,** Effect of chemical speciation on the uptake of copper by *Chironomus tentans, Environ. Sci. Technol.,* 13, 1287, 1979.
385. **Doe, K. G., Parker, W. R., Ernst, W. R., and Horne, W. H.,** Studies on the toxicity of Matacil 180F concentrate, its formulations, and their components to aquatic fauna, in Environmental Protection Service Surveillance Report, EPS-5-AR-83-1, Dartmouth, 1983, 1.
386. **Doelman, P., Nieboer, G., Schrooten, J., and Visser, M.,** Antagonistic and synergistic toxic effects of Pb and Cd in a simple foodchain: nematodes feeding on bacteria or fungi, *Bull. Environ. Contam. Toxic.,* 32, 717, 1984.
387. **Dohmen, G. P.,** Secondary effects of air pollution: enhanced aphid growth, *Environ. Pollut. Ser. A,* 39, 227, 1985.
388. **Dohmen, G. P.,** Secondary effects of air pollution: ozone decreases brown rust disease potential in wheat, *Environ. Pollut.,* 43, 189, 1987.
389. **Dohmen, G. P.,** Indirect effects of air pollutants: changes in plant/parasite interactions, *Environ. Pollut.,* 53, 197, 1988.
390. **Dohmen, G. P., McNeill, S., and Bell, J. N. B.,** Air pollution increases *Aphis fabae* pest potential, *Nature,* 307, 52, 1984.
391. **Dojmi di Delupis, G. and Rotondo, V.,** Phototaxis in aquatic invertebrates: possible use in ecotoxicology tests, *Ecotoxicol. Environ. Safety,* 16, 189, 1988.
392. **Dolin, V. G., Stovbchaty, V. N., and Titar, V. M.,** Biological consequences of the Chernobyl radiation accident as indicated by the insect fauna of the region, Fourth Int. Congress Entomol., Gödöllö, Hungary, Abstract volume, 1991, 46.

393. **Dominguez, T. M., Calabrese, E. J., Kostecki, P. T., and Coler, R. A.,** The effects of tri- and dichloroacetic acids on the oxygen consumption of the dragonfly nymph *Aeschna umbrosa*, *J. Environ. Sci. Health,* A23, 251, 1988.
394. **Donker, M. H. and Bogert, C. G.,** Adaptation to cadmium in three populations of the isopod *Porcellio scaber*, *Comp. Biochem. Physiol.,* 100C, 143, 1991.
395. **Dosdall, L. M.,** Survival of selected aquatic insects exposed to metoxychlor treatment of the Saskatchewan river system, *Water Pollut. Res. J. Can.,* 26, 27, 1991.
396. **Dosdall, L. M. and Lehmkuhl D. M.,** Drift of aquatic insects following methoxychlor treatment of the Saskatchewan River system, *Can. Entomol.,* 121, 1077, 1989.
397. **Dosdall, L. M. and Lehmkuhl, D. M.,** The impact of methoxychlor treatment of the Saskatchewan River system on artificial substrate populations of aquatic insects, *Environ. Pollut.,* 60, 209, 1989.
398. **Douwes, P., Mikkola, K., Petersen, B., and Vestergren, A.,** Melanism in *Biston betularia* from north-west Europe (Lepidoptera: Geometridae)., *Entomol. Scand.,* 7, 261, 1976.
399. **Drablos, D. and Tollan, A., Eds.,** Ecological impact of acid precipitations, Proc. Int. Conference, SNSF Project, Ås-NLH, Norway, 1980.
400. **Dreyer, D. L. and Jones, K. G.,** Feeding deterrency of flavonoids and related phenolics towards *Schizaphis graminum* and *Myzus persicae:* aphid feeding deterrents in wheat, *Phytochemistry,* 20, 2489, 1981.
401. **Driscoll, C. T. and Newton, R. M.,** Chemical characteristics of Adirondack Lakes, *Environ. Sci. Technol.,* 19, 1018, 1985.
402. **Driscoll, C. T., Jr., Baker, J. P., Bisogni, J. J., Jr., and Schofield, C. L.,** Effect of aluminium on fish in dilute acidified waters, *Nature,* 284, 161, 1980.
403. **Duda, A. M., Lenat, D. R., and Penrose, D. L.,** Water quality in urban streams — what we can expect, *J. Water Pollut. Contr. Fed.,* 54, 1139, 1982.
404. **Duffy, D. C.,** Environmental uncertainty and commercial fishing: effects of Peruvian guano birds, *Biol. Conserv.,* 26, 227, 1983.
405. **Dumnicka, E. and Kownacki, A.,** A regulated river ecosystem in a polluted section of the Upper Vistula, *Acta Hydrobiol.,* 30, 81, 1988.
406. **Dumnicka, E., Kasza, H., Kownacki, A., Krzyzanek, E., and Kuflikowski, T.,** Effects of regulated stream on the hydrochemistry and zoobenthos in differently polluted parts of the upper Vistula River (southern Poland), *Hydrobiologia,* 169, 183, 1988.
407. **Dunger, W., Dunger, H.-D., Eugelmann, H.-D. , and Schneider, R.,** Untersuchungen zur Langzeitwirkung von Industrie-Emissionen auf Böden, Vegetation und Bödenfauna des Neissetales bei Ostritz/Oberlausitz, *Abh. Ber. Naturkundemus. Görlitz.,* 4, 405, 1974.
408. **Durzan, D. J.,** Nitrogen metabolism of *Picea glauca.* II. Diurnal changes of free amino acids, amides, and guanidino compounds in roots, buds, and leaves during the onset of dormancy of white spruce saplings, *Can. J. Bot.,* 46, 921, 1968.
409. **Du Toit, R.,** Trypanosomiasis in Zululand and the control of tsetse flies by chemical means, *Onderstepoort J. Vet. Res.,* 26, 317, 1954.
410. **Eamus, D. and Jarvis, P. G.,** The direct effects of increase in the global atmospheric CO_2 concentration on natural and commercial temperate trees and forests, *Adv. Ecol. Res.,* 19, 1, 1989.
411. **Ebeling, W.,** Sorptive dusts for pest control, *Ann. Rev. Entomol.,* 16, 123, 1971.
412. **Eberhardt, L. L. and Thomas, J. M.,** Designing environmental field studies, *Ecol. Monogr.,* 61, 53, 1991.
413. **Edmunds, G. F., Jr.,** Ecology of black pineleaf scale (Homoptera: Diaspididae), *Environ. Entomol.,* 2, 765, 1973.
414. **Edmunds, G. F., Jr. and Allen, R. K.,** Comparison of black pineleaf scale population density on normal ponderosa pine and those weakened by other agents, *Proc. 10th Int. Congr. Entomol. (Montreal),* 4, 391, 1956.
415. **Edwards, C. A.,** *Persistant Pesticide in Environment,* CRC Press, Cleveland, OH, 1973.
416. **Egglishaw, H. J.,** The quantitative relationship between bottom fauna and plant detritus in streams of different calcium concentrations, *J. Appl. Ecol.,* 5, 731, 1968.

417. **Ehnström, B. and Waldén, H. W.,** *Faunavård i skogsbruket. Del 2 - Den lägre faunan,* Skogsstyrelsen, Jönköping, Sweden, 1986.
418. **Ehrlich, P. R. and Raven, P. H.,** Butterflies and plants: a study in coevolution, *Evolution,* 18, 586, 1964.
419. **Eidt, D. C.,** The effect of fenitrothion from large-scale forest spraying on benthos in New Brunswick headwaters streams, *Can. Entomol.,* 107, 743, 1975.
420. **Eidt, D. C.,** Recovery of aquatic arthropod population in a woodland stream after depletion by fenitrothion treatment, *Can. Entomol.,* 113, 303, 1981.
421. **Eidt, D. C. and Sundaram, K. M.,** The insecticide fenitrothion in headwaters streams from large-scale forest spraying, *Can. Entomol.,* 107, 735, 1975.
422. **Eidt, D. C. and Weaver, C. A. A.,** Threshold concentrations of aminocarb that causes drift of stream insects, *Can. Entomol.,* 115, 715, 1983.
423. **Eilers, J. M., Lien, G. J., and Berg, R. G.,** Aquatic organisms in acidic environments: a literature review, *Wis. Nat. Res. Tech. Bull.,* 150, 1, 1984.
424. **Eisler, R.,** Polycyclic aromatic hydrocarbon hazards to fish, wildlife, and invertebrates: a synoptic review, *U.S. Fish Wildl. Serv. Biol. Rep.,* 85(1.11), 1, 1987.
425. **Eisler, R.,** Boron hazards to fish, wildlife, and invertebrates: a synoptic review, Biol. Rep 85(1.20), Contam. Hazard Rew., Rep. 20, U.S. Department of the Interior, Fish and Wildlife Service, 1990.
426. **Eisler, R.,** Chlordane hazards to fish, wildlife, and invertebrates: a synoptic review, Biol. Rep 85(1.21), Contam. Hazard Rew., Rep. 21, U.S. Department of the Interior, Fish and Wildlife Service, 1990.
427. **Eisler, R.,** Paraquat hazards to fish, wildlife, and invertebrates: a synoptic review, Biol. Rep 85(1.22), Contam. Hazard Rew., Rep. 22, U.S. Department of the Interior, Fish and Wildlife Service, 1990.
428. **Elden T. C., Howell, R. K., and Webb, R. E.,** Influence of ozone on pea aphid resistance in selected alfalfa strains, *J. Econ. Entomol.,* 71, 283, 1978.
429. **Ellenberg, H.,** Ökologische Veränderungen in Biozönosen durch Stickstoffeintrag, *Berichte Ökol. Forsch. Forschungszentrum Jülich Gmbh,* 4, 75, 1991.
430. **Elliger, C. A., Wong, Y., Chan, B. G., and Waiss, A. C., Jr.,** Growth inhibitors in tomato (*Lycopersicon*) to tomato fruitworms (*Heliothis zea*), *J. Chem. Ecol.,* 7, 753, 1981.
431. **El Sayed, E. J., Graves, J. B., and Bonner, F. L.,** Chlorinated hydrocarbon insecticide residues in selected insects and birds found in association with cotton fields, *J. Agric. Food Chem.,* 15, 1014, 1967.
432. **Elton, C. S.,** *The Ecology of Invasions by Animals and Plants,* Methuen and Co., London, 1958.
433. **Elzen, G. W.,** Sublethal effects of pesticides on beneficial parasitoids, in *Pesticides and Non-target Invertebrates,* Jepson, P. C., Ed., Intercept, Wimborne, Dorset, England, 1989, 129.
434. **Elzen, G. W., O'Brien, P. J., and Powell, J. E.,** Toxic and behavioral effects of selected insecticides on the parasitoid *Microplitis croceipes, Entomophaga,* 34, 87, 1988.
435. **Emanuel, W. R., Shugart, H. H., and Stevenson, M. P.,** Climate change and the broadscale distribution of terrestrial ecosystem complexes, *Climate Change,* 7, 29, 1985.
436. **Emanuel, W. R., Shugart, H. H., and Stevenson, M. P.,** Response to comment: "climatic change and the broad-scale distribution of terrestrial complexes," *Clim. Change,* 7, 457, 1985.
437. **van Emden, H. F.,** Plant resistance to *Myzus persicae* induced by a plant regulator and measured by aphid growth rate, *Entomol. Exp. Appl.,* 12, 125, 1969.
438. **van Emden, H. F., Ed.,** *Aphid Technology,* Academic Press, London, 1972.
439. **van Emden, H. F. and Bashford, M. A.,** A comparison of the reproduction of *Brevicoryne brassicae* and *Myzus persicae* in relation to soluble nitrogen concentration and leaf age (leaf position) in the Brussels sprout plant, *Entomol. Exp. Appl.,* 12, 351, 1969.
440. **Endress, A. G. and Post, S. L.,** Altered feeding preference of Mexican bean beetle *Epilachna varivestis* for ozonated soybean foliage, *Environ. Pollut. Ser A.,* 39, 9, 1985.

441. **Engelmann, F.,** Reproduction in insects, in *Ecological Entomology,* Huffaker, C. B. and Rabb, R. L., Eds., John Wiley & Sons, New York, 1984, 113.
442. **Eriksson, M. O. G., Henrikson, L., Nilsson, B.-I., Nyman, G., Oscarson, H. G., and Stenson, A. E.,** Predator-prey relations important for the biotic changes in acidified lakes, *Ambio,* 9, 248, 1980.
443. **Ernsting, G. and Joosse, E. N. G.,** Predation on two species of surface dwelling Collembola. A study with radio-isotope labelled prey, *Pedobiologia,* 14, 222, 1974.
444. **Esher, R. J.,** Simulated acid rain effects on fine roots, ectomycorrhizae, microorganisms, and invertebrates in pine forests of the southern United States, *Water Air Soil Pollut.,* 61, 269, 1992.
445. **Evans, R. A.,** Response of limnetic insect populations of two acidic, fishless lakes to liming and brook trout (*Salvelinus fontinalis*), *Can. J. Fish Aquat. Sci.,* 46, 342, 1989.
446. **Evenden, J. C.,** Memorandum for files relative to the Northport smelter area, USDA Forest Insect Field St., Coeur d'Alene, ID, 1929.
447. **Evenden J. C.,** The role of forest insects in respect to timber damage in the smelter fume area near Northport, Washington, USDA Forest Insect Field St., Coeur d'Alene, ID, 1931.
448. **Everard, L. B. and Swain, R.,** Isolation, characterization and induction of metallothionein in the stonefly *Eusthenia spectabilis* following exposure to cadmium, *Comp. Biochem. Physiol.,* 75C, 275, 1983.
449. **Everts, J. W., Aukema, B., Hengeveld, R., and Koeman, J. H.,** Side-effects of pesticides on ground-dwelling predatory arthropods in arable ecosystems, *Environ. Pollut.,* 59, 203, 1989.
450. **Everts, J. W., van Frankenhuyzen, K., Roman, B., and Koeman, J. H.,** Side-effects of experimental pyrethroid applications for the control of tsetse flies in a riverine forest habitat (Africa), *Arch. Environ. Contam. Toxicol.,* 12, 91, 1983.
451. **Extence, C. A. and Ferguson, A. J. D.,** Aquatic invertebrate surveys as a water quality management tool in the Anglian Water region, *Reg. Rivers Res. Manage.,* 4, 139, 1989.
452. **Eyre, M. D. and Rushton, S. P.,** Quantification of conservation criteria using invertebrates, *J. Appl. Ecol.,* 26, 159, 1989.
453. **Faegri, K. and van der Pijl, L.,** *Principles of Pollination Ecology,* Pergamon, Oxford, 1978.
454. **Fairchild, F. J., Boyle, T., English, R. W., and Rabeni, C.,** Effects of sediment and contaminated sediment on structural and functional components of experimental stream ecosystems, *Water Air Soil Pollut.,* 36, 271, 1987.
455. **Fajer, E. D.,** The effects of enriched CO_2 atmospheres on plant-insect herbivore interactions: growth responses of larvae of the specialist butterfly, *Junonia coenia* (Lepidoptera: Nymphalidae), *Oecologia,* 81, 514, 1989.
456. **Fajer, E. D., Bowers, M. D., and Bazzaz, F. A.,** The effects of enriched carbon dioxide atmospheres on plant-insect herbivore interactions, *Science,* 243, 1198, 1989.
457. **Fajer, E. D., Bowers, M. D., and Bazzaz, F. A.,** The effects of enriched CO_2 atmospheres on the buckeye butterfly, *Junonia coenia, Ecology,* 72, 751, 1991.
458. **Falkengren-Grerup, U.,** Effect of stemflow on beech forest soils and vegetation in southern Sweden, *J. Appl. Ecol.,* 26, 341, 1989.
459. **Falkengren-Grerup, U. and Björk, L.,** Reversibility of stemflow-induced soil acidification in Swedish beech forest, *Environ. Pollut.,* 74, 31, 1991.
460. **Federenko, A.,** Instars and species-specific diets in two species of *Chaobrus, Limnology and Oceanography,* 20, 238, 1975.
461. **Fehrman, H., von Tiedemann, A., and Fabian, P.,** Predisposition of wheat and barley to fungal leaf attack by preinoculative treatment with ozone and sulphur dioxide, *Z. Pflanzenkr. Pflanzenschutz,* 93, 313, 1986.
462. **Feir, D. and Hale, R.,** Growth and reproduction of an insect model in controlled mixtures of air pollutants, *Int. J. Environ. Stud.,* 20, 223, 1983.
463. **Feir, D. and Hale, R.,** Responses of large milkweed bug, *Oncopeltus fasciatus* (Hemiptera: Lygaeidae) to high levels of air pollutants, *Int. J. Environ. Stud.,* 20, 269, 1983.

464. **Felkl, G.,** First investigations on the abundance of epigeal arthropods, cereal aphids and stenophagous aphid predators in herbicide-free border strips of winter wheat fields in Hesse, *Gesunde Pflanzen,* 40, 483, 1988.
465. **Felton, G. W. and Dahlman, D. L.,** Allelochemical induced stress: effects of L-canavanine on the pathogenity of *Bacillus thuringiensis* in *Manduca sexta, J. Invertebr. Pathol.,* 44, 187, 1984.
466. **Ferenbaugh, R. W.,** Effects of simulated acid rain on *Phaseolus vulgaris* L. (Fabaceae), *Am. J. Bot.,* 63, 283, 1976.
467. **Ferens, M. C. and Beyers, R. J.,** Studies of a simple laboratory microsystem: effects of stress, *Ecology,* 53, 709, 1976.
468. **Ferrario, J. B., Lawler, G. C., DeLeon, L. R., and Laseter, J. L.,** Volatile organic pollutants in biota and sediments of Lake Pontchartrain, *Bull. Environ. Contam. Toxicol.,* 34, 246, 1985.
469. **Ferrell, G. T.,** Moisture stress threshold of susceptibility to fir engraver beetles in pole-size white fir, *For. Sci,* 24, 85, 1978.
470. **Ferry, B. M., Baddeley, M. S., and Hawksworth, D. L.,** *Air Pollution and Lichens,* Athlone Press, London, 1973.
471. **Fiance, S. B.,** Effects of pH on the biology and distribution of *Ephemerella funeralis* (Ephemeroptera), *Oikos,* 31, 332, 1978.
472. **Fischer, P. and Führer, E.,** Effect of soil acidity on the entomophilic nematode *Steinernema kraussei* Steiner, *Biol. Fertil. Soils,* 9, 174, 1990.
473. **Fisher, D. J., Burton, D. T., and Paulson, R. L.,** Aqute toxicity of a complex mixture of synthetic hexachloroethane (HC) smoke combustion products. 1. Comparative toxicity to freshwater aquatic organisms, *Environ. Toxicol. Chem.,* 9, 745, 1990.
474. **Fisher, E., Filip, J., and Molnar, L.,** The effect of bivalent heavy metals on the oxygen-dependent nuclear volume alterations of the chloragocytes in *Tubifex tubifex* Müll., *Environ. Pollut. Ser A.,* 23, 261, 1980.
475. **Fisher, M.,** The effect of previously infested spruce needles on the growth of the green spruce aphid, *Elatobium abietinum,* and the effect of the aphid on the amino acid balance of the host plant, *Ann. Appl. Biol.,* 111, 33, 1987.
476. **Fisher, S. W.,** Effects of pH on the toxicity and uptake of (^{14}C)lindane in the midge, *Chironomus riparius, Ecotoxicol. Environ. Safety,* 10, 202, 1985.
477. **Fisher, S. W.,** Changes in the toxicity of three pesticides as a function of environmental pH and temperature, *Bull. Environ. Contam. Toxicol.,* 46, 197, 1991.
478. **Fisher, S. W. and Wadleigh, R. W.,** Effects of pH on the acute toxicity and uptake of (^{14}C) pentachlorophenol in the midge, *Chironomus riparius, Ecotoxicol. Environ. Safety,* 11, 1, 1986.
479. **Fishman, J., Fakhruzzaman, K., Cros, B., and Nganga, D.,** Identification of widespread pollution in the Southern Hemisphere deduced from satellite analyses, *Science,* 252, 1693, 1991.
480. **Fix, L. A. and Plapp, F. W., Jr.,** Effect of parasitism on the susceptibility of tobacco budworm (Lepidoptera: Noctuidae) to methyl parathion and permethrin, *Environ. Entomol.,* 12, 976, 1983.
480a. **Fjellheim, A. and Raddum, G.,** Recovery of acid-sensitive species of Ephemeroptera, Plecoptera, and Trichoptera in River Audna after liming, *Environ. Pollut.,* 78, 173, 1992.
481. **Flannagan, J. F., Townsend, B. E., and DeMarch, B. G. E.,** Acute and long term effects of methoxychlor larviciding on the aquatic invertebrates of the Athabasca River, Alberta, in *Control of Black Flies in the Athabasca River,* Haufe, W. O. and Croome, G. C. R., Eds., Technical Rep., Alberta Environment, Edmonton, 1980, 151.
482. **Flannagan, J. F., Townsend, B. E., DeMarch, B. G. E., Friesen, M. K., and Leonhard, S. L.,** The effects of an experimental injection of methoxychlor on aquatic invertebrates: accumulation, standing crop, and drift, *Can. Entomol.,* 111, 73, 1979.
483. **Flemming, C. A. and Trevors, J. T.,** Copper toxicity and chemistry in the environment: a review, *Water Air Soil Pollut.,* 44, 1989, 143.

484. **Floravårdskommittén för svampar,** Annoterad lista över hotade svampar i Sverige [An annotated red data list of fungi from Sweden], *Windahlia,* 19, 87, 1991.
485. **Flückiger, W. and Braun, S.,** Effects of air pollutants on insects and hostplant/insect relationships, in proceedings of a workshop jointly organized within the framework of the Concerted Action. Effects of Air Pollution on Terrestrial and Aquatic Ecosystems. Working Party III: How are the effects of air pollutants on agricultural crops influenced by the interaction with other limiting factors? Commission of the European Communities and the National Agency of Environmental Protection, Risö National Laboratory, Denmark, 1986.
486. **Flückiger, W., Oertli, J. J., and Baltensweiler, W.,** Observations of an aphid infestation on hawthorn in the vicinity of a motorway, *Naturwissenschaften,* 65, 654, 1978.
487. **Folmar, L. C.,** Avoidance chamber responses of mayfly nymphs exposed to eight herbicides, *Bull. Environ. Contam. Toxicol.,* 19, 312, 1978.
488. **Folmar, L. C., Sanders, H. O., and Julin, A. M.,** Toxicity of the herbicide glyphosate and several of its formulations to fish and aquatic invertebrates, *Arch. Environ. Contam. Toxicol.,* 8, 269, 1979.
489. **Ford, B. C., Jester, W. A., Griffith, S. M., Morse, R. A., Zall, R. R., Burgett, D. M., Bodyfelt, F. W., and Lisk, D. J.,** Cesium-134 and cesium-137 in honey bees and cheese samples collected in the U.S. after the Chernobyl accident, *Chemosphere,* 17, 1153, 1988.
490. **Ford, M. J.,** *The Changing Climate,* Allen and Unwin, London, 1982.
491. **Förstner, U. and Wittman, G. T.W.,** *Metal Pollution in the Aquatic Environment,* Springer-Verlag, New York, 1979.
492. **Fox, C. J. S.,** The effects of five herbicides on the numbers of certain invertebrate animals in grassland soil, *Can. J. Plant D'Sci.,* 44, 405, 1964.
493. **Frank, K. D.,** Impact of outdoor lightning on moths: an assessment, *J. Lepidopt. Soc.,* 42, 63, 1988.
494. **van Frankenhuyzen, K. and Geen, G. H.,** Microbemediated effect of low pH on availability of detrital energy to a shredder, *Clistoronia magnifica* (Trichoptera, Limnephilidae), *Can. J. Zool.,* 64, 42, 1986.
495. **Frankie, G. W. and Koehler, L. E.,** Ecology of insects in urban environments, *Ann. Rev. Entomol.,* 23, 367, 1978.
496. **Franklin, J. F., Shugart, H. H., and Harmon, M. E.,** Tree death as an ecological process, *BioScience,* 37, 550, 1987.
497. **Fraser, J.,** Acclimation to lead in the freshwater isopod, *Asellus aquaticus, Oecologia,* 45, 419, 1980.
498. **Fratello, B., Bertolani, R., Sabatini, M. A., Mola, L., and Rassu, M. A.,** Effects of atrazine on soil microarthropods in experimental maize fields, *Pedobiologia,* 28, 161, 1985.
499. **Fredeen, F. J. H.,** Tests with single injections of methoxychlor black fly (Diptera: Simuliidae) larvicides in large rivers, *Can. Entomol.,* 106, 285, 1974.
500. **Fredeen, F. J. H.,** Effects of a single injection of methoxychlor black fly larvicide on insect larvae in a 161 km (100 mile) section of the North Saskatchewan River, *Can. Entomol.,* 107, 807, 1975.
501. **Fredeen, F. J. H.,** Trends in numbers of aquatic invertebrates in a large Canadian river during four years of black fly larviciding with methoxychlor (Diptera: Simuliidae), *Quaest. Entomol.,* 19, 53, 1983.
502. **Freedman, B.,** *Environmental Ecology. The Impacts of Pollution and Other Stresses on Ecosystem Structure and Function,* Academic Press, San Diego, 1989.
503. **Freedman, B. and Hutchinson, T. C.,** Effects of smelter pollutants on forest leaf litter decomposition near a nickel-copper smelter at Sudbury, Ontario, Canada, *Can. J. Bot.,* 58, 1722, 1980.
504. **Freitag, R. and Hastings, L.,** Kraft mill fallout and ground beetle populations, *Atmos. Environ.,* 7, 587, 1973.

505. **Freitag, R. and Poulter, F.,** The effects of the insecticides sumithion and phosphamidon on populations of five species of carabid beetles and two species of lycosid spiders in northwestern Ontario, *Can. Entomol.,* 102, 1307, 1970.
506. **Freitag, R., Hastings, L., Mercer, W., and Smith, A.,** Ground beetle populations near a kraft mill, *Can. Entomol.,* 105, 299, 1973.
507. **Fremling, C. R.,** Factors influencing the distribution of burrowing mayflies along the Mississippi River, in *Proc. First Int. Conf. Ephemeroptera,* Peters, W. L. and Peters, J. G., Eds., E. J. Brill, Leiden, 1973, 12.
508. **Fremling, C. R.,** *Hexagenia* mayflies: biological monitors of water quality in the upper Mississippi River, *J. Minn. Acad. Sci.,* 55, 139, 1989.
509. **Frenzel, S. A.,** Effects of municipal wastewater discharges on aquatic communities, Boise River, ID, *Water Resour. Bull.,* 26, 279, 1990.
510. **Fri, R. W.,** Sustainable development. Can we put these principles into practice? *J. For.,* 89, 24, 1989.
511. **Friberg, F., Otto, C., and Svensson, B. S.,** Effects of acidification on the dynamics of allochtonous leaf material and benthic invertebrate communities in running waters, in *Ecological Impact of Acid Precipitations,* Drablos, D. and Tollan, A., Eds., Proc. Int. Conf., SNSF Project, Ås-NLH, Norway, 1980, 304.
512. **Frick, K. G. and Herrmann, J.,** Aluminium accumulation in a lotic mayfly at low pH — A laboratory study, *Ecotoxicol. Environ. Safety,* 19, 81, 1990.
513. **Friend, W. G.,** The nutritional requirement of Diptera, in *Radiation, Radioisotopes and Rearing Methods in the Control of Insect Pests,* Krippner, M., Ed., Proc. IAEA, Vienna, Austria, 1968, 41.
514. **Fritz, R. S., Gaud, W. S., Sacchi, C. F., and Price, P. W.,** Patterns of intra- and interspecific association of gall-forming sawflies in relation to shoot size on their willow host plant, *Oecologia,* 73, 159, 1987.
515. **Fuhremann, T. W. and Liechtenstein, E. P.,** Increase in the toxicity of organophosphorus insecticides to house flies due to polychlorinated biphenyl compounds, *Toxicol. Appl. Pharmacol.,* 22, 628, 1972.
516. **Führer, E.,** Immissionen und Forstschädlinge, *Allg. Forstz. Wien,* 94, 164, 1983.
517. **Führer, E.,** Air pollution and the incidence of forest insect problems, *Z. Angew. Entomol.,* 99, 371, 1985.
518. **Führer, E.,** Forest decline in central Europe: additional aspects of its cause, *For. Ecol. Manage.,* 37, 249, 1990.
519. **Fujii, M.,** Alkaline phosphatase in the alimentary canal of silkworm, *Bombyx mori,* L., larvae, poisoned by the oral application of florine compound, *J. Sericult. Sci. Japan,* 44, 337, 1975.
520. **Fujii, M. and Hayashi, H.,** Fluorides contained in mulberry leaves and silkworms in the area around a tile factory, *J. Sericult. Sci. Japan,* 41, 150, 1972.
521. **Fujii, M. and Honda, S.,** The relative oral toxicity of fluorine compounds for silkworm larvae, *J. Sericult. Sci. Japan,* 41, 104, 1972.
522. **Futuyama, D. J.,** *Evolutionary Biology,* Sinauer, Sunderland, MA, 1979.
523. **Galecka, B.,** Structure and functioning of community of Coccinellidae (Coleoptera) in industrial and agricultural-forest regions, *Pol. Ecol. Stud.,* 6, 717, 1980.
524. **Galecka, B.,** Phenological development of *Frangula alnus* Mill. in an industrial region and the number of *Aphis frangulae* Kalt., *Pol. Ecol. Stud.,* 10, 141, 1984.
525. **Garber, E. D.,** A nutrition-inhibition hypothesis of pathogenity, *Am. Nat.,* 90, 183, 1956.
526. **Gardarsson A., Gislason M. G., and Einarsson A.,** Long term changes in the lake Myvatn ecosystem, *Aqua Fenn.,* 18, 125, 1988.
527. **Gärdenfors, U.,** Impact of airborne pollution on terrestrial invertebrates with particular reference to molluscs, *Nat. Swedish Environ. Prot. Board Rep.,* 3362, 1987, 115.
528. **Garrison, P. J. and Webster, K. E.,** The effects of acidification on the invertebrate portion of the aufwuchs in a mesocosm experiment, *Verh. Internat. Verein. Limnol.,* 23, 2273, 1988.

529. **Gartner, E. J.,** Forest decline in the Federal Republic of Germany — appearance, extent, potential causes, *Geojournal,* 17, 165, 1988.
530. **Gauss, J. D., Woods, P. E., Winner, R. W., and Skillings, J. H.,** Acute toxicity of copper to three life stages of *Chironomus tentans* as affected by water hardness-alkalinity, *Environ. Pollut. Ser. A,* 37, 149, 1985.
531. **Geckler, J. R., Horning, W. B., Neiheisel, T. M., Pickering, Q. H., Robinson, E. L., and Stephan, C. E.,** Validity of laboratory tests for predicting copper toxicity in streams, Ecol. Res. Ser., EPA-600/3-76-116, U.S. Environmental Protection Agency, Duluth, MN, 1976.
532. **Gehring, C. A. and Whitham, T. G.,** Herbivore-driven mycorrhizal mutualism in insect-susceptible pinyon pine, *Nature,* 353, 556, 1991.
533. **Geldmacher-v. Mallinckrodt, M.,** Ecogenetics, in *Pollutants and Their Ecotoxicological Significance,* Nürnberg, H. W., Ed., John Wiley & Sons, Chichester, 1985, 441.
534. **Georghiou, G. P.,** The evolution of resistance to pesticides, *Ann. Rev. Ecol. Syst.,* 3, 133, 1972.
535. **Gepp, J.,** Programmrahmen für einer umfassenden Lepidopterenschutz. Eine Synopsis der Beitrüge, Diskussionen und Anregungen des II. Europäischen Kongresses für Lepidopterologie in Karlsruhe 1980 zum Thema "Europas Schmetterlinge sind bedroht!" *Beih. Veröff. Naturschutz Landschaftspflege Baden-Württemberg,* 21, 191, 1981.
536. **Gerdes, R. A., Smith, J. D., and Applegate, H. G.,** The effects of atmospheric hydrogen fluoride upon *Drosophila melanogaster.* I. Differential genotypic response, *Atmos. Environ.,* 5, 113, 1971.
537. **Gerdes, R. A., Smith, J. D., and Applegate, H. G.,** The effects of atmospheric hydrogen fluoride upon *Drosophila melanogaster.* II. Fecundity, hatchability and fertility, *Atmos. Environ.,* 5, 117, 1971.
538. **Gerhardt, A.,** Effects of subacute doses of cadmium on pH-stressed *Leptophlebia marginata* (L.) and *Baetis rhodani* Pictet (Insecta: Ephemeroptera), *Environ. Pollut.,* 67, 29, 1990.
539. **Gershenzon, J.,** Changes in the levels of plant secondary metabolites under water and nutrient stress, in *Phytochemical Adaptations to Stress,* Timmermann, B. N., Steelink, C., and Loewus, F. A., Eds., *Rec. Adv. Phytochem.,* Vol. 18, Plenum Press, New York, 1984, 273.
540. **Gibbs, K. E., Brautigam, F. C., Stubbs, C. S., and Zibilske, L. M.,** Experimental applications of *B. t. i.* for larval black fly control: persistence and down-stream carry, efficacy, impact on nontarget invertebrates and fish feeding, *Maine Life Sci. Agric. Exp. Stn. Tech. Bull.,* 123, 1, 1986.
541. **Giesy, J. P. and Hoke, R. A.,** Freshwater sediment toxicity bioassessment: rationale for species selection and test design, *J. Great Lakes Res.,* 15, 539, 1989.
542. **Giesy, J. P., Bowling, J. W., Kania, H. J., Knight, R. L., and Mashburn, S.,** Fates of cadmium introduced into channel microcosms, *Environ. Int.,* 5, 159, 1981.
543. **Giesy, J. P., Rosiu, C. J., Graney, R. L., and Henry, M. G.,** Benthic invertebrate bioassays with toxic sediment and pore water, *Environ. Toxicol. Chem.,* 9, 233, 1990.
544. **Gilbert, M. D. and Lisk, D. J.,** Honey as an environmental indicator of radionuclide contamination, *Bull. Environ. Contam. Toxicol.,* 19, 32, 1978.
545. **Gilbert, O. L.,** Some indirect effects of air pollution on bark-living invertebrates, *J. Appl. Ecol.,* 8, 77, 1970.
546. **Gillet, J. W. and Witt, J. M., Eds.,** *Terrestrial Microcosms,* Workshop on Terrestrial Microcosms, Otter Crest, The Foundation, Washington, D.C., 1978.
547. **Ginevan, M. E. and Lane, D. D.,** Effects of SO_2 in air on the fruit fly *Drosophila melanogaster, Environ. Sci. Technol.,* 12, 828, 1978.
548. **Gish, C. D. and Christensen, R. E.,** Cadmium, nickel, lead and zinc in earthworms from roadside soil, *Environ. Sci. Technol.,* 7, 1060, 1973.
549. **Glynn, P.,** Widespread coral mortality and the 1982-83 El Niño warming event, *Environ. Conserv.,* 11, 133, 1984.

550. **Godzik, S. and Linskens, H. F.,** Concentration changes of free amino acid in primary bean leaves after continuous and interrupted SO_2 fumigation and recovery, *Environ. Pollut.*, 7, 25, 1974.
551. **Goldstein H. S. and Babich H.,** Differential effects of arsenite and arsenate to *Drosophila melanogaster* in a combined adult/developmental toxicity assay, *Bull. Environ. Contam. Toxicol.*, 42, 276, 1989.
552. **Goodnight, C. J.,** The use of aquatic macroinvertebrates as indicators of stream pollution, *Trans. Am. Microsc. Soc.*, 92, 1, 1973.
553. **Gordon, H. T.,** Growth and development of insects, in *Ecological Entomology*, Huffaker, C. B. and Rabb, R. L., Eds., John Wiley & Sons, New York, 1984, 53.
554. **Goriup, P. D.,** Acidic air pollution and birds in Europe, *Oryx*, 23, 82, 1989.
555. **Goudie, A.,** *Human Impact on the Natural Environment*, 2nd ed., MIT Press, Cambridge, MA, 1986.
556. **Gower, A. M. and Darlington, S. T.,** Relationships between copper concentrations in larvae of *Plectrocnemia conspersa* (Curtis) (Trichoptera) and in mine drainage streams, *Environ. Pollut.*, 65, 155, 1990.
557. **Granett, A. L. and Musselman, R. C.,** Simulated acid fog injures lettuce, *Atmos. Environ.*, 18, 887, 1984.
558. **Granett, J. and Weseloh, R. H.,** Dimilin toxicity to the gypsy moth larval parasitoid, *Apanteles melanoscelus*, *J. Econ. Entomol.*, 68, 577, 1975.
559. **Grant, I. F.,** Monitoring insecticide side-effects in large-scale treatment programmes: tsetse spraying in Africa, in *Pesticides and Non-target Invertebrates*, Jepson, P. C., Ed., Intercept, Wimborne, Dorset, England, 1989, 43.
560. **Gray, J.,** Do bioassays adequately predict ecological effects of pollutants? *Hydrobiologia*, 188/189, 397, 1989.
561. **Green, L. R. and Dorough, H. W.,** House fly age as a factor in their response to certain carbamates, J. *Econ. Entomol.*, 61, 88, 1968.
562. **Greenfield, J.P. and Ireland, M. P.,** A survey of the macrofauna of a coal-waste polluted Lancashire fluvial system, *Environ. Pollut.*, 16, 105, 1978.
563. **Greville, R. W. and Morgan, A. J.,** A comparison of (Pb, Cd and Zn) accumulation in terrestrial slugs maintained in microcosms: evidence for metal tolerance, *Environ. Pollut.*, 74, 115, 1991.
564. **Griswold, M. J. and Trumble, J. T.,** Consumption and utilization of celery, *Apium graveolus*, by the beet armyworm (*Spodoptera exigua*), *Entomol. Exp. Appl.*, 38, 73, 1985.
565. **Grosch, D. S.,** Reproductive performance of *Bracon hebetor* after sublethal doses of carbaryl, *J. Econ. Entomol.*, 68, 659, 1975.
566. **Grue, C. E., DeWeese, L. R., Mineau, P., Swanson, G. A., Foster, J. R., Arnold, P. M., Huckins, J. N., Sheehan, P. J., Marshall, W. K., and Ludden, A. P.,** Potential impacts of agricultural chemicals on waterfowl and other wildlife inhabiting prairie wetland: an evaluation of research needs and approaches, *Trans. North Am. Wildl. Nat. Resour. Conf.*, 51, 357, 1986.
567. **Gruttke, H., Kratz, W., Weigmann, G., and Haque, A.,** Terrestrial model food chain and environmental chemicals. I. Transfer of sodium (^{14}C)Pentachlorophenate between springtails and carabids, *Ecotoxicol. Environ. Safety*, 15, 253, 1988.
568. **Grzybkowska, M.,** Production estimates of the dominant taxa of Chironomidae (Diptera) in the modified River Widawka and the natural River Grabia, central Poland, *Hydrobiologia*, 179, 245, 1989.
569. **Guderian, R.,** *Air Pollution*, Springer-Verlag, Berlin, 1977.
570. **Guderian, R., Ed.,** *Air Pollution by Photochemical Oxidants: Formation, Transport, Control, and Effects on Plants*, Ecol. Stud., Vol. 52, Springer-Verlag, New York, 1985.
571. **Guiney, P. D., Sykora, J. L., and Keleti, G.,** Environmental impact of an aviation kerosene spill on stream water quality in Cambria County, PA, *Environ. Toxicol. Chem.*, 6, 977, 1987.

572. **Gunn, J. M. and Keller, W.,** Biological recovery of an acid lake after reductions in industrial emissions of sulfur, *Nature,* 345, 437, 1990.
573. **Gunnarsson, B.,** Spruce-living spiders and forest decline; the importance of needle-loss, *Biol. Conserv.,* 43, 309, 1988.
574. **Gunnarsson, B. and Johnsson, J.,** Effects of simulated acid rain on growth rate in a spruce-living spider, *Environ. Pollut.,* 56, 311, 1989.
575. **Gunnison, A. F.,** Sulfite toxicity: a critical review of in vitro and in vivo data, *Food Cosmet. Toxicol.,* 19, 667, 1981.
576. **Gupta, A. P.,** Arthropod immunocytes. Identification, structure, functions, and analogies to the functions of vertebrate B- and T-lymphocytes, in *Hemocytic and Humoral Immunity in Arthropods,* Gupta, A. P., Ed., John Wiley & Sons, New York, 1986, 3.
577. **Hagen, K. S., Dadd, R. H., and Reese, J.,** The food of insects, in *Ecological Entomology,* Huffaker, C. B. and Rabb, R. L., Eds., John Wiley & Sons, New York, 1984, 79.
578. **Hagler, J. R., Waller, G. D., and Lewis, B. E.,** Mortality of honeybees (Hymenoptera: Apidae) exposed to permethrin and combinations of permethrin with piperonyl butoxide, *J. Apic. Res.,* 28, 208, 1989.
579. **Hågvar, S.,** Effects of liming and artificial acid rain on Collembola and Protura in coniferous forest, *Pedobiologia,* 27, 341, 1984.
580. **Hågvar, S.,** Atmospheric deposition: impact via soil biology, in *Acidification and Its Policy Implications,* Schneider, T., Ed., *Stud. Environ. Sci.,* 30, 153, 1986.
581. **Hågvar, S.,** Reactions to soil acidification in microarthropods: is competition a key factor? *Biol. Fertil. Soils,* 9, 178, 1990.
582. **Hågvar, S. and Abrahamsen, G.,** Eksperimentelle forsuringsforsök i skog 5. Jordbunnszoologiske undersökelser [Acidification experiments in conifer forest 5. Studies on the soil fauna], SNSF-prosjektet, IR 32/77, NISK, Oslo-Ås, 1977, 1.
583. **Hågvar, S. and Abrahamsen, G.,** Colonisation by Enchytraeidae, Collembola and Acari in sterile soil samples with adjusted pH levels, *Oikos,* 34, 245, 1980.
584. **Hågvar, S. and Abrahamsen, G.,** Microarthropoda and Enchytraeidae (Oligochaeta) in naturally lead-contaminated soil: a gradient study, *Environ. Entomol.,* 19, 1263, 1990.
585. **Hågvar, S. and Amundsen, T.,** Effects of liming and artificial acid rain on the mite (Acari) fauna in coniferous forest, *Oikos,* 37, 7, 1981.
586. **Hågvar, S. and Kjöndal, B. R.,** Effects of artificial acid rain on the microarthropod fauna in decomposing birch leaves, *Pedobiologia,* 22, 409, 1981.
587. **Hågvar, S., Abrahamsen, G., and Bakke, A.,** Angrep av furuskuddmøll (*Exoteleia dodecella* L.) på Sörlandet. Mulig sammenheng med sur nedbør [Attack by the pine bud moth (*Exoteleia dodecella* L.) in southernmost Norway: possible effect of acid precipitation], SNSF-prosjektet, IR 15/76, NISK, Oslo-Ås, 1976.
588. **Hågvar. S., Fjellberg, A., Klausen, F. E., Kvamme, T., and Refseth, D.,** Experimentelle forsuringsforsøk i skog. 7. Virkning av syrebehandling på insekter og edderkopper i skogbunnern, SNSF-project IR68/80, Oslo-Ås, 1980.
589. **Hain, F. P.,** Interactions of insects, trees and air pollutants, *Tree Physiol.,* 3, 93, 1987.
590. **Hain, F. P. and Arthur, F. H.,** The role of atmospheric deposition in the latitudinal variation of fraser fir mortality caused by balsam woolly adelgids, *Adelges piceae* Ratz. (Hemiptera: Adelgidae): a hypothesis, *Z. Angew. Entomol.,* 99, 145, 1985.
591. **Haines, T.,** Acidic precipitation and its consequences for aquatic ecosystems: a review, *Trans. Am. Fish. Soc.,* 110, 669, 1981.
592. **Hairston, N. G., Smith, F. E., and Slobodkin, L. B.,** Community structure, population control, and competition, *Am. Nat.,* 94, 421, 1960.
593. **Hale, J. G.,** Toxicity of metal mining wastes, *Bull. Environ. Contam. Toxicol.,* 17, 66, 1977.
594. **Hall, R. J. and Ide, F. P.,** Evidence of acidification effects on stream insect communities in central Ontario between 1937 and 1985, *Can. J. Fish. Aquat. Sci.,* 44, 1652, 1987.

595. **Hall, R. J., Driscoll, C. T., and Likens, G. E.,** Importance of hydrogen ions and aluminium in regulating the structure and function of stream ecosystems: an experimental test, *Freshwater Biol.,* 18, 17, 1987.
596. **Hall, R. J., Driscoll, C. T., Likens, G. E., and Pratt, J. M.,** Physical, chemical and biological consequences of episodic aluminium additions to a stream, *Limnol. Oceanogr.,* 30, 212, 1985.
597. **Hall, R. J., Likens, G. E., Fiance, S. B., and Hendrey, G. R.,** Experimental acidification of a stream in the Hubbard Brook experimental forest, N. H., *Ecology,* 61, 976, 1980.
598. **Hall, R. J., Pratt, J. M., and Likens, G. E.,** Effects of experimental acidification on macroinvertebrate drift diversity in Aam, *Water Air Soil Pollut.,* 18, 273, 1982.
599. **Hall, R. W., Barger, J. H., and Townsend, A. M.,** Effects of simulated acid rain, ozone and sulfur dioxide on suitability of elms for elm leaf beetle. *J. Arboricult.,* 14, 61, 1988.
600. **Hämäläinen, H. and Huttunen, P.,** Estimation of acidity in streams by means of benthic invertebrates: evaluation of two methods, *Acidification in Finland,* Kauppi, P., Anttila, P., and Kenttämies, K., Eds., Springer-Verlag, Berlin, 1990, 1051.
601. **Hamilton, A. L. and Saether, O. A.,** The occurrence of characteristic deformities in the chironomid larvae of several Canadian lakes, *Can. Entomol.,* 103, 363, 1971.
602. **Hammer, U. T., Huang, P. M., and Liaw, W.,** Bioaccumulation of mercury in aquatic ecosystems, *Can. Techn. Rep. Fish. Aquat. Sci.,* 1163, 69, 1982.
603. **Hanazato, T.,** Effects of repeated applications of carbaryl on zooplankton communities in experimental ponds with or without the predator *Chaoborus, Environ. Pollut.,* 74, 309, 1991.
604. **Hanazato, T. and Yasuno, M.,** Effects of a carbamate insecticide, carbaryl, on the summer phyto- and zooplankton communities in ponds, *Environ. Pollut.,* 48, 145, 1987.
605. **Hanazato, T. and Yasuno, M.,** Effects of carbaryl on the spring zooplankton communities in ponds, *Environ. Pollut.,* 56, 1, 1989.
606. **Hanazato, T. and Yasuno, M.,** Influence of *Chaoborus* density on the effects of an insecticide on zooplankton communities in ponds, *Hydrobiologia,* 194, 183, 1990.
607. **Haney, A. and Lipsey, R. L.,** Accumulation and effects of methyl mercury hydroxide in a terrestrial food chain under laboratory conditions, *Environ. Pollut.,* 5, 305, 1973.
608. **Hansen, J., Fung, I., Lacis, A., Lebedeff, S., Rind, D., Ruedy, R., and Russel, G.,** Prediction of near-term climate evolution: what can we tell decision-makers now? In *Proc. First North American Conf., Preparing for Climate Change: A Cooperative Approach,* Topping, J. C., Jr., Ed., Government Inst., Rockville, MD, 1988, 35.
609. **Hansen, J., Lacis, A., Rind, D., Russell, G., Fung, I., and Lebedeff, S.,** Evidence for future warming: how large and when, in *The Greenhouse Effect, Climate Change, and U.S. Forests,* Shands, W. E. and Hoffman, J. S., Eds., Conservation Foundation, Washington, D.C., 1987, 57.
610. **Hansen, S. R. and Garton, R. R.,** The effects of diflubenzuron on a complex laboratory stream community, *Arch. Environ. Contam. Toxicol.,* 11, 1, 1982.
611. **Hanski, I.,** Metapopulation dynamics: does it help to have more of the same? *Trends Evol. Ecol.,* 4, 113, 1989.
612. **Hanski, I.,** Single-species metapopulation dynamics: concepts, models and observations, *Biol. J. Linnean Soc.,* 42, 17, 1991.
613. **Hanski, I. and Gilpin, M.,** Metapopulation dynamics: brief history and conceptual domain, *Biol. J. Linnean Soc.,* 42, 3, 1991.
614. **Haque, A., Gruttke, H., Kratz, W., Kielhorn, U., Weigmann, G., Meyer, G., Bornkamm, R., Schuphan, I., and Ebing, W.,** Environmental fate and distribution of sodium (^{14}C)-pentachlorophenate in a section of urban wasteland ecosystem, *Sci. Total Environ.,* 68, 127, 1988.
615. **Hare, J. D.,** Seasonal variation in plant-insect associations: utilization of *Solanum dulcamara* by *Leptinotarsa decemlineata, Ecology,* 64, 345, 1983.
616. **Hare, L. and Carter, J. C. H.,** The distribution of *Chironomus* (s.s.)? *cucini* (*salinarius* group) larvae (Diptera: Chironomidae) in Parry Sound, Georgian Bay, with particular reference to structural deformities, *Can. J. Zool.,* 54, 2129, 1976.

617. **Hare, L., Campbell, P. G. C., Tessier, A., and Belzile, N.,** Gut sediments in a burrowing mayfly (Ephemeroptera, *Hexagenia limbata*): their contribution to animal trace element burdens, their removal, and the efficacy of a correction for their presence, *Can. J. Fish. Aquat. Sci.,* 46, 451, 1989.
618. **Hare, L., Saouter, E., Campbell, P. G. C., Tessier, A., Ribeyre, F., and Boudou, A.,** Dynamics of cadmium, lead, and zinc exchange between nymphs of the burrowing mayfly *Hexagenia rigida* (Ephemeroptera) and the environment, *Can. J. Fish. Aquat. Sci.,* 48, 39, 1991.
619. **Harp, G. L. and Campbell, R. J.,** The distribution of *Tendipes plumosus* in mineral acid water, *Limnol. Oceanogr.,* 12, 260, 1967.
620. **Harper, J. L.,** *Population Biology of Plants,* Academic Press, London, 1980.
621. **Harrel, R. C.,** Effects of crude oil spill on water quality and macrobenthos of a southeast Texas stream, *Hydrobiologia,* 124, 223, 1985.
622. **Harrewijn, P.,** Reproduction of the aphid *Myzus persicae* related to mineral nutrition of potato plants, *Entomol. Exp. Appl.,* 13, 307, 1970.
623. **Harriman, R. and Morrison, B. R. S.,** Ecology of streams draining forested and nonforested catchments in an area of central Scotland subject to acidid precipitation, *Hydrobiologia,* 88, 251, 1982.
624. **Harrison, W., Winnik, M. A., Kwong, T. Y., and Mackay, D.,** Disappearance of aromatic and aliphatic components from small sea-surface slicks, *Environ. Sci. Technol.,* 9, 231, 1975.
625. **Hartman, W. A. and Martin, D. B.,** Effects of four agricultural pesticides on *Daphnia pulex, Lemna minor,* and *Potamogeton pectinatus, Bull. Environ. Contam. Toxicol.,* 35, 646, 1985.
626. **Hartman, W. L.,** Lake Erie, effects of exploitation, environmental changes and new species on the fishery resources, *J. Fish. Res. Board Can.,* 29, 899, 1972.
627. **Hartnett, D. C. and Bazzaz, F. A.,** Leaf demography and plant-insect interactions: goldenrods and phloem-feeding aphids, *Am. Nat.,* 124, 137, 1984.
628. **Hassan, S. A.,** Testing methodology and the concept of the IOBC/WPRS Working Group, in *Pesticides and Non-target Invertebrates*, Jepson, P. C., Ed., Intercept, Wimborne, Dorset, England, 1989, 1.
629. **Hassell, M. P., Comins, H. N., and May, R. M.,** Spatial structure and chaos in insect population dynamics, *Nature* 353, 255, 1991.
630. **Hastings, F. L., Brady, U. E., and Jones, A. S.,** Lindane and fenitrothion reduce soil and litter mesofauna on Piedmont and Appalachian sites, *Environ. Entomol.,* 18, 245, 1989.
631. **Hatakeyama, S.,** Chronic effects of Cd on reproduction of *Polypedilum nubifer* (Chironomidae) through water and food, *Environ. Pollut.,* 48, 249, 1987.
632. **Hatakeyama, S.,** Effect of copper and zinc on the growth and emergence of *Epeorus latifolium* (Ephemeroptera) in an indoor model stream, *Hydrobiologia,* 174, 17, 1989.
633. **Hatakeyama, S. and Yasuno, M.,** A method for assessing chronic effects of toxic substances on the midge *Paratanytarsus parthenogeneticus* — effects of copper, *Arch. Environ. Contam. Toxicol.,* 10, 705, 1981.
634. **Hatakeyama, S. and Yasuno, M.,** Chronic effects of Cd on the reproduction of the guppy (*Poecilia reticulata*) through Cd-accumulated midge larvae (*Chironomus yoshimatsui*), *Ecotoxicol. Environ. Safety,* 14, 191, 1987.
635. **Hatakeyama, S., Satake, K., and Fukushima, S.,** Flora and fauna in heavy metal polluted rivers. I. Density of *Epeorus latifolium* (Ephemeroptera) and heavy metal concentration of *Baetis* (Ephemeroptera) relating to Cd, Cu and Zn concentrations, *Res. Rep. Nat. Inst. Environ. Stud. Jpn.,* 99, 15, 1986.
636. **Hatakeyama, S., Shiraishi, H., and Kobayashi, N.,** Effects of aerial spraying of insecticides on nontarget macrobenthos in a mountain stream, *Ecotoxicol. Environ. Safety,* 19, 254, 1990.
637. **Haufe, W. O., Depner, K. R., and Charnetski, W. A.,** Impact of methoxychlor on drifting aquatic invertebrates, in *Control of Black Flies in the Athabasca River,* Haufe, W. O. and Croome, G. C. R., Eds., Alberta Environment, Edmonton, 1980, 159.

638. **Havas, M.**, Physiological response of aquatic animals to low pH, in *Effects of Acidic Precipitation on Benthos*, Singer, R., Ed., Proc. Symp., North American Benthol. Society, Springfield, IL, 1981, 49.
639. **Havas, M.**, Aluminium bioaccumulation and toxicity to *Daphnia magna* in soft water at low pH, *Can. J. Fish. Aquat. Sci.*, 42, 1741, 1985.
640. **Havas, M. and Hutchinson, T. C.**, Aquatic invertebrates from the Smoking Hills, N.W.T.: effects of pH and metals on mortality, *Can. J. Fish. Aquat.*, 39, 890, 1982.
641. **Havas, M. and Likens, G. E.**, Toxicity of aluminium and hydrogen ions to *Daphnia catawba, Holopedium gibberum, Chaoborus punctipennis* and *Chironomus anthracinus* from Mirror Lake, New Hampshire, *Can. J. Zool.*, 63, 1114, 1985.
642. **Havas, P.**, Injury to pines in the vicinity of a chemical processing plant in northern Finland, *Acta For. Fenn.*, 121, 1, 1971.
643. **Hawkins, B. A. and Lawton, J. H.**, Species richness for parasitoids of British phytophagous insects, *Nature*, 326, 788, 1987.
644. **Hay, C. J.**, Bibliography of Arthropoda and air pollution, USDA Forest Service General Tech. Rep., NE-24, USDA-FS, Upper Darby, PA, 1977, 1.
645. **Haynes, K. F. and Baker, T. C.**, Sublethal effects of permethrin on the chemical communication system of the pink bollworm moth, *Pectinophora gossypiella, Arch. Insect Biochem. Physiol.*, 2, 283, 1985.
646. **Heagle, A. S.**, Interactions between air pollutants and plant parasites, *Ann. Rev. Phytopathol.*, 11, 365, 1973.
647. **Heagle, A. S.**, Response of three obligate parasites to ozone, *Environ. Pollut.*, 9, 91, 1975.
648. **Heath, R. L.**, Air pollutant effects on biochemicals derived from metabolism: organic, fatty, and amino acids, in *Gaseous Air Pollutants and Plant Metabolism*, Koziol, M. J. and Whatley, F. R., Eds., Butterworths, London, 1984, 275.
649. **Hegazi, E. M., Rawash, I. A., El-Gayar, F. H., and Kares, E. A.**, Effect of parasitism by *Microplitis rufiventris* Kok. on the susceptibility of *Scodoptera littoralis* (Boisd.) larvae to insecticides, *Acta Phytopathol. Acad. Sci. Hungaricae*, 17, 115, 1982.
650. **Heil, G. W. and Diemont, W. M.**, Raised nutrient levels change heathland into grassland, *Vegetatio*, 53, 113, 1983.
651. **Heliövaara, K.**, Occurrence of *Petrova resinella* (Lepidoptera, Tortricidae) in a gradient of industrial air pollutants, *Silva Fenn.*, 20, 83, 1986.
652. **Heliövaara, K. and Väisänen, R.**, Industrial air pollution and the pine bark bug, *Aradus cinnamomeus* Panz. (Het., Aradidae), *J. Appl. Entomol.*, 101, 469, 1986.
653. **Heliövaara, K. and Väisänen, R.**, Parasitization in *Petrova resinella* (Lepidoptera, Tortricidae) galls in relation to industrial air pollutants, *Silva Fenn.*, 20, 233, 1986.
654. **Heliövaara, K. and Väisänen, R.**, Interactions among herbivores in three polluted pine stands, *Silva Fenn.*, 22, 283, 1988.
655. **Heliövaara, K. and Väisänen, R.**, Invertebrates of young Scots pine stands in the industrialized town of Harjavalta, *Silva Fenn.*, 23, 13, 1988.
656. **Heliövaara, K. and Väisänen, R.**, Between-species differences in heavy metal levels in four pine diprionids (Hymenoptera) along an air pollutant gradient, *Environ. Pollut.*, 62, 253, 1989.
657. **Heliövaara, K. and Väisänen, R.**, Reduced cocoon size of diprionids (Hymenoptera) reared on pollutant affected pines, *J. Appl. Entomol.*, 107, 32, 1989.
658. **Heliövaara, K. and Väisänen, R.**, Heavy-metal contents in pupae of *Bupalus piniarius* (Lepidoptera: Geometridae) and *Panolis flammea* (Lepidoptera: Noctuidae) near an industrial source, *Environ. Entomol.*, 19, 481, 1990.
659. **Heliövaara, K. and Väisänen, R.**, Concentrations of heavy metals in the food, faeces, adults, and empty cocoons of *Neodiprion sertifer* (Hymenoptera, Diprionidae), *Bull. Environ. Contam. Toxicol.*, 45, 13, 1990.
660. **Heliövaara, K. and Väisänen, R.**, Air pollution levels and abundance of forest insects, in *Acidification in Finland*, Kauppi, P., Anttila, P., and Kenttämies, K., Eds., Springer-Verlag, Berlin, 1990, 447.

661. **Heliövaara, K. and Väisänen, R.,** Changes in population dynamics of pine insects induced by air pollution, in *Population Dynamics of Forest Insects,* Watt, A. D., et al., Eds., Intercept, Andover, 1990, 209.
662. **Heliövaara, K. and Väisänen, R.,** Prolonged development in *Diprion pini* (Hymenoptera, Diprionidae) reared on pollutant affected pines, *Scand. J. For. Res.,* 5, 127, 1990.
663. **Heliövaara, K. and Väisänen, R.,** Bark beetles and associated species with high heavy metal tolerance, *J. Appl. Entomol.,* 111, 397, 1991.
664. **Heliövaara, K., Terho, E., and Koponen, M.,** Parasitism in the eggs of the pine bark bug, *Aradus cinnamomeus* (Heteroptera, Aradidae), *Ann. Entomol. Fenn.,* 48, 31, 1982.
665. **Heliövaara, K., Väisänen, R., Braunschweiler, H., and Lodenius, M.,** Heavy metal levels in two biennial pine insects with sap-sucking and gall-forming life-styles, *Environ. Pollut.,* 48, 13, 1987.
666. **Heliövaara, K., Väisänen, R., and Kemppi, E.,** Change of pupal size of *Panolis flammea* (Lepidoptera; Noctuidae) and *Bupalus piniarius* (Geometridae) in response to concentration of industrial pollutants in their food plant, *Oecologia,* 79, 179, 1989.
667. **Heliövaara, K., Väisänen, R., and Varama, M.,** Fecundity and egg viability in relation to female body size in *Neodiprion sertifer* (Hymenoptera: Diprionidae), *Holarctic Ecol.,* 13, 166, 1990.
667a. **Heliövaara, K., Väisänen, R., and Varama, M.,** Larval mortality of pine sawflies (Hymenoptera: Diprionidae) in relation to pollution level: a field experiment, *Entomophaga,* 36, 315, 1991.
668. **Heliövaara, K., Väisänen, R., and Varama, M.,** Acidic precipitation increases egg survival in *Neodiprion sertifer, Entomol. Exp. Appl.,* 62, 55, 1992
669. **Hellawell, J. M.,** *Biological Indicators of Freshwater Pollution and Environmental Management,* Elsevier, London, 1986.
670. **Hendrey, G. and Wright, R.,** Acid precipitation in Norway: effects on aquatic fauna, *J. Great Lakes Res.,* 2 (Suppl. 1), 192, 1976.
671. **Hendry, G. R., Yan, N. D., and Baumgartner, B. J.,** Responses of freshwater plants and invertebrates to acidification, in *Restoration of Lakes and Inland Rivers,* EPA 440/5-81/010, U.S. Environmental Protection Agency, Washington D.C., 1980, 457.
672. **Hennig, W.,** *Insect Phylogeny,* John Wiley & Sons, Chichester, U.K., 1981, 514.
673. **Henrikson, L., Oscarson, H. G., and Stenson, J. A. E.,** Does the change of predator system contribute to the biotic development in acidic lakes, in *Ecological Impact of Acid Precipitations,* Drablos, D. and Tollan, A., Eds., Proc. Int. Conf., SNSF Project, Ås-NLH, Norway, 1980, 316.
674. **Heppner, J. B.,** Faunal regions and the diversity of Lepidoptera, *Tropic. Lepidopt.,* 2 (Suppl. 1), 1, 1991.
675. **Herricks, E. E. and Cairns, J.,** The recovery of stream macrobenthos from low pH stress, *Rev. Biol. (Lisboa),* 10, 1, 1974.
676. **Herrmann, J.,** Sodium levels of lotic mayfly nymphs being exposed to aluminium at low pH — a preliminary report, *Ann. Soc. R. Zool. Belg.,* 117, 181, 1987.
677. **Herrmann, J.,** Aluminium impact on freshwater invertebrates at low pH — a review, in *Speciation of Metals in Water, Sediment and Soil System, Lecture Notes in Earth Sciences,* Vol. 11, Landner, L., Ed., Springer-Verlag, Berlin, 1987, 157.
678. **Herrmann, J. and Andersson, G.,** Aluminium impact on respiration of lotic mayflies at low pH, *Water Air Soil Pollut.,* 30, 703, 1986.
679. **Hershey, A. E.,** Tubes and foraging behavior in larval Chironomidae: implications for predator avoidance, *Oecologia,* 73, 236, 1987.
680. **Heynen, C. von,** Untersuchungen zum Einfluss von diflubenzuron (Dimilin) auf das Wirt-parasit-system *Scodoptera littoralis* Boisd. (Lep., Noctuidae)/*Microplitis rufiventris* Kok. (Hym., Braconidae), *Z. Angew. Entomol.,* 100, 113, 1985.
681. **Hienton, T. E.,** Summary of investigations of electric traps, *U.S. Dept. Agric. Tech. Bull.,* 1498, 1974, 1.

682. **Hildebrand, L. D., Sullivan, D. S., and Sullivan, T. P.,** Effects of Roundup® herbicide on populations of *Daphnia magna* in a forest pond, *Bull. Environ. Contam. Toxicol.,* 25, 353, 1980.
683. **Hildrew, A. G., Townsend, C. R., Francis, J., and Finch, K.,** Cellulolytic decomposition in streams of contrasting pH and its relationship with invertebrate community structure, *Freshwater Biol.,* 14, 323, 1984.
684. **Hill, I. R.,** Aquatic organisms and pyrethroids, *Pest. Sci.,* 27, 429, 1989.
685. **Hill, L. and Michaelis, F. B.,** Conservation of insects and related wildlife, Occ. Paper 13, Australian National Parks and Wildlife Service, Canberrra, 1988.
686. **Hillmann, R. C. and Benton, A. W.,** Biological effects of air pollution on insects, emphasizing the reactions of the honeybee (*Apis mellifera* L.) to sulphur dioxide, *J. Elisha Mitchell Sci. Soc.,* 88, 195, 1972.
687. **Hiltbrunner, E. and Flückiger, W.,** Altered feeding preference of beech weevil *Rhynchaenus fagi* L. for beech foliage under ambient air pollution, *Environ. Pollut.,* 75, 333, 1992.
688. **Hiltunen, J. K. and Manny, B. A.,** Distribution and abundance of macrozoobenthos in the Detroit River and Lake St. Clair, U.S. Fish and Wildlife Service, Administrative Report 82-2, National Great Lakes Fisheries Center, Ann Arbor, MI, 1982.
689. **Hiltunen, J. K. and Schloesser, D. W.,** The occurrence of oil and the distribution of *Hexagenia* (Ephemeroptera: Ephemeridae) nymphs in the St. Marys River, Michigan and Ontario, *Freshwater Invertebr. Biol.,* 2, 199, 1983.
690. **Hinton, H. E.,** *Biology of Insect Eggs,* Vols. I, II, and III, Pergamon Press, Oxford, 1981.
691. **Hirano, M.,** Characteristics of pyrethroids for insect pest control in agriculture, *Pest. Sci.,* 27, 353, 1989.
692. **Hobbie, J. E., Cole, J., Dungan, J., Houghton, R. A., and Peterson, B. J.,** Role of biota in global CO_2 balance: the controversy, *BioScience,* 34, 492, 1984.
693. **Hodkinson, I. D. and Hughes, M. K.,** *Insect Herbivory,* Chapman and Hall, London, 1982.
694. **Hodson, P. V., Borgman, U., and Shear, H.,** Toxicity of copper to aquatic biota, in *Copper in the Environment, Part II: Health Effects,* Nriagu, J. O., Ed., John Wiley & Sons, New York, 1979, 308.
695. **Hoehn, R. C., Stauffer, J. R., Masnik, M. T., and Hocutt, C. H.,** Relationships between sediment oil concentrations and the macroinvertebrates present in a small stream following an oil spill, *Environ. Lett.,* 7, 345, 1974.
696. **Hoffman, M. R.,** Comment on acid fog, *Environ. Sci. Technol.,* 18, 61, 1984.
697. **Hoffman, W. A., Jr., Lindgren, S. E., and Turner, R. R.,** Precipitation acidity: the role of canopy in acid exchange, *J. Environ. Qual.,* 9, 95, 1980.
698. **Holcombe, G. W., Phipps, G. L., Sulaiman, A. H., and Hoffman, A. D.,** Simultaneous multiple species testing: acute toxicity of 13 chemicals to 12 diverse freshwater amphibian, fish, and invertebrate families, *Arch. Environ. Contam. Toxicol.,* 16, 697, 1987.
699. **Holdgate, M. W.,** *A Perspective of Environmental Pollution,* Cambridge University Press, Cambridge, U.K., 1979.
700. **Holopainen, J. K., Kainulainen, E., Oksanen, J., Wulff, A., and Kärenlampi, L.,** Effect of exposure to fluoride, nitrogen compounds and SO_2 on the numbers of spruce shoot aphids on Norway spruce seedlings, *Oecologia,* 86, 51, 1991.
701. **Holopainen, J. K., Mustaniemi, A., Kainulainen, P., Satka, H., and Oksanen, J.,** Conifer aphids in an air-polluted environment. I. Aphid density, growth and accumulation of sulphur and nitrogen by Scots pine and Norway spruce seedlings, *Environ. Pollut.,* in press.
702. **Honék A.,** Regulation of body size in a heteropteran bug, *Pyrrhocoris apterus, Entomol. Exp. Appl.,* 44, 257, 1987.
703. **Honék, A and Kocourek, F.,** Thermal requirements for development of aphidophagous Coccinellidae (Coleoptera), Chrysopidae, Hemerobiidae (Neuroptera), and Syrphidae (Diptera): some general trends, *Oecologia,* 76, 455, 1988.

704. **van Hook, R. I. and Yates, A. J.,** Transient behavior of cadmium in a grassland arthropod food chain, *Environ. Res.,* 9, 76, 1975.
705. **Hopkin, S. P.,** Ecophysiological strategies of terrestrial arthropods for surviving heavy metal pollution, in *Proc. Third Eur. Congress Entomol.,* Velthuis, H. H. W., Ed., Nederl. Entomol. Ver., Amsterdam, 1986, 263.
706. **Hopkin, S. P.,** *Ecophysiology of Metals in Terrestrial Invertebrates,* Elsevier, London, 1989.
707. **Hopkin, S. P. and Martin, M. H.,** Assimilation of zinc, cadmium, lead and copper by the centipede *Lithobius variegatus* (Chilopoda), *J. Appl. Ecol.,* 21, 535, 1984.
708. **Hopkin, S. P. and Martin, M. H.,** Transfer of heavy metals from leaf litter to terrestrial invertebrates, *J. Sci. Food Agric.,* 36, 538, 1985.
709. **Hopkin, S. P. and Martin, M. H.,** Assimilation of zinc, cadmium, lead, copper and iron by the spider *Dysdera crocata,* a predator of woodlice, *Bull. Environ. Contam. Toxicol.,* 34, 183, 1985.
710. **Hopkin, S. P., Watson, K., Martin, M. H., and Mould, M. L.,** The assimilation of heavy metals by *Lithobius variegatus* and *Glomeris marginata* (Chilopoda; Diplopoda), *Bijdragen tot de Dierkunde,* 55, 88, 1985.
711. **Hopkins, P. S., Kratz, K. W., and Cooper, S. D.,** Effects of an experimental acid pulse on invertebrates in a high altitude Sierra Nevada stream, *Hydrobiologia,* 171, 45, 1989.
712. **Horn, H. S.,** Succession, in *Theoretical Ecology,* May, R. M., Ed., Blackwell Scientific, Oxford, U.K., 1976, 187.
713. **Hothem, R. L. and Ohlendorf, H. M.,** Contaminants in foods of aquatic birds at Kesterson Reservoir, California, 1985, *Arch. Environ. Contam. Toxicol.,* 18, 773, 1989.
714. **Houghton, R. A., Hobbie, J. E., Melillo, J. M., Moore, B., Peterson, B. J., Shaver, G. R., and Woodwell, G. M.,** Changes in the carbon content of terrestrial biota and soils between 1860 and 1980: a net release of CO_2 to the atmosphere, *Ecol. Monogr.,* 53, 235, 1983.
715. **Houlden, G., McNeill, S., Aminu-Kano, M., and Bell, J. N.,** Air pollution and agricultural aphid pests. I Fumigation experiments with SO_2 and NO_2, *Environ. Pollut.,* 67, 305, 1990.
716. **Houlden, G., McNeill, S., Craske, A., and Bell, J. N. B.,** Air pollution and agricultural aphid pests. II Chamber filtration experiments, *Environ. Pollut.,* 72, 45, 1991.
717. **Hovmand, M. F., Tjell, J. C., and Mosbaek, H.,** Plant uptake of airborne cadmium, *Environ. Pollut. Ser. A,* 30, 27, 1983.
718. **Howells, E. J., Howells, M. E., and Alabaster, J. S.,** A field investigation of water quality, fish and invertebrates in the Mawddach river system, *Wales J. Fish. Biol.,* 22, 447, 1983.
719. **Howlett, R. J. and Majerus, M. E. N.,** The understanding of industrial melanism in the peppered moth (*Biston betularia*) (Lepidoptera: Geometridae), *Biol. J. Linnean Soc.,* 30, 31, 1987.
720. **Hoy, J. B.,** Ecological impact of lindane on a pine plantation soil microarthropod community, *Environ. Entomol.,* 9, 164, 1981.
721. **Hoy, J. B. and Shea, P. J.,** Effects of lindane, chlorpyrifos, and carbaryl on a California pine forest soil arthropod community, *Environ. Entomol.,* 10, 732, 1981.
722. **Huffaker, C. B., Gordon, H. T., and Rabb, R. L.,** Meaning of ecological entomology — the ecosystem, in *Ecological Entomology,* Huffaker, C. B. and Rabb, R. L., Eds., John Wiley & Sons, New York, 1984, 3.
723. **Hughes, M. K., Lepp, N. W., and Phipps, D. A.,** Aerial heavy metal pollution and terrestrial ecosystems, *Adv. Ecol. Res.,* 11, 218, 1980.
724. **Hughes P. R. and Laurence, J. A.,** Relationship of biochemical effects of air pollutants on plants to environmental problems: insect and microbial interactions, in *Gaseous Air Pollutants and Plant Metabolism,* Koziol, M. J. and Whatley, E. R., Eds., Butterworths, London, 1984, 361.

725. **Hughes, P. R., Chiment, J. J., and Dickie, A. I.,** Effects of pollutant dose on the response of Mexican bean beetle (Coleoptera: Coccinellidae) to SO_2-induced changes in soybean, *Environ. Entomol.,* 14, 718, 1985.
726. **Hughes, P. R., Dickie, A. I., and Penton, M. A.,** Increased success of the Mexican bean beetle on field-grown soybeans exposed to SO_2, *J. Environ. Qual.,* 12, 565, 1983.
727. **Hughes, P. R., Potter, J. E., and Weinstein, L. H.,** Effects of air pollution on plant-insect interactions: reactions of the Mexican bean beetle to SO_2-fumigated Pinto beans, *Environ. Entomol.,* 10, 741, 1981.
728. **Hughes, P. R., Potter, J. E., and Weinstein, L. H.,** Effects of air pollution on plant-insect interactions: increased susceptibility of greenhouse-grown soybeans to the Mexican bean beetle after plant exposure to SO_2, *Environ. Entomol.,* 11, 173, 1982.
729. **Hughes, P. R., Weinstein, L. H., Scott, H. W., Chiment, J. J., Doss, G. J., Culliney, T. W., Gutenmann, W. H., Bache, C. A., and Lisk, D. J.,** Effect of fertilization with municipal sludge on the glutathione, polyamine, and cadmium content of cole crops and associated loopers (*Trichoplusia ni*), *J. Agric. Food Chem.,* 35, 50, 1987.
730. **Huhta, V., Ikonen, E., and Vilkamaa, P.,** Succession of invertebrate populations in artificial soil made of sewage sludge and crushed bark, *Ann. Zool. Fennici,* 16, 223, 1979.
731. **Humbert, W.,** The mineral concretion in the midgut of *Tomocerus minor* (Collembola): microprobe analysis and physioecological significance, *Rev. Ecol. Biol. Sol,* 14, 71, 1977.
732. **Humbert, W.,** Cytochemistry and X-ray microprobe analysis of the midgut of *Tomocerus minor* Lubbock (Insecta, Collembola) with special reference to the physiological significance of the mineral concretions, *Cell Tiss. Res.,* 187, 397, 1978.
733. **Hunter, B. A. and Johnson, M. S.,** Food chain relationships of copper and cadmium in contaminated grassland ecosystems, *Oikos,* 38, 108, 1982.
734. **Hunter, B. A., Hunter, L. M., Johnson, M. S., and Thompson, D. J.,** Dynamics of metal accumulation in the grasshopper *Chorthippus brunneus* in contaminated grasslands, *Arch. Environ. Contam. Toxicol.,* 16, 711, 1987.
735. **Hunter, B. A., Johnson, M. S., and Thompson, D. J.,** Ecotoxicology of copper and cadmium in a contaminated grassland ecosystem. I. Soil and vegetation contamination, *J. Appl. Ecol.,* 24, 573, 1987.
736. **Hunter, B. A., Johnson, M. S., and Thompson, D. J.,** Ecotoxicology of copper and cadmium in a contaminated grassland ecosystem. II. Invertebrates, *J. Appl. Ecol.,* 24, 587, 1987.
737. **Hunter, B. A., Johnson, M. S., and Thompson, D. J.,** Ecotoxicology of copper and cadmium in a contaminated grassland ecosystem. III. Small mammals, *J. Appl. Ecol.,* 24, 601, 1987.
738. **Hurlbert, S. H., Mulla, M. S., and Wilson, H. R.,** The effects of an organophosphorous insecticide on the phytoplankton, zooplankton, and insect populations of freshwater ponds, *Ecol. Monogr.,* 42, 269, 1972.
739. **Hynes, H. B. N.,** *The Biology of Polluted Waters,* Liverpool University Press, Liverpool, U.K., 1960.
740. **Ide, F. P.,** Effects on forest spraying with DDT on aquatic insects of salmon streams in New Brunswick, *J. Fish. Res. Board Can.,* 24, 769, 1967.
741. **Imai, S. and Sato, S.,** On the black spots observed in the integument of silkworms poisoned by fluorine compounds, *Jpn. Soc. Air Pollut.,* 9, 401, 1974.
742. **Deleted in proof.**
743. **Inglesfield, C.,** Pyrethroids and terrestrial non-target organisms, *Pest. Sci.,* 27, 387, 1989.
744. **Inman, J. C. and Parker, G. R.,** Decomposition and heavy metal dynamics of forest litter in northwestern Indiana, *Environ. Pollut.,* 17, 39, 1978.
745. **Inouye, D. W.,** Resource partitioning and community structure: a study of bumblebees in the Colorado Rocky Mountains, Ph.D. thesis, University of North Carolina, 1976.
746. **Isensee, A. R., Kearney, P. C., and Jones, G. E.,** Modeling aquatic ecosystems for metabolic studies, *Am. Chem. Soc. Symp. Ser.,* 99, 195, 1979.

747. **Islam, A. and Roy, S.,** Effects of $CdCl_2$ on the quantitative variations of carbohydrate, protein, amino acid and cholesterol in *Chrysochoris stolli* Wolf (Insecta: Hemiptera), *Curr. Sci.,* 52, 215, 1983.

748. **Ito, T. and Arai, N.,** Nutritive effects of alanine, cystine, glycine, serine, and tyrosine on the silkworm, *Bombyx mori, J. Insect Physiol.,* 13, 1813, 1967.

749. **IUCN East European Programme,** *Environmental Status Reports: 1990,* Vol. 3, U.S.S.R., Page Brothers, Norwich, U.K., 1991.

750. **IUCN East European Programme,** The environment in Eastern Europe: 1990, *Environ. Res. Ser.,* 3, 1, 1991.

751. **IUCN East European Programme,** The lowland grasslands of central and eastern Europe, *Environ. Res. Ser.,* 4, 1, 1991.

752. **IUCN East European Programme,** Lake Baikal: on the brink? *Environ. Res. Ser.,* 2, 1, 1991.

753. **IUCN, UNEP, and WWF,** *Caring for the Earth. A Strategy for Sustainable Living,* Gland, Switzerland, 1991.

754. **Jackim, E. J.,** Influence of lead and other metals on fish gamma-aminolevulinate dehydrase activity, *Fish. Res. Board Can.,* 30, 560. 1970.

755. **Jackson, D. R. and Watson, A. P.,** Disruption of nutrient pools and transport of heavy metals in a forested watershed near a lead smelter, *J. Environ. Qual.,* 6, 331, 1977.

756. **Jackson, J. B. C., Cubit, J. D., Keller, B. D., Batista, V., Burns, K., Caffey, H. M., Caldwell, R. L., Garrity, S. D., Getter, C. D., Gonzalez, C., Guzman, H. M., Kaufmann, K. W., Knap, A. H., Levings, S. C., Marshall, M. J., Steger, R., Thompson, R. C., and Weil, E.,** Ecological effects of a major oil spill on Panamanian coastal marine communities, *Science,* 243, 37, 1989.

757. **Jackson, D. R., Selvidge, W. J., and Ausmus, B. S.,** Behavior of heavy metals in forest microcosms. II. Effects on nutrient cycling processes, *Water Air Soil Pollut.,* 10, 13, 1978.

758. **Jäger, H. J. and Grill, D.,** Einfluss von SO_2 und HF auf freie Aminosäuren der Fichte (*Picea abies*, Karsten), *Eur. J. For. Pathol.,* 5, 279, 1975.

759. **Jäger, J.,** Climate: approaches to projecting temperature and moisture changes, in *Toward Ecological Sustainability in Europe,* Solomon, A. M. and Kauppi, L., Eds., IIASA, Laxenburg, Austria, 1990, 7.

760. **James, R.,** Lepidopterological notes for 1914, *Entomol. Rec. J. Variat.,* 27, 3, 1915.

761. **Jamil, K. and Jyothi, K. N.,** Enhanced reproductive potential of *Neochetina bruchi* Hostache fed on water hyacinth plants from polluted water bodies, *Curr. Sci.,* 57, 195, 1988.

762. **Janeva, I. J.,** The zoobenthos of the river Vit. II. Dynamics of the benthic zoocoenoses under anthropogenic influence, *Hydrobiology,* 32, 3, 1988, (in Russian; English summary).

763. **Janssen, M. P. M. and Bedaux, J. J. M.,** Importance of body-size for cadmium accumulation by forest litter arthropods, *Neth. J. Zool.,* 39, 194, 1989.

764. **Janssen, M. P. M., Bruins, A., De Vries, T. H., and Van Straalen, N. M.,** Comparison of cadmium kinetics in four soil arthropod species, *Arch. Environ. Contam. Toxicol.,* 20, 305, 1991.

765. **Janssen, M. P. M., Joosse, E. N. G., and Van Straalen, N. M.,** Seasonal variation in concentration of cadmium in litter arthropods from a metal contaminated site, *Pedobiologia,* 34, 257, 1990.

766. **Jansson, A.,** Distribution of Micronectidae (Heteroptera, Corixidae) in Lake Päijänne, central Finland: correlation with eutrophication and pollution, *Ann. Zool. Fenn.,* 14, 105, 1977.

767. **Jansson, A.,** *Micronectae* (Heteroptera, Corixidae) as indicators of water quality in two lakes in southern Finland, *Ann. Zool. Fenn.,* 14, 118, 1977.

768. **Jansson, A.,** *Micronectinae* (Heteroptera, Corixidae) as indicators of water quality in Lake Vesijärvi, southern Finland, during the period of 1976–1986, *Biol. Res. Rep. Univ. Jyväskylä,* 10, 119, 1987.

769. **Janzen, D. H.,** Host-plants as islands. II. Competition in evolutionary and contemporary time, *Am. Nat.,* 107, 786, 1973.
770. **Jarvis, S. C., Jones, L. H. P., and Hopper, M. J.,** Cadmium uptake from solutions by plants and its transport from roots to shoots, *Plant Soil,* 44, 179, 1976.
771. **Jeffords, M. R. and Endress, A. G.,** Possible role of ozone in tree defoliation by the Gypsy Moth (Lepidoptera: Lymantriidae), *Environ. Entomol.,* 13, 1249, 1984.
772. **Jennings, M. D. and Reganold, J. P.,** Hierarchy and subidy-stress as a theoretical basis for managing environmentally sensitive areas, *Landscape Urban Plan.,* 21, 31, 1991.
773. **Jensen, J.,** Report on organochlorines, *Arctic Centre Publ.,* University of Lapland, Rovaniemi, 2, 335, 1991.
774. **Jensen, J. W.,** Diversity of Ephemeroptera and Plecoptera in Norway relative to size and qualities of catchment area, *Fauna Norv. Ser. B.,* 37, 67, 1990.
775. **Jensen, S.,** Report of a new chemical hazard, *New Sci.,* 32, 612, 1966.
776. **Jensen, S., Johnels, A. G., Olssen, M., and Otterlind, G.,** DDT and PCB in marine animals from Swedish waters, *Nature,* 224, 247, 1969.
777. **Jepson, P. C.,** The temporal and spatial dynamics of pesticide side-effects on non-target invertebrates, in *Pesticides and Non-target Invertebrates,* Jepson, P. C., Ed., Intercept, Wimborne, Dorset, England, 1989, 95.
778. **Jepson, P. C., Ed.,** *Pesticides and Non-target Invertebrates,* Intercept, Wimborne, Dorset, England, 1989.
779. **Jepson, P. C. and Sherratt, T. N.,** Predicting the long-term impact of pesticides on predatory invertebrates, Brighton Crop Protection Conference, Weeds 7B-5, U.K., 1991, 911.
780. **Johansen C. A.,** Honeybee poisoning by chemicals: signs, contributing factors, current problems and prevention, *Bee World,* 60, 109, 1979.
781. **Johansen, C. A., Coffey, M. D., and Quist, J. A.,** Effect of insecticide treatments to alfalfa on honey bees, including insecticidal residue and honey flavor analyses, *J. Econ. Entomol.,* 50, 721, 1957.
782. **Johansson, T. S. K. and Johansson, M. P.,** Sublethal doses of sodium fluoride affecting fecundity of confused flour beetles, *J. Econ. Entomol.,* 65, 356, 1972.
783. **Johnson, B. T.,** Potential impact of selected agricultural chemical contaminants on a northern prairie wetland: a microcosm evaluation, *Environ. Toxicol. Chem.,* 5, 473, 1986.
784. **Johnson, D. K. and Aasen, P. W.,** The Metropolitan Wastewater Treatment Plant and the Mississippi River: 50 years of improving water quality, *J. Minn. Acad. Sci.,* 55, 143, 1989.
785. **Johnson, G. M., McNeil, C. O., and George, E. S.,** Benthic macroinvertebrate associations in relation to environmental factors in Georgian Bay, *J. Great Lakes Res.,* 13, 310, 1987.
786. **Johnson, N. E. and Lawrence, W. H.,** Role of pesticides in the management of American forests, in *Pesticides in the Environment,* White-Stevens, R., Ed., Vol. 3, Dekker, New York, 1977.
787. **Johnson, R. H. and Lincoln, D. E.,** Sagebrush carbon allocation patterns and grasshopper nutrition: the influence of CO_2 enrichment and soil mineral limitation, *Oecologia,* 87, 127, 1991.
788. **Johnson, W. W. and Finley, M. T.,** Handbook of acute toxicity of chemicals to fish and aquatic invertebrates, U.S. Fish Wildlife Serv., Res. Publ., 137, Columbia, MO, 1980.
789. **Jones, C. G.,** Microorganisms as mediators of plant resource exploitation by insect herbivores, in *A New Ecology: Novel Approaches to Interactive Systems,* Price, P. W., Slobodchikoff, C. N., and Gaud, W. S., Eds., John Wiley & Sons, New York, 1984, 51.
790. **Jones, C. G. and Coleman, J. S.,** Plant stress and insect behavior: cottonwood, ozone and the feeding and oviposition preference of a beetle, *Oecologia,* 76, 51, 1988.
791. **Jones, C. G. and Coleman, J. S.,** Plant stress and insect herbivory: toward an integrated perspective, in *Response of Plants to Multiple Stresses,* Mooney, H. A., Winner, W. E., and Pell, E. J., Eds., Academic Press, New York, 1991, 422.

792. **Jones, D. S. and Macfadden, B. J.,** Induced magnenization in the monarch butterfly *Danaus plexippus* (Insecta, Lepidoptera), *J. Exp. Biol.,* 96, 1, 1982.
793. **Jones, J. R. E.,** The fauna of four streams in the Black Mountain district of south Wales, *J. Animal Ecol.,* 17, 51, 1984.
794. **Jones, K. C.,** Contaminant trends in soils and crops, *Environ. Pollut.,* 69, 311, 1991.
795. **Jones, R.,** Honey as an indicator of heavy metal contamination, *Water Air Soil Pollut.,* 33, 179, 1987.
796. **Jones, R. C. and Clark, C. C.,** Impact of watershed urbanization on stream insect communities, *Water Resour. Bull.,* 23, 1047, 1987.
797. **Jones, R. H., Luedke, A. J., Walton, T. E., and Metcalf, H. E.,** Bluetongue in the United States, and entomological perspective toward control, *World Anim. Rev.,* 38, 2, 1981.
798. **Joosse, E. N. G.,** New developments in the ecology of Apterygota, *Pedobiologia,* 25, 217, 1983.
799. **Joosse, E. N. G. and Buker, J. B.,** Uptake and excretion of lead by litter-dwelling Collembola, *Environ. Pollut.,* 18, 235, 1979.
800. **Joosse, E. N. G. and van Straalen, N. M.,** Development and present status of terrestrial ecotoxicology, in *Ecological Responses to Environmental Stresses,* Rozema, J. and Verkleij, J. A. C., Eds., Kluwer, the Netherlands, 1991, 210.
801. **Joosse, E. N. G. and Verhoef, S. C.,** Lead tolerance in Collembola, *Pedobiologia,* 25, 11, 1983.
802. **Joosse, E. N. G. and Verhoef, H. A.,** Developments in ecophysiological research on soil invertebrates, *Adv. Ecol. Res.,* 16, 175, 1987.
803. **Joosse, E. N. G. and van Vliet, L. H. H.,** Iron, manganese and zinc inputs in soil and litter near a blast-furnace plant and the effects on the respiration of woodlice, *Pedobiologia,* 26, 249, 1984.
804. **Jordan, M. J. and Lechvalier, M. P.,** Effects of zinc-smelter emissions on forest soil microflora, *Can. J. Microbiol.,* 21, 1855, 1975.
805. **Jördens-Röttger, D.,** Das Verhalten der schwarzen Bohnenblattlaus, *Aphis fabae* Scop. gegenüber chemischen Reizen von Pflanzenoberflachen, *Z. Angew. Entomol.,* 88, 158, 1979.
806. **Joy, V. C. and Chakravorty, P. P.,** Impact of insecticides on nontarget microarthropod fauna in agricultural soil, *Ecotoxicol. Environ. Safety,* 22, 8, 1991.
807. **Jungreis, A. M.,** Physiology of moulting in insects, *Adv. Insect Physiol.,* 14, 109, 1979.
808. **Kainulainen, P., Satka, H., Mustaniemi, A., Holopainen, J. K., and Oksanen, J.,** Conifer aphids in an air-polluted environment. II. Host plant quality, *Environ. Pollut.,* in press.
809. **Kaisila, J.,** Immigrationen un Expansion der Lepidopteren in Finnland in den Jahren 1869–1960, *Acta Entomol. Fenn.,* 18, 1, 1962.
810. **Kansanen, P. H. and Aho, J.,** Changes in the macrozoobenthos associations of polluted Lake Vanajavesi, southern Finland, over a period of 50 years, *Ann. Zool. Fenn.,* 18, 73, 1981.
811. **Kansanen, P. H. and Venetvaara, J.,** Comparison of biological collectors of airborne heavy metals near ferrochrome and steel works, *Water Air Soil Pollut.,* 60, 337, 1991.
812. **Kansanen, P. H., Aho, J., and Paasivirta, L.,** Testing the benthic lake type concept based on chironomid associations in some Finnish lakes using multivariate statistical methods, *Ann. Zool. Fenn.,* 21, 55, 1984.
813. **Karban, R.,** Interspecific competition between folivorous insects on *Erigeron glaucus, Ecology,* 67, 1063, 1986.
814. **Karban, R., Adamchak, R., and Schnathorst, W. C.,** Induced resistance and interspecific competition between spider mites and a vascular wilt fungus, *Science,* 235, 678, 1987.
815. **Karlsson, B. and Wiklund, C.,** Egg weight variation and lack of correlation between egg weight and offspring fitness in the wall brown butterfly *Lasiommata megera, Oikos,* 43, 376, 1984.

816. **Karnak, R. E. and Collins, W. J.,** The susceptibility to selected insecticides and acetylcholinesterase activity in a laboratory colony of midge larvae, *Chironomus tentans* (Diptera: Chironomidae), *Bull. Environ. Contam. Toxicol.,* 12, 62, 1974.
817. **Karr, R. A.,** Greenhouse skeptic out of the cold, *Science,* 246, 1118, 1989.
818. **Katayev (Kataev), O. A., Golutvin, G. I., and Selikhovkin, A. V.,** Changes in arthropod communities of forest biocoenoses with atmospheric pollution, *Entomol. Rev.,* 62, 20, 1983.
819. **Kauppi, P., Anttila, P., and Kenttämies, K., Eds.,** *Acidification in Finland,* Springer-Verlag, Berlin, 1990.
820. **Keane, K. D. and Manning, W. J.,** Effects of ozone and sulphur dioxide on mycorrhizal formation in paper birch and white pine, in *Acid Rain: Scientific and Technical Advances,* Perry, R., Harrison, R. M., Bell, J. N. B., and Lester, J. N., Eds., Selper, London, 1987, 608.
821. **Keating, S. T. and Yendol, W. G.,** Influence of selected host plants on gypsy moth (Lepidoptera: Lymantridae) larval mortality caused by a baculovirus, *Environ. Entomol.,* 16, 459, 1987.
822. **Keith, L. H. and Telliard, W. A.,** Priority pollutants. I. A perspective view, *Environ. Sci. Technol.,* 13, 416, 1979.
823. **Kelly, J. R., Rudnick, D. T., Morton, R. D., Buttel, L. A., Levine, S. N., and Carr, K. A.,** Tributyltin and invertebrates of a seagrass ecosystem: exposure and response of different species, *Mar. Environ. Res.,* 29, 245, 1990.
824. **Kemp, W. J.,** Temperature and western spruce budworm development, in *Recent Advances in Spruce Budworms Research,* Sanders, C. J., Stark, R. W., Mullins, E. J., and Murphy, J., Eds., Proc. CANUSA Spruce Budworms Research Symposium, Ottawa, Ontario, 1985, 78.
825. **Kereks, J. and Freedman, B.,** Physical, chemical, and biological characteristics of three watersheds in Kejimkujik National Park, Nova Scotia, *Arch. Environ. Contam. Toxicol.,* 18, 183, 1988.
826. **Kereks, J., Freedman, B., Howell, G., and Clifford, P.,** Comparison of the characteristics of an acidic eutrophic and acidic oligotrophic lake near Halifax, Nova Scotia, *Water Pollut. Res. J. Can.,* 19, 1, 1984.
827. **Kerswill, C. J.,** Studies on effects of forest spraying with insecticides, 1952–63, on fish and aquatic invertebrates in New Brunswick streams: Introduction and summary, *J. Fish. Res. Board Can.,* 24, 701, 1967.
828. **Kettle, A., Port, G., and Davison, A.,** Preliminary survey of the impact of fluoride pollution on invertebrates, *Bull. Br. Ecol. Soc.,* 14, 110, 1983.
829. **Kettlewell, B.,** *The Evolution of Melanism,* Clarendon Press, Oxford, 1973.
830. **Kettlewell, H. B. D.,** Selection experiments on industrial melanism in the Lepidoptera, *Heredity,* 9, 323, 1955.
831. **Kettlewell, H. B. D.,** Recognition of appropriate backgrounds by the pale and black phases of Lepidoptera, *Nature,* 175, 943, 1955.
832. **Kettlewell, H. B. D.,** A survey of the frequencies of *Biston betularia* (L.) (Lepidoptera) and its melanic forms in Great Britain, *Heredity,* 12, 51, 1958.
833. **Kettlewell, H. B. D. and Heard, M. J.,** Accidental radioactive labelling of a migrating moth, *Nature,* 189, 676, 1961.
834. **Kevan, P. G.,** Forest application of the insecticide fenitrothion and its effects on wild bee pollinators (Hymenoptera: Apoidea) of low-bush blueberries (*Vaccinium* spp.) in southern New Brunswick, Canada, *Biol. Conserv.,* 7, 301, 1975.
835. **Kevan, P. G.,** Blueberry crops in Nova Scotia and New Brunswick — pesticides and crop reductions, *Can. J. Agric. Econ.,* 25, 61, 1977.
836. **Kevan, P. G. and Baker, H. G.,** Insect on flowers, in *Ecological Entomology,* Huffaker, C. B. and Rabb, R. L., Eds., John Wiley & Sons, New York, 1984, 607.
837. **Kickert, R. N. and Krupa, S. V.,** Forest responses to tropospheric ozone and global climate change: an analysis, *Environ. Pollut.,* 68, 29, 1990.

838. **Kickert, R. N. and Krupa, S. V.,** Modeling plant response to tropospheric ozone: a critical review, *Environ. Pollut.*, 70, 271, 1991.
839. **Kidd, N. A. C.,** The role of the host plant in the population dynamics of the large pine aphid, *Cinara pinea, Oikos,* 44, 114, 1985.
840. **Kidd, N. A. C.,** The effects of simulated acid mist on the growth rates of conifer aphids and the implications for tree health, *J. Appl. Entomol.*, 110, 524, 1990.
841. **Kidd, N. A. C. and Thomas, M. B.,** The effects of acid mist on conifer aphids and their implications for tree health, in *Air Pollution and Ecosystems,* Mathy, P., Ed., Reidel, Dordrecht, Holland, 1988, 780.
842. **Kimball, B. A.,** CO_2 stimulation of growth and yield under environmental constraints, in *Carbon Dioxide Enrichment of Greenhouse Crops, Vol. II, Physiology, Yield and Economics,* Enoch, H. Z. and Kimball, B. A., Eds., CRC Press, Boca Raton, FL, 1986, 53.
843. **Kimball, K. D. and Levin, S. A.,** Limitations of laboratory bioassays: the need for ecosystem level testing, *BioScience,* 35, 165, 1985.
844. **Kimmel, W. G., Murrphey, D. J., Sharpe, W. E., and DeWalle, D. R.,** Macroinvertebrate community structure and detritus processing rates in two southwestern Pennsylvania streams acidified by atmospheric deposition, *Hydrobiologia,* 124, 97, 1985.
845. **Kirschik, V. A., Barbosa, P., and Reichelderfer, C. F.,** Three trophic level interactions: allelochemicals, *Manduca sexta* (L.), and *Bacillus thuringiensis* var. *kurstaki* Berliner, *Environ. Entomol.*, 17, 476, 1988.
846. **Kirschvink, J. L., Jones, D. S., and Macfadden, B. J., Eds.,** *Magnetite Biomineralization and Magnetoreception in Organisms. Topics in Geobiology,* Vol. 5, Plenum Press, New York, 1985.
847. **Kis, B.,** Changes of the Romanian Orthoptera fauna in the last four decades, Fourth Eur. Congr. Entomol., Gödöllö, Hungary, Abstract vol., 1991, 107.
848. **Kitayama, K., Stinner, R. E., and Labb, R. L.,** Effects of temperature, humidity and soybean maturity on longevity and fecundity of the adult Mexican bean beetle, *Epilachna varivestis, Environ. Entomol.*, 8, 458, 1979.
849. **Klausnitzer, B. and Schummer, R.,** Zum Vorkommen der Formen von *Adalia bipunctata* L. in der DDR (Insecta, Coleoptera), *Entomol. Nachrichten Berichte,* 27, 159, 1983.
850. **Kleinert, J.,** Bystruskovité (Coleoptera, Carabidae) v alkalickej imisnej oblasti Jelsava, Lubenik. [Synusiae of ground beetles (Coleoptera, Carabidae) in the environs of Jelsava and Lubenik affected by alkaline emissions], *Biológia (Bratislava),* 43, 159, 1988.
851. **Klerks, P. L. and Weis, J. S.,** Genetic adaptation to heavy metals in aquatic organisms: a review, *Environ. Pollut. Ser. A,* 45, 173, 1987.
852. **Kline, D. L. and Greiner, E. C.,** Observations on larval habitats of suspected *Culicoides* vectors of bluetongue virus in Florida, *Proc. Symp. Bluetongue Relat. Viruses, Prog. Clin. Biol. Res.*, 178, 221, 1985.
853. **Kline, E. R., Mattson, V. R., Pickering, Q. H., Spehar, D. L., and Stephan, C. E.,** Effects of pollution on freshwater organisms, *Res. J. WPCF,* 59, 539, 1987.
854. **Kloke, A.,** Orientierungsdaten fur tolerierbare Gesamtgehalte einiger Elemente in Kulturböden, *Mitt. Verband Dt. landwirtsch. Unters.Forstschunganst.*, H. 2, 32, 1977.
855. **Knezovich, J. P. and Harrison, F. L.,** The bioavailability of sediment-sorbed chlorobenzenes to larvae of the midge, *Chironomus decorus., Ecotoxicol. Environ. Safety,* 15, 226, 1988.
856. **Knezovich, J. P., Harrison, F. L., and Wilhelm, R. G.,** The bioavailability of sediment-sorbed organic compounds: a review, *Water Air Soil Pollut.*, 32, 233, 1987.
857. **Knight, A. L and Norton, G. W.,** Economics of agricultural pesticide resistance in arthropods, *Ann. Rev. Entomol.*, 34, 293, 1989.
858. **Knopf, F. L. and Sedgwick, J. A.,** Latent population responses of summer birds to a catastrophic climatological event, *Condor,* 89, 869, 1987.
859. **Koehn, T. and Frank, C.,** Effect of thermal pollution on the chironomid fauna in an urban channel, in *Chironomidae, Ecology, Systematics, Cytology and Physiology,* Murray, D. A., Ed., Pergamon Press, Oxford, 1980, 187.

860. **Koeman, J.H., Den Boer, W.M.J., Feith, A.F., de Iongh, H.H., Spliethoff, P.C., Na'Isa, B. K., and Spielberger, U.,** Three years' observation on side effects of helicopter applications on insecticides used to exterminate *Glossina* species in Nigeria, *Environ. Pollut. Ser. A,* 15, 31, 1978.
861. **Kognitzki, S.,** Die Libellenfauna des Landkreises Erlangen-Höchstadt: Biotope-Gefährdung-Förderungsmassnahmen. Scriftenreihe Bayerischen Landesamt für Umweltschutz, Heft 79, *Beitr. Artenschutz,* 4, 75, 1988.
862. **König, F.,** Allgemeine Betrachtungen uber die Vergangenheit, Gegenwart und Zukunft der rumänischen Lepidopterenfauna, *Beih. Veröff. Naturschutz Landschaftspflege Baden-Württemberg,* 21, 73, 1981.
863. **Kooijman, S. A. L. M.,** Toxicity at population level, in *Multispecies Toxicity Testing,* Cairns, J., Jr., Ed., Pergamon Press, New York, 143, 1985.
864. **Korganova, G. A.,** Influence of experimental soil pollution with ^{90}Sr on the fauna of soil protists, *Zool. J.,* 52, 939, 1973, (in Russian).
865. **Koryak, M., Shapiro, M. A., and Sykora, J. L.,** Riffle zoobenthos in streams receiving acid mine drainage, *Water Res.,* 6, 1239, 1972.
866. **Kosalwat, P. and Knight, A. W.,** Acute toxicity of aqueous and substrate-bound copper to the midge, *Chironomus decorus, Arch. Environ. Contam. Toxicol.,* 16, 275, 1987.
867. **Kosalwat, P. and Knight, A. W.,** Chronic toxicity of copper to a partial life cycle of the midge, *Chironomus decorus, Arch. Environ. Contam. Toxicol.,* 16, 283, 1987.
868. **Koussouris, T. S, Diapoulis, A. C., Bertahas, I. T., and Gritzalis, K. C.,** Self-purification processes along a polluted river in Greece, *Water Sci. Techol.,* 21, 1869, 1989.
869. **Kovats, Z. E. and Ciborowski, J. J. H.,** Aquatic insect adults as indicators of organochlorine contamination, *J. Great Lakes Res.,* 15, 623, 1989.
870. **Koziol, M. J. and Jordan, C. F.,** Changes in carbohydrate levels in red kidney bean (*Phaseolus vulgaris* L.) exposed to sulphur dioxide, *J. Exp. Bot.,* 29, 1037, 1978.
871. **Kozlowski, T. T.,** Impacts of air pollution on forest ecosystems, *BioScience,* 30, 88, 1980.
872. **Kraft, K. J. and Sypniewski, R. H.,** Effect of sediment copper on the distribution of benthic macroinvertebrates in the Keweenaw waterway, Lake Superior, *J. Great Lakes Res.,* 3, 258, 1981.
873. **Kramer, P. J. and Sionit, N.,** Effects of increasing carbon dioxide concentration on the physiology and growth of forest trees, in *The Greenhouse Effect, Climate Change, and U.S. Forests,* Shanas, W. L. and Hoftman, J. S., Eds., Conservation Foundation, Washington, D.C., 1987, 219.
874. **Krantzberg, G.,** The influence of bioturbation on physical, chemical and biological parameters in aquatic environments: a review, *Environ. Pollut. Ser. A,* 39, 99, 1985.
875. **Krantzberg, G.,** Metal accumulation by chironomid larvae: the effects of age and body weight on metal body burdens, *Hydrobiologia,* 188/189, 497, 1989.
876. **Krantzberg, G. and Stokes, P. M.,** The importance of surface adsorption and pH in metal accumulation by chironomids, *Environ. Toxicol. Chem.,* 7, 653, 1988.
877. **Krantzberg, G. and Stokes, P. M.,** Metal concentrations and tissues distribution in larvae of *Chironomus* with reference to X-ray microprobe analysis, *Arch. Environ. Contam. Toxicol.,* 19, 84, 1990.
878. **Krawczyk, A.,** Sensitivity of aphids to sulphur dioxide, *Acta Biol. Katowice,* 17, 103, 1985.
879. **Krebs, C. J.,** *Ecology. The Experimental Analysis of Distribution and Abundance,* Harper & Row, New York, 1985.
880. **Kreutzweiser, D. P. and Kingsbury, P. D.,** Permethrin treatments in Canadian forests. II. Iimpact on stream invertebrates, *Pest. Sci.,* 19, 49, 1987.
881. **Kreutzweiser, D. P. and Sibley, P. K.,** Invertebrate drift in a headwater stream treated with permethrin, *Arch. Environ. Contam. Toxicol.,* 20, 330, 1991.
882. **Kreutzweiser, D. P., Kingsbury, P. D., and Feng, J. C.,** Drift response of stream invertebrates to aerial applications of glyphosate, *Bull. Environ. Contam. Toxicol.,* 42, 331, 1989.

883. **Krivolutsky, D. A.,** The effect of an increased Ra content on the soil animals, Proc. VII Int. Coll. Soil Zool., Syracuse, NY, 1980, 391.
884. **Krivolutsky, D. A.,** Radiation ecology of soil animals, *Biol. Fertil. Soils,* 3, 51, 1987.
885. **Krivolutsky, D. A., Tichomirova, A. L., and Turchaninova, V. A.,** Strukturänderungen des Tierbesatzes (Land- und Bodenwirbellosen) unter dem Einfluss der Kontamination des Bodens mit ^{90}Sr, *Pedobiologia,* 12, 374, 1973.
886. **Krivolutsky, D. A., Usachev, V. L., Kozhevnikova, T. L., and Bakurov, A. S.,** Effect of ^{241}Am soil pollution on mesofauna of a meadow biogeocoenosis, *Dokl. Akad. Nauk SSSR,* 305, 241, 1989, (in Russian); *Dokl. Biol. Sci.,* 305, 164, 1989.
887. **Krupa, S. V. and Manning, W. J.,** Atmospheric ozone: formation and effects on vegetation, *Environ. Pollut.,* 50, 101, 1988.
888. **Kudrna, O.,** *Aspects of the Conservation of Butterflies in Europe. Butterflies of Europe,* Vol. 8, Aula-Verlag, Wiesbaden, 1986.
889. **Kukkonen, J.,** Effects of pH and natural humic substances on the accumulation of organic pollutants into two freshwater invertebrates, in *Humic Substances in the Aquatic and Terrestrial Environment,* Allard, B., Borén, H., and Grimvall, Eds., Proc. Int. Symp., Lecture Notes in Earth Science, 33, 413, 1991.
890. **Kulfan, J.,** Húsenice motyl^{3}ov (Lepidoptera) na troch druhoch listnatych drevín ovplyvnovanych cementárenskymi imisiami [Larvae of moths on three species of broadleaved trees exposed to cement factory pollution], *Lesnictví,* 34, 537, 1988.
891. **Kuribayashi, S.,** Environmental pollution effects on sericulture and its countermeasures, *Sericult. Sci. Technol.,* 10, 48, 1971.
892. **Kuribayashi, S.,** Influence of air pollution with fluorine on sericulture, *J. Sericult. Soc. Jpn.,* 41, 316, 1972.
893. **Kuribayashi, S., Yatomi, K., and Kadota, M.,** Effects of hydrogen fluoride and sulfur dioxide on mulberry trees and silkworms, *J. Jpn. Soc. Air Pollut.,* 6, 155, 1976.
894. **Kushner, D. J. and Harvey, G. T.,** Antibacterial substances in leaves: their possible role in insect resistances to disease, *J. Invertebr. Pathol.,* 4, 155, 1962.
895. **Kuussaari, M., Mikkola, K., and Vakkari, P.,** Perhosten teollisuusmelanismi pääkaupunkiseudun ilmanlaadun osoittajana. Vuodet 1988–1989, *Pääkaupunkiseudun Julkaisusarja C,* 1990 (3), 1, 1990.
896. **Lacey, L. A. and Mulla, M. S.,** Larvicidal and ovicidal activity of Dimilin® against *Simulium vittatum, J. Econ. Entomol.,* 70, 369, 1977.
897. **Lacey, L. A. and Mulla, M. S.,** Safety of *Bacillus thuringiensis* var. *israelensis* and *Bacillus sphaericus* to nontarget organisms in the aquatic environment, in *Safety of Microbial Insecticides,* Laird, M., Lacey, L. A., and Davidson, E. W., Eds., CRC Press, Boca Raton, FL, 1990, 169.
898. **Lamborg, M. R., Hardy, R. W. F., and Paul, E. A.,** Microbial effects, in *CO_2 and Plants: The Response of Plants to Rising Levels of Atmospheric CO_2,* Lemon, E. R., Ed., Westview, Boulder, CO, 1983, 131.
899. **Lancaster, J. L. and Tugwell, N. P.,** Mosquito control from applications made for control of rice water weevil, *J. Econ. Entomol.,* 62, 1511, 1969.
900. **Lane, P. A.,** Vertebrate and invertebrate predation intensity on freshwater zooplankton communities, *Nature,* 280, 391, 1979.
901. **Lang, C., l'Eplattenier, G., and Raymond, O.,** Water quality in rivers of western Switzerland: application of an adaptable index based on benthic invertebrates, *Aquat. Sci.,* 51, 224, 1989.
902. **Langford, T. E.,** A comparative assessment of thermal effects in some British and North American rivers, in *River Ecology and Man,* Oglesby, R. T., Carlson, C. A., and McCann, J. A., Eds., Academic Press, New York, 1972, 318.
903. **LaPoint, T. W., MeLancan, S. M., and Morris, M. K.,** Relationships among observed metal concentrations, criteria, and benthic structural community responses in 15 streams, *J. Water Pollut. Control Fed.,* 56, 1030, 1984.

904. **Larsson, P.,** Transport of PCBs from aquatic to terrestrial environments by emerging chironomids, *Environ. Pollut. Ser. A,* 34, 283, 1984.
905. **Larsson S.,** Stressful times for the plant stress — insect performance hypothesis, *Oikos,* 56, 277, 1989.
906. **Larsson, S. and Tenow, O.,** Areal distribution of a *Neodiprion sertifer* (Hym., Diprionidae) outbreak on Scots pine as related to stand condition, *Holarctic Ecol.,* 7, 81, 1984.
907. **Larsson, S., Björkman, C., and Gref, R.,** Responses of *Neodiprion sertifer* (Hym., Diprionidae) larvae to variation in needle resin acid concentration in Scots pine, *Oecologia,* 70, 77, 1986.
908. **Larsson, S., Oren, R., Waring, R. H., and Barrett, J. W.,** Attacks of mountain pine beetle as related to tree vigor of ponderosa pine, *For. Sci,* 29, 395, 1983.
909. **Larsson, S., Wirén, A., Lundgren, L., and Ericsson, T.,** Effects of light and nutrient stress on a leaf phenolic chemistry in *Salix dasyclados* and susceptibility to *Galerucella lineola* (Coleoptera), *Oikos,* 47, 205, 1986.
910. **Laws, R. M.,** Animal conservation in the Antarctic, *Symp. Zool. Soc. Lond.,* 54, 3, 1985.
911. **Lawton, J. H. and Hassell, M. P.,** Interspecific competition in insects, in *Ecological Entomology,* Huffaker, C. B. and Rabb, R. L., Eds., John Wiley & Sons, New York, 1984, pp. 451–495.
912. **Lawton, J. H. and McNeill, S.,** Between the devil and the deep blue sea: on the problem of being a herbivore, in *Population Dynamics,* Anderson, R. M., Turner, B. D., and Taylor, L. R., Eds., Blackwell Scientific, Oxford, 1979, 223.
913. **Lechleitner, R. A., Cherry, D. S., Cairns, J., Jr., and Stetler, D. A.,** Ionoregulatory and toxicological responses of stonefly nymphs (Plecoptera) to acidic and alkaline pH, *Arch. Environ. Contam. Toxicol.,* 14, 179, 1985.
914. **Lechowicz, M. J.,** Resource allocation by plants under air pollution stress: implications for plant-pest-pathogen interactions, *Bot. Rev.,* 53, 281, 1987.
915. **Leckie, J. O. and Davis, J. A., III,** Aqueous environmental chemistry in copper, in *Copper in the Environment, Part I: Ecological Cycling,* Nriagu, J. O., Ed., John Wiley & Sons, New York, 1979, 89.
916. **Lee, K. J. and Kim, T. W.,** Air pollution and forest declines in Far-East Asia, B Report IUFRO, XIX World Congr., Montreal, Canada, 1990, 37.
917. **Lees, D. R.,** The distribution of melanism in the Pale Brindled Beauty Moth, *Phigalia pedaria,* in Great Britain, in *Ecological Genetics and Evolution,* Creed, E. R., Ed., Blackwell Scientific, Oxford, 1971, 152.
918. **Lees, D. R.,** Industrial melanism: genetic adaptation of animals to air pollution, in *Genetic Consequences of Man-made Change,* Bishop, J. A. and Cook, L. M., Eds., Academic Press, London, 1981, 129.
919. **Lees, D. R. and Creed, E. R.,** Industrial melanism in *Biston betularia*: the role of selective predation, *J. Anim. Ecol.,* 44, 67, 1975.
920. **Lees, D. R. and Dent C. S.,** Industrial melanism in the spittlebug *Philaneus spumarius* (L.) (Homoptera: Aphrophoridae), *Biol. J. Linnean Soc.,* 19, 115, 1983.
921. **Lehmkuhl, D. N., Danks, H. V., Behan-Pelletier, V. M., Larson, D. J., Rosenberg, D. M., and Smith, I. M.,** The black cutworm *Agrotis ipsilon* (Hufnagel) (Lepidoptera, Noctuidae), *Bull. Entomol. Soc. Can.,* 16, 1, 1984.
922. **Lehtiö, H.,** Effect of air pollution on the volatile oil in needles of Scots pine (*Pinus sylvestris*), *Silva Fenn.,* 15, 122, 1981.
923. **Leivestad, H., Hendry, G., Muniz, I. P., and Snevik, E.,** Effects of acid precipitation on freshwater organisms, in *Impact of Acid Precipitation on Forest and Freshwater Ecosystems in Norway,* Brakke, F. H., Ed., Res. Rep. FR 6/76, SNSF Project, Oslo, Norway, 1976, 87.
924. **Leland, H. V., Fend, S. V., Dudley, T. L., and Carter, J. L.,** Effects of copper on species composition of benthic insects in a Sierra Nevada, California stream, *Freshwater Biol.,* 21, 163, 1989.

925. **Lemly, D. A.,** Modification of benthic insect communities in polluted streams: combined effects of sedimentation and nutrient enrichment, *Hydrobiology,* 87, 229, 1982.
926. **Leonard, A. and Lauwerys, R. R.,** Carcinogenicity, teratogenicity, and mutagenicity of arsenic, *Mutat. Res.,* 75, 49, 62.
927. **Lesniak, A.,** Wplyw niektórych czynników antropogenicznych na owady lesne, *Prace Inst. Badawczego Lesnictwa* (Warszawa), 1979, 542, 1979.
928. **Levin, S. A., Ed.,** *New Perspectives in Ecotoxicology,* ERC Rep. No. 14, Ecosystem Research Center, Cornell University, Ithaca, NY, 1982.
929. **Levy, R., Chui, Y. J., and Cromroy, H. L.,** Effects of ozone on three species of Diptera, *Environ. Entomol.,* 1, 608, 1972.
930. **Levy, R., Jouvenaz, D. P., and Cromroy, H. L.,** Tolerance of three species of insects to prolonged exposures to ozone, *Environ. Entomol.,* 3, 184, 1974.
931. **Lewis, R. G., Martin, B. E., Sgontz, D. L., and Howes, J. E., Jr.,** Measurement of fugitive atmospheric emissions of polychlorinated biphenyls from hazardous waste landfills, *Environ. Sci. Technol.,* 19, 986, 1985.
932. **Lhonoré, D.,** Donneés histophysiologiques sur le développement post-embryonnaire d'un insecte Trichoptère, *Ann. Limnol.,* 9, 157, 1973.
933. **Liang, T. T. and Liechtenstein, E. P.,** Synergism of insecticides by herbicides: effect of environmental conditions, *Science,* 186, 1128, 1974.
934. **Liebert, T. G. and Brakefield, P. M.,** Behavioural studies on the peppered moth *Biston betularia* and a discussion of the role of pollution and lichens in industrial melanism, *Biol. J. Linnean Soc.,* 31, 129, 1987.
935. **Liechtenstein, E. P., Liang, T. T., and Anderegg, B. N.,** Synergism of insecticides by herbicides under various environmental conditions, *Environ. Conserv.,* 2, 148, 1975.
936. **Likens, G. E., Wright, R. F., Galloway, J. N., and Butler, T. J.,** Acid rain, *Sci. Am.,* 241, 43, 1979.
937. **Lincoln, D. E. and Couvet, D.,** The effect of carbon supply on allocation to allelochemicals and caterpillar consumption of peppermint, *Oecologia,* 78, 112, 1989.
938. **Lincoln, D. E., Couvet, D., and Sionit, N.,** Response of an insect herbivore to host plants grown in carbon dioxide enriched atmospheres, *Oecologia,* 69, 556, 1986.
939. **Lincoln, D. E., Sionit, N., and Strain, B. R.,** Growth and feeding response of *Pseudoplusia includens* (Lepidoptera: Noctuidae) to host plants grown in controlled carbon dioxide atmospheres, *Environ. Entomol.,* 13, 1527, 1984.
940. **Lindqvist, L.,** Transport av kadmium från växter till insekter i en terrester miljö [Transport of cadmium from herbs to insects in a terrestrial environment], *Entomol. Tidskrift,* 109, 119, 1988.
941. **Linn, C. E. and Roelofs, W. L.,** Modulatory effects of octopamine and serotonin on male sensitivity and periodicity of response to sex pheromone in the cabbage looper moth, *Trichoplusia ni, Arch. Insect Biochem. Physiol.,* 3, 161, 1986.
942. **Lis, J. A.,** The influence of industrial pollutions on associations of true-bugs (Heteroptera) in selected plant communities in the zincwork "Miasteczko Slaskie" region (Upper Silesia, Poland), Abstract vol., 4th Eur. Congr. Entomol., Gödöllö, Hungary, 1991, 128.
943. **Livingston, R. J. and Meeter, D. A.,** Correspondence of laboratory and field results: what are the criteria for verification, in *Multispecies Toxicity Testing,* Cairns, J., Ed., Pergamon Press, Oxford, 1985, 76.
944. **Lock, M. A., Wallace, R. R., and Barton, D. R.,** The effects of synthetic crude oil on microbial and macroinvertebrate benthic river communities: Part 1: colonization of synthetic crude oil contaminated substrata, *Environ. Pollut. Ser. A,* 24, 207, 1981.
945. **Lockwood, D. F., Labb, R. L., Stinner, R. E., and Sprenkel, R. K.,** The effects of two host plant species and phenology on three population parameters of adult Mexican bean beetle in North Carolina, *J. Ga. Entomol. Soc.,* 14, 220, 1979.
946. **Lodenius, M.,** Mercury contents of dipterous larvae feeding on macrofungi, *Ann. Entomol. Fenn.,* 47, 63, 1981.

947. **Lodenius, M.,** Sorption of mercury in soils, in *Encyclopedia of Environmental Control Technology, Vol. 4, Hazardous Waste Containment and Treatment,* Cheremisinoff, P. N., Ed., Gulf Publishing, Houston, TX, 1990, 339.
948. **Lohner, T. W. and Collins, W. J.,** Determination of uptake rate constants for six organochlorines in midge larvae, *Environ. Toxicol. Chem.,* 6, 137, 1987.
949. **Lubin, Y. D.,** Changes in the native fauna of the Galápagos Islands following invasion by the little red fire ant, *Wasmannia auropunctata, Biol. J. Linnean Soc.,* 21, 229, 1984.
950. **Luckey, T. D.,** *Hormesis with Ionizing Radiation,* CRC Press, Boca Raton, FL, 1980.
951. **Luczak, J.,** Spiders of industrial areas, *Pol. Ecol. Stud.,* 10, 157, 1984.
952. **Luxmoore, R. J.,** CO_2 and phytomass, *BioScience,* 31, 626, 1981.
953. **Lynch, T. R., Popp, C. J., and Jacobi, G. Z.,** Aquatic insects as environmental monitors of trace metal contamination: Red River, New Mexico, *Water Air Soil Pollut.,* 42, 19, 1988.
954. **Macdonald, D. R.,** Biological assessment of aerial forest spraying against spruce budworm in New Brunswick. III. Effects on two overwintering parasites, *Can. Entomol.,* 91, 330, 1959.
955. **Mackay, R. J. and Kersey, K. E.,** A preliminary study of aquatic insect communities and leaf decomposition in acid streams near Dorset, Ontario, *Hydrobiologia,* 122, 3, 1985.
956. **MacLean, A. J.,** Cadmium in different plant species, *Can. J. Soil Sci.,* 56, 129, 1975.
957. **MacLeod, A. R.,** Effects of open-air fumigation with sulphur dioxide on the occurence of fungal pathogens in winter cereals, *Phytopathology,* 78, 88, 1988.
958. **Mackie, D. W.,** A melanic form of *Salticus scenicus* (Clerck), *Bull. Br. Spider Stud. Group,* 8, 3, 1964.
959. **Macrì, A. and Sbardella, E.,** Toxicological evaluation of nitrofurazone and furaltadone on *Selenastrum carpicornutum, Daphnia magna* and *Musca domestica, Ecotoxicol. Environ. Safety,* 8, 101, 1984.
960. **Macrì, A., Stazi, A. V., and Dojmi Di Delupis, G.,** Acute toxicity of furazolidone on *Artemia salina, Daphnia magna,* and *Culex pipiens molestus* larvae, *Ecotoxicol. Environ. Safety,* 16, 90, 1988.
961. **Maier, K. J. and Knight, A. W.,** The toxicity of waterborne boron to *Daphnia magna* and *Chironomus decorus* and the effects of water hardness and sulfate on boron toxicity, *Arch. Environ. Contam. Toxicol.,* 20, 282, 1991.
962. **Maki, A. W. and Johnson, H. E.,** Kinetics of lampricide (TFM: 3-trifluoromethyl-4-nitrophenol) residues in model stream communities, *J. Fish. Res. Board Can.,* 34, 276, 1977.
963. **Maki, A. W., Geissel, L., and Johnson, H. E.,** Comparative toxicity of the larval lampricide TFM (3 trifluoromethyl-4-nitrophenol) to selected benthic macroinvertebrates, *J. Fish. Res. Board Can.,* 32, 1455, 1975.
964. **Maksymiuk, B.,** Occurrence and nature of antibacterial substances in plants affecting *Bacillus thuringiensis* and other entomogenous bacteria, *J. Invertebr. Pathol.,* 15, 356, 1970.
965. **Malhotra, S. S. and Khan, A. A.,** Biochemical and physiological impact of major pollutants, in *Air Pollution and Plant Life,* Treshow, M., Ed., John Wiley & Sons, New York, 1984, 113.
966. **Malhotra, S. S. and Sarkar, S. K.,** Effects of sulfur dioxide on sugar and free amino acid content of pine seedlings, *Physiol. Plant.,* 47, 223, 1979.
967. **Malicky, H.,** Köcherfliegen-Lichtfallenfang am Donauufer in Linz (Trichoptera), *Linzer Biol. Beitr.,* 10, 135, 1978.
968. **Malicky, H.,** Lichtfallenuntersuchungen über die Köcherfliegen (Insecta, Trichoptera) des Rheins, *Mainzer Naturw. Archiv,* 18, 71, 1980.
969. **Malicky, H.,** Der Indikatorwert von Köcherfliegen (Trichoptera) in grossen Flüssen, *Mitt. Dtsch. Ges. Allg. Angew. Entomol.,* 3, 135, 1981.
970. **Malisch, V. R., Schulte, E., and Acker, L.,** Chlororganische Pestizide, Polychlorierte Biphenyls und Phthalate in Sedimenten von Rhein und Neckar, *Chem. Zeitung,* 105, 187, 1981.

971. **Mallow, D., Snider, R. J., and Robertson, L. S.,** Effects of different management practices on Collembola and Acarina in corn production systems. II. The effects of moldboard plowing and atrazine, *Pedobiologia,* 28, 115, 1985.
972. **Malueg, K. W., Schuytema, G. S., Gakstatter, J. H., and Krawczyk, D. F.,** Effect of *Hexagenia* on *Daphnia* response in sediment toxicity tests, *Environ. Toxicol. Chem.,* 2, 73, 1983.
973. **Mani, G. S.,** Theoretical models of melanism in *Biston betularia* — a review, *Biol. J. Linnean Soc.,* 39, 355, 1990.
974. **Mankovska, B.,** Vplyv imisii fuoru z hlinikarne na jeho obsah v roznych vyvojovych stadiach obalovaca mladikovecho *Rhyacionia buoliana* Den. et Schiff. (Lepidoptera) [Influence of fluorine emission from an aluminium factory plant and its content in different developmental stages of European pine shoot moth *Rhyacionia buoliana* (Den. & Schiff.) (Lepidoptera)], *Biologia (Bratislava) Ser. B,* 30, 355, 1975.
975. **Mann, K. H. and Clark, R. B.,** Long-term effects of oil spills on marine intertidal communities, *J. Fish. Res. Board Can.,* 35, 791, 1977.
976. **Manning, W. J., Feder, W. A., Pappia, P. M., and Perkins, I.,** Influence of foliar ozone injury on root development and root surface fungi of pinto bean plants, *Environ. Pollut.,* 1, 305, 1971.
977. **Manning, W. J., Feder, W. A., Perkins, I., and Glickman, M.,** Ozone injury and infection of potato leaves by *Botrytis cinerea, Plant Dis. Rep.,* 53, 412, 1969.
978. **Mansour, S. A., Ali, A. D., and Al-Jalili, M. K.,** The residual toxicity of honeybees of some insecticides on clover flowers: laboratory studies, *J. Apicult. Res.,* 23, 213, 1984.
979. **Manton, S. M.,** *The Arthropoda: Habits, Functional Morphology and Evolution,* Clarendon Press, Oxford, 1977.
980. **Maroni, G. and Watson, D.,** Uptake and binding of cadmium, copper and zinc by *Drosophila melanogaster* larvae, *Insect Biochem.,* 15, 55, 1985.
981. **Marshall, A. T.,** X-Ray microanalysis of copper and sulphur-containing granules in the fatbody cells of homopteran insects, *Tissue Cell,* 15, 311, 1983.
982. **Marshall, E.,** Recalculating the cost of Chernobyl, *Science,* 236, 658, 1987.
983. **Martin, M. H. and Coughtrey, P. J.,** *Biological Monitoring of Heavy Metal Pollution, Land and Air,* Applied Science, London, 1982.
984. **Martin, M. H., Coughtrey, P. J., and Young, E. W.,** Observation on the availability of lead, zinc and cadmium by the woodlouse *Oniscus asellus, Chemosphere,* 5, 313, 1976.
985. **Martin, M. H., Duncan, E. M., and Coughtrey, P. J.,** The distribution of heavy metals in a contaminated woodland ecosystem, *Environ. Pollut. Ser. B,* 3, 147, 1982.
986. **Martin, P. A., Lasenby, D. C., and Evans, R. D.,** Fate of dietary cadmium at two intake levels in the odonate nymph, *Aeshna canadensis, Bull. Environ. Contam. Toxicol.,* 44, 54, 1990.
987. **Martinek, V.,** Zum Problem der Übervermehrung der Gemeinen Fichtengespinstblattwespe (*Cephalcia abietis* L.) (Hym., Pamphiliidae) in Böhmen, *Rozprcesk. Akad. Ved. Rada. Mat. Prir. Ved.,* 90, 1, 1980.
988. **Martoja, R., Bouquegnau, J., and Verthe, C.,** Toxicological effects of cadmium and mercury in an locust *Locusta migratoria* (Orthoptera), *J. Inv. Pathol.,* 42, 17, 1983.
989. **Masnik, M. T., Stauffer, J. R., Hocutt, C. H., and Wilson, J. H.,** The effects of an oil spill on the macroinvertebrates and fish in a small southwestern Virginia creek, *J. Environ. Sci. Health,* 4, 281, 1976.
990. **Mason, G. A. and Johnson, M. W.,** Tolerance to permethrin and fenvalerate in hymenopterous parasitoids associated with *Liriomyza* spp. (Diptera: Agromyzidae), *J. Econ. Entomol.,* 81, 123, 1988.
991. **Masui, H. and Matsubara F.,** Utilization and distribution of nickel in the larvae and pupae of the silkworm, *Bombyx mori. J. Sericult. Sci. Jpn.,* 53, 331, 1984.
992. **Mathova, A.,** Biological effects and biochemical alterations after long-term exposure of *Galleria mellonella* (Lepidoptera, Pyralidae) larvae to cadmium containing diet, *Acta Entomol. Bohemoslov.,* 87, 241, 1990.

993. **Matsubara, F., Nakayama, Y., and Masui, H.,** Influence of heavy metals on aseptically reared silkworm larvae (VI). Amount of accumulated Cd in the tissues and organs by the administration of Cd, Cd.Zn and Cd.EDTA, *Environ. Control Biol.,* 20, 35, 1982.
994. **Matthews, E. G.,** *Insect Ecology,* University of Queensland Press, St. Lucia, Queensland, 1976.
995. **Matthiessen, P. and Douthwaite, B.,** The impact of tsetse fly campaign on African wildlife, *Oryx,* 19, 202, 1985.
996. **Mattson, W. J.,** *The Role of Arthropods in Forest Ecosystems,* Springer-Verlag, New York, 1977.
997. **Mattson, W. J.,** Herbivory in relation to plant nitrogen content, *Ann. Rev. Ecol. Syst.,* 11, 119, 1980.
998. **Mattson, W. J. and Haack, R. A.,** The role of drought stress in provoking outbreaks of phytophagous insects, in *Insect Outbreaks,* Barbosa, P. and Schultz, J. C., Eds., Academic Press, San Diego, 1987, 365.
999. **Mauck, W. L. and Olson, L. E.,** Polychlorinated biphenyls in adult mayflies (*Hexagenia bilineata*) from the upper Mississippi River, *Bull. Environ. Contam. Toxicol.,* 17, 387, 1977.
1000. **Maund, S. J., Peither, A., Taylor, E. J., Jüttner, I., Beyerle, R., Lay, J. P., and Pascoe, D.,** Toxicity of lindane to freshwater insect larvae in compartments of an experimental pond, *Ecotoxicol. Environ. Safety,* 23, 76, 1992.
1001. **Maurizio, A.,** Factors affecting the toxicity of fluorine compounds to bees, *Bee World,* 38, 314, 1957.
1002. **Maurizio, A.,** Bestimmung der letalen Dosis einiger Fluorverbindungen für Bienen. Zugleich ein Beitrag zur Methodik der Giftwertbestimmung in Bienenversuchen, Verhadl. IV Int. Pflanzenschutz-Kongr., Hamburg 1957, Bd. 2, 1709, 1960.
1003. **Maurizio, A. and Staub, M.,** Bienenvergiftung mitt Fluorhaltigen in der Schweiz, *Schweiz. Bienen-Ztg.,* 79, 476, 1956.
1004. **May, R. M.,** How many species are there on Earth? *Science,* 241, 1441, 1988.
1005. **Mayer, D. F., Lunden, J. D., and Weinstein, L. H.,** Evaluation of fluoride levels and effects on honey bees (*Apis mellifera* L.) (Hymenoptera: Apidae), *Am. Bee J.,* 126, 832, 1986.
1006. **Mayer, F. L., Jr. and Sanders, H. O.,** Toxicology of phthalatic acid esters in aquatic organisms, *Environ. Health Perspect.,* 3, 153, 1973.
1007. **Mayr, E.,** *Animals, Species and Evolution,* Belknap Press, Cambridge, MA., 1963.
1008. **Mayr, E.,** *Populations, Species, and Evolution. An Abridgement of Animal Species and Evolution,* Belknap Press, Cambridge, MA., 1970.
1009. **McCahon, C. P. and Pascoe, D.,** Culture techniques for three freshwater macroinvertebrate species and their use in toxicity tests, *Chemosphere,* 17, 2471, 1988.
1010. **McCahon, C. P. and Pascoe, D.,** Use of *Gammarus pulex* (L.) in safety evaluations tests — culture and selection of a sensitive life stage, *Ecotoxicol. Environ. Safety,* 15, 245, 1988.
1011. **McCahon, C. P. and Pascoe, D.,** Short-term experimental acidification of a Welsh stream: toxicity of different forms of aluminium at low pH to fish and invertebrates, *Arch. Environ. Contam. Toxicol.,* 18, 233, 1989.
1012. **McCahon, C. P. and Pascoe, D.,** Episodic pollution: causes, toxicological effects and ecological significance, *Funct. Ecol.,* 4, 375, 1990.
1013. **McCahon, C. P. and Pascoe, D.,** Brief-exposure of first and fourth instar *Chironomus riparius* larvae to equivalent assumed doses of cadmium: effects on adult emergence, *Water Air Soil Pollut.,* 60, 395, 1991.
1014. **McCahon, C. P., Brown, A. F., Poulton, M. J., and Pascoe. D.,** Effects of acid, aluminium and lime additions on fish and invertebrates in a chronically acidic Welsh stream, *Water Air Soil Pollut.,* 45, 345, 1989.

1015. **McCahon, C. P., Pascoe, D., and McKavanagh, C.,** Histochemical observations on the salmonids *Salmo salar* L. and *Salmo trutta* L. and ephemeropterans *Baetis rhodani* (Pict.) and *Ecdyonurus venosus* (Fabr.) following a simulated episode of acidity in an upland stream, *Hydrobiologia,* 153, 3, 1987.

1016. **McCahon, C. P., Whiles, A. J., and Pascoe, D.,** The toxicity of cadmium to different larval instars of the trichopteran larvae *Agapetus fuscipes* Curtis and the importance of life cycle information to the design of toxicity tests, *Hydrobiologia,* 185, 153, 1989.

1017. **McConnochie, K. and Likens, G. E.,** Some Trichoptera of the Hubbard Brook Experimental Forest in central New Hampshire, *Can. Field Nat.,* 83, 147, 1969.

1018. **McDonald, R. C., Solomon, K. R., Surgeoner, G. A., and Harris, C. R.,** Laboratory studies on the mechanics of resistance to permethrin of a field selected strain of house flies, *Pest. Sci.,* 16, 10, 1985.

1019. **McDowell, W. M.,** Potential effects of acid deposition on tropical terrestrial ecosystems, in *Acidification in Tropical Countries,* SCOPE 36, Rodhe, H. and Herrera, R., Eds., John Wiley & Sons, Chichester, U.K., 1988, 117.

1020. **McEwen, F. L. and Stephenson, G. R.,** *The Use and Significance of Pesticides in the Environment,* John Wiley & Sons, Toronto, 1979.

1021. **McKee, M. J. and Knowles, C. O.,** Effects of fenvalerate on biochemical parameters, survival, and reproduction of *Daphnia magna, Ecotoxicol. Environ. Safety,* 12, 70, 1986.

1022. **McKin, J. M.,** Evaluation of tests with early life stages of fish for predicting long-term toxicity, *J. Fish. Res. Board Can.,* 34, 1148, 1977.

1023. **McLaughlin, S. B.,** Effects of air pollution on forests, *J. Air Pollut. Contr. Assoc.,* 35, 512, 1985.

1024. **McLeese, D. W.,** Fenitrothion toxicity to the freshwater crayfish *Orconectes limosus, Bull. Environ. Contam. Toxicol.,* 16, 411, 1976.

1025. **McLeod, A. R., Holland, M. R., Shaw, P. J. A., Sutherland, P. M., Darrall, N. M., and Skeffington, R. A.,** Enhancement of nitrogen deposition to forest trees exposed to SO_2, *Nature,* 347, 277, 1990.

1026. **McNary, T. J., Milchunas, D. G., Leetham, J. W, Lavenroth, W. K., and Dodd, J. L.,** Effect of controlled low levels of SO_2 on grasshopper densities on a northern mixed-grass prairie, *J. Econ. Entomol.,* 74, 91, 1981.

1027. **McNeill, A.,** The effects of a timber preservative spillage on the ecology of the River Lossie, *J. IWEM.,* 3, 496, 1989.

1028. **McNeill, S. and Southwood, T. R. E.,** The role of nitrogen in the development of insect/plant relationship, in *Biochemical Aspects of Plant and Animal Coevolution,* Harbourne, J. B., Ed., Academic Press, New York, 1978, 77.

1029. **McNeill, S. and Whittaker, J. B.,** Air pollution and tree-dwelling aphids, in *Population Dynamics of Forest Insects,* Watt, A. D., Leather, S. R., Hunter, M. D., and Kidd, N. A.C., Eds., Intercept, Andover, Hampshire, U.K., 1990, 195.

1030. **McNeill, S., Aminu-Kano, M., Houlden, G., Bullock, J. M., Citrone, S., and Bell, J. N. B.,** The interactions between air pollution and sucking insects, in *Acid Rain: Scientific and Technical Advances,* Perry, R., Harrison, R. M., Bell, J. N. B., and Lester, J. N., Eds., Selper, London, 602, 1987.

1031. **McNeill, S., Bell, J. N. B., Aminu-Kano, M., and Mansfield, P.,** SO_2, plant, insect and pathogen interactions, in *How are the Effects of Air Pollutants on Agricultural Crops Influenced by the Interaction with Other Limiting Factors?* C.E.C., Brussels, 1986, 108.

1032. **McNicol, D. K., Bendell, B. E., and Ross, R. K.,** Studies on the effects of acidification on aquatic wildlife in Canada: waterfowl and trophic relationships in small lakes in northern Ontario, Occasional Paper, No: 62, Canadian Wildlife Service, Minister of Supply and Services Canada, Catalogue No. CW69-1/62E, 1987.

1033. **McNicol, R. E. and Scherer, E.,** Behavioural responses of *Acroneuria lycorias* (Ins., Plecopt.) larvae to acute and chronic acid exposure, *Can. Tech. Rep. Fish. Aquat. Sci.,* 1512, 1,1987.

1034. **Medici, J. C. and Taylor, M. W.,** Interrelationships among copper, zinc and cadmium in the diet of the confused flour beetle, *J. Nutr.,* 93, 307, 1967.

1035. **Melillo, J. M., Aber, J. D., and Muratore, J. F.,** Nitrogen and lignin control of hardwood leaf litter decomposition dynamics, *Ecology,* 63, 621, 1982.

1036. **Menhinick, E. F. and Crossely, D. A., Jr.,** A comparison of radiation profiles of *Acheta domesticus* and *Tenebrio molitor, Ann. Entomol. Soc. Am.,* 61, 1359, 1968.

1037. **Menzel, D. B.,** Oxidation of biologically active reducing substances by ozone, *Arch. Environ. Health,* 23, 149, 1971.

1038. **Meriläinen, J. J. and Hynynen, J.,** Benthic invertebrates in relation to acidity in Finnish forest lakes, in *Acidification in Finland,* Kauppi, P., Anttila, P., and Kenttämies, K., Eds., Springer-Verlag, Berlin, 1990, 1029.

1039. **Merritt, R. W., Walker, E. D., Wilzbach, M. A., Cummins, K. W., and Morgan, W. T.,** A broad evaluation of *B.t.i.* for black fly (Diptera: Simuliidae) control in a Michigan River: efficacy, carry and nontarget effects on invertebrates and fish, *J. Am. Mosq. Contr. Assoc.,* 5, 397, 1989.

1040. **Metcalf, R. L.,** Model ecosystem approach to insecticide degradation, *Annu. Rev. Entomol.,* 22, 241, 1977.

1041. **Metcalf, R. L.,** Insect resistance to insecticides, *Pest. Sci.,* 26, 333, 1989.

1042. **Metcalf, R. L., Booth, G. M., Schuth, C. K., Hansen, D. J., and Lu, P. Y.,** Uptake and fate of di-2-ethylhexyl phthalate in aquatic organisms and in a model ecosystem, *Environ. Health Perspect.,* 4, 27, 1973.

1043. **Miguel, A. H.,** Environmental pollution research in South America, *Environ. Sci. Technol.,* 25, 591, 1991.

1044. **Migula, P.,** Wrazliwosc wybranych gatunkow owadow na skazenia powietrza gazami i pylami przemyslowymi oraz tolerancja przez nie termicznych zmian srodowiska [Sensitivity of some insect species to gaseous and dust pollution and their tolerance of ambient temperature changes], University of Silesia Press, Katowice, Poland, 765, 1, 1985.

1045. **Migula, P. and Karpinska, B.,** The effect of atmospheric pollution on α-glycerophosphate dehydrogenase activity in the satin moth (*Leucoma salicis* (L.)). *Environ. Monit. Assess.,* 11, 69, 1988.

1046. **Migula, P., Kafel, A., Kedziorski, A., and Nakonieczny, M.,** Combined and separate effects of cadmium, lead and zinc on growth and feeding in the house cricket (*Acheta domesticus*). *Biologia (Bratislava),* 44, 911, 1989.

1047. **Mikkola, K.,** Behavioural and electrophysical responses of night-flying insects, especially Lepidoptera, to near-ultraviolet and visible light, *Ann. Zool. Fenn.,* 9, 225, 1972.

1048. **Mikkola, K.,** Frequencies of melanic forms of *Oligia* moths (Lepidoptera, Noctuidae) as a measure of atmospheric pollution in Finland, *Ann. Zool. Fenn.,* 12, 197, 1975.

1049. **Mikkola, K.,** Origin and genetics of industrial melanism of *Oligia strigilis* (L.) in Finland (Lepidoptera: Noctuidae), *Entomol. Scand.,* 11, 1, 1980.

1050. **Mikkola, K.,** Dominance relationships among the melanic forms of *Biston betularius* and *Odontopera bidentata* (Lepidoptera, Geometridae), *Heredity,* 52, 9, 1984.

1051. **Mikkola, K.,** On the selective forces acting in the industrial melanism of *Biston* and *Oligia* moths (Lepidoptera: Geometridae and Noctuidae), *Biol. J. Linnean Soc.,* 21, 409, 1984.

1052. **Mikkola, K.,** The first case of industrial melanism in the subarctic lepidopteran fauna: *Xestia gelida* f. *inferna* f. n. (Noctuidae), *Notulae Entomol.,* 69, 1, 1989.

1053. **Mikkola, K. and Albrecht, A.,** Radioactivity in Finnish night-flying moths (Lepidoptera) after the Chernobyl accident, *Notulae Entomol.,* 66, 153, 1986.

1054. **Mikkola, K. and Albrecht, A.,** The melanism of *Adalia bipunctata* around the Gulf of Finland as an industrial phenomenon (Coleoptera, Coccinellidae), *Ann. Zool. Fenn.,* 25, 177, 1988.

1055. **Mikkola, K., Jalas, I., and Peltonen, O., Eds.,** *Suomen perhoset. Mittarit 2.* Suomen Perhostutkijain Seura, Recallmed, Hanko, Finland, 1989.

1056. **Miller, M. C. and Stout, J. R.,** Effects of a controlled under-ice oil spill on invertebrates of an arctic and subarctic stream, *Environ. Pollut. Ser. A,* 42, 99, 1986.

1057. **Miller, P. R.,** Ozone effects in the San Bernardino National Forest, in *Air Pollution and the Productivity of the Forest,* Davis, D, D., Miller, A. A., and Dochinger, L., Eds., Izaak Walton League of America, Arlington, VA, 1983, 161.

1058. **Miller, P. R., Cobb, F. W., Jr., and Zavarin, E.,** Effect of injury upon oleoresin composition, phloem carbohydrates, and phloem pH, *Hilgardia,* 39, 135, 1968.

1059. **Miller, T. J., Crowder, L. B., Rice, J. A., and Marschall, E. A.,** Larval size and recruitment mechanisms in fishes: toward a conceptual framework, *Can. J. Fish. Aquat. Sci.,* 45, 1657, 1988.

1060. **Minshall, G. W. and Minshall, J. N.,** Further evidence of the role of chemical factors in determining the distribution of benthic invertebrates in the river Duddon, *Arch. Hydrobiol.,* 83, 324, 1978.

1061. **Mitterböck, F. and Führer, E.,** Wirkungen fluorbelasteter Fichtennadeln auf Nonnenraupen, *Lymantria monacha* L. (Lep., Lymantriidae) [Effects of fluoride-polluted spruce leaves on nun moth caterpillars (*Lymantria monacha*)], *J. Appl. Entomol.,* 105, 19, 1988.

1062. **Miura, T. and Takahashi, R. M.,** Insect development inhibitors: effects of candidate mosquito control agents on nontarget aquatic organisms, *Environ. Entomol.,* 3, 631, 1974.

1063. **Miura, T. and Takahashi, R. M.,** Impact of fenoxycarb, a carbamate insect growth regulator, on some aquatic invertebrates abundant in mosquito breeding habitats, *J. Am. Mosq. Contr. Assoc.,* 3, 476, 1987.

1064. **Mochida, M. and Yoshida, M.,** Symptoms of fluorine intoxication on silkworms, especially the abnormal arthroidal membrane, 41st Jpn. Soc. Sericult. Ann. Meet., Tsukyba, Ibaraki, Japan, 1974, 24.

1065. **Modde, T. and Drewes, H. G.,** Comparison of biotic index values for invertebrate collections from natural and artificial substrates, *Freshwater Biol.,* 23, 171, 1990.

1066. **Mohsen, Z. H. and Mulla, M. S.,** Field evaluation of *Simulium* larvicide: effects on target and non-target insects, *Environ. Entomol.,* 11, 390, 1982.

1067. **Mola, L., Sabatini, M. A., Fratello, B., and Bertolani, R.,** Effects of atrazine on two species of Collembola (Onychiuridae) in laboratory tests, *Pedobiologia,* 30, 145, 1987.

1068. **Moller, D.,** Estimation of the global man-made sulfur emission, *Atmos. Environ.,* 18, 19, 1984.

1069. **Mooney, H. A. and Drake, J. A.,** *Ecology of Biological Invasions of North America and Hawaii,* Springer-Verlag, New York, 1986.

1070. **Moore, J. C., Snider, R.J., and Robertson, L. S.,** Effects of different management practices on Collembola and Acarina in corn production systems. I. The effects of notillage and atrazine, *Pedobiologia,* 26, 143, 1984.

1071. **Moore, J. M., Beaubien, V. A., and Sutherland, D. J.,** Comparative effects of sediment and water contamination on benthic invertebrates in four lakes, *Bull. Environ. Contam. Toxicol.,* 23, 840, 1979.

1072. **Moore, J. W. and Ramamoorthy, S.,** *Heavy Metals in Natural Waters: Applied Monitoring and Impact Assessment,* Springer-Verlag, New York, 1984.

1073. **Moreau, J.-P.,** Pucerons et pollution: une association catastrophique? *Phytoma,* 398, 22, 1988.

1074. **Morgan, E. L., Young, R. C., Crane, C. N., and Armitage, B. J.,** Developing automated multispecies biosensing for contaminant detection, *Water Sci. Technol.,* 19, 73, 1987.

1075. **Moriarty, F.,** *Ecotoxicology. The Study of Pollutants in Ecosystems,* 2nd ed., Academic Press, London, 1990.

1076. **Moriarty, F. and Walker, C. H.,** Bioaccumulation in food chains, a rational approach, *Ecotoxicol. Environ. Safety,* 13, 208, 1987.

1077. **Morrill, P. K. and Neal, B. R.,** Impact of deltamethrin insecticide on Chironomidae (Diptera) of prairie ponds, *Can. J. Zool.,* 68, 289, 1990.

1078. **Morris, O. N.,** Inhibitory effects of foliage extracts of some forest trees on commercial *Bacillus thuringiensis, Can. Entomol.,* 104, 1357, 1972.

1079. **Morrison, I. K.,** Acid rain: a review of literature on acid deposition effects in forest ecosystems, *For. Abstr.,* 45, 483, 1984.

1080. **Morse, R. A., van Campen, D. R., Gutenmann, W. H., Lisk, D. J., and Collison, C.,** Analysis of radioactivity in honeys produced near Three-Mile Island nuclear power plant, *Nutr. Rep. Int.* 22, 319, 1980.

1081. **Morse, R. A., Culliney, T. W., Gutenmann, W. H., Littman, C. B., and Lisk, D. J.,** Polychlorinated biphenyls in honey bees, *Bull. Environ. Contam. Toxicol.,* 38, 271, 1987.

1082. **Morse, R. A., St. John, L. E., and Lisk, D. J.,** Residue analysis of Sevin in bees and pollen, *J. Econ. Entomol.,* 56, 415, 1963.

1083. **Mossberg, P. and Nyberg, P.,** Bottom fauna of small and acid forest lakes, *Inst. Freshwater Res. Drottningholm Rep.,* 58, 77, 1979.

1084. **Mozolevskaya, Ye. G. and Pechenzhskaya, M. N.,** Trunk pests in the zone of action of industrial wastes, *Ekologiya i Zashcita Lesa,* 5, 110, 1980, (in Russian).

1085. **Mudd, J. B. and Freeman, B. A.,** Reaction of ozone with biological membranes, in *Biochemical Effects of Environmental Pollutants,* Lee, S. D., Ed., Ann Arbor Science, Ann Arbor, MI, 1977, 97.

1086. **Muggleton, M., Londsdale, D., and Benham, B. R.,** Melanism in *Adalia bipunctata* L. (Col., Coccinellidae) and its relationships to atmospheric pollution, *J. Appl. Ecol.,* 12, 451, 1975.

1087. **Mueller, B. and Worseck, M.,** Damage to bees caused by arsenic- and fluorine-containing industrial flue gas, *Monatsh. Veterinaermed.,* 25, 554, 1970.

1088. **Muir, D. C. G., Townsend, B. E., and Lockhart, W. L.,** Bioavailability of six organic chemicals to *Chironomus tentans* larvae in sediment and water, *Environ. Toxicol. Chem.,* 2, 269, 1983.

1089. **Muirhead-Thomson, R. C.,** *Pesticides and Freshwater Fauna,* Academic Press, London, 1971.

1090. **Muirhead-Thomson, R. C.,** Lethal and behavioral impact of permethrin (NRDC-143) on selected stream macroinvertebrates, *Mosq. News,* 38, 829, 1978.

1091. **Muirhead-Thomson, R. C.,** *Pesticide Impact on Stream Fauna with Special Reference to Macroinvertebrates,* Cambridge University Press, Cambridge, U.K.,1987.

1092. **Mulder, R. and Gijswijt, M. J.,** The laboratory evaluation of two new insecticides which interfere with cuticle deposition, *Pest. Sci.,* 4, 737, 1973.

1093. **Mulla, M. S. and Khasawinah, A. M.,** Laboratory and field evaluation of larvicides against chironomid midges, *J. Econ. Entomol.,* 62, 37, 1969.

1094. **Mulla, M. S., Darwazeh, H. A., and Norland, R. L.,** Insect growth regulators: evaluation procedures and activity against mosquitoes, *J. Econ. Entomol.,* 67, 329, 1974.

1095. **Mullens, B. A. and Rodriguez, J. L.,** Colonization and response of *Culicoides variipennis* (Diptera: Ceratopogonidae) to pollution levels in experimental dairy wastewater ponds, *J. Med. Entomol.,* 25, 441, 1988.

1096. **Muller, P., Nagel, P., and Flacke, W.,** Ecological side effects of dieldrin application against tsetse flies in Adamaoua, Cameroon, *Oecologia,* 50, 187, 1981.

1097. **Mullié, W. C., Verwey, P. J., Berends, A. G., Sene, F., Koeman, J. H., and Everts, J. W.,** The impact of Furadan 3 G (Carbofuran) applicatins on aquatic macroinvertebrates in irrigated rice in Senegal, *Arch. Environ. Contam. Toxicol.,* 20, 177, 1991.

1098. **Mullin, C. A. and Croft, B. A.,** An update on development of selective pesticides favoring arthropod natural enemies, in *Biological Control in Agricultural IPM Systems,* Hoy, M. A. and Herzog, D. C., Eds., Academic Press, New York, 1985, 123.

1099. **Mullock, P. and Christiansen, E.,** The threshold of successful attack by *Ips typographus* on *Picea abies*: a field experiment, *For. Ecol. Manage.,* 14, 125, 1986.

1100. **Mumma, R. O., Raupach, D. C., Waldman, J. P., Tong, S. S. C., Jacobs, M. L., Babish, J. G., Hotchkiss, J. H., Wszolek, P. C., Gutenmann, W. H., Bache, C. A., and Lisk, D. J.,** National survey of elements and other constituents in municipal sewage sludges, *Arch. Environ. Contam. Toxicol.,* 13, 75, 1984.

1101. **Münster-Swendsen, M.,** The effect of precipitation on radial increment in Norway spruce (*Picea abies* Karst.) and on the dynamics of a lepidopteran pest insect, *J. Appl. Ecol.,* 24, 563, 1987.

1102. **Murphy, D. D. and Weiss, S. B.,** The effects of climate change on biological diversity in western North America: species losses and mechanisms, in *Proc. World Wildlife Fund Conf., Consequences of the Greenhouse Effect for Biological Diversity,* Peters, R. L. and Lovejoy, T. E., Eds., Yale University Press, New Haven, CT, in press.

1103. **Murphy, T. J., Formanski, L. J., Brownawell, B., and Meyer, J. A.,** Polychlorinated biphenyl emissions to the atmosphere in the Great Lakes region. Municipal landfills and incinerators, *Environ. Sci. Technol.,* 19, 942, 1985.

1104. **Muskett, C. J. and Jones, M. P.,** The dispersal of lead, cadmium and nickel from motor vehicles and effects on roadside invertebrate macrofauna, *Environ. Pollut. Ser. A,* 23, 231, 1980.

1105. **Mutanen, R. M., Siltanen H. T., Kuukka, V. P., Annila, E. A., and Varama, M. M. O.,** Residues of diflubenzuron and two of its metabolites in a forest ecosystem after control of the pine looper moth, *Bupalus piniarius* L., *Pest. Sci.,* 23, 131, 1988.

1106. **Myers, J. H.,** Effects of physiological condition of the host plant on the ovipositional choice of the cabbage white butterfly *Pieris rapae, J. Anim. Ecol.,* 54, 193, 1985.

1107. **Myers, J. H.,** Can a general hypothesis explain population cycles of forest Lepidoptera? *Adv. Ecol. Res.,* 18, 179, 1988.

1108. **Myers, J. H. and Post, B. J.,** Plant nitrogen and fluctuation of insect populations: a test with the cinnabar moth — tansy ragwort system, *Oecologia,* 48, 151, 1981.

1109. **Myers, J. M.,** Interactions between western tent caterpillars and wild rose: a test of some plant herbivore hypotheses, *J. Anim. Ecol.,* 50, 11, 1981.

1110. **Nair, K. K. and Subramanyam, G.,** Effects of variable dose-rates on radiation damage in the rust-red flour beetle, *Tribolium castaneum* Herbst., in *Radiation and Radioisotopes Applied to Insects of Agricultural Importance,* Proc. Symp. Use and Application of Radioisotopes and Radiation in the Control of Plant and Animal Insect Pests, International Atomic Energy Agency/Food and Agriculture Organization, IAEA, Vienna, Austria, 1963, 425.

1111. **National Academy of Sciences,** *Eutrophication, Causes, Consequences, Correctives,* Proc. Symp., National Academy Press, Washington, D.C., 1969.

1112. **National Academy of Sciences,** *Petroleum in the Marine Environment,* National Academy Press, Washington, D.C., 1975.

1113. **National Research Council of Canada,** Pesticide-pollinator interactions, NRCC Publ., 18471, Ottawa, Ontario, 1985.

1114. **National Research Council of Canada,** TFM and Bayer 73: Lampricides in the Aquatic Environment, NRCC Publ., 22488, Ottawa, Ontario, 1981.

1115. **National Research Council of Canada,** Pyrethroids: Their effects on aquatic and terrestrial ecosystems, NRCC Publ., 24376, Ottawa, Ontario, 1986.

1116. **Nebeker, A. V., Cairns, M. A., and Wise, C. M.,** Relative sensitivity of *Chironomus tentans* life stages to copper, *Environ. Toxicol. Chem.,* 3, 151, 1984.

1117. **Nehring, R. B.,** Aquatic insects as biological monitors of heavy metal pollution, *Bull. Environ. Contam. Toxicol.,* 15, 147, 1976.

1118. **Nehring, R. B., Nisson, R., and Minasian, G.,** Reliability of aquatic insects versus water samples as measures of aquatic lead pollution, *Bull. Environ. Contam. Toxicol.,* 22, 103, 1979.

1119. **Nelson, B. C.,** Ecology of medically important arthropods in urban environments, in *Perspectives in Urban Entomology,* Frankie, G. W. and Koehler, C. S., Eds., Academic Press, New York, 1978, 87.

1120. **Nenonen, M.,** Report on acidification in the Arctic countries: man-made acidification in a world of natural extremes, *Arctic Centre Publ. (Univ. Lapland, Rovaniemi)*, 2, 7, 1991.
1121. **Neuvonen, S. and Lindgren, M.,** The effect of simulated acid rain on performance of the aphid *Euceraphis betulae* (Koch) on silver birch, *Oecologia*, 74, 77, 1987.
1122. **Neuvonen, S., Saikkonen, K., and Haukioja, E.,** Simulated acid rain reduces the susceptibility of the European pine sawfly (*Neodiprion sertifer*) to its nuclear polyhedrosis virus, *Oecologia*, 83, 209, 1990.
1123. **Neuvonen, S., Saikkonen, K., and Suomela, J.,** Effect of simulated acid rain on the growth performance of the European pine sawfly (*Neodiprion sertifer*), *Scand. J. For. Res.*, 5, 541, 1990.
1124. **New, T. R.,** *Insect Conservation. An Australian Perspective*, Ser. Entomol., 32, Dr. W. Junk, Dordrecht, Netherlands, 1984.
1125. **Nicolas, G. and Sillans, D.,** Immediate and latent effects of carbon dioxide on insects, *Annu. Rev. Entomol.*, 34, 97, 1989.
1126. **Nilssen, J. P.,** Acidification of a small watershed in southern Norway and some characteristics of acidic aquatic environments, *Int. Rev. Ges. Hydrobiol.*, 65, 177, 1980.
1127. **Nilsson, J. and Grennfelt, P.,** Critical loads for sulphur and nitrogen. Report from a workshop held at Skokloster, Sweden, 1988, organized by UN-ECE and the Nordic Council of Ministers, NORD, 1988, 15.
1128. **Nopp, H., Pruscha, H., Künig, M., Vogel, W., and Schopf, A.,** Die Entwicklung von *Glyptapanteles liparidis* (Bouché) (Braconidae) in Abhängigkeit von der Cd- und Cu-Belastung des Wirtes *Lymantria dispar* (L.) (Lymantriidae), *Anz. Österr. Akad. Wiss. Mathem.-Naturwiss. Klasse*, 1990(2), 9, 1990.
1129. **Norby, R. J. and Kozlowski, T. T.,** Interactions of SO_2-concentration and post-fumigation temperature on growth of five species of woody plants, *Environ. Pollut. Ser. A*, 25, 27, 1981.
1130. **Norby, R. J. and Kozlowski, T. T.,** Relative sensitivity of three species of woody plants to SO_2 at high or low exposure temperature, *Oecologia*, 51, 33, 1981.
1131. **Norby, R. J. and Kozlowski, T. T.,** Response of SO_2-fumigated *Pinus resinosa* seedlings to postfumigation temperature, *Can. J. Bot.*, 59, 470, 1981.
1132. **Nordgren, A., Bååth, E., and Söderström, B.,** Microfungi and microbial activity along a heavy metal gradient, *Appl. Environ. Microbiol.*, 45, 1829, 1983.
1133. **Nordman, A. F.,** The significance for insects of climatic changes, *Fennia*, 75, 60, 1952.
1134. **Notman, P. A.,** Review of invertebrate poisoning by compound 1080, *N.Z. Entomol.*, 12, 67, 1989.
1135. **Nottrot, F., Joosse, E. N. G., and van Straalen, N. M.,** Sublethal effects of iron and manganese soil pollution on *Orchesella cincta* (Collembola), *Pedobiologia*, 30, 45, 1987.
1136. **Novák, I. and Spitzer, K.,** Industrial melanism in *Biston betularia* (Lepidoptera, Geometridae) in Czechoslovakia, *Acta Entomol. Bohemoslov.*, 83, 185, 1986.
1137. **Novak, M. A., Reilly, A. A., Bush, B., and Shane, L.,** *In situ* determination of PCB congener-specific firts order absorption/desorption rate constants using *Chironomus tentans* larvae (Insecta: Diptera: Chironomidae), *Water Res.*, 24, 321, 1990.
1138. **Novak, M. A., Reilly, A. A., and Jackling, S. J.,** Long-term monitoring of polychlorinated biphenyls in the Hudson River (New York) using caddisfly larvae and other macroinvertebrates, *Arch. Environ. Contam. Toxicol.*, 17, 699, 1988.
1139. **Nowakowski, E.,** Influence of urbanization on the structure of wireworm communities (Coleoptera, Elateridae), in *Animals in Urban Environment*, Luniak, M. and Pisarski, B., Eds., Zaklad Narodowy im. Ossolinskich, Wroclaw, 1982, 79.
1140. **Nowakowski, J. T.,** Influence of urban pressure on communities of Diptera-Acalyptrata, in *Animals in Urban Environment*, Luniak, M. and Pisarski, B., Eds., Zaklad Narodowy im. Ossolinskich, Wroclaw, 1982, 91.
1141. **Nriagu, J. O. and Pacyna, J. M.,** Quantitative assessment of worldwide contamination of air, water, and soils by trace metals, *Nature*, 333, 134, 1988.

1142. **Nuorteva P.,** Tutkimuksia metallien osuudesta metsiä tuhoavassa monistressisairaudessa [The role of metals in the multistress disease killing forests in Europe], *Lounais-Hämeen Luonto,* 75, 62, 1988.
1143. **Nuorteva, P., Häsänen, E., and Nuorteva, S.L.,** Bioaccumulation of mercury in sarcosaprophagous insects, *Norw. J. Entomol.,* 25, 79, 1987.
1144. **Nuorteva, P., Witkowski, Z., and Nuorteva, S.-L.,** Chronic damage by *Tortrix viridana* L. (Lepidoptera, Tortricidae) related to the content of iron, aluminium, zinc, cadmium and mercury in oak leaves in Niepolomice forest, *Ann. Entomol. Fenn.,* 53, 36, 1987.
1145. **Nuttall, P. M.,** The effects of sand deposition upon the macro-invertebrate fauna of the river Camel, Cornwall, *Freshwater Biol.,* 2, 181, 1972,
1146. **O'Brien, P. J., Elzen, G. W., and Vinson, S. B.,** Toxicity of azinphos methyl and chlordimefon to parasitoid *Bracon mellitor* (Hymenoptera: Braconidae): lethal and reproductive effects, *Environ. Entomol.,* 14, 891, 1985.
1147. **Odenkirchen, E. W. and Eisler, R.,** Chlorpyrifos hazards to fish, wildlife, and invertebrates: a synoptic review, *U.S. Fish Wildlife Serv. Biol. Rep.,* 85(1.13), 1, 1988.
1148. **Odum, E. P.,** *Fundamentals of Ecology,* 3rd ed., W. B. Saunders, Philadelphia, 1971.
1149. **Odum, E. P., Finn, J. T., and Franz, E. H.,** Perturbation theory and the subsidy-stress gradient, *BioScience,* 29, 349, 1979.
1150. **Ogg, C. L. and Gold, R. E.,** Exposure and field evaluation of fenoxycarb for German cockroach (Orthoptera: Blattellidae) control, *J. Econ. Entomol.,* 81, 1408, 1988.
1151. **Ojima, D. S., Kittel, T. G. F., Rosswall, T., and Walker, B. H.,** Critical issues for understanding global change effects on terrestrial ecosystems, *Ecol. Appl.,* 1, 316, 1991.
1152. **Oke, T. R.,** *Boundary Layer Climates,* Methuen, London, 1978.
1153. **Økland, J. and Økland, K. A.,** pH level and food organisms for fish: studies of 1000 lakes in Norway, in *Ecological Impact of Acid Precipitations,* Drablos, D. and Tollan, A., Eds., Proc. Int. Conf., SNSF Project, Ås-NLH, Norway, 1980, 326.
1154. **Økland, J. and Økland, K. A.,** The effects of acid deposition on benthic animals in lakes and streams, *Experientia,* 42, 471, 1986.
1155. **Oladimeji, A. A. and Offem, B. O.,** Toxicity of lead to *Clarias lazera, Oreochromis niloticus, Chironomus tentans* and *Benacus* sp., *Water Air Soil Pollut.,* 44, 191, 1989.
1156. **Oliver, B. G. and Nicol, K. D.,** Chlorobenzenes in sediments, water, and selected fish from lakes Superior, Huron, Erie, and Ontario, *Environ. Sci. Technol.,* 16, 532, 1982.
1157. **Olson, W. P. and O'Brien, R. D.,** The relation between physical properties and the penetration of solutes into the cockroach cuticle, *J. Insect Physiol.,* 9, 777, 1963.
1158. **Onuf, C. P.,** Nutritive value as a factor in plant-insect interactions with an emphasis on field studies, in *The Ecology of Arboreal Folivores,* Montgomery, G., Ed., Smithsonian Institution Press, Washington, D.C., 1978, 85.
1159. **Oppermann, T. A.,** Rinden- und holzbrütenden Insekten an immissiongeschädigten Fichten und Kiefern, *Holz-Zentralblatt,* 111, 213, 1985.
1160. **Ormerod, S. J. and Edwards, R. W.,** Modelling the ecological impact of acidification: problems and possibilities, *Verh. Int. Verein. Limnol.,* 24, 1738, 1991.
1161. **Ormerod, S. J. and Wade, K. R.,** The role of acidity in the ecology of Welsh lakes and streams, in *Acid Waters in Wales,* Edwards, R. W., Stoner, J. H., and Gee, A. S., Eds., Kluwer, The Hague, 1990, 93.
1162. **Ormerod, S. J., Boole, P., McCahon, C. P., Weatherley, N. S., Pascoe, D., and Edwards, R. W.,** Short-term experimental acidification of a Welsh stream: comparing the biological effects of hydrogen ions and aluminium, *Freshwater Biol.,* 17, 341, 1987.
1163. **Ormerod, S. J., O'Halloran, J. O., Gribbin, S. D., and Tyler, S. J.,** The ecology of dippers *Cinclus cinclus* in relation to stream acidity in upland Wales: breeding performance, calcium physiology and nestling growth, *J. Appl. Ecol.,* 28, 419, 1991.
1164. **Ormerod, S. J., Wade, K. R., and Gee, A. S.,** Macrofloral assemblages in upland Welsh streams in relation to acidity, and their importance to invertebrates, *Freshwater Biol.,* 18, 545, 1987.

1165. **Ormerod, S. J., Weatherley, N. S., Merrett, W. J., Gee, A. S., and Whitehead, P. G.,** Restoring acidified streams in upland Wales: a modelling comparison of the chemical and biological effects of liming and reduced sulphate deposition, *Environ. Pollut.,* 64, 67, 1990.
1166. **Ormrod, D. P., Black, V. J., and Linsworth, M. H.,** Depression of net photosynthesis in *Vicia faba* L. exposed to sulphur dioxide and ozone, *Nature,* 291, 585, 1981.
1167. **O'Rourke, M. J., Loomis, E. C., and Smith, D. W.,** Observations of some *Culicoides variipennis* (Diptera: Ceratopogonidae) larval habitats in areas of bluetongue virus outbreaks in California, *Mosq. News,* 43, 147, 1983.
1168. **Ortel, J.,** Effects of lead and cadmium on chemical composition and total water content of the pupal parasitoid, *Pimpla turionellae, Entomol. Exp. Appl.,* 59, 93, 1991.
1169. **Ortel, J. and Vogel, W. R.,** Effects of lead and cadmium on oxygen consumption and life expectancy of the pupal parasitoid, *Pimpla turionellae, Entomol. Exp. Appl.,* 52, 83, 1989.
1170. **Osborne, L. L., Davies, R. W., and Linton, K. J.,** Effects of limestone strip mining on benthic macroinvertebrate communities, *Water Res.,* 13, 1285, 1979.
1171. **Otto, C. and Svensson, B. S.,** Properties of acid brown water streams in south Sweden, *Archiv. Hydrobiol.,* 99, 15, 1983.
1172. **Owen, D. F.,** Industrial melanism in North American moths, *Am. Nat.,* 95, 227, 1961.
1173. **Owen, D. F.,** The evolution of melanism in six species of North American geometrid moths, *Ann. Entomol. Soc. Am.,* 55, 699, 1962.
1174. **Owens, E. H.,** Estimating and quantifying oil contamination on the shoreline, *Mar. Pollut. Bull.,* 18, 110, 1987.
1175. **Paasivirta, L.,** Macrozoobenthos of Lake Pyhäjärvi (Karelia), *Finnish Fish. Res.,* 8, 27, 1987.
1176. **Parker, D. M., Lodola, A., and Holbrook, J. J.,** Use of sulphite adduct of nicotinamide-adenine dinucleotide to study ionizations and the kinetics of lactate dehydrogenase and malate dehydrogenase, *Biochem. J.,* 173, 959, 1978.
1177. **Parker, S. P.,** *Synopsis and Classification of Living Organisms,* McGraw-Hill, New York, 1982.
1178. **Parsons, J. D.,** The effects of acid strip-mine effluents on the ecology of a stream, *Arch. Hydrobiol.,* 65, 25, 1968.
1179. **Parsons, P. A.,** Environmental stresses and conservation of natural populations, *Annu. Rev. Ecol. Syst.,* 20, 29, 1989.
1180. **Pascoe, D.,** The role of aquatic toxicity tests in predicting and monitoring pollution effects, Proc. Int. Union of Biol. Sciences Conference on Bioindicators, Cairo, *Acta Biol. Hungarica,* 38, 47, 1987.
1181. **Pascoe, D., Brown, A. F., Evans, B. M. J., and McKavanagh, C.,** Effects and fate of cadmium during toxicity tests with *Chironomus riparius* — the influence of food and artificial sediment, *Arch. Environ. Contam. Toxicol.,* 19, 872, 1990.
1182. **Pascoe, D., Williams, K. A., and Green, D. W. J.,** Chronic toxicity of cadmium *Chironomus riparius* Meigen — effects upon larval development and adult emergence, *Hydrobiologia,* 175, 109, 1989.
1183. **Pastor, J. and Post, W. M.,** Responses of northern forests to CO_2-induced climate change, *Nature,* 34, 55, 1988.
1184. **Pavoni, B., Duzzin, B., and Donazzolo, R.,** Contamination by chlorinated hydrocarbons (DDT, PCBs) in surface sediment and macrobenthos of the river Adige (Italy), *Sci. Tot. Environ.,* 65, 21, 1987.
1185. **Pearson, R. G. and Penridge, L. K.,** The effects of pollution by organic sugar mill effluent on the macro-invertebrates of a stream in tropical Queensland, Australia, *J. Environ. Manage.,* 24, 205, 1987.
1186. **Peleg, B. A.,** Effect of a new phenoxy juvenile hormone analog on California red scale (Homoptera: Diaspididae), Florida wax scale (Homoptera: Coccidae) and the ectoparasite *Aphytis holoxanthus* DeBach (Hymenoptera: Aphelinidae), *J. Econ. Entomol.,* 81, 88, 1988.

1187. **Pell, E. J., Weissberger, W. C., and Spermi, J. J.,** Impact of ozone on quantity and quality of greenhouse grown potato plants, *Environ. Sci. Technol.,* 14, 568, 1980.
1188. **Perfecto, I.,** Indirect and direct effects in a tropical agroecosystem: the maize-pest-ant system in Nicaragua, *Ecology,* 71, 2125, 1990.
1189. **Persson, C., Rodhe, H., and De Geer, L.E.,** The Chernobyl accident — a meteorological analysis of how radionuclides reached and were deposited in Sweden, *Ambio,* 16, 20, 1987.
1190. **Persson, T., Bååth, E., Clarholm, M., Lundkvist, H., Söderström, B. E., and Sohlenius, B.,** Trophic structure, biomass dynamics and carbon metabolism of soil organisms in a Scots pine forest, in *Structure and Function of Northern Coniferous Forest — An Ecosystem Study,* Persson, T., Ed., *Ecol. Bull.,* 32, 419, 1980.
1191. **Petal, J.,** The effects of industrial pollution of Silesia on populations of ants, *Pol. Ecol. Stud.,* 6, 665, 1980.
1192. **Peters, R. L.,** Effects of global warming on forests, *For. Ecol. Manage.,* 35, 13, 1990.
1193. **Petersen, H. and Luxton, M.,** A comparative analysis of soil fauna populations and their role in decomposition processes, *Oikos,* 39, 287, 1982.
1194. **Petersen R. C., Jr., Landner, L., and Blanck, H.,** Assessment of the impact of the Chernobyl reactor accident on the biota of Swedish streams and lakes, *Ambio,* 15, 327, 1986.
1195. **Peterson, L. B.-M. and Peterson, R. C., Jr.,** Anomalies in hydropsychid capture nets from polluted streams, *Freshwater Biol.,* 13, 185, 1983.
1196. **Peterson, R. H., Gordon, D. J., and Johnston, D. J.,** Distribution of mayfly nymphs (Insecta: Ephemeroptera) in some streams of eastern Canada as related to stream pH, *Can. Field Nat.,* 99, 490, 1985.
1197. **Petters, R. M. and Mettus, R. V.,** Reproductive performance of *Bracon hebetor* females following acute exposure to sulphur dioxide in air, *Environ. Pollut. Ser. A,* 27, 155, 1982.
1198. **Pettigrove, V.,** Larval mouthpart deformities in *Procladius paludicola* Skuse (Diptera: Chironomidae) from the Murray and Darling Rivers, Australia, *Hydrobiologia,* 179, 111, 1989.
1199. **Pfadt, R. E.,** *Fundamentals of Applied Entomology,* MacMillan, New York, 1985.
1200. **Pfeffer, A.,** Insektenschädlinge an Tannen im Bereich der Gasexhalationen, *Z. Angew. Entomol.,* 51, 203, 1962.
1201. **Phillippi, A. M.,** Aquatic macroinvertebrates as indicators of water quality in the Cache River drainage, Transact. Illinois State Acad. Sci., 80th Annual Meeting, Southern Illinois University, Carbondale, 1987, 52.
1202. **Phillippi A. M. and Coltharp B. G.,** Posttreatment effects of forest fertilization on the predominant benthic community of a headwater stream in eastern Kentucky, *Trans. Ky. Acad. Sci.,* 51, 18, 1990.
1203. **Pickering, Q., Carle, D. O., Pilli, A., Willingham, T., and Lazorchak, J. M.,** Effects of pollution on freshwater organisms, *Res. J. WPCF,* 61, 998, 1989.
1204. **Pihlajamäki, J., Väänänen, V. M., Koskinen, P., and Nuorteva, P.,** Metal levels in *Laothoe populi* and *Sphinx pinastri* (Lepidoptera, Sphingidae) in Finland, *Ann. Entomol. Fennici,* 55, 17, 1989.
1205. **Pimentel, D. and Edwards, C. A.,** Pesticides and ecosystems, *BioScience,* 32, 595, 1982.
1206. **Pimentel, D. and Levitan, L.,** Pesticides: amounts applied and amounts reaching pests, *BioScience,* 36, 86, 1986.
1207. **Pistrang, L. A. and Burger, J. F.,** Effect of *Bacillus thuringiensis* var. *israelensis* on a genetically-defined population of black flies (Diptera: Simuliidae) and associated insects in a montane New Hampshire stream, *Can. Entomol.,* 116, 975, 1984.
1208. **Pitt, R. and Bozeman, M.,** Sources of urban runoff pollution and its effects on an urban creek, EPA-600/S2-82-090, U.S. Environmental Protection Agency, 1983.
1209. **Plapp, F. W.,** Biochemical genetics of insecticide resistance, *Annu. Rev. Entomol.,* 21, 79, 1976.

1210. **Plowright, R. C.,** The effect of fenitrothion on forest pollinators in New Brunswick, Proc. N. R. C. Symp. Fenitrothion, NRCC/CNRC 16073, Ottawa, 335, 1977.
1211. **Plowright, R. C., Pendrel, B. A., and McLaren, I. A.,** The impact of aerial fenitrothion spraying upon the population biology of bumble bees (*Bombus* Latr.: Hym.) in southwestern New Brunswick, *Can. Entomol.,* 110, 1145, 1978.
1212. **Poirier, D. G. and Surgeoner, G. A.,** Laboratory flow-through bioassays of four forestry insecticides against stream invertebrates, *Can. Entomol.,* 119, 755, 1987.
1213. **Poirier, D. G. and Surgeoner, G. A.,** Evaluation of a field bioassay technique to predict the impact of aerial applications of forestry insecticides on stream invertebrates, *Can. Entomol.,* 120, 627, 1988.
1214. **Pomeroy, D. and Service. M. W.,** *Tropical Ecology,* Longman Scientific, Burnt Mill, Harlow, Essex, U.K., 1986.
1215. **Pontasch, K. W. and Brusven, M. A.,** Macroinvertebrate and periphyton response to streambed agitation for release of substrate-trapped hydrocarbons, *Arch. Environ. Contam. Toxicol.,* 18, 545, 1989.
1216. **Pontasch, K. W., Smith, E. P., and Cairns, J., Jr.,** Diversity indices, community comparison indices and canonical discriminant analysis: interpreting the results of multispecies toxicity tests, *Water Res.,* 23, 1229, 1989.
1217. **Popescu, C.,** Natural selection in the industrial melanic psocid, *Mesopsocus unipunctatus* (Müll.) (Insecta: Psocoptera), in northern England, *Heredity,* 42, 133, 1979.
1218. **Popovici, I., Stan, G., Stefan, V., Tomescu, R., Dumela, A., Tatra, A., and Dan, F.,** The influence of atrazine on soil fauna, *Pedobiologia,* 17, 209, 1977.
1219. **Port, G. R.,** Auchenorrhyncha on roadside verges. A preliminary survey, *Acta Entomol. Fennica,* 38, 456, 1981.
1220. **Port, G. R. and Hooton, C.,** Some effects of pollution on roadside fauna, in *Insect-Plant Relationships,* Visser, J. H. and Minks, A. K., Eds., Pudoc, Wageningen, 1982, 449.
1221. **Port, G. R. and Thompson, J. R.,** Outbreaks of insect herbivores on plants along motorways in the United Kingdon, *J. Appl. Ecol.,* 17, 649, 1980.
1222. **Posthuma, L.,** Genetic differentiation between populations of *Orchesella cincta* (Collembola) from heavy metal contaminated sites, *J. Appl. Ecol.,* 27, 609, 1990.
1223. **Posthuma, L., Hogervorst, L. R., and van Straalen, N. M.,** Adaptation to soil pollution by cadmium excretion in natural populations of *Orchesella cincta* (L.) (Collembola), *Arch. Environ. Contam. Toxicol.,* 22, 146, 1992.
1224. **Potts, G. R. and Wickerman, G. P.,** Studies of the cereal ecosystem, *Adv. Ecol. Res.,* 8, 107, 1974.
1225. **Powell, W., Dean, G. J., and Dewar, A.,** The influence of weeds on polyphagous arthropod predators in winter wheat, *Crop Protect.,* 4, 298, 1985.
1226. **Powlesland, C. and George, J.,** Acute and chronic toxicity of nickel to larvae of *Chironomus riparius* (Meigen), *Environ. Pollut. Ser. A,* 42, 47, 1986.
1227. **Pratt, C. R. and Sikorski, R. S.,** Lead content of wildflowers and honey bees (*Apis mellifera*) along a roadway: possible contamination of a simple food chain, *Proc. Penn. Acad. Sci.,* 56, 151, 1982.
1228. **Prestidge, R. A.,** Instar duration, adult consumption, oviposition and nitrogen utilization efficiences of leafhoppers feeding on different quality food (Auchenorrhyncha: Homoptera), *Ecol. Entomol.,* 7, 91, 1982.
1229. **Prestidge, R. A. and McNeill, S.,** The role of nitrogen in the ecology of grassland Auchenorrhyncha, in *Nitrogen as an Ecological Factor,* Lee, J. A., McNeill, S., and Rorison, I. H., Eds., Blackwell Scientific, Oxford, U.K., 1983, 257.
1230. **Preszler, R. W. and Price, P. W.,** Host quality and sawfly populations: a new approach to life table analysis, *Ecology,* 69, 2012, 1988.
1231. **Price, P. W.,** *Insect Ecology,* John Wiley & Sons, New York, 1975.
1232. **Price, P. W.,** Hypotheses on organization and evolution in herbivorous insect communities, in *Variable Plants and Herbivores in Natural and Managed Systems,* Denno, R. F. and McClure, M. S., Eds., Academic Press, New York, 1983, 559.

1233. **Price, P. W.,** The concept of ecosystem, in *Ecological Entomology,* Huffaker, C. B. and Rabb, R. L., Eds., John Wiley & Sons, New York, 1984, 19.

1234. **Price, P. W., Bouton, C. E., Gross, P., McPheron, B. A., Thompson, J. N., and Weis, A. E.,** Interaction among three trophic levels: influence of plants on interactions between insect herbivores and natural enemies, *Annu. Rev. Ecol. Syst.,* 11, 41, 1980.

1235. **Price, P. W., Rathcke, B. J., and Gentry, D. A.,** Lead in terrestrial arthropods: evidence for biological concentration, *Environ. Entomol.,* 3, 370, 1974.

1236. **Price, P. W., Roininen, H., and Tahvanainen, J.,** Why does the bud-galling sawfly, *Euura mucronata,* attack long shoots? *Oecologia,* 74, 1, 1987.

1237. **Priebe, A., Klein, H., and Jager, H. J.,** The role of polyamines in SO_2-polluted pea plants, *J. Exp. Bot.,* 29, 1045, 1978.

1238. **Prinz, B., Krause, G. H. M., and Stratmann, H.,** Vorläufiger Bricht der Landesanstalt für Immissionschutz über Untersuchungen zur Aufklärung der Waldschäden in der Bundesrepublik Deutschland, *LIS Rep.,* 28, 1, 1982.

1239. **Proctor, M. and Yeo, P.,** *The Pollination of Flowers,* Collins, London, 1973.

1240. **Przybylski, Z.,** Results of consecutive observations of effect of SO_2, SO_3 and H_2SO_4 on fruit trees and some harmful insects near the sulfur mine and sulfur processing plant at Machow near Tarnobrzega, *Adv. Agric. Sci. Warsaw,* 14, 111, 1967.

1241. **Przybylski, Z.,** The effect of automobile exhaust gases on the arthropods of cultivated plants, meadows and orchards, *Environ. Pollut. Ser. A,* 19, 157, 1979.

1242. **Przybylski, Z.,** The occurrence of *Sitobion avenae* (F.) on wheat in agroecosystems polluted by sulphur compounds in 1987–1988, *Arch. Phytopathol.,* 26, 473, 1990.

1243. **Pyle, R. M., Bentzien, M., and Opler, P.,** Insect conservation, *Annu. Rev. Entomol.,* 26, 233, 1981.

1244. **Qian, D., Li, Z., Wang, J., and Gao, X.,** A study on fluoride and mulberry-silkworm ecosystem, *J. Environ. Sci. China,* 5, 7, 1984.

1245. **Quarles, H. D., Hanawalt, R. B., and Odum, W. E.,** Lead in small mammals, plants and soil at varying distances from a highway, *J. Appl. Ecol.,* 11, 937, 1974.

1246. **Rabe, R. and Kreeb, K. H.,** Enzyme activities and chlorophyll and protein content in plants as indicators of air pollution, *Environ. Pollut. Ser. A,* 19, 119, 1979.

1247. **Raddum, G.,** Invertebrates: quality and quantity as fish food, in *Limnological Aspects of Acid Precipitation,* Hendry, G. R., Ed., Rep. No. 51074, Brookhaven National Laboratory, Brookhaven, NY, 1978, 1.

1248. **Raddum, G. G.,** Comparison of benthic invertebrates in lakes with different acidity, in *Ecological Impact of Acid Precipitations,* Drablos, D. and Tollan, A., Eds., Proc. Int. Conf., SNSF Project, Ås-NLH, Norway, 1980, 330.

1249. **Raddum, G. G. and Fjellheim, A.,** Acidification and early warning organisms in freshwater in western Norway, *Verh. Int. Ver. Theor. Angew. Limnol.,* 22, 1973, 1984.

1250. **Raddum, G. G. and Saether, O. E.,** Chironomid communities in Norwegian lakes with different degrees of acidification, *Verh. Int. Ver. Theor. Angew. Limnol.,* 21, 367, 1981.

1251. **Rae, J. G.,** Chironomid midges as indicators of organic pollution in the Scioto River Basin, Ohio, *Ohio J. Sci.,* 89, 5, 1989.

1252. **Raffa, K.,** The mountain pine beetle in Western North America, in *Dynamics of Forest Insect Populations. Patterns, Causes, Implications,* Berryman, A. A., Ed., Plenum Press, New York, 1988, 505.

1253. **Raffa, K. F. and Berryman, A. A.,** Physiological differences between lodgepole pines resistant and susceptible to the mountain pine beetle and associated microorganisms, *Environ. Entomol.,* 11, 486, 1982.

1254. **Rammell, C. G. and Fleming, P. A.,** Compound 1080. Properties and Use of Sodium Monofluoroacetate in New Zealand, Animal Health Div., Ministry Agric. Fisheries, Wellington, 1978.

1255. **Ramusino, M. C., Pacchetti, G., and Lucchese, A.,** Influence of chromium (IV) upon Ephemeroptera in the Pre-Alps, *Bull. Environ. Contam. Toxicol.,* 26, 228, 1981.

1256. **Rands, M. R. W.,** Pesticide use on cereals and the survival of grey partridge chicks: a field experiment, *J. Appl. Ecol.,* 22, 49, 1985.

1257. **Rasmussen, K. and Lindegaard, C.,** Effects of iron compounds on macroinvertebrate communities in a Danish lowland river system, *Water Res.,* 22, 1101, 1988.

1258. **Rathore, R. S., Sanghui, P. K., and Swarup, H.,** Toxicity of cadmium chlorid and lead nitrate to *Chironomus tentans* larvae, *Environ. Pollut.,* 18, 173, 1979.

1259. **Ratsep, R.,** Review of biological variables used for long-term monitoring. Preliminary report prepared for the Working Group for Environmental Monitoring in the Nordic Countries, Lund, Sweden, 1991.

1260. **Raupp, M. J. and Denno, R. F.,** Leaf age as a predictor of herbivore distribution and abundance, in *Variable Plants and Herbivores in Natural and Managed Systems,* Denno, R. F. and McClure, M. S., Eds., Academic Press, New York, 1983, 91.

1261. **Raupp, M. J., Milan, F. R., Barbosa, P., and Leonhard, B. A.,** Methylcyclopentanoid monoterpenes mediate interactions among insect herbivores, *Science,* 232, 1408, 1986.

1262. **Raven, J. A.,** Phytophages of xylem and phloem: a comparison of animal and plant sap-feeders, *Ann. Rev. Ecol. Syst.,* 13, 135, 1983.

1263. **Raven, P. J. and George, J. J.,** Recovery by riffle macroinvertebrates in a river after a major accidental spillage of chlorpyrifos, *Environ. Pollut.,* 59, 55, 1989.

1264. **Ravetto, P., Cavaglia, D., Colombo, V., and Peila, V.,** Proposta per l'utilizzazione dell'ape mellifera come efficiente indicatore di contaminazioni radioattive [Proposal for the utilisation of honey bee as an efficient indicator of radioactive contamination], *Apicult. Mod.,* 78, 187, 1987.

1265. **van Rhee, J. A.,** Development of earthworm populations in orchard soils, in *Progress in Soil Biology,* Satchell, J. E. and Graff, O., Eds., North Holland, Amsterdam, 1967, 360.

1266. **Reekie, E. G. and Bazzaz, F. A.,** Competition and patterns of resource use among seedlings of five tropical trees grown at ambient and elevated CO_2, *Oecologia,* 79, 212, 1989.

1267. **Reese, J. C. and Field, M. D.,** Defense against insect attack in susceptible plants: black cutworm (Lepidoptera: Noctuidae) growth on corn seedlings and artificial diet, *Ann. Entomol. Soc. Am.,* 79, 372, 1986.

1268. **Reeves, R. G., Woodham, D. W., Ganyard, M. C., and Bond, C. A.,** Preliminary monitoring of agricultural pesticides in a cooperative tobacco management project in North Carolina, 1971 — first year study, *Pest. Monit. J.,* 11, 99, 1977.

1269. **Rehfeldt G. V.,** Wirkung von Talsperren und Gewässerbelastung auf Invertebratengesellschaften in Fliessgewässern und Auen des Harzes, *Arch. Hydrobiol.,* 111, 255, 1987.

1270. **Rehwoldt, R., Lasko, L., Shaw, C., and Wirhowski, E.,** The acute toxicity of some heavy metal ions toward benthic organisms, *Bull. Environ. Contam. Toxicol.,* 10, 291, 1973.

1271. **Reich, M. and Kuhn, K.,** Stand der Libellenerfassung in Bayern und Anwendbarkeit der Ergebnisse in Arten- und Biotopschutzprogrammen. Scriftenreihe Bayerischen Landesamt fur Umweltschutz, Heft 79, *Beitr. Artenschutz,* 4, 27, 1988.

1272. **Reich P. B.,** Quantifying plant response to ozone: a unifying theory, *Tree Physiol.,* 3, 63, 1987.

1273. **Reich, P. B. and Amundson, R. G.,** Ambient levels of ozone reduce net photosynthesis in tree and crop species, *Science,* 230, 566, 1985.

1274. **Reid, G. K.,** *Ecology of Inland Waters and Estuaries,* Van Nostrand, New York, 1961.

1275. **Reynolds, S. E.,** Integration of insect behaviour and physiology in ecdysis, *Adv. Ins. Physiol.,* 15, 475, 1980.

1276. **Reynoldson, B. T., Schloesser, W. D., and Manny, A. B.,** Development of a benthic invertebrate objective for mesotrophic Great Lakes waters, *J. Great Lakes Res.,* 15, 669, 1989.

1277. **Rhoades, D. F.,** Evolution of plant chemical defense against herbivores, in *Herbivores: Their Interaction with Secondary Plant Metabolites,* Academic Press, London, 1979, 3.

1278. **Rhoades, D. F.,** Herbivore population dynamics and plant chemistry, in *Variable Plants and Herbivores in Natural and Managed Systems,* Denno, R. F. and McClure, M. S., Eds., Academic Press, New York, 1983, 155.
1279. **Rhoades, D. and Cates, R. G.,** Toward a general theory of plant antiherbivore chemistry, *Recent Adv. Phytochem.,* 10, 168, 1976.
1280. **Ricci, E. D., Hubert, W. A., and Richard, J. J.,** Organochlorine residues in sediment cores of a midwestern reservoir, *J. Environ. Qual.,* 12, 418, 1983.
1281. **Rice, A. D., Gibson, R. W., and Stribley, M. F.,** Effect of deltamethrin on walking, flight and potato virus Y-transmission by pyrethroid-resistant *Myzus persicae, Ann. Appl. Biol.,* 102, 229, 1983.
1282. **Richter, K. and Klausnitzer, B.,** Zum Einfluss ausgewahlter antropogener Noxen auf NAD-abhängige Malatdehydrogenase und Transaminasen aus *Aphis sambuci* (Hom. Aphidina) und einigen anderen Blattlausen, *Wiss. Z. Karl-Marx-Univ., Leipzig, Math. Naturwiss.,* 29, 611, 1980.
1283. **Riechers, G. D. and Strain, B. R.,** Growth of blue grama (*Bouteloua gracilis*) in response to atmospheric carbon dioxide enrichment, *Can. J. Bot.,* 66, 1570, 1988.
1284. **Riemer, J. and Whittaker, J. B.,** Air pollution and insect herbivores: observed interactions and possible mechanisms, in *Insect-Plant Interactions, Vol. I,* Bernays, E. A., Ed., CRC Press, Boca Raton, FL, 1989, 73.
1285. **Rind, D.,** A character sketch of greenhouse, *EPA J.,* 15, 4, 1989.
1286. **Riskallah, M. P.,** Influence of posttreatment temperature on the toxicity of pyrethroid insecticides to susceptible and resistant larvae of the Egyptian cotton leafworm, *Spodoptera littoralis, Experientia,* 40, 188, 1984.
1287. **Roback, S. S. and Richardson, J. W.,** The effects of acid mine drainage on aquatic insects, *Proc. Natl. Acad. Sci. U.S.A.,* 121, 81, 1969.
1288. **Roberts, R. D. and Johnson, M. S.,** Dispersal of heavy metals from abandoned mine workings and their transference through terrestrial food chains, *Environ. Pollut.,* 16, 294, 1978.
1289. **Roberts, R. D., Johnson, M. S., and Hutton, M.,** Lead contamination of small mammals from abandoned metalliferous mines, *Environ. Pollut.,* 15, 61, 1978.
1290. **Rockwood, J. P., Coler, R. A., and Yin, C.-M.,** The effect of aluminium in soft water at low pH on osmoregulation and ionic balance in the dragonfly *Libellula julia* Uhler, *Comp. Biochem. Physiol.,* 91C, 499, 1988.
1291. **Rockwood, J. P., Jones, D. S., and Coler, R. A.,** The effect of aluminium in soft water at low pH on oxygen consumption by the dragonfly *Libellula julia* Uhler, *Hydrobiologia,* 190, 55, 1990.
1292. **Rodecap, K. D. and Tingey, D. T.,** Ozone-induced ethylene release from leaf surfaces, *Plant Sci.,* 44, 73, 1986.
1293. **Rodhe, H. and Herrera, R.,** *Acidification in Tropical Countries,* SCOPE 36, Rodhe, H. and Herrera, R., Eds., John Wiley & Sons, Chichester, U.K., 1988.
1294. **Rodhe, H., Cowling, E., Galbally, I., Galloway, J., and Herrera, R.,** Acidification and regional air pollution in the tropics, in *Acidification in Tropical Countries,* SCOPE 36, Rodhe, H. and Herrera, R., Eds., John Wiley & Sons, Chichester, U.K., 1988, 3.
1295. **Rodrigues, C. S. and Wright, R. E.,** Evaluation of the insect growth regulators methoprene and diflubenzuron against floodwater mosquitoes (Diptera: Culicidae) in southwestern Ontario, *Can. Entomol.,* 110, 319, 1978.
1296. **Roff, J. C. and Kwiatkowski, R. E.,** Zooplankton and zoobenthos communities of selected northern Ontario lakes of different acidities, *Can. J. Zool.,* 55, 899, 1977.
1297. **Roline, R. A.,** The effects of heavy metals pollution of the upper Arkansas River on the distribution of aquatic macroinvertebrates, *Hydrobiologia,* 160, 3, 1988.
1298. **Root, M.,** Biological monitors of pollution, *BioScience,* 40, 83, 1990.
1299. **Rose, A. F., Jones, K. C., Haddon, W. F., and Dreyer, D. L.,** Grindelane diterpene acids from *Grindelia humilis*: feeding deterrency of diterpene acids towards aphids, *Phytochemistry,* 20, 2249, 1981.

1300. **Rosenberg, D. M. and Wiens, A. P.,** Community and species responses of Chironomidae (Diptera) to contamination of fresh waters by crude oil and petroleum products, with special reference to the Trail River, Northwest Territories, *J. Fish. Res. Board Can.*, 33, 1955, 1976.

1301. **Rosenheim, J. A. and Hoy, M. A.,** Sublethal effects of pesticides on the parasitoid *Aphytis melinus* (Hymenoptera: Aphelinidae), *J. Econ. Entomol.*, 81, 476, 1988.

1302. **Rosiu, J. C., Giesy, P. J., and Kreis, R. G., Jr.,** Toxicity of vertical sediments in the Trenton channel, Detroit River, Michigan, *J. Great Lakes Res.*, 15, 570, 1989.

1303. **Roth-Holzapfel, M. and Funke, M.,** Element content of bark-beetles (*Ips typographus* Linne, *Trypodendron lineatum* Olivier; Scolytidae): a contribution to biological monitoring, *Biol. Fertil. Soils*, 9, 192, 1990.

1304. **Rotty, R. M. and Barland, G.,** Fossil fuel consumption: recent amounts, patterns, and trends of CO_2, in *The Changing Carbon Cycle: A Global Analysis*, Trabalka, J. R. and Reichle, D. E., Eds., Springer-Verlag, New York, 1986.

1305. **Rubinstein, N. I., Lores, E., and Gregory, N. R.,** Accumulation of PCBs, mercury and cadmium by *Nereis virens*, *Mercenaria mercenaria* and *Palaemonetes pugio* from contaminated harbor sediments, *Aquat. Toxicol.*, 3, 249, 1983.

1306. **Rudd, R. L. and Genelly, R. E.,** Pesticides: their use and toxicity in relation to wildlife, *State of Calif., Dept. Fish Game Manage. Branc. Game Bull.*, 7, 1, 1956.

1307. **Rühling, Å. and Tyler, G.,** Heavy metal pollution and decomposition of spruce needle litter, *Oikos*, 24, 402, 1973.

1308. **Rühling, Å., Bååth, E., Nordgren, A., and Söderström, B.,** Fungi in metal contaminated soil near the Gusum brass mill, Sweden, *Ambio*, 13, 34, 1984.

1309. **Rühling, Å., Rasmussen, L., Pilegaard, K., Mäkinen, A., and Steinnes, E.,** *Survey of Atmospheric Heavy Metal Deposition*, Graphic Systems AB, Göteborg, Sweden, 1987.

1310. **Rushton, S. P., Luff, M. L., and Eyre, M. D.,** Effects of pasture improvement and management on the ground beetle and spider communities of upland grasslands, *J. Appl. Ecol.*, 26, 489, 1989.

1311. **Ruszcyk, A.,** Mortality of *Papilio scamander* scamander (Lep., Papilionidae) pupae in four districts of Porto Alegre (S. Brazil) and the causes of the superabundance of some butterflies in urban areas, *Rev. Bras. Biol.*, 46, 567, 1986.

1312. **Rutt, G. P., Weatherley, N. S., and Ormerod, S. J.,** Relationships between the physicochemistry and macroinvertebrates of British upland streams: the development of modelling and indicator systems for predicting fauna and detecting acidity, *Freshwater Biol.*, 24, 463, 1990.

1313. **Ryabinin, N. A., Ganin, G. N., and Pankov, A. N.,** Influence of sulphur-acid production waste upon complexes of soil invertebrates, *Ekologya (Sverdlovsk)*, 6, 29, 1988.

1314. **Ryck, F. M. and Duchrow, R. M.,** Oil pollution and the aquatic environment in Missouri, 1960–1972, *Trans. Mo. Acad. Sci.*, 7–8, 164, 1973–1974.

1315. **Rydell, J.,** Seasonal use of illuminated areas by foraging northern bats *Eptesicus nilssoni*, *Holarctic Ecol.*, 14, 203, 1991.

1316. **Sabatini, M. A., Pederzoli, A., Fratello, B., and Bertolani, R.,** Microarthropod communities in soil treated with atrazine, *Boll. Zool.*, 46, 333, 1979.

1317. **Sala, G. M., Gibson, G. E., Jr., and Harrel, R. C.,** Physicochemical conditions and benthic macroinvertebrates of a tertiary sewage treatment system, *Hydrobiologia*, 52, 161, 1977.

1318. **Salo, J. and Pyhälä, M.,** *Amazonia*, Otava, Helsinki, Finland, 1991.

1319. **Sander, K.,** Specification of the basic body pattern in insect embryogenesis, *Adv. Ins. Physiol.*, 12, 125, 1976.

1320. **Sanders, H. O. and Cope, O. B.,** The relative toxicities of several pesticides to naiads of three species of stonefly, *Limnol. Oceanogr.*, 33, 112, 1968.

1321. **Sanders, H. O., Mayer, F. L., and Walsh, D. F.,** Toxicity residue dynamics and reproductivity effects of phthalate esters in aquatic invertebrates, *Environ. Res.*, 6, 84, 1973.

1322. **Sanders, C. J., Stark, R. W., Mullins, E. J., and Murphy, J.,** Recent advances in spruce budworm research, Proc. of the CANUSA Spruce Budworms Research Symposium, 1984, Ottawa, Ontario, 1985.

1323. **Santamarina, J. and Guitian, J.,** Quelques données sur le régime alimentaire du desman (*Galemys pyrenaicus*) dans le nord-ouest de l'Espagne, *Mammalia*, 52, 301, 1988.

1324. **Saouter, E., Ribeyre, F., and Boudou, A.,** Bioaccumulation of mercury compounds ($HgCl_2$ and CH_3HgCl) by *Hexagenia rigida* (Ephemeroptera), in *Proceedings of the International Conference: Heavy Metals in the Environment*, Vernet, J. P., Ed., CEP Consultant, Edinburgh, 1989, 378.

1325. **Saouter, E., Ribeyre, F., Boudou, A., and Maury-Brachet, R.,** *Hexagenia rigida* (Ephemeroptera) as a biological model in aquatic ecotoxicology: experimental studies on mercury transfers from sediment, *Environ. Pollut.*, 69, 51, 1991.

1326. **Sardi, K.,** Changes in the soluble protein content of soybean *Glycine max* L. and pea *Pisum sativum* L. under continuous SO_2 and soot pollution, *Environ. Pollut. Ser. A*, 25, 181, 1981.

1327. **Sargent, T.,** Melanism in moths of Central Massachusetts (Noctuidae, Geometridae), *J. Lepidopt. Soc.*, 28, 145, 1972.

1328. **Sasa, M.,** Recent advances in the environmental and medical sciences achieved by the biosystematic studies — with special reference to the chironomids, *Pure Appl. Chem.*, 59, 505, 1987.

1329. **Sassen, B.,** The effect of two pyrethroids on the feeding behaviour of three aphid species and on transmission of two different viruses, *J. Plant Dis. Prot.*, 90, 119, 1983.

1330. **Satake, K. N. and Yasuno, M.,** The effects of diflubenzuron on invertebrates and fishes in a river, *Jpn. J. Sanit. Zool.*, 38, 303, 1987.

1331. **Schaefer, C. H. and Dupras, E. F., Jr.,** Factors affecting the stability of Dursban in polluted waters, *J. Econ. Entomol.*, 63, 701, 1970.

1332. **Scheffer, T. C. and Hedgcock, G. G.,** Injury to northwestern forest trees by sulfur dioxide from smelters, *USDA For. Serv. Techn. Bull.*, 1117, 1955.

1333. **Schindler, D. W.,** Effects of acid rain on freshwater ecosystems, *Science*, 239, 149, 1988.

1334. **Schindler, D. W. and Turner, M. A.,** Biological, chemical and physical responses of lakes to experimental acidification, *Water Air Soil Pollut.*, 18, 259, 1982.

1335. **Schindler, D. W., Kasian, S. E. M., and Hesslein, R. H.,** Losses of biota from American aquatic communities due to acid rain, *Environ. Monit. Assess.*, 12, 269, 1989.

1336. **Schindler, D. W., Mills, K. H., Malley, D. F., Findlay, D. L., Shearer, J. A., Davies, I. J., Turner, M. A., Lindsey, G. A., and Cruikshank, O. R.,** Long-term ecosystem stress: the effects of years of experimental acidification on a small lake, *Science*, 228, 1395, 1985.

1337. **Schloesser, D. W.,** Zonation of mayfly nymphs and caddisfly larvae in the St. Marys River, *J. Great Lakes Res.*, 14, 227, 1988.

1338. **Schmidt, G. H.,** Use of grasshoppers as test animals for the ecotoxicological evaluation of chemicals in the soil, *Agric. Ecosyst. Environ.*, 16, 175, 1986.

1339. **Schmidt, G. H. and Fielbrand, B.,** Wirkung einer simulierten Dauerbelastung durch $HgCl_2$ auf die Generationsfolge der Feldheuschrecke *Acrotylus patruelis* (H.-S.) (Orthoptera, Acrididae), *Anz. Schädlingskde., Pflanzenschutz, Umweltschutz*, 60, 84, 1987.

1340. **Schmidt, G. H. and Fielbrand, B.,** Quecksilber-Dauerbelastung und Feldheuschreckenentwicklung, *Mitt. Dtsch. Ges. Allg. Angew. Entomol.*, 6, 464, 1988.

1341. **Schmidt, G. H., Ibrahim, N. M. M., and Abdallah, M. D.,** Toxicological studies on the long-term effects of heavy metals (Hg, Cd, Pb) in soil on the development of *Aiolopus thalassinus* (Fabr.) (Saltatoria: Acrididae), *Sci. Total Environ.*, 107, 109, 1991.

1342. **Schmidtmann, E. T., Mullens, B. A., Schwager, S. J., and Spear, S.,** Distribution, abundance, and a probability model for larval *Culicoides variipennis* (Diptera: Ceratopogonidae) on dairy farms in New York State, *Environ. Entomol.*, 12, 768, 1983.

1343. **Schneider, S. H.,** The greenhouse effect: science and policy, *Science*, 243, 771, 1989.

1344. **Schoones, J. and Giliomee, J. H.,** The toxicity of methidathion to parasitoids of red scale, *Aonidella aurantii* (Hemiptera: Diaspididae), *J. Entomol. Soc. South Africa,* 45, 261, 1982.

1345. **Schoonhoven, L. M. and Derksen-Koppers, I.,** Effects of some allelochemics on food uptake and survival of a polyphagous aphid, *Myzus persicae, Entomol. Exp. Appl.,* 19, 52, 1976.

1346. **Schöpfer, W. und Hradetzky, J.,** Der Indizienbeweis: Luftverschmutzung massgebliche Ursache der Walderkrankung, *Forstw. Centralbl.,* 103, 231, 1984.

1347. **Schultz, J. C.,** Impact of variable plant defensive chemistry on susceptibilty of insects to natural enemies, in *Plant Resistance to Insects,* Hedin, P. A., Ed., American Chemical Society, Washington, D.C., 1983, 37.

1348. **Schütt, P. and Cowling, E. B.,** Waldsterben - a general decline of forest in central Europe: symptoms, development and possible causes, *Plant Dis.,* 67, 548, 1985.

1349. **Schütt, S. and Nuorteva, P.,** Metylkvicksilvrets inverkan på aktivitet hos *Tenebrio molitor* (L.) (Col. Tenebrionidae), *Acta Entomol. Fenn.,* 42, 78, 1983.

1350. **Schwenke, W., Braun, G., and Maschning, E.,** Situation und Prognose des Forstschädlingsbefalls in Bayern 1983/84, *Allg. Forstzeit.,* 19, 477, 1984.

1351. **Scorgie, H. R. A.,** Ecological effects of the aquatic herbicide cyanatryn on a drainage channel, *J. Appl. Ecol.,* 17, 207, 1980.

1352. **Scullion, J. and Edwards, R. W.,** The effects of coal industry pollutants on the macroinvertebrate fauna of a small river in the South Wales coalfield, *Freshwater Biol.,* 10, 141, 1980.

1353. **Seastedt, T. R.,** The role of microarthropods in decomposition and mineralization processes, *Ann. Rev. Entomol.,* 29, 25, 1984.

1354. **Seaward, M. R. D. and Hitch, C. J. B.,** *Atlas of the Lichens of the British Isles,* 1, Institute of Terrestrial Ecology, Cambridge, U.K., 1982.

1355. **Sebastien, R. J., Brust, R. A., and Rosenberg, D. M.,** Impact of methoxychlor on selected nontarget organisms in a riffle of the Souris River, Manitoba, *Can. J. Fish. Aquat. Sci.,* 46, 1047, 1989.

1356. **Seidman, L. A., Bergstrom, G., Gingrich, D. J., and Remsen, C. C.,** Accumulation of cadmium by the fourth instar larva of the fly *Chironomus thummi, Tissue Cell,* 18, 395, 1986.

1357. **Selikhovkin, A. V.,** The effects of sulfur dioxide on the development of the gypsy moth, *Ekologiya i Zashcita Lesa,* 5, 114, 1980 (in Russian).

1358. **Sergievsky, S. O.,** The multifunctionality and plasticity of genetical polymorphism (the population mechanism of *Adalia bipunctata* taken as an example), *J. Gen. Biol.,* 156, 491, 1985 (in Russian).

1359. **Sheedy, B. R., Lazorchak, J. M., Grunwald, D. J., Pickering, Q. H., Pilli, A., Hall, D., and Webb, R.,** Effects of pollution on freshwater organisms, *Res. J. WPCF,* 63, 619, 1991.

1360. **Sheehan, P. J., Baril, A., Mineau, P., Smith, D. K., Harfenist, A., and Marshall, W. K.,** The impact of pesticides on the ecology of prairie nesting ducks, *Can. Wildlife Serv. Tech. Rep. Ser.,* 19, 1, 1987.

1361. **Sheehan, P. J., Miller, D. R., Butler, G. C., and Bourdeau, P.,** *Effects of Pollutants at the Ecosystem Level,* SCOPE 22, John Wiley & Sons, Chichester, U.K., 1984.

1362. **Shirt, D. B.,** *British Red Data Books: 2. Insects,* Nature Conservancy Council, Peterborough, U.K., 1987.

1363. **Shriner, D. S.,** Terrestrial ecosystems, wet deposition, in *Air Pollutants and Their Effects on the Terrestrial Ecosystem. Advances in Environmental Science and Technology,* Vol. 18, Legge, A. H. and Krupa, S. V., Eds., John Wiley & Sons, New York, 1986, 365.

1364. **Shugart, H. H. and Emanuel, W. R.,** Carbon dioxide increase: the implications at the ecosystem level, *Plant Cell Environ.,* 8, 381, 1985.

1365. **Sibley, P. K., Kaushik, N. K., and Kreutzweiser, D. P.,** Impact of a pulse application of permethrin on the macroinvertebrate community of a headwater stream, *Environ. Pollut.,* 70, 35, 1991.

1366. **Sierpinski, Z.,** Schädliche Insekten an jungen Kiefernbeständen in Rauchschadengebieten in Oberschlesien, *Arch. Forstwesen,* 15, 1105, 1966.
1367. **Sierpinski, Z.,** Einfluss von industriellen Luftverunreinigungen auf die Populationsdynamik einiger primärer Kiefernschädlinge, XIV IUFRO-Kongress, 1967, Referate, Bind 5, Section 24, 1967.
1368. **Sierpinski, Z.,** Szkodniki wtórne sosny w drzewostanach znajdujacych sie w zasiegu dzialania emisji przemyslowych zawierajacych zwiazki azotowe [Summary: secondary noxious insects of pine in stands growing on areas with industrial air pollution containing nitrogen compounds], *Sylwan,* 115, 11, 1971.
1369. **Sierpinski, Z.,** The significance of secondary pine insects in areas of chronic exposure to industrial pollution, *Mitt. Forstl. Bundes Versuchsanst. (Wien),* 97, 609, 1972.
1370. **Sierpinski, Z.,**Über den Einfluss von Luftverunreinigungen auf Schadinsekten in polnischen Nadelbaumbeständen, *Forstwiss. Centralbl.,* 103, 83, 1984.
1371. **Sierpinski, Z.,** Luftverunreinigungen und Forstschädlinge, *Z. Angew. Entomol.,* 99, 1, 1985.
1372. **Siewert, H. F., Miller, C. J., and Torke, B. G.,** Water quality and macroinvertebrate populations before and after a hazardous waste cleanup, *Water Resour. Bull.,* 25, 685, 1989.
1373. **Simonet, D. E., Knausenberger, W. I., Townsend, L. H., Jr., and Turner, E. C., Jr.,** A biomonitoring procedure utilizing negative phototaxis of first instar *Aedes aegypti* larvae, *Arch. Environ. Contam. Toxicol.,* 7, 339, 1978.
1374. **Simpson, K. W.,** Abnormalities in the tracheal gills of aquatic insects collected from streams receiving chlorinated or crude oil wastes, *Freshwater Biol.,* 10, 581, 1980.
1375. **Simpson, K. W.,** Communities of Chironomidae (Diptera) from an acid-stressed stream in the Adirondack Mountains, New York, *Mem. Am. Entomol. Soc.,* 34, 315, 1983.
1376. **Simpson, S. J. and Absigold, J. D.,** Compensation by locust for changes in dietary nutrients: behavioral mechanisms, *Physiol. Entomol.,* 10, 443, 1985.
1377. **Simpson, K. W. and Bode, R. W.,** Common larvae of Chironomidae (Diptera) from New York State streams and rivers, *NY State Mus. Bull.,* 439, 1980, 1.
1378. **Simpson, K. W., Bode, R. W., and Colquhoun, J. R.,** The macroinvertebrate fauna of an acid-stressed headwater stream system in the Adirondack Mountains, New York, *Freshwater Biol.,* 15, 671, 1985.
1379. **Sinha, S. N., Lakhani, K. H., and Davis B. N. K.,** Studies on the toxicity of insecticidal drift to the first instar larvae of the large white butterfly *Pieris brassicae* (Lepidoptera: Pieridae), *Ann. Appl. Biol.,* 116, 27, 1990.
1380. **Skärby, L.,** Changes in the nutritional quality of crops, in *Gaseous Air Pollutants and Plant Metabolism,* Koziol, M. J. and Whatley, F. R., Eds., Butterworths, London, 1984, 351.
1381. **Skeffington, R. A. and Roberts, T. M.,** The effects of ozone and acid mist on Scots pine saplings, *Oecologia,* 65, 201, 1985.
1382. **Skelly, J. M., Yang, Y.-S., Chevone, B. I., Long, S. J., Nellessen, J. E., and Winner, W. E.,** Ozone concentrations and their influence on forest species in the Blue Ridge mountains of Virginia, in *Air Pollution and the Productivity of the Forest,* Davis, D, D., Miller, A. A., and Dochinger, L., Eds., Izaak Walton League of America, Arlington, VA, 1983, 143.
1383. **Skinner, W. D. and Arnold, D. E.,** Short term biotic response before and during the treatment of an acid mine drainage with sodium carbonate, *Hydrobiologia,* 199, 229, 1990.
1384. **Skriver, J.,** Okkers invirkning pa° invertebratfaunaens forekomst og maengde i midt- og vestjyske hedeslettevandlöb, Bilag 9 til Okkerredegörelsen, Nat. Agency Environ. Protect., Copenhagen, 1984.
1385. **Skrzypczynska, M.,** Preliminary studies on entomofauna of cones of *Abies alba* in Ojcowski and Tatranski national parks in Poland, *Z. Angew. Entomol.,* 98, 375, 1984.
1386. **Sloof, W.,** Benthic macroinvertebrates and water quality assessment; some toxicological considerations, *Aquat. Toxicol.,* 4, 73, 1983.

1387. **Sloof, W. and Canton, J. H.,** Comparison of the susceptibility of 11 freshwater species to 8 chemical compounds. II. (Semi)chronic toxicity tests, *Aquat. Toxicol.,* 4, 271, 1983.
1388. **Sloof, W., Canton, J. H., and Hermens, J. L. M.,** Comparison of the susceptibility of 22 freshwater species to 15 chemical compounds. I. (Sub)acute toxicity tests, *Aquat. Toxicol.,* 4, 113, 1983.
1389. **Sloof, W., De Zwart, D., and Van de Kerkhoff, J. F. J.,** Monitoring the Rivers Rhine and Meuse in the Netherlands for toxicity, *Aquat. Toxicol.,* 4, 189, 1983.
1390. **Smies, M., Evers, R. H. J., Peijnenburg, F. H. M., and Koeman, J. H.,** Environmental aspects of field trials with pyrethroids to eradicate tsetse fly in Nigeria, *Ecotoxicol. Environ. Safety,* 4, 114, 1980.
1391. **Smith, G. E. and Isom, B. G.,** Investigations on effects of large-scale applications of 2,4-D on aquatic fauna and water quality, *Pest. Monit. J.,* 1, 16, 1967.
1392. **Smith, M. E. and Kaster, J. L.,** Effect of rural highway on stream benthic macroinvertebrates, *Environ. Pollut.,* 32, 157, 1983.
1393. **Smith, M. E., Wyskowski, B. J., Brooks, C. M., Driscoll, C. T., and Cosentini, C. C.,** Relationships between acidity and benthic invertebrates of low-order woodland streams in the Adirondack Mountains, New York, *Can. J. Fish. Aquat. Sci.,* 47, 1318, 1990.
1394. **Smith, S. P., Strain, B. R., and Sharkey, T. D.,** Effects of CO_2 enrichment on four Great Basin grasses, *Funct. Ecol.,* 1, 139, 1987.
1395. **Smith, W. H.,** *Air Pollution and Forest Interactions Between Air Contaminants and Forest Ecosystems,* Springer-Verlag, New York, 1981.
1396. **Smith, W. H.,** Effects of regional air pollutants on forests in the USA, *Forstwiss. Centralb.,* 103, 48, 1984.
1397. **Smith, W. H.,** Forest and air quality, *J. For.,* February, 83, 1985.
1398. **Smith, W. H.,** *Air Pollution and Forests. Interactions between Air Contaminants and Forest Ecosystems,* Springer-Verlag, New York, 1990.
1399. **Smith, W. H., Geballe, G., and Fubner, J.,** Effects of acidic deposition on forest vegetation: interaction with insect and microbial agents of stress, in *Direct and Indirect Effects of Acidic Deposition on Vegetation,* Linthurst, R. A., Ed., Acid Precip. Ser. Vol. 5, Proc. of a Symp. on Acid Precipitation, Am. Chem. Soc., 1982, Butterworths, Boston, MA, 1984.
1400. **Smock, L. A.,** Relationship between metal concentrations and organism size in aquatic insects, *Freshwater Biol.,* 13, 313, 1983.
1401. **Södergren, A.,** Significance of interfaces in the distribution and metabolism of di-2-ethylhexyl phthalate in aquatic laboratory model ecosystem, *Environ. Pollut.,* 27, 263, 1982.
1402. **Sohal, R. S., Peters, P. D., and Hall, T. A.,** Fine structure and X-ray microanalysis of mineralized concretions in the Malpighian tubules of the housefly, *Musca domestica, Tissue Cell,* 8, 447, 1976.
1403. **Sokolov, V. E., Krivolutzky, D. A., Ryabov, I. N., Taskaev, A. I., and Shevchenko, V. A.,** Bioindication of biological after-effects of the Chernobyl Atomic Power Station accident in 1986–1987, *Biol. Int.,* 18, 6, 1989.
1404. **Sotherton, N. W.,** Effects of herbicides on the chrysomelid beetle *Gastrophysa polygoni* (L.) in laboratory and field, *Z. Angew. Entomol.,* 94, 446, 1982.
1405. **Sotherton, N. W. and Moreby, S.,** Contact toxicity of some foliar fungicides sprayed to three species of polyphagous predators found in cereal fields. Tests of Agrochem. and Cultiv. 5, *Ann. Appl. Biol. Suppl.,* 105, 423, 1984.
1406. **Sotherton, N. W. and Moreby, S. J.,** The effects of foliar fungicides on beneficial arthropods in wheat fields. *Entomophaga,* 33, 87, 1988.
1407. **Sotherton, N. W., Moreby, S. J., and Langley, M. G.,** The effects of the foliar fungicide pyrazophos on beneficial arthropods in barley fields, *Ann. Appl. Biol.,* 111, 75, 1987.
1408. **Southward, A. J. and Southward, E. C.,** Recolonization of rocky shores in Cornwall after use of toxic dispersants to clean up the Torrey Canyon Spill, *J. Fisheries Res. Board Can.,* 35, 682, 1978.

1409. **Spahr, H.-J.,** Die bodenbiologische Bedeutung von Collembolen und ihre Eignung als Testorganismen für die Ökotoxicologie, *Anz. Schädlingskde. Pflanzenschutz Umweltschutz,* 54, 27, 1981.

1410. **Spaic', I.,** Promjene u sastavu sumske entomofaune uzrokovane uporabom insekticida [Synopsis: changes in the composition of forest entomofauna due to insecticide application], *Acta Entomol. Jugoslavica,* 13, 61, 1977.

1411. **Sparks, T. C., Lockwood, J. A., Byford, R. L., Graves, J. B., and Leonard, B. R.,** The role of behavior in insecticide resistance, *Pest. Sci.,* 26, 383, 1989.

1412. **Sparks, T. C., Pavloff, A. M., Rose, R. L., and Clower, D. F.,** Temperature-toxicity relationships of pyrethroids on *Heliothis virescens* (F.) (Lepidoptera: Noctuidae) and *Anthonomus grandis grandis* Boheman (Coleoptera: Curculionidae), *J. Econ. Entomol.,* 76, 243, 1983.

1413. **Spear, P. A. and Pierce, R. C.,** Copper in the aquatic environment: chemistry, distribution, and toxicology, National Council of Canada, NRCC No. 16454, Ottawa, Canada, 1979.

1414. **Spehar, R. L., Anderson, R. L., and Fiandt, J. T.,** Toxicity and bioaccumulation of cadmium and lead in aquatic invertebrates, *Environ. Pollut.,* 15, 195, 1978.

1415. **Spencer, H. J. and Port, G. R.,** Effects of roadside conditions on plants and insects. II. Soil conditions, *J. Appl. Ecol.,* 25, 709, 1988.

1416. **Spencer, H. J., Scott, N. E., Port, G. R., and Davison, A. W.,** Effects of roadside conditions on plants and insects. I. Atmospheric conditions, *J. Appl. Ecol.,* 25, 639, 1988.

1417. **Sprague, J. B., Elson, P. F., and Saunders, R. L.,** Sublethal copper-zinc pollution in a salmon river — a field laboratory study, *Air Water Soil Pollut.,* 9, 531, 1965.

1418. **Srivastava, H. S. and Ormrod, D. P.,** Effects of nitrogen dioxide and nitrate on growth and nitrate assimilation in bean leaves, *Plant Physiol.,* 76, 418, 1984.

1419. **Stachurska-Hagen, T.,** Acidification experiments in conifer forest. 8. Effects of acidification and liming on some soil animals: Protozoa, Rotifera and Nematoda, SNSF-Project, IR 74/80, 1980, 1.

1420. **Stark, R. W., Miller, P. R., Cobb, F. W., Jr., Wood, D. L., and Parmeter, J. R., Jr.,** I. Incidence of bark beetle infestation in injured trees, *Hilgardia,* 39, 121, 1968.

1421. **Stary, P. and Kubiznakova, J.,** Content and transfer of heavy metal air pollutants in populations of *Formica* spp. wood ants (Hym., Formicidae), *J. Appl. Entomol.,* 104, 1, 1987.

1422. **Staxang, B.,** Acidification of bark on some deciduous trees, *Oikos,* 20, 224, 1969.

1423. **Stebbing, A. R. D.,** Hormesis — the stimulation of growth by low levels of inhibitors, *Sci. Total Environ.,* 22, 213, 1982.

1424. **Stegeman, L. C.,** The effects of the carbamate insecticide carbaryl upon forest soil mites and Collembola, *J. Econ. Entomol.,* 70, 119, 1964.

1425. **Steinwascher, K.,** Egg size variation in *Aedes aegypti*: relationship to body size and other variables, *Am. Midl. Nat.,* 112, 76, 1984.

1426. **Steward, R. C.,** Industrial and nonindustrial melanism in the peppered moth *Biston betularia* (L.), *Ecol. Entomol.,* 2, 231, 1977.

1427. **Stinner, D. H., Stinner, B. R., and McCartney, D. A.,** Effects of simulated acidic precipitation on plant-insect interactions in agricultural systems: corn and black cutworm larvae. *J. Environ. Qual.,* 17, 371, 1988.

1428. **Stoner, A., Wilson, W. T., and Harvey, J.,** Dimethoate (Cygon®): effect of long-term feeding of low doses on honey bees in standard-size field colonies, *Southwest. Entomol.,* 8, 174, 1983.

1429. **Stoner, A., Wilson, W. T., and Rhodes, H. A.,** Carbofuran: effects of long-term feeding of low doses in sucrose syrup on honey bees in standard-size field colonies, *Environ. Entomol.,* 11, 53, 1982.

1430. **Stoner, J. H., Gee, A. S., and Wade, K. R.,** The effects of acidification on the ecology of streams in the upper Tywi catchment in West Wales, *Environ. Pollut. Ser. A,* 35, 125, 1984.

1431. **van Straalen, N. M.,** Size-specific mortality patterns in two species of forest floor Collembola, *Oecologia,* 67, 220, 1985.

1432. **van Straalen, N. M.,** Production and biomass turnover in two populations of forest floor Collembola, *Neth. J. Zool.,* 39, 156, 1989.

1433. **van Straalen, N. M.,** Soil and sediment quality criteria derived from invertebrate toxicity data, in *Ecotoxicology of Metals in Invertebrates,* Dallinger, R. and Rainbow, P. S., Eds., Lewis Publishers, Chelsea, MI, 1992, in press.

1434. **van Straalen, N. M. and Ernst, W. H. O.,** Metal biomagnification may endanger species in critical pathways, *Oikos,* 62, 255, 1991.

1435. **van Straalen, N. M. and de Goede, R. G. M.,** Productivity as a population performance index in life-cycle toxicity tests. *Water Sci. Technol.,* 19, 13, 1987.

1436. **van Straalen, N. M. and van Meerendonk, J. H.,** Biological half-lives of lead in *Orchesella cincta* (L.) (Collembola), *Bull. Environ. Contam. Toxicol.,* 38, 213, 1987.

1437. **van Straalen, N. M. and van Wensem, J.,** Heavy metal content of forest litter arthropods as related to body-size and trophic level, *Environ. Pollut. Ser. A,* 42, 209, 1986.

1438. **van Straalen, N. M., Burghouts, T. B. A., and Doornhof, M. J.,** Dynamics of heavy metals in populations of Collembola in a contaminated pine forest soil, in Proc. Int. Conf., *Heavy Metals in the Environment,* Vol. 1, CEP Consultants, Edinburgh, 1985, 613.

1439. **van Straalen, N. M., Burghouts, T. B. A., Doornhof, M. J., Groot, G. M., Janssen, M. P. M., Joosse, E. N. G., van Meerendonk, J. H., Theeuwen, J. P. J. J., Verhoef, H. A., and Zoomer, H. R.,** Efficiency of lead and cadmium excretion in populations of *Orchesella cincta* (Collembola) from various contaminated forest soils, *J. Appl. Ecol.,* 24, 953, 1987.

1440. **van Straalen, N. M., Schobben, J. H. M., and de Goede, R. G. M,.** Population consequences of cadmium toxicity in soil micro-arthropods, *Ecotoxicol. Environ. Safety,* 17, 190, 1989.

1441. **Strain, B. R.,** Physiological and ecological controls on carbon sequestering in ecosystems, *Biochemistry,* 1, 219, 1985.

1442. **Strain, B. R. and Bazzaz, F. A.,** Terrestrial plant communities, in *CO_2 and Plants: The Response of Plants to Rising Levels of Atmospheric Carbon Dioxide,* Lemon, E. R., Ed., Westview Press, Washington D.C. (Ass. Adv. Sci.), 1983, 280.

1443. **Strek, H. J. and Weber, J. B.,** Behaviour of polychlorinated biphenyls (PCBs) in soils and plants, *Environ. Pollut. Ser. A,* 28, 291, 1982.

1444. **Strojan, C. L.,** The impact of zinc smelter emissions on forest litter arthropods, *Oikos,* 31, 41, 1978.

1445. **Strong, C. R. and Luoma, S. N.,** Variations in the correlation of body size with concentrations of copper and silver in the bivalve *Macoma baltica, Can. J. Fish. Aquat. Sci.,* 38, 1059, 1981.

1446. **Strong, D. R., Lawton, J. H., and Southwood, R.,** *Insects on Plants. Community Patterns and Mechanisms,* Blackwell Scientific, Oxford, 1984.

1447. **Subagja, J. and Snider, R. J.,** The side effects of the herbicide atrazine and paraquat upon *Folsomia candida* and *Tullbergia granulata* (Insecta, Collembola), *Pedobiologia,* 22, 141, 1981.

1448. **Sumi, Y., Fukuoka, H. R., Murakami, T., Suzuki, T., Hatakeyama, S., and Suzuki, K. T.,** Histochemical localization of copper, iron and zinc in the larvae of the mayfly *Baetis thermicus* inhabiting a river polluted with heavy metals, *Zool. Sci.,* 8, 287, 1991.

1449. **Sumi, Y., Suzuki, T., Yamamura, M., Hatakeyama, S., Sugaya, Y., and Suzuki, K. T.,** Histochemical staining of cadmium taken up by the midge larva *Chironomus yoshimatsui* (Diptera, Chironomidae), *Comp. Biochem. Physiol.,* 79A, 353, 1984.

1450. **Surber, E. W.,** *Cricotopus bicinctus,* a midgefly resistant to electroplating wastes, *Trans. Am. Fish. Soc.,* 89, 111, 1959.

1451. **Sutcliffe, D. W. and Carrick, T. R.,** Studies on mountain streams in the English Lake District. 1. pH, calcium and the distribution of invertebrates in the river Duddon, *Freshwater Biol.,* 3, 437, 1973.

1452. **Sutcliffe, D. W. and Hildrew, A. G.,** Invertebrate communities in acid streams, in *Acid Toxicity and Aquatic Animals,* Morris, R., Taylor, E. W., Brown, D. J. A., and Brown, J. A., Eds., Cambridge University Press, Cambridge, U.K., 1989, 13.

1453. **Suttman, C. E. and Barrett, G. W.,** Effects of Sevin on arthropods in an agricultural and an old-field plant community, *Ecology,* 60, 628, 1979.

1454. **Suzuki, H.,** Defenses triggered by previous invaders: fungi, in *Plant Disease,* Vol. V, Horsfall, J. G. and Cowling, E. B., Eds., Academic Press, New York, 1980, 319.

1455. **Suzuki, K. T., Aoki, Y., Nishikawa, M., Masui, H., and Matsubara, F.,** Effect of cadmium-feeding on tissue concentrations of elements in germ-free silkworm (*Bombyx mori*) larvae and distribution of cadmium in the alimentary canal, *Comp. Biochem. Physiol.,* 79C, 249, 1984.

1456. **Suzuki, K. T., Sunaga, H., Aoki, Y., Hatakeyama, S., Sugaya, Y., Sumi, Y., and Suzuki, T.,** Binding of cadmium and copper in the mayfly *Baetis thermicus* larvae that inhabit a river polluted with heavy metals, *Comp. Biochem. Physiol.,* 91C, 487, 1988.

1457. **Suzuki, K. T., Sunaga, H., Hatakeyama, S., Sumi, Y., and Suzuki, T.,** Differential binding of cadmium and copper to the same protein in a heavy metal tolerant species of mayfly (*Baetis thermicus*) larvae, *Comp. Biochem. Physiol.,* 94C, 99, 1989.

1458. **Suzuki, K. T., Yamamura, M., Hatakeyama, S., Aoki, Y., Masui, H., Matsubara, F., Sumi, Y., and Suzuki, T.,** Difference in tolerance mechanism to cadmium among three insect larvae (midge, fleshfly and silkworm). *J. Pharmacobio-Dyn.,* 8, 25, 1985.

1459. **Swanson, S. M.,** Food-chain transfer of U-series radionuclides in a northern Saskatchewan aquatic system, *Health Phys.,* 49, 747, 1985.

1460. **Swift, M. J., Heal, O. W., and Anderson, J. M.,** *Decomposition in Terrestrial Ecosystems,* Blackwell Scientific, Oxford, 1979.

1461. **Symons, P. E. K.,** Dispersal and toxicology of the insecticide fenitrothion: predicting hazards of forest spraying, *Residue Rev.,* 68, 1, 1977.

1462. **Symons, P. K. E. and Metcalfe, J. L.,** Mortality, recovery and survival of larval *Brachycentrus numerosus* (Trichoptera) after exposure to insecticide fenitrothion, *Can. J. Zool.,* 56, 1284, 1978.

1463. **Tabaru, Y.,** Study on the chemical control of a nuisance chironomid midge (Diptera: Chironomidae). 4. Efficacy of two insect growth regulators to *Chironomus yoshimatsui* in laboratory and field, *Jpn. J. Sanit. Zool.,* 36, 309, 1985.

1464. **Tabashnik, B. E.,** Responses of pest and non-pest *Colias* butterfly larvae to intraspecific variation in leaf nitrogen and water content, *Oecologia,* 55, 389, 1982.

1465. **Takken, W., Balk, F., Jansen, R.C., and Koeman, J. H.,** The experimental application of insecticides from a helicopter for the control of riverine populations of *Glossina tachinoides* in West Africa. VI. Observations on side effects. *Pest Articles News Summ.,* 24, 455, 1978.

1466. **Tanabe, S.,** PCB problems in the future: foresight from current knowledge, *Environ. Pollut.,* 50, 5, 1988.

1467. **Tapp, R. L. and Hockaday, A.,** Combined histochemical and X-ray microanalytical studies on the copper-accumulating granules in the mid-gut of larval *Drosophila, J. Cell Sci.,* 26, 201, 1977.

1468. **Tarzwell, C. M.,** Toxicity of oil and oil dispersant mixtures to aquatic life, in *Water Pollution by Oil,* Hepple, P., Ed., Proc. 1970 Aviemore Seminar, Inst. Petroleum, London, 1971, 263.

1469. **Taylor, E. J., Maund, S. J., and Pascoe, D.,** Toxicity of four common pollutants to the freshwater macroinvertebrates *Chironomus riparius* Meigen (Insecta: Diptera) and *Gammarus pulex* (L.) (Crustacea: Amphipoda), *Arch. Environ. Contam. Toxicol.,* 21, 371, 1991.

1470. **Taylor, L. R. and Carter, C. I.,** The analysis of numbers and distribution in an aerial population of Macrolepidoptera, *Trans. R. Entomol. Soc. Lond.,* 113, 369, 1961.

1471. **Taylor, O. C. and Eaton, F. M.,** Suppression of plant growth by nitrogen dioxide, *Plant Physiol.,* 41, 132, 1966.

1472. **Templin, E.,** Zur Populationsdynamik einiger Kiefernschadinsekten in rauchgeschädigten Beständen, *Wiss. Zeitschr. Techn. Univ. Dresden,* 11, 631, 1962.
1473. **Thalenhorst, W.,** Untersuchungen über den Einfluss fluorhaltiger Abgase auf die Disposition der Fichte für den befall Durch die Gallenlaus *Sacchiphantes abietis* (L.), *Z. Pflanzenkr. Pflanzenschuzt,* 81, 717, 1975.
1474. **Theiling, K. M. and Croft, B. A.,** Toxicity, selectivity and sublethal effects of pesticides on arthropod natural enemies: a data-base summary, in *Pesticides and Non-target Invertebrates,* Jepson, P. C., Ed., Intercept, Wimborne, Dorset, U.K., 1989, 213.
1475. **Thomas, A. G. B.,** L'application de l'étude d'impact sur l'environnement dans la pratique: bien plus une question de choix judicieux des bioindicateurs qu'un défi à la science (deux exemples pris dans les écosystemès lotiques), *Rev. Suisse Zool.,* 94, 503, 1987.
1476. **Thomas, A. G. B.,** La valeur de bioindicateurs des Coléoptères (Elmidae en particulier) comparativement aux autres insectes, face à une violente pollution industrielle (Kraft) en rivière, *Ann. S.S.N.A.T.V.,* 40, 163, 1988.
1477. **Thomas, A. T. and Hodkinson, I. D.,** Nitrogen, water stress and the feeding efficiency of lepidopteran herbivores, *J. Appl. Ecol., 28,* 703, 1991.
1478. **Thomas, C. F. G., Hol, E. H. A., and Everts, J. W.,** Modelling the diffusion component of dispersal during recovery of a population of linyphiid spiders from exposure to an insecticide, *Funct. Ecol.,* 4, 357, 1990.
1479. **Thomas, J. A.,** The conservation of butterflies in temperate countries: past efforts and lessons for the future, in *The Biology of Butterflies,* Vol. 11, Vane-Wright, R. I. and Ackery, P. R., Eds., Academic Press, London, 1984, 333.
1480. **Thomlinson, G. H., Ed.,** *Effects of Acidic Deposition on the Forests of Europe and North America,* CRC Press, Boca Raton, FL, 1990.
1481. **Thompson, A. J., Shepherd, R. F., Harris, J. W. E., and Silversides, R. H.,** Relating weather to outbreaks of western spruce budworm, *Choristoneura occidentalis* (Lepidoptera: Tortricidae), in British Columbia, *Can. Entomol.,* 116, 375, 1984.
1482. **Thompson, A. R. and Gore, F. L.,** Toxicity of twenty-nine insecticides to *Folsomia candida.* Laboratory studies, *J. Econ. Entomol.,* 65, 1255, 1972.
1483. **Thompson, J. N.,** Insect diversity and trophic structure of communities, in *Ecological Entomology,* Huffaker, C. B. and Rabb, R. L., Eds., John Wiley & Sons, New York, 1984, 591.
1484. **Thornley, S. and Hamdy, Y.,** An assessment of the bottom fauna and sediment of the Detroit River, Ontario Ministry of the Environment, Techn. Rep., ISBN 0-7743-8474-3, London, Ontario, 1984.
1485. **Thorp, V. J. and Lake, P. S.,** Toxicity bioassays of cadmium on selected freshwater invertebrates and the interaction of cadmium and zinc on the freshwater shrimp *Paratya tasmaniensis* Riek., *Aust. J. Mar. Freshwater Res.,* 25, 97, 1974.
1486. **Thurén, A.,** Determination of phthalates in aquatic environments, *Bull. Environ. Contam. Toxicol.,* 36, 33, 1986.
1487. **Ting, I. P. and Mukerji, S. K.,** Leaf ontogeny as a factor in susceptibility to ozone: amino acid and carbohydrate changes during expansion, *Am. J. Bot.,* 58, 497, 1971.
1488. **Tingey, D. T., Wilhour, R. G., and Standley, C.,** The effect of chronic ozone exposures on the metabolite content of ponderosa pine seedlings, *For. Sci.,* 22, 234, 1976.
1489. **Tomkiewicz, S. M. and Dunson, W. A.,** Aquatic insect diversity and biomass in a stream marginally polluted by acid strip mine drainage, *Water Res.,* 11, 397, 1977.
1490. **Tomlinson, H. and Rich, S.,** Metabolic changes in free amino acids of bean leaves exposed to ozone, *Phytopathology,* 57, 972, 1967.
1491. **Tong, S. S. C., Morse, R. A., Bache, C. A., and Lisk, D. J.,** Elemental analysis of honey as an indicator of pollution, *Arch. Environ. Health,* 30, 329, 1975.
1492. **Tooby, T. E., Thompson, A. N., Rycroft, R. J., Black, I. A., and Hewson, R. T.,** A pond study to investigate the effects on fish and aquatic invertebrates of deltamethrin applied directly onto water, Aquatic Environ. Prot. 2, Fisheries Laboratory, Burnham on Crouch, Essex, U.K., 1981.

1493. **Torup, V. J. and Lake, P. S.,** Pollution of a Tasmanian river by mine effluents. II. Distribution of macroinvertebrates, *Int. Rev. Ges. Hydrobiol. Hydrogr.*, 58, 885, 1973.

1494. **Toshkov, A. S., Shabanov, M. M., and Ibrishimov, N. I.,** Attempts to use bees to prove impurities in the environment, *Dokl. Bolg. Akad. Nauk.*, 27, 699, 1974.

1495. **Townsend, C. R., Hildrew, A. G., and Francis, J.,** Community structure in some southern English streams: the influence of physicochemical factors, *Freshwater Biol.*, 13, 521, 1983.

1496. **Tranvik, L. and Eijsackers, H.,** On the advantage of *Folsomia fimetarioides* over *Isotomiella minor* (Collembola) in a metal polluted soil, *Oecologia*, 80, 195, 1989.

1497. **Treshow, M. and Anderson, F. K.,** *Plant Stress from Air Pollution*, John Wiley & Sons, Chichester, U.K., 1991, 283.

1498. **Troiano, J. J. and Leone, I. A.,** Changes in growth rate and nitrogen content of tomato plants after exposure to NO_2, *Phytopathology*, 67, 1130, 1977.

1499. **Truhaut, R.,** Ecotoxicology: objectives, principles and perspectives, *Ecotoxicol. Environ. Safety*, 1, 151, 1977.

1500. **Trumble, J.T. and Hare, J.D.,** Acidic fog-induced changes in host-plant suitability. Interactions of *Trichoplusia ni* and *Phaseolus lunatus*, *J. Chem. Ecol.*, 15, 2379, 1989.

1501. **Trumble, J. T., Dercks, W., Quiros, C. F., and Beier, R. C.,** Host plant resistance and linear furanocoumarin content of *Apium* accessions, *J. Econ. Entomol.*, 83, 519, 1990.

1502. **Trumble, J. T., Hare, J. D., Musselman, R. C., and McCool, P. M.,** Ozone-induced changes in host-plant suitability: interactions of *Keiferia lycopersicella* and *Lycopersicon esculentum*, *J. Chem. Ecol.*, 13, 1, 1987.

1503. **Trumble, J. T., Moar, W. J., Brewer, M. J., and Carson, W. G.,** Impact of UV radiation on activity of linear furanocoumarins and *Bacillus thuringiensis* var. *kurstaki* against *Spodoptera exigua*: implications for tritrophic interactions, *J. Chem. Ecol.*, 17, 973, 1991.

1504. **Tutt, J. W.,** *British Moths*, George Routledge & Sons, England, 1896.

1505. **Tyler, G.,** Heavy metal pollution and soil enzymatic activity, *Plant Soil*, 41, 303, 1973.

1506. **Tyler, G.,** Heavy metal pollution, phosphatase activity and mineralization of organic phosphorus in forest soils, *Soil Biol. Biochem.*, 8, 327, 1976.

1507. **Tyler, G.,** Edaphical distribution patterns of macrofungal species in deciduous forest of south Sweden, *Acta Oecol; Oecol. Gener.*, 10, 309, 1986.

1508. **Tyler, G., Palsberg Påhlsson, M., Bengtsson G., Bååth, E., and Tranvik, L.,** Heavy-metal ecology of terrestrial plants, microorganisms and invertebrates, *Water Air Soil Pollut.*, 47, 189, 1989.

1509. **Tyus, H. M. and Minckley, W. L.,** Migrating mormon crickets, *Anabrus simplex* (Orthoptera: Tettigonidae), as food for stream fishes, *Great Basin Nat.*, 48, 25, 1988.

1510. **Udvardy, M. F. D.,** Notes on the ecological concepts of habitat, biotope and niche, *Ecology*, 40, 725, 1959.

1511. **Ulrich, B.,** Effects of air pollution on forest ecosystems and water — the principles demonstrated at a case study in central Europe, *Atmos. Environ.*, 18, 621, 1984.

1512. **Ultsch, G. R. and Gros, G.,** Mucus as a diffusion barrier to oxygen: possible role in O_2 uptake at low pH in carp (*Cyprinus carpio*) gills, *Comp. Biochem. Physiol.*, 62A, 685, 1979.

1513. **Umbach, D. M. and Davis, D. D.,** Severity and frequency of SO_2-induced leaf necrosis on seedlings of 57 tree species, *For. Sci.*, 30, 587, 1984.

1514. **UNEP (United Nations Environment Programme),** *Environmental Data Report*, 2nd ed., 1989/1990, Blackwell Reference, Alden Press, Oxford, 1989.

1515. **Urbahn, E.,** Zunahme von Melanismus-Beobachtungen bei Makrolepidopteren Europas in neuerer Zeit, *Mitt. Münchener Entomol. Ges.*, 1, 1, 1972.

1516. **van Urk, G. and Kerkum, F. C. M.,** Chironomid mortality after the Sandoz accident and deformities in *Chironomus* larvae due to sediment pollution in the Rhine, *Aqua*, 4 191, 1987.

1517. **U.S. Environmental Protection Agency (USEPA),** Fish kills caused by pollution, fifteen-year summary 1961–1975, Office of Water Planning and Standards (WH-551), Washington, D.C., EPA-440/4-78-011, 1978.

1518. **Uutala, A. J.,** *Chaoborus* (Diptera: Chaoboridae) mandibles — paleolimnological indicators of the historical status of fish populations in acid-sensitive lakes, *J. Paleolimnol.,* 4, 139, 1990.

1519. **Uzunov, Y. and Kovachev, S.,** The macrozoobenthos of Struma River: an example of a recovered community after the elimination of a heavy industrial impact with suspended materials, *Arch. Hydrobiol. Suppl. (Monogr. Beitr.),* 76, 169, 1987.

1520. **Vanderplank, J. E.,** *Disease Resistance in Plants,* 2nd ed., Academic Press, New York, 1984.

1521. **Vangenechten, J. H. D., Witters, H., and Vanderborght., O. L. J.,** Laboratory studies on invertebrate survival and physiology in acid waters, in *Acid Toxicity and Aquatic Animals,* Morris, R., Taylor, E. W., Brown, D. J. A., and Brown, J. A., Eds., Cambridge University Press, Cambridge, U.K., 1989, 151.

1522. **Varty, I. W.,** Hazard to non-target insects, in *1977 Environmental Surveillance of Insecticide Spray Operations in New Brunswick's Budworm-infested Forests,* Varty, I. W., Ed., Committee for Environmental Monitoring of Forest Insect Control Operations, Maritimes Forest Research Centre, Fredericton, NB Canada, Rep. M-X-87, 1978, 14.

1523. **Varty, I. W. and Carter, N. E.,** A baseline inventory of litter dwelling arthropods and airborne insects including pollinators in two fir-spruce stands with dissimilar histories of insecticide treatment, Can. For. Serv., Marit. For. Res. Centre Info. Rep. M-X-48, 1974, 1.

1524. **Vepsäläinen, K. and Pisarski, B.,** The structure of urban ant communities along a geographical gradient from North Finland to Poland, in *Animals in Urban Environment,* Luniak, M. and Pisarski, B., Eds., Zaklad Narodowy im. Ossolinskich, Wroclaw, Poland, 1982, 155.

1525. **Verhoef, H. A. and de Goede, R. G. M.,** Effects of collembolan grazing on nitrogen dynamics in a coniferous forest, in *Ecological Interactions in Soil,* Fitter, A. H., Ed., Special Publication No. 4 of the British Ecol. Soc., Blackwell Scientific, Oxford, 1985, 367.

1526. **Vickerman, G. P. and Sotherton, N. W.,** Effects of some foliar fungicides on the chrysomelid beetle *Gastrophysa polygoni* (L.), *Pest. Sci.,* 14, 405, 1983.

1527. **Vighi, M., Garlanda, M. M., and Calamari, D.,** QSARs for toxicity of organophosphorous pesticides to *Daphnia* and honeybees, *Sci. Total Environ.,* 109/110, 605, 1991.

1528. **Villemant, C.,** Influence de la pollution atmosphérique sur les microlépidoptères du pin en forêt de Roumare (Seine-Maritime), *Acta Oecol; Oecol. Appl.,* 1, 291, 1980.

1529. **Villemant, C.,** Modification du complexe entomophage de la tordeuse des pousses de pin *Rhyacionia buoliana* Schiff. (Lépidoptère Tortricidae) en liaison avec la pollution atmosphérique en forêt de Roumare (Seine-Maritime), *Acta Oecol; Oecol. Appl.,* 1, 139, 1980.

1530. **Villemant, C.,** Influence de la pollution atmospherique sue les populations d'aphides du pin sylvestre en Forêt de Roumaire (Seine Maritime), *Environ. Pollut. Ser. A,* 24, 245, 1981.

1531. **Vogel, W. R.,** Zur Schwermetallbelastung der Borkenkäfer, *Entomol. Exp. Appl.,* 42, 259, 1986.

1532. **Vogel, W. R.,** Die Belastung von Arthropoden mit Blei und Cadmium in unterschiedlich schadstoffexponierten Waldgebieten, *Mitt. Schweiz. Entomol. Gesell.,* 61, 205, 1988.

1533. **Vogel, W. R,.** Zur Aufnahme und Auswirkung der Schwermetalle Zinc und Cadmium beim Mehlkäfer *Tenebrio molitor* L. (Col., Tenebrionidae) unter Berücksichtigung möglicher Wechselwirkungen, *Zool. Anz.,* 220, 25, 1988.

1534. **Vossbrinck, C. R., Coleman, D. C., and Woolley, T. A.,** Abiotic and biotic factors in litter decomposition in a semi-arid grassland, *Ecology,* 60, 265, 1979.

1535. **Wade, K. R., Ormerod, S. J., and Gee, A. S.,** Classification and ordination of macroinvertebrate assemblages to predict stream acidity in upland Wales, *Hydrobiologia,* 171, 59, 1989.
1536. **Waggoner, P. E.,** Agriculture and carbon dioxide, *Am. Sci.,* 72, 179, 1984.
1537. **Wagner, R. M. and Frantz P. D.,** Influence of induced water stress in ponderosa pine on pine sawflies, *Oecologia,* 83, 452, 1990.
1538. **Waite, M. E., Evans, K. E., Thain, J. E., and Waldock, M. J.,** Organotin concentrations in the Rivers Bure and Yare, Norfolk Broads, England, *Appl. Organomet. Chem.,* 3, 383, 1989.
1539. **Wallace, J. B. and Brady, U. E.,** Residue levels of dieldrin in aquatic invertebrates and effect of prolonged exposure on populations, *Pest. Monit. J.,* 5, 295, 1974.
1540. **Wallace, J. B., Cuffney, T. F., Lay, C. C., and Vogel, D.,** The influence of an ecosystem-level manipulation on prey consumption by a lotic dragonfly, *Can. J. Zool.,* 65, 35, 1987.
1541. **Wallace, R. R. and Hynes, H. B. N.,** The catastrophic drift of stream insects after treatment with methoxychlor (1,1,1-trichloro-2. 2-bis (p-methoxyphenyl) ethane), *Environ. Pollut.,* 8, 255, 1975.
1542. **Wallace, R. R. and Hynes, H. B. N.,** The effect of chemical treatments against blackfly larvae on the fauna of running waters, in *Blackflies: The Future for Biological Methods in Integrated Control,* Laird, M., Ed., Academic Press, London, 1981, 237.
1543. **Wallace, R. R., Hynes, H. B. N., and Merritt, W. F.,** Laboratory and field experiments with methoxychlor as a larvicide for Simuliidae (Diptera), *Environ. Pollut.,* 10, 251, 1976.
1544. **Wallace, R. R., West, A. S., Downe, A. E. R., and Hynes, H. B. N.,** The effects of experimental blackfly (Diptera: Simuliidae) larviciding with Abate, Dursban and Methoxychlor on stream invertebrates, *Can. Entomol.,* 105, 817, 1973.
1545. **Wallwork-Barber, M. K., Ferenbaugh, R. W., and Gladney, E. S.,** The use of honey bees as monitors of environmental pollution, *Am. Bee J.,* 122, 770, 1982.
1546. **Walter, G.,** Ökologische Untersuchungen über die Wirkung Fe-II-haltiger Braunkohlengruben-Abwässer auf Vorflutorganismen, *Wiss. Z. Karl-Marx-Univ., Leipzig,* 15, 247, 1966.
1547. **Walter, H.,** *Vegetation of the Earth and Ecological Systems of the Geo-biosphere,* 3rd revised and enlarged ed., Springer-Verlag, Berlin, 1985.
1548. **Wang, D., Karnosky, D. F., and Bormann, F. H.,** Effects of ambient ozone on the productivity of *Populus tremuloides* Michx. grown under field conditions, *Can. J. For. Res.,* 16, 47, 1986.
1549. **Wang, J. and Bian, Y.,** Fluoride effects on the mulberry-silkworm system, *Environ. Pollut.,* 52, 11, 1988.
1550. **Wang, J., Li, Z., Qian, D., Gao, X., Ma, D., and Li, R.,** Accumulation and translocation of fluoride in mulberry-silkworm ecosystem, *Environ. Qual. China,* 1, 22, 1980.
1551. **Wang, J., Qian, D., Li, Z., Gao, X., Ma, D., and Li, R.,** Effects of fluoride in mulberry leaves on growth and development of the silkworm, *Environ. Qual. China,* 2, 33, 1980.
1552. **Waring, R. H. and Pitman, G. B.,** Modifying lodgepole pine stands to change susceptibility to mountain pine beetle attack, *Ecology,* 66, 889, 1985.
1553. **Waringer, J. A.,** Phenology and the influence of meteorological parameters on the catching success of light-trapping for Trichoptera, *Freshwater Biol.,* 25, 307, 1991.
1554. **Warnick, S. L. and Bell, H. L.,** The acute toxicity of some heavy metals to different species of aquatic insect, *J. Water Pollut. Cont. Fed.,* 41, 280, 1969.
1555. **Deleted in proof.**
1556. **Warren, C. E.,** *Biology and Water Pollution Control,* Saunders, Philadelphia, 1971.
1557. **Warrington, S.,** Relationship between SO_2 dose and growth of the pea aphid, *Acyrthosiphon pisum,* on peas, *Environ. Pollut.,* 43, 155, 1987.
1558. **Warrington, S.,** Ozone enhances the growth rate of cereal aphids, *Agric. Ecosyst. Environ.,* 26, 65, 1989.
1559. **Warrington, S. and Whittaker, J. B.,** Interactions between Sitka spruce, the green spruce aphid, sulfur dioxide pollution and drought, *Environ. Pollut.,* 65, 363, 1990.

1560. **Warrington, S., Cottam, D. A., and Whittaker, J. B.,** Effects of insect damage on photosynthesis, transpiration and SO_2 uptake by sycamore, *Oecologia,* 80, 155, 1989.
1561. **Warrington, S., Mansfield, T. A. M., and Whittaker, J. B.,** Effect of SO_2 on the reproduction of pea aphids, *Acyrthosiphon pisum,* and the impact of SO_2 and aphids on the growth and yield of peas, *Environ. Pollut.,* 48, 285, 1987.
1562. **Warwick, W. F.,** Pasqua Lake, southeastern Saskatchewan: a preliminary assessment of trophic status and contamination based on the Chironomidae (Diptera), in *Chironomidae: Ecology, Systematics, Cytology and Physiology,* Murray, D. A., Ed., Pergamon Press, Oxford, 1980, 255.
1563. **Warwick, W. F.,** Palaeolimnology of the Bay of Quinte, Lake Ontario: 2800 years of cultural influence, *Can. Bull. Fish. Aquat. Sci.,* 206, 1, 1980.
1564. **Warwick, W. F.,** Morphological abnormalities in Chironomidae (Diptera) larvae as measures of toxic stress in freshwater ecosystems: indexing antennal deformities in *Chironomus* Meigen, *Can. J. Fish. Aquat. Sci.,* 42, 1881, 1985.
1565. **Warwick, W. F.,** Morphological deformities in larvae of *Procladius* Skuse (Diptera: Chironomidae) and their biomonitoring potential, *Can. J. Fish. Aquat. Sci.,* 46, 1255, 1989.
1566. **Warwick, W. F., Fitchko, J., McKee, P. M., Hart, D. R., and Burt, A. J.,** The incidence of deformities in *Chironomus* spp. from Port Hope Harbour, Lake Ontario, *J. Great Lakes Res.,* 13, 88, 1987.
1567. **Waterhouse, J. C. and Farrell, M. P.,** Identifying pollution related changes in chironomid communities as a function of taxonomic rank, *Can. J. Fish. Aquat. Sci.,* 42, 406, 1985.
1568. **Watson, A. P., van Hook, R. I., Jackson, D. R., and Reichle, D. E.,** Impact of a lead mining-smelting complex on the forest-floor litter arthropod fauna in the New Lead Belt region of southwest Missouri, Oak Ridge National Laboratory, ORNL/NSF/EATC-30, Oak Ridge, TN, 1976.
1569. **Watson, J. A. L., Arthington, A. H., and Conrick, D. L.,** Effect of sewage effluent on dragonflies (Odonata) of Bulimba Creek, Brisbane, *Aust. J. Mar. Freshwater Res.,* 33, 517, 1982.
1570. **Watson, M. A.,** Integrated physiological units in plants, *Trends Evol. Ecol.,* 1, 119, 1986.
1571. **Way, J. M., Newman, J. F., Moore, N. W., and Knaggs, F. W.,** Some ecological effects of the use of paraquat for the control of weeds in small lakes, *J. Appl. Ecol.,* 8, 509, 1971.
1572. **Wayland, M.,** Effect of carbofuran on selected macroinvertebrates in a prairie parkland pond: an enclosure approach, *Arch. Environ. Contam. Toxicol.,* 21, 270, 1991.
1573. **Wayland, M. and Boag, D. A.,** Toxicity of carbofuran to selected macroinvertebrates in prairie ponds, *Bull. Environ. Contam. Toxicol.,* 45, 74, 1990.
1574. **Webber, E. C., Bayne, D. R., and Seesock, W. C.,** DDT contamination of benthic macroinvertebrates and sediments from tributaries of Wheeler Reservoir, Alabama, *Arch. Environ. Contam. Toxicol.,* 18, 728, 1989.
1575. **Weber, W. J.,** *Diseases Transmitted by Rats and Mice,* Thomson Publ., Fresno, CA, 1982.
1576. **Wegner, G. S. and Hamilton, R. W.,** Effect of calcium sulphide on *Chironomus riparius* (Diptera: Chironomidae) egg hatchability, *Environ. Entomol.,* 5, 256, 1976.
1577. **Weidlich M.,** Lepidopterologische und coleopterologische Beobachtungen aus den mittleren und nördlichen Teilen des Bezirkes Halle/S. unter besonderer Berucksichtigung von Gefährdungsursachen, *Faun. Abhand. Staat. Museum Tierk. Dresden,* 14, 131, 1987.
1578. **Weinstein, L.,** Fluoride and plant life, *J. Occup. Med.,* 19, 49, 1977.
1579. **Weir, C. F. and Walter, W. M.,** Toxicity of cadmium in the freshwater snail *Physa gyrina* (Say), *J. Environ. Qual.,* 5, 359, 1976.
1580. **Weiss, S. B., Murphy, D. D., and White, R. B.,** Sun, slope, and butterflies: topographic determinants of habitat quality for *Euphydryas editha, Ecology,* 69, 1486, 1988.
1581. **Wellburn, A. R.,** Effects of SO_2 and NO_2 on metabolic function, in *Effects of Gaseous Air Pollution in Agriculture and Horticulture,* Unsworth, M. H. and Ormrod, D. P., Eds., Butterworths, London, 1982, 169.

1582. **Wellings, P. W. and Dixon, A. F. G.,** Sycamore aphid numbers and population density. III. The role of aphid induced changes in plant quality, *J. Anim. Ecol.,* 56, 161, 1987.
1583. **Wellington, W. G.,** Air mass climatology of Ontario north of Lake Huron and Lake Superior before outbreaks of spruce budworm, *Choristoneura fumiferana* (Clem.), and forest tent caterpillar, *Malacosoma disstria* (Hbn.) (Lepidoptera: Tortricidae; Lasiocampidae), *Can. J. Zool.,* 3, 114, 1952.
1584. **Wells, S. M., Pyle, R. M., and Collins, N. M.,** *The IUCN Invertebrate Red Data Book,* IUCN, Gland, Switzerland, 1983, 632.
1585. **Wentsel, R., McIntosh, A., and McCafferty, W. P.,** Emergence of the midge *Chironomus tentans* when exposed to heavy metal contaminated sediment, *Hydrobiologia,* 57, 195, 1978.
1586. **Wentzel, K. F.,** Insekten als Immissionfolgeschädlinge, *Naturwissenschaften,* 52, 10, 1965.
1587. **Wentzel, K. F. and Ohnesorge, B.,** Zum auftreten von Schadinsekten bei Luftverunreinigung, *Forstarchiv,* 32, 177, 1961.
1588. **Werner, M. D. and Adams, V. D.,** Consequences of oil pollution on the decomposition of vascular plant litter in freshwater lakes: Part 2: nutrient exchange between litter and the environment, *Environ. Pollut. Ser. A,* 34, 101, 1984.
1589. **West, C.,** Factors underlying the late seasonal appearance of the lepidopterous leaf-mining guild on oak, *Ecol. Entomol.,* 10, 111, 1985.
1590. **West, R. L. and Snyder-Conn, E.,** Effects of Prudhoe Bay reserve pit fluids on water quality and macroinvertebrates of arctic tundra ponds in Alaska, Biological Report, 87, Fish and Wildlife Service, U.S. Department of Interior, Washington, D.C., 1987, 48.
1591. **Wetzel, R. G.,** *Limnology,* Saunders, Toronto, Ontario, 1975.
1592. **Whale, G., Sheahan, D., and Matthiessen, P.,** The toxicity of tecnazene, a potato sprouting inhibitor, to freshwater fauna, *Chemosphere,* 17, 1205, 1988.
1593. **Whicker, F. W., Pinder, J. E., III, Bowling, J. W., Alberts, J. J., and Brisbin, I. L., Jr.,** Distribution of ^{137}Cs, ^{90}Sr, ^{238}Pu, ^{239}Pu, ^{241}Am and ^{244}Cm in Pond B, Savannah River Site, Report No.: SREL-35 Savannah River Ecology Laboratory, Aiken, SC, 1989, 79.
1594. **White, G. C., Hakonson, T. E., and Ahlquist, A. J.,** Factors affecting radionuclide availability to vegetables grown at Los Alamos, *Environ. Qual.,* 10, 294, 1981.
1595. **White, T. C. R.,** A hypothesis to explain outbreaks of looper caterpillars, with special reference to populations of *Selidosema suavis* in a plantation of *Pinus radiata* in New Zealand, *Oecologia,* 16, 279, 1974.
1596. **White, T. C. R.,** The abundance of invertebrate herbivores in relation to the availability of nitrogen in stressed food plants, *Oecologia,* 63, 90, 1984.
1597. **Whitehurst, T. I. and Lindsey, I. B.,** The impact of organic enrichment on the benthic macroinvertebrate communities of a lowland river, *Water Res.,* 24, 625, 1990.
1598. **Whitham, T. G.,** Habitat selection by *Pemphigus* aphids in response to resource limitation and competition, *Ecology,* 59, 1164, 1978.
1599. **Whittaker, J. B. and Warrington, S.,** Effects of atmospheric pollutants on interactions between insects and their food plants, in *Pests, Pathogens and Plant Communities,* Burdon, J. J. and Leather, S. R., Eds., Blackwell Scientific, Oxford, 1990, 97.
1600. **Whittaker, J. B., Kristiansen, L. W., Mikkelsen, T. N., and Moore, R.,** Responses to ozone of insects feeding on a crop and a weed species, *Environ. Pollut.,* 62, 89, 1989.
1601. **Whittaker, R. H.,** *Communities and Ecosystems,* MacMillan, New York, 1975.
1602. **Whittaker, R. H., Levin, S. A., and Rott, R. B.,** Niche, habitat, and ecotope, *Am. Nat.,* 107, 321, 1973.
1603. **Wiackowski, S.,** Impact of industrial air pollution upon parasites of pine bud moth (*Exoteleia dodecella*), aphid predators, and certain other insects occurring on pine in vicinity of Tomaszow Maz, *Folia For. Polonica Ser. A,* 23, 175, 1978.
1604. **Wickham, P., van de Walle, E., and Planas, D.,** Comparative effects of mine wastes on the benthos of an acid and an alkaline pond, *Environ. Pollut.,* 44, 83, 1987.

1605. **Wiederholm, T.,** Responses of aquatic insects to environmental pollution, in *The Ecology of Aquatic Insects,* Resh, V. H. and Rosengren, D. M., Eds., Praeger, New York, 1984, 508.
1606. **Wiederholm, T. and Eriksson, L.,** Benthos of an acid lake, *Oikos,* 29, 261, 1977.
1607. **Wiens, J. A.,** Spatial scaling in ecology, *Funct. Ecol.,* 3, 385, 1989.
1608. **Wigglesworth, V. B.,** *The Principles of Insect Physiology,* 6th ed., Methuen, London, 1965.
1609. **Wiklund, C.,** The evolutionary relationship between adult oviposition preferences and larval host plant range in *Papilio machaon* L., *Oecologia,* 18, 185, 1975.
1610. **Wild, H.,** Termites and the serpentines of the Great Dyke of Rhodesia, *Trans. Rhod. Sci. Assoc.,* 57, 1, 1975.
1611. **Wildish, D. J. and Phillips, R. L.,** Acute lethality of fenitrothion to freshwater aquatic invertebrates, *Fish Res. Board Can. M. S. Rep.,* 2010, 1, 1972.
1612. **Williams, D. D. and Hynes, H. B. N.,** The recolonization mechanisms of stream benthos, *Oikos,* 27, 265, 1976.
1613. **Williams, G.,** Invertebrate conservation, in *Australia's Endangered Species,* Kennedy, M., Ed., Simon & Schuster Australia, Brookvale, New South Wales, 1990.
1614. **Williams, K. A., Green, D. W. J., Pascoe, D., and Gower, D. E.,** The acute toxicity of cadmium to different larval stages of *Chironomus riparius* (Diptera: Chironomidae) and its ecological significance for pollution regulation, *Oecologia,* 70, 362, 1986.
1615. **Williams, K. A., Green, D. W. J., Pascoe, D., and Gower, D. E.,** Effect of cadmium on oviposition and egg viability in *Chironomus riparius* (Diptera: Chironomidae), *Bull. Environ. Contam. Toxicol.,* 38, 86, 1987.
1616. **Williams, K. A., Green, W. J., and Pascoe, D.,** Toxicity testing with freshwater macroinvertebrates: methods and application in environmental management, in *Freshwater Biological Monitoring. Advances in Water Pollution Control,* Pascoe, D. and Edwards, R. W., Eds., Pergamon Press, Oxford, 1984, 81.
1617. **Williams, M. W., Hoeschele, J. D., Turner, J. E., Jacobson, K. B., Christie, N. T., Paton, C. L., Smith, L. H., Witschi, H. R., and Lee, E. H.,** Chemical softness and acute metal toxicity in mice and *Drosophila, Toxicol. Appl. Pharmacol.,* 63, 461, 1982.
1618. **Williams, S. T., McNeilly, T., and Wellington, E. M. H.,** The decomposition of vegetation growing on metal mine waste, *Soil Biol. Biochem.,* 9, 271, 1977.
1619. **Williams, W. E., Garbutt, K., Bazzaz, F. A., and Vitousek, P. M.,** The response of plants to elevated CO_2. IV. Two deciduous-forest tree communities, *Oecologia,* 69, 454, 1986.
1620. **Williams, W. E., Garbutt, K., and Bazzaz, F. A.,** The response of plants to elevated CO_2. V. Performance of an assemblage of serpentine grassland herbs, *Environ. Exp. Bot.,* 28, 123, 1988.
1621. **Williamson, P.,** Comparison of metal levels in invertebrate detrivores and their natural diets: concentration factors reassessed, *Oecologia,* 44, 75, 1979.
1622. **Willoughby, L. G.,** The ecology of *Baetis muticus* and *Baetis rhodani* (Insecta, Ephemeroptera) with special emphasis on acid backgrounds, *Int. Rev. Ges. Hydrobiol.,* 73, 259, 1988.
1623. **Wilson, C. L. and Graham, C. L.,** *Exotic Plant Pests of North America and Hawaii,* Academic Press, New York, 1986.
1624. **Wilson, D. C. and Bond, C. E.,** The effects of herbicides diquat and dichlobenil (Casaron) on pond invertebrates. Part I. Acute toxicity, *Trans. Am. Fish. Soc.,* 3, 438, 1969.
1625. **Wilson, R. S.,** A survey of the zinc-polluted River Nent (Cumbria) and the East and West Allen (Northumberland), England, using chironomid pupal exuviae, *Spixiana,* 14, 167, 1988.
1626. **van Wingerden, W. K. R. E., Musters, J. C. M., and Maaskamp, F. I. M.,** The influence of temperature on the duration of egg development in West European grasshoppers (Orthoptera: Acrididae), *Oecologia,* 87, 417, 1991.

1627. **Winner, R. W., Boesel, M. W., and Farrell, M. P.,** Insect community structure as an index of heavy-metal pollution in lotic ecosystem, *Can. J. Fish. Aquat. Sci.,* 37, 647, 1980.

1628. **Winner, R. W., Scott van Dyke, J., Caris, N., and Farrell, M. P.,** Response of macroinvertebrate fauna to a copper gradient in an experimentally-polluted stream, *Int. Ver. Theor. Angew. Limnol. Verh.,* 19, 2121, 1975.

1629. **Winner, W. E. and Mooney, H. A.,** Ecology of SO_2 resistance. II. Photosynthetic changes in relation to SO_2 absorption and stomatal behavior, *Oecologia,* 44, 296, 1980.

1630. **Winterbourn, M. J., Hildrew, A. G., and Box, A.,** Structure and grazing of stone surface organic layers in some acid streams of southern England, *Freshwater Biol.,* 15, 363, 1985.

1631. **Witkowski, Z. and Borusiewicz, K.,** Ecology, energetics and the significance of phytophagous insects in deciduous and coniferous forests, in *Forest Ecosystems in Industrial Regions,* Grodzinski, W., Weiner, J., and Maycock, P. F., Eds., Springer-Verlag, New York, 1984, 103.

1632. **Witkowski, Z., Madziara-Borusiewicz, K., Plonka, P., and Zurek, Z.,** Insect outbreaks in mountain national parks in Poland — their causes, course and effects, *Ekol. Pol.,* 35, 465, 1987.

1633. **Witt, P. N.,** Drugs alter web-building of spiders: a review and evaluation, *Behav. Sci.,* 16, 98, 1971.

1634. **Wittassek, R.,** *Untersuchungen zur Verteilung des Kupfers in Boden, Vegetation und Bodenfauna eines Weinbergökosystems,* Inaugural-Dissert. Dr. Agrarwiss., Rheinische Friedrich-Wilhelms-Univ., Bonn, Germany, 1987, 180.

1635. **Wittig, R. and Neite, H.,** Acid indicators around the trunk base of *Fagus sylvatica* in limestone and loess beechwoods: distribution patterns and phytosociological problems, *Vegetatio,* 64, 113, 1985.

1636. **Wittmann, D.,** Tracer-versuche zur Passage von Insektiziden durch Ammenbienen als Basis für die Abschätzung von Intoxikationswegen der Bienenbrut, *Apidologie,* 13, 328, 1982.

1637. **Wittwer, S. H.,** Rising atmospheric CO_2 and crop productivity, *Hortscience,* 18, 667, 1983.

1638. **Woin, P. and Larsson, P.,** Phthalate esters reduce predation efficiency of dragonfly larvae, Odonata, *Aeshna, Bull. Environ. Contam. Toxicol.,* 38, 220, 1987.

1639. **Woltering, D. M.,** Population responses to chemical exposure in aquatic multispecies systems, in *Multispecies Toxicity Testing,* Cairns, J., Jr., Ed., Pergamon Press, New York, 1985, 61.

1640. **Wolters, V.,** Die Wirkung der Bodenversauerung auf Protura, Diplura und Collembola (Insecta, Apterygota) — Untersuchungen am Stammfuss von Buchen, *Jber. Naturwiss. Ver. Wuppertal,* 42, 45, 1989.

1641. **Wong, P. T. S.,** Toxicity of cadmium to freshwater microorganisms, phytoplankton, and invertebrates, in *Cadmium in the Aquatic Environment, Vol. 19, Advances in Environmental Science and Technology,* Nriagu, J. O. and Sprague, J. B., Eds., John Wiley & Sons, New York, 1987.

1642. **Wood, C. W. and Nash, T. N., III,** Copper smelter effluent on the Sonoran desert vegetation, *Ecology,* 57, 1311, 1976.

1643. **Wood, G. W.,** Recuperation of native bee populations in blueberry fields exposed to drift of fenitrothion from forest spray operations in New Brunswick, *J. Econ. Entomol.,* 72, 36, 1979.

1644. **Wood, L. W., Rhee, G.-Y., Bush, B., and Barnard, E.,** Sediment desorption of PCB congeners and their bio-uptake by dipteran larvae, *Water Res.,* 21, 875, 1987.

1645. **Wood, T. G.,** The role of termites (Isoptera) in decomposition processes, in *The Role of Terrestrial and Aquatic Organisms in Decomposion Processes,* Anderson, J. M. and Macfadyen, A., Eds., Blackwell Scientific, Oxford, 1976, 145.

1646. **Woodham, D. W., Robinson, H. F., Reeves, R. G., Bond, A. B., and Richardson, H.,** Monitoring agricultural insecticides in the cooperative cotton pest management program in Arizona, 1971 — first year study, *Pest. Monit. J.,* 10, 159, 1977.

1647. **Woodring, J. P., Clifford, C. W., and Beckman, B. R.,** Food utilization and metabolic efficiency in larval and adult house crickets, *J. Insect Physiol.,* 25, 903, 1979.

1648. **Woodward, D. F. and Riley, R. G.,** Petroleum hydrocarbon concentrations in a salmonid stream contaminated by oil field discharge water and effects on macrobenthos, *Arch. Environ. Contam. Toxicol.,* 12, 327, 1983.

1649. **Woodward, D. F., Little, E. E.,, and Smith, L. M.,** Toxicity of five shale oils to fish and aquatic invertebrates, *Arch. Environ. Contam. Toxicol.,* 16, 239, 1987.

1650. **Woodward, D. F., Mehrle, P. M., and Mauck, W. L.,** Accumulation and sublethal effects of a Wyoming crude oil in the cutthroat trout, *Trans. Am. Fish. Soc.,* 110, 437, 1981.

1651. **Woodwell, G. M., Craig, P. P., and Johnson, H. A.,** DDT in the biosphere: where does it go, *Science,* 1974, 1101, 1971.

1652. **Woodwell, G. M., Hobbie, J. E., Houghton, R. A., Melillo, J. M., Moore, B., Peterson, B. J., and Shaver, G. R.,** Global deforestation: contribution to atmospheric carbon dioxide, *Science,* 222, 1081, 1983.

1653. **Worth, C. B. and Muller, J.,** Captures of large moths by an ultraviolet light trap, *J. Lepid. Soc.,* 33, 261, 1979.

1654. **Wren, C. D. and Stephenson, G. L.,** The effect of acidification on the accumulation and toxicity of metals to freshwater invertebrates, *Environ. Pollut.,* 71, 205, 1991.

1655. **Wright, J. H., Moss, D., Armitage, P. D., and Furse, M. T.,** A preliminary classification of running water sites in Great Britain based on macroinvertebrate species and the prediction of community type using environmental data, *Freshwater Biol.,* 14, 221, 1984.

1656. **Wright, M. A. and Stringer, A.,** Lead, zinc and cadmium content of earthworms from pasture in the vicinity of an industrial smelting complex, *Environ. Pollut. Ser. A,* 23, 313, 1980.

1657. **Wulff, A., Ropponen, L., and Kärenlampi, L.,** Causes of conifer injuries in some industrial environments, in *Acidification in Finland,* Kauppi, P., Anttila, P., and Kenttämies, K., Eds., Springer-Verlag, Berlin, 1990, 1237.

1658. **Wüstemann, O.,** Gewässergütebestimmung mittels Indikatorenorganismen, *Z. Binnenfisch. DDR,* 36, 237, 1989.

1659. **Yamamura, M., Suzuki, K. T., Hatakeyama, S., and Kubota, K.,** Tolerance to cadmium and cadmium-binding proteins induced in the midge larva, *Chironomus yoshimatsui* (Diptera, Chironomidae), *Comp. Biochem. Physiol.,* 75C, 21, 1983.

1660. **Yanovskii, V. M.,** Influence of waste material from thermal electric power stations on forest insect activity. *Ékologiya,* 2, 74, 1989.

1661. **Yasuno, M., Fukushima, S., Hasegawa, J., Shioyama, F., and Hatakeyama, S.,** Changes in the benthic fauna and flora after application of temephos to a stream on Mt. Tsukuba, *Hydrobiologia,* 89, 205, 1982.

1662. **Yasuno, M., Hatakeyama, S., and Sugaya, Y.,** Characteristic distribution of chironomids in the rivers polluted with heavy metals, *Verh. Int. Ver. Limnol.,* 22, 2371, 1985.

1663. **Yasuno, M., Okita, J., Saito, K., Nakamura, Y., Hatakeyama, S., and Kasuga, K.,** Effects of fenitrothion on benthic fauna in small streams of Mt. Tsukuba, Japan, *Jpn. J. Ecol.,* 31, 237, 1981.

1664. **Yasuno, M., Shioyama, H., and Hasegawa, J.,** Field experiment on susceptibility of macrobenthos in streams to temephos, *Jpn. J. Sanit. Zool.,* 32, 229, 1981.

1665. **Yasuno, M., Sugaya, Y., and Iwakuma, T.,** Effects of insecticides on the benthic community in a model stream, *Environ. Pollut.,* 38, 31, 1985.

1666. **Yevtushenko, N. Yu., Bren, N. V., and Sytnik, Yu. M.,** Heavy metal contents in invertebrates of the Danube river, *Water Sci. Techol.,* 22, 119, 1990.

1667. **Young, A. L.,** Minimizing the risk associated with pesticide use: an overview, in *Pesticides. Minimizing the Risks,* Ragsdale, N. and Kuhr, R. J., Eds., ACS Symp. Ser., 336, American Chemical Society, Washington, D.C., 1987, 183.

1668. **Young, A. L., Cockerham, L. G., and Thalken, C. E.,** A long-term study of ecosystem contamination with 2,3,7,8-tetrachlorodibenzo-p-dioxin, *Chemosphere,* 16, 1791, 1987.

1669. **Yu, C. C., Booth, G. M., Hansen, D. J., and Larsen, J. R.,** Fate of carbofuran in a model ecosystem, *J. Agric. Food. Chem.*, 22, 431, 1974.
1670. **Zabecki, W.,** Rola owadów kambio- i ksylofagicznych w procesie zamierania jodlowych drzewostanów Ojcowskiego Parku narodowego, znajdujacych sie pod wplywem imisji przemyslowych. *Acta Agraria Silvestria (S. silvestris)*, 27, 17, 1988.
1671. **Zera, A. J., Koehn, R. K., and Hall, J. G.,** Allozymes and biochemical adaptation, in *Comprehensive Insect Physiology, Biochemistry and Pharmacology*, Kerkut, G. A. and Gilbert, L. I., Eds., Pergamon Press, Oxford, 1985, 633.
1672. **Zimakowska-Gnoinska, D.,** The effects of industrial pollution on bioenergetic indices and on chemical composition of polyphagous predators — Araneae, *Pol. Ecol. Stud.*, 7, 61, 1981.
1673. **Zimmerman, E. C.,** *Insects of Hawaii I*, University of Hawaii, Honolulu, 1948.
1674. **Zimmerman, E. C.,** Adaptive radiation on Hawaii with special reference to insects, *Biotropica*, 2, 32, 1970.
1675. **Zischke, J. A., Arthur, J. H., Nordlie, K. J., Hermanutz, R. O., Standen, D. A., and Henry, T. P.,** Acidification effects on macroinvertebrates and feathed minnows (*Pimephales promelas*) in outdoor experimental channels, *Water Res.*, 17, 47, 1983.
1676. **Zumr, V. and Landa, M.,** Zu den Auswirkungen der Immissionsschäden in der CSSR, *Allg. Forstzeit.*, 39, 364, 1984.
1677. **Zwiazek, J. J. and Shay, J. M.,** The effects of sodium fluoride on cytoplasmic leakage and the lipid and fatty acid composition of jack pine (*Pinus banksiana*) seedlings, *Can. J. Bot.*, 66, 535, 1988.

INDEX

B

C

E

F

H

U

V

W